CYTOKINE-INDUCED TUMOR IMMUNOGENICITY

From Exogenous Molecules To Gene Therapy

EDITED BY

GUIDO FORNI
CNR Center for Immunogenetics and Experimental Oncology
University of Turin, Turin, Italy

ROBIN FOA
Department of Biological Sciences and Human Oncology
University of Turin, Turin, Italy

ANGELA SANTONI
Department of Experimental Medicine
University La Sapienza, Rome, Italy

LUIGI FRATI
Department of Experimental Medicine
University La Sapienza, Rome, Italy

ACADEMIC PRESS
Harcourt Brace & Company, Publishers
London San Diego New York
Boston Sydney Tokyo Toronto

ACADEMIC PRESS LIMITED
24–28 Oval Road
LONDON NW1 7DX

U.S. Edition Published by
ACADEMIC PRESS INC.
San Diego, CA 92101

This book is printed on acid free paper

A catalogue record for this book is available from the British Library

ISBN 0–12–261520–4

Printed in Great Britain by T.J. Press (Padstow) Ltd, Padstow, Cornwall

PREFACE

When one of us suggested the possibility of preparing a book on the use of cytokines to increase tumor immunogenicity, we all agreed that it was a sound and timely idea. We also thought that word processing technology would have made it possible to edit the work of our numerous colleagues in this field in a relatively easy fashion, with speed and reasonable "elegance". This prompted us to favor a "camera ready" realization. It has been very gratifying to see the enthusiastic response to our invitation. We are pleased and excited by the contributions to this volume and we believe that it provides a representative picture of the various facets of the current work in gene immunotherapy.

On the other hand, we had a lot to learn about ease and speed. We have come to appreciate thoroughly why so many books are heartily dedicated to the patience and understanding of spouses and family members. Word processing may be a dream, but collecting the work from well over a hundred contributors has its pitfalls. Some authors vanished from sight; some sent incompatible computer files; some were enmeshed in legal problems; some were wedded to idiosyncratic stylistics; some were untraceable by means of faxes or phone calls, and others refused to be bothered with references. This last issue, in fact, leads us to refer the reader to the authors of the individual chapters for any specific questions regarding references.

Finally, we wish to acknowledge the financial support by the special project "Gene Therapy" of the Italian Association for Cancer Research (AIRC) and by the National Research Council (CNR). We wish to thank Massimiliano Sammarco and Monica Raviola for their patience in solving computer compatibility problems, and Marcia Vitiello for her careful reading of the manuscripts.

The editors

CONTENTS

Experimental models

Conclusions

CONTRIBUTORS

Alessandra Allione
CNR Center for Immunogenetics and Experimental Oncology,
University of Turin,
Via Santena 19,
I-10126 Turin, Italy

Andrea Anichini
Division of Experimental Oncology D,
Istituto Nazionale Tumori,
Via Venezian 1
I-20133 Milan, Italy

Ron N. Apte
Department of Microbiology and Immunology,
Faculty of Health Sciences,
Ben-Gurion University of the Negev,
Beer-Sheva 84105, Israel

Flavio Arienti
Division of Experimental Oncology D,
Istituto Nazionale Tumori,
Via Venezian 1,
I-20133 Milan, Italy

Timothy C. Back
Biological Carcinogenesis and Development Program,
PRI/DynCorp, Inc,
Frederick, MD 21702-1201, USA

Filippo Belardelli
Laboratory of Virology,
Istituto Superiore di Sanità,
Viale Regina Elena 299,
I-00161 Rome, Italy

Filiberto Belli
Division of Clinical Oncology B,
Istituto Nazionale Tumori,
Via Venezian 1,
I-20133 Milan, Italy

Sabrina Bertini
CNR Center for Immunogenetics and Experimental Oncology,
University of Turin,
Via Santena 19,
I-10126 Turin, Italy

Thomas Blankenstein
Institute of Immunology,
Klinikum Steglitz,
Freie Universitat Berlin,
Hindenburgdamm 27,
D-1000 Berlin 45, FRG

Thierry Boon
Ludwig Institute for Cancer Research,
Brussels Branch,
74 Av. Hippocrate,
B-1200 Brussels, Belgium

Maria Carla Bosco
Macrophage Cell Biology Section,
Biological Response Modifiers Program,
National Cancer Institute,
Frederick Cancer Research and Development Center,
Frederick, MD 21702-1201, USA

Vincent Brichard
Ludwig Institute for Cancer Research,
Brussels Branch,
74 Av. Hippocrate,
B-1200 Brussels, Belgium

Jan Bubenik
Institute of Molecular Genetics,
Academy of Sciences of the Czech Republic,
Flemingovo nam. 2,
16637 Prague 6, Czech Republic

Natale Cascinelli
Division of Clinical Oncology B,
Istituto Nazionale Tumori,
Via Venezian 1,
I-20133 Milan, Italy

Federica Cavallo
CNR Center for Immunogenetics and Experimental Oncology,
University of Turin,
Via Santena 19,
I-10126 Turin, Italy

Martin A. Cheever
Division of Medical Oncology,
University of Washington,
Seattle, WA 98195, USA

Wei Chen
Division of Medical Oncology,
University of Washington,
Seattle, WA 98195, USA

Alessandro Cignetti
Department of Biomedical Sciences and Human Oncology,
University of Turin,
Via Genova 3,
I-10126, Turin, Italy

Marco Cippitelli
Department of Experimental Medicine,
University La Sapienza,
Viale Regina Elena 324,
I-00161 Rome, Italy

Mario P. Colombo
Division of Experimental Oncology D,
Istituto Nazionale Tumori,
Via Venezian 1,
I-20133 Milan, Italy

Manuela Consalvo
CNR Center for Immunogenetics and Experimental Oncology,
Institute of Microbiology,
University of Turin,
Via Santena 9,
I-10126 Turin, Italy

Giorgio Cortesina
Otorhinolaryngology Clinic,
University of Turin,
Via Genova 3,
I-10126 Turin, Italy

Katherine Cronin
Division of Hematologic Oncology,
Memorial Sloan-Kettering Cancer Center,
New York, NY 10021, USA

Carla De Giovanni
Cancerology Institute,
University of Bologna,
Viale Filopanti 22,
I-40126 Bologna, Italy

Ruggero De Maria
Department of Experimental Medicine,
University La Sapienza,
Viale Regina Elena 324,
I-00161 Rome, Italy

Antonella De Stefani
Otorhinolaryngology Clinic,
University of Turin,
Via Genova 3,
I-10126 Turin, Italy

Sylvie Delassus
U.277 - Biologie Moleculaire du Gène,
Institut Pasteur,
25, rue du Dr. Roux,
F-75724 Paris cédex 15, France

Tibor Diamantstein
Institute of Immunology,
Klinikum Steglitz,
Freie Universitat Berlin,
Hindenburgdamm 27,
D-1000 Berlin 45, FRG

Mary L. Disis
Division of Medical Oncology,
University of Washington,
Seattle, WA 98195, USA

Marion Dorsch
Institute of Immunology,
Klinikum Steglitz,
Freie Universitat Berlin,
Hindenburgdamm 27,
D-1000 Berlin 45, FRG

Amos Douvdevani
Department of Microbiology and Immunology,
Faculty of Health Sciences,
Ben-Gurion University of the Negev,
Beer-Sheva 84105, Israel

Tatyana Dvorkin
Department of Microbiology and Immunology,
Faculty of Health Sciences,
Ben-Gurion University of the Negev,
Beer-Sheva 84105, Israel

Jos Even
U.277 - Biologie Moleculaire du Gène,
Institut Pasteur,
25, rue du Dr. Roux,
F-75724 Paris cédex 15, France

Robert G. Fenton
Biological Response Modifiers Program,
National Cancer Institute,
Frederick Cancer Research and Development Center,
Frederick, MD 21702-1201, USA

Maria Ferrantini
Laboratory of Virology,
Istituto Superiore di Sanità,
Viale Regina Elena 299,
I-00161 Rome, Italy

Eyal Fima
Department of Microbiology and Immunology,
Faculty of Health Sciences,
Ben-Gurion University of the Negev,
Beer-Sheva 84105, Israel

Lynda Finch
Departments of Medicine/Oncology and Immunology,
University of Washington,
Seattle, WA 89195, USA

Robin Foa
Department of Biomedical Sciences and Human Oncology,
University of Turin,
Via Genova 3,
I-10126, Turin, Italy

Guido Forni
CNR Center for Immunogenetics and Experimental Oncology,
University of Turin,
Via Santena 19,
I-10126 Turin, Italy

Jose L. Franco
Laboratory of Experimental Immunology,
National Cancer Institute,
Frederick Cancer Research and Development Center,
Frederick, MD 21702-1201, USA

Luigi Frati
Department of Experimental Medicine,
University La Sapienza,
Viale Regina Elena 324,
I-00161 Rome, Italy

Lucia Gabriele
Laboratory of Virology,
Istituto Superiore di Sanità,
Viale Regina Elena 299,
I-00161 Rome, Italy

Emanuela Galeazzi
Otorhinolaryngology Clinic,
University of Turin,
Via Genova 3,
I-10126 Turin, Italy

Bernd Gansbacher
Division of Hematologic Oncology,
Memorial Sloan-Kettering Cancer Center,
New York, NY 10021, USA

Mark Gilbert
Departments of Medicine/Oncology and Immunology,
University of Washington,
Seattle, WA 89195, USA

Eli Gilboa
Department of Experimental Surgery,
Duke University Medical Center,
Durham, NC 27710, USA

Anna Gillio Tos
Department of Biomedical Sciences and Human Oncology,
University of Turin,
Via Genova 3,
I-10126, Turin, Italy

Mirella Giovarelli
CNR Center for Immunogenetics and Experimental Oncology,
Institute of Microbiology,
University of Turin,
Via Santena 9,
I-10126 Turin, Italy

Anna Maria Giuffrida
Department of Experimental Medicine,
University La Sapienza,
Viale Regina Elena 324,
I-00161 Rome, Italy

Evelyne Goillot
MGH Cancer Center,
Massachusetts General Hospital and Harvard Medical School,
Charlestown, MA 02129, USA

Philip D. Greenberg
Departments of Medicine/Oncology and Immunology,
University of Washington,
Seattle, WA 89195, USA

Ion Gresser
Laboratory of Viral Oncology,
Institut de Recherches Scientifiques sur le Cancer,
F-94802 Villejuif, France

M. Eilene Gruys
Laboratory of Experimental Immunology,
National Cancer Institute,
Frederick Cancer Research and Development Center,
Frederick, MD 21702-1201, USA

Anna Guarini
Department of Biomedical Sciences and Human Oncology,
University of Turin,
Via Genova 3,
I-10126, Turin, Italy

Moshe Hacham
Department of Microbiology and Immunology,
Faculty of Health Sciences,
Ben-Gurion University of the Negev,
Beer-Sheva 84105, Israel

Hanno Hock
Institute of Immunology,
Klinikum Steglitz,
Freie Universitat Berlin,
Hindenburgdamm 27,
D-1000 Berlin 45, FRG

Alan N. Houghton
Division of Clinical Immunology,
Memorial Sloan-Kettering Cancer Center,
New York, NY 10021, USA

Thomas Kaido
Laboratory of Viral Oncology,
Institut de Recherches Scientifiques sur le Cancer,
F-94802 Villejuif, France

Kristin L. Komschlies
Biological Carcinogenesis and Development Program,
PRI/DynCorp, Inc,
Frederick, MD 21702-1201, USA

Philippe Kourilsky
U.277 - Biologie Moleculaire du Gène,
Institut Pasteur,
25, rue du Dr. Roux,
F-75724 Paris cédex 15, France

Lorena Landuzzi
Cancerology Institute,
University of Bologna,
Viale Filopanti 22,
I-40126 Bologna, Italy

Jonathan J. Lewis
Department of Surgery,
Memorial Sloan-Kettering Cancer Center,
New York, NY 10021, USA

Pier Luigi Lollini
Cancerology Institute,
University of Bologna,
Viale Filopanti 22,
I-40126 Bologna, Italy

Luciano Lombardi
Division of Experimental Oncology A,
Istituto Nazionale Tumori,
Via Venezian 1,
I-20133 Milan, Italy

Dan L. Longo
Biological Response Modifiers Program,
National Cancer Institute,
Frederick Cancer Research and Development Center,
Frederick, MD 21702-1201, USA

Cristina Maccalli
Division of Experimental Oncology D,
Istituto Nazionale Tumori,
Via Venezian 1,
I-20133 Milan, Italy

Arabella Mazzocchi
Division of Experimental Oncology D,
Istituto Nazionale Tumori,
Via Venezian 1,
I-20133 Milan, Italy

Craig S. McCune
Missoula Medical Oncology, P.C.,
615 West Adler,
Missoula, MO 59802, USA

N. Avrion Mitchison
Deutsches Rheuma Forschungszentrum Berlin,
Robert Koch Institute,
D-13353 Berlin, FRG

Andrea Modesti
Institute of Experimental and Social Medicine,
University of Chieti,
I-66100 Chieti, Italy

Franca Moretti
Department of Experimental Medicine,
University La Sapienza,
Viale Regina Elena 324,
I-00161 Rome, Italy

Roberta Mortarini
Division of Experimental Oncology D,
Istituto Nazionale Tumori,
Via Venezian 1,
I-20133 Milan, Italy

Craig A. Mullen
Metabolism Branch,
Division of Cancer Biology, Diagnosis and Centers,
National Cancer Institute,
National Institutes of Health,
Bethesda, MD 20892, USA

Piero Musiani
Institute of Experimental and Social Medicine,
University of Chieti,
I-66100 Chieti, Italy

Gary J. Nabel
Departments of Internal Medicine and Biological Chemistry,
Howard Hughes Medical Institute,
University of Michigan Medical Center,
Ann Arbor, MI 48109-0684, USA

Satoshi Nagoya
Departments of Medicine/Oncology and Immunology,
University of Washington,
Seattle, WA 89195, USA

Patrizia Nanni
Cancerology Institute,
University of Bologna,
Viale Filopanti 22,
I-40126 Bologna, Italy

Brad Nelson
Departments of Medicine/Oncology and Immunology,
University of Washington,
Seattle, WA 89195, USA

Giordano Nicoletti
Cancerology Institute,
University of Bologna,
Viale Filopanti 22,
I-40126 Bologna, Italy

Franco Novelli
CNR Center for Immunogenetics and Experimental Oncology,
University of Turin,
Via Santena 19,
I-10126 Turin, Italy

Gabriella Palmieri
Department of Experimental Medicine,
University La Sapienza,
Viale Regina Elena 324,
I-00161 Rome, Italy

Drew M. Pardoll
Departments of Oncology, Medicine and Molecular Biology, and Genetics,
Johns Hopkins University,
School of Medicine,
Baltimore, MD 21205, USA

Giorgio Parmiani
Division of Experimental Oncology D,
Istituto Nazionale Tumori,
Via Venezian 1,
I-20133 Milan, Italy

David J. Peace
Division of Hematology/Oncology,
Loyola University Chicago,
2160 South First Avenue,
Maywood, IL 60153, USA

Manuela Peretti
Laboratory of Virology,
Istituto Superiore di Sanità,
Viale Regina Elena 299,
I-00161 Rome, Italy

Gregory E. Plautz
Department of Pediatrics and Communicable Diseases,
Howard Hughes Medical Institute,
University of Michigan Medical Center,
Ann Arbor, MI 48109-0684, USA

Enrico Proietti
Laboratory of Virology,
Istituto Superiore di Sanità,
Viale Regina Elena 299,
I-00161 Rome, Italy

Antonello Punturieri
Institute of Human Pathology and History of Medicine,
University of Chieti,
I-66100 Chieti, Italy

Nicholas P. Restifo
Surgery Branch,
Division of Cancer Treatment,
National Cancer Institute,
National Institutes of Health,
Bethesda, MD 20892, USA

Carlo Riccardi
Institute of Medical Pharmacology,
University of Perugia,
Via del Giochetto,
I-06100 Perugia, Italy

Stanley R. Riddell
Departments of Medicine/Oncology and Immunology,
University of Washington,
Seattle, WA 89195, USA

Ludovica Riera
Department of Biomedical Sciences and Human Oncology,
University of Turin,
Via Genova 3,
I-10126, Turin, Italy

Licia Rivoltini
Division of Experimental Oncology D,
Istituto Nazionale Tumori,
Via Venezian 1,
I-20133 Milan, Italy

Felicia Rosenthal
Division of Hematologic Oncology,
Memorial Sloan-Kettering Cancer Center,
New York, NY 10021, USA

Claude Roth
U.277 - Biologie Moleculaire du Gène,
Institut Pasteur,
25, rue du Dr. Roux,
F-75724 Paris cédex 15, France

Manuela Sacchi
Otorhinolaryngology Clinic,
University of Turin,
Via Genova 3,
I-10126 Turin, Italy

Stefania Salvi
Division of Experimental Oncology D,
Istituto Nazionale Tumori,
Via Venezian 1,
I-20133 Milan, Italy

Laura Santodonato
Laboratory of Virology,
Istituto Superiore di Sanità,
Viale Regina Elena 299,
I-00161 Rome, Italy

Angela Santoni
Department of Experimental Medicine,
University La Sapienza,
Viale Regina Elena 324,
I-00161 Rome, Italy

Shraga Segal
Department of Microbiology and Immunology,
Faculty of Health Sciences,
Ben-Gurion University of the Negev,
Beer-Sheva 84105, Israel

Maria Luisa Sensi
Division of Experimental Oncology D,
Istituto Nazionale Tumori,
Via Venezian 1,
I-20133 Milan, Italy

Noa Shimoni
Department of Microbiology and Immunology,
Faculty of Health Sciences,
Ben-Gurion University of the Negev,
Beer-Sheva 84105, Israel

Antonella Stoppacciaro
Department of Experimental Medicine,
University La Sapienza,
Viale Regina Elena 324,
I-00161 Rome, Italy

Robert I. Tepper
MGH Cancer Center,
Massachusetts General Hospital and Harvard Medical School,
Charlestown, MA 02129, USA

Andrea Tubaro
Urology Clinic,
University of L'Aquila,
I-67100 L'Aquila, Italy

Stefania Vai
CNR Center for Immunogenetics and Experimental Oncology,
University of Turin,
Via Santena 19,
I-10126 Turin, Italy

Benoit Van den Eynde
Ludwig Institute for Cancer Research,
Brussels Branch,
74 Av. Hippocrate,
B-1200 Brussels, Belgium

Aline Van Pel
Ludwig Institute for Cancer Research,
Brussels Branch,
74 Av. Hippocrate,
B-1200 Brussels, Belgium

Francesca Velotti
Laboratory of Molecular Pathology,
University of Tuscia,
I-01100 Viterbo, Italy

Elena Voronov
Department of Microbiology and Immunology,
Faculty of Health Sciences,
Ben-Gurion University of the Negev,
Beer-Sheva 84105, Israel

Beth Walter
Departments of Medicine/Oncology and Immunology,
University of Washington,
Seattle, WA 89195, USA

Yoshihiko Watanabe
Department of Molecular Microbiology,
Faculty of Pharmaceutical Sciences,
Kyoto University,
Sakyo-ku,
Kyoto 606, Japan

Rosalyn M. White
Department of Microbiology and Immunology,
Faculty of Health Sciences,
Ben-Gurion University of the Negev,
Beer-Sheva 84105, Israel

Robert H. Wiltrout
Laboratory of Experimental Immunology,
National Cancer Institute,
Frederick Cancer Research and Development Center,
Frederick, MD 21702-1201, USA

Thomas Wolfel
I Medizinische Klinik der Johannes Gutenberg-Universitat,
Langenbeckstrasse 1,
D-6500 Mainz, FRG

Cassia Yee
Departments of Medicine/Oncology and Immunology,
University of Washington,
Seattle, WA 89195, USA

Karen Zier
Department of Medicine,
Mount Sinai School of Medicine,
New York, NY 10029, USA

Margot Zoller
Institute for Radiology and Pathophysiology,
German Cancer Research Center,
Im Neuenheimer Feld 280,
D-6900, Heidelberg, FRG

LIST OF ABBREVIATIONS

ADCC	antibody dependent cell cytotoxicity
AFP	alpha fetoprotein
APC	antigen presenting cell(s)
BCG	bacillus Calmette-Guerin
BRM	biological response modifier(s)
CEA	carcinoembrionic antigen
CR	complete remission/response(s)
CTL	cytotoxic T lymphocyte(s)
DTH	delayed type hypersensitivity
EBV	Epstein-Barr virus
FCS	fetal calf serum
G-CSF	granulocyte colony stimulanting factor
GM-CSF	granulocyte/macrophage colony stimulating factor
h	hour(s)
HLA	human leukocyte antigen
IFN	interferon(s)
Ig	immunoglobulin(s)
IL	interleukin
i.p.	intraperitoneal/intraperitoneally
i.v.	intravenous/intravenously
IU	international units
LAK	lymphokine activated killer

mAb	monoclonal antibody (or antibodies)
MHC	major histocompatibility complex
MR	minor remission/response(s)
NK	natural killer
PBL	peripheral blood lymphocytes
PCR	polymerase chain reaction
PMN	polymorphonuclear leukocytes
PR	partial remission/response(s)
SD	stable disease
s.c.	subcutaneous/subcutaneously
TCR	T-cell receptor
TGF	transforming growth factor
Th	T helper
TIL	tumor infiltrating lymphocytes/leukocytes
TNF	tumor necrosis factor
U	unit(s)

1

Prefatory Chapter: CYTOKINES, GENES AND CANCER THERAPY

D. L. Longo
Biological Response Modifiers Program, NCI-Frederick Cancer Research and Development Center, Frederick, MD, USA

Anyone who has seen a patient whose joints have been destroyed by severe long-standing rheumatoid arthritis, or examined a rejected kidney allograft grossly and microscopically, cannot help being impressed by the awesome destructive power of the immune system. In the first instance, the mechanisms that enable the immune system to distinguish between self and nonself have gone haywire, and the immune system relentlessly attacks the self. In the second instance, the physicians have performed a breach of the immunologic integrity of the host the likes of which has not occurred in nature (with the exception of the conception and gestation of an infant) since corals started to compete with each other for the same space on a rock. This breach of immunologic integrity is rapidly detected by the host and the response is specific, swift, strong, diverse, and relentless. In fact, we are still learning amazing things about the capabilities of the immune system. And most workers agree that it would be wonderful (and is an appropriate goal for researchers) to be able to harness the destructive power of the immune system against cancer. The immune system has several distinctive attributes:

- It is specific
- It employs diverse effector mechanisms
- It can be expanded and regulated.

None of the currently employed cancer treatments encompasses all these attributes; in fact, the lack of specificity of radiation therapy, chemotherapy and surgery is responsible for much suffering, disability and even mortality among patients with cancer.

Yet, the immune system has seemed largely indifferent to cancer cells. It has developed numerous mechanisms for killing environmental parasites such as bacteria and viruses. Indeed, the sophisticated genetic mechanisms that give rise to antibodies and the TCR seem designed to accommodate nearly any pathogen, including those that do not yet exist. Even though about 8-10% of human cancer may be attributable to viral causes, and it is clear that the immune system is extremely capable of controlling most viral infections (particularly with the help of immunization), histologic examination of most human cancers reveals

CYTOKINE-INDUCED TUMOR IMMUNOGENICITY
ISBN 0-12-261520-4

little to support the notion that lymphocytes have noticed something awry.

Research over the last several years has elucidated reasons for this apparent immunological blindness to cancer, the insidious destroyer that originates within rather than outside the host. Insights into the requirements for antigen presentation to T cells have led to the recognition that tumor cells are frequently deficient in many of the essentials of this process. For example, in general, they lack the costimulatory molecules, such as B7 (and other members of this growing family) which are required to activate rather than anergize T cells. They often down-regulate their class I MHC molecules (either directly or indirectly by failing to express β2-microglobulin, which is required for surface expression of class I MHC) and do not express class II MHC molecules. Class II molecules are crucial for presenting peptides that elicit $CD4^+$ helper phenotype T cells, which are important for promoting the maturation of cytotoxic $CD8^+$ T cells. These CTL destroy the tumor by recognizing a peptide presented by class I MHC molecules. Thus, tumor cells without class I or class II MHC molecules must have their antigenic peptides presented by other APC, and those cells are not normally found within a growing tumor mass. Even if tumor cells express the appropriate MHC molecules, they may lack critical components of the intracellular pathway for processing proteins into peptides and getting such peptides loaded into MHC molecules for expression on the cell surface.

In addition to failing to present antigen and failing to provide T-cell costimulation, there is another set of hurdles erected by tumors. Most tumors produce or elicit the production of immunosuppressive factors that reduce the capacity of the host to respond not only to the tumor, but to other antigenic challenges. Prostaglandin E_2 and TGF-β are only two such immunosuppressive factors, and there may be several others. We have recently described a novel form of tumor induced immunosuppression involving the alteration of the structure of the T-cell antigen receptor, the signal transduction pathway of T-cell activation, and the transcription factors that mediate expression of effector molecules critical to T-cell action.

T cells from tumor bearing mice and people have an abnormal T-cell receptor isoform in which the ζ chain is replaced by another ζ family member, the Fc_ε γ-chain. Neither *lck* nor *fyn* tyrosine kinases are associated with the altered receptor, and the pattern of tyrosine phosphorylation after activation is markedly altered. Calcium release is decreased. Upon activation neither *c-rel* nor NF-κB p65 proteins translocate to the nucleus. The NF-κB p50 protein does move to the nucleus, but it is truncated at its aminoterminal end to a p48 molecule that homodimerizes and binds to NF-κB DNA sequences. It does not drive transcription. Indeed, it blocks transcription by authentic NF-κB p65-p50 heterodimers. This defect can be induced in vitro by culturing normal T cells with a cell free malignant pleural effusion. Thus, we expect to isolate a factor responsible for these changes. In the course of our studies, we have found that this same change occurs in antigen specific T cells exposed to their cognate antigen for 4 days in culture. Therefore, this alteration may be the functional equivalent of receptor desensitization. The alteration can also be induced by IL-1 and, there are therefore examples of both homologous and heterologous receptor desensitization similar to those reported for the β adrenergic receptor. We have also found this defect in T cells from aged mice, a finding that may explain in part the relationship between aging and cancer. Once the tumor related factor that produces these changes is isolated, strategies to overcome this particular form of tumor induced immunosuppression should be conceivable. The defect has been found in over 90% of specimens of TIL from renal cell carcinomas and in the peripheral blood T cells of roughly half of the patients that we have tested with melanoma, breast cancer, Hodgkin's disease, and myeloma. Defects such as this, and perhaps others not yet detected, may help explain the exceedingly poor response rates to biological therapy in patients with cancer.

Given these many problems that we know about and the probability that there are others that we do not yet know about, coupled with the admittedly poor results that have been achieved to date with immunologic approaches to cancer treatment in general and immunization approaches in particular, it would seem that only fools would persist in the notion that patients will one day be able to receive an effective immunization against cancer. However, a number of recent findings encourage us to persist in the believe that the notion is not only feasible, but likely to work. In fact, this book is filled with promising data and ideas that are ready for clinical testing in man. It seems to me that two of the many recent observations have been crucial in converting the pursuit of cancer vaccines from foolish persistence by a few die-hards to a high priority endeavor pursued by the finest minds in immunology and cancer research.

The first critical finding was that intracellular proteins can be targets of an immune attack. Of course, this result, mainly attributable to Thierry Boon and his colleagues, was built upon the novel findings that elucidated the antigen processing and presentation pathways. However, it allowed us to stop the largely futile search for tumor specific cell surface antigens and to begin focusing on any features that distinguish cancer cells form normal cells. Thus, there are now successful vaccine approaches to cancer in animal models that include as targets mutated oncogenes (*ras, abl*), mutated tumor suppressor genes (p53), viral gene products (papilloma virus E7), and intracellular proteins (MAGE, tyrosinase). The enormous effort to find cell surface proteins that are tumor specific and immunogenic has not been entirely futile. For example, the idiotype of the Ig on B-cell lymphomas, the mucins on breast and colon cancer, and glycolipids like the Lewis blood group antigen and GD2 and GD3 gangliosides which were defined serologically can also be the targets of cellular immunity. However, the potential spectrum of targets is now quite broad.

The second critical finding was that cytokines can compensate for many and perhaps all of the defects in tumor antigen presentation and can amplify the immune response to tumors by both antigen specific and antigen nonspecific T cells. Observations initially reported by Bob Tepper in Philip Leder's laboratory for IL-4 and by Eric Fearon and Drew Pardoll for IL-2 began an avalanche of studies using many different cytokines to correct the defects in tumor cell antigen presentation and immune response amplification. The mechanisms by which all these cytokines are able to elicit an immune response are in some cases quite complicated and multifactorial (e.g. IL-2), in other cases seemingly more straightforward (e.g. GM-CSF), and in yet other cases undefined (e.g. IL-1, IL-7). However, it is clear that the local production of a cytokine within a tumor is a key component to successful therapy. Direct injection of tumor nodules with a cytokine is less effective, probably because the elevated interstitial pressures within tumor nodules pump the free cytokine out of the tumor too fast for it to elicit the immune response. To solve this serious (but largely pharmacologic) problem, many investigators have resorted to introducing cytokine genes into tumor cells or normal cells and introducing the gene transfected cells into tumors. This technique works best when the nontransfected tumor cells and the transfected tumor or normal cells are mixed together ex vivo and coinjected, a circumstance that does not adequately reproduce the usual clinical scenario. Indeed, one of the major goals for this field of research is to improve the efficacy of this technique to the point that established tumor masses can be injected with transfected tumor or normal cells and produce tumor regression.

There is some evidence the gene transfected cells may be capable of doing more than acting as a depot cytokine preparation. On the other hand, there are also data that chemically modified cytokines (e.g. PEGylated) or cytokines incorporated into liposomes of various sizes and compositions may also be effective. The idea of using gene transfection techniques has also led to novel strategies for tumor cell killing. Nabel and his colleagues are hoping to bring about tumor rejection by making tumor cells express a foreign MHC antigen by transfer of naked DNA specifying the synthesis of a foreign MHC protein into a tumor (without a viral vector). The hope is to elicit an allogeneic response analogous to the rejection of an organ allograft. Since it is clear in all T-cell responses that only a minority of the T cells in a lesion are antigen specific and most of the cells were recruited into the response by antigen nonspecific mechanisms, it is hoped that an initial allospecific response will recruit in other T cells among which may be cells capable of recognizing and rejecting the tumor in other body sites where the tumor does not express the alloantigen. This interesting strategy is described in detail in this book.

Another exciting prospect for gene therapy is the introduction into tumors of genes whose expression would make the cells vulnerable to a particular drug. An example of this is the introduction of herpes virus thymidine kinase into tumor cells to make them vulnerable to ganciclovir. A surprising finding emerged from the studies by Blaese and his colleagues at the NCI; they found that noninfected tumor cells in the general vicinity of cells that had been infected with a thymidine kinase containing virus were also killed by ganciclovir. The basis for this innocent bystander effect is not yet clear, but clinical trials are proceeding.

There are many other prospects for gene therapy that are either beyond the scope of this book or insufficiently developed to include here at this point. For example, one may use the thymidine kinase-ganciclovir system to eliminate effector cells when their job is done. Some of the most convincing data that T cells can kill tumors in vivo in man are derived from the decreased recurrence rate in allogeneic bone marrow transplant patients experiencing graft-vs-tumor reactions. One could introduce alloreactive T cells from healthy donors, perhaps immunized against the patient's tumor, into the tumor bearing host as effector cells. If such cells are transduced with thymidine kinase, one can envisage eliciting a vigorous allogeneic response analogous to the graft-vs-tumor response (as well as graft-vs-host disease) and then killing off the effector lymphocytes with ganciclovir after the tumor is gone if and when graft-vs-host disease manifestations appear clinically. Also not covered in this book, because there are not yet published data, is the use of chemokine genes to orchestrate the migration into tumors of particular cell subsets. It has been shown that IL-8 attracts neutrophils, and that the β-chemokines MIP-1α and MIP-1β are selective in chemoattracting $CD8^+$ and $CD4^+$ T cells, respectively. Genes for these chemokines, and perhaps even other cytokines yet to be discovered, used alone or in various combinations, may provide even more powerful antitumor effects.

This remarkable book contains ample evidence for the persistent and waxing enthusiasm that exists for vaccine approaches to cancer. It includes chapters from nearly every research group actively contributing to the field. The contributors have captured the excitement that pervades researchers in biologic therapy at this moment because of the confluence of advances from basic research in genetics, immunology and cancer cell biology. This confluence of ideas makes introduction of cytokine genes into cells and administration of cells to patients feasible. Of course, translational research such as this is not usually characterized by enormous success with the first clinical application of the preclinical idea. It is highly likely that our initial clinical efforts will be based upon some false assumption or another that will require further clarification with more laboratory experiments. Then, with a refined clinical hypothesis in hand, we can go back into patients, and more likely than not, we shall find yet another obstacle to overcome. Translational research must be iterative in nature. The laboratory and the clinic must work together very closely in the application of new ideas to cancer treatment. As each new barrier is overcome, new ones appear that require ingenuity and new ideas to meet.

The overwhelming message of this book is that our level of understanding has got much deeper in the last five years; indeed, sufficiently deep that our clinical ideas seem at last to be more than empirical leaps of faith. The information that has been obtained in many laboratories all over the world and is presented here has illuminated our immediate next steps on the tortuous path to successful therapy. And it is quite possible that the light we see is the first light of the dawn of a new day in cancer treatment.

CONCEPTS

2

THE IMMUNOGENICITY OF TUMORS

N. A. Mitchison
Deutsches Rheuma Forschungszentrum Berlin, Robert Koch Institute, Berlin, Federal Republic of Germany

Introduction

This historical introduction takes the form of a brief general statement of the function of the immune system, followed by a critical survey of the theory of immunosurveillance of cancer. It goes on to exclude viral cancer from the condemnation of surveillance, and then to present a view of tumor specific transplantation antigens that accords with modern ideas about mutation in oncogenes and elsewhere in the genome. Presentation of endogenous peptides by class I MHC molecules is cited as providing fresh inspiration for tumor immunology. The relevance of evolutionary considerations is discussed.

Throughout its history the immune system has served the purpose of defense against infection. This is demonstrated today by those rare experiments of nature in which infants are born with a nonfunctional immune system because one of the key genes has mutated. If left untreated, these infants will die from infection and can survive only in an isolator. These infants, and their animal counterparts such as mice with a congenital thymus defect, do not have a high incidence of most types of cancer and do make normal numbers of red blood cells. If kept alive in a semi-sterile environment, mice with a congenital thymus defect reproduce quite well. So any idea that the immune system is necessary to prevent cancer, or to help red blood cells develop, or to facilitate reproduction must be mistaken. This does not necessarily mean that the immune system could not be made to attack cancer cells, exert a trophic influence on erythropoiesis or facilitate reproduction, as some immunologists hope. But one has to be realistic: it will not be easy to persuade the system to perform tasks to which it is unaccustomed in the course of evolution.

By the standards of later chapters of this book this introduction may sound somewhat negative in tone. More optimistic assessments are readily available: the chapters on tumor associated antigens (James and Sikora), and on the immune responses to tumors (Johnson and Riethmuller) in the new edition of *Clinical Aspects of Immunology*, 1993 fall

CYTOKINE-INDUCED TUMOR IMMUNOGENICITY
ISBN 0-12-261520-4

eminently into this category, and at the same time, provide comprehensive coverage.

Immune surveillance

The concept of immune surveillance of tumors was proposed in the early 1960's, independently by Richard Prehn, Lewis Thomas and Macfarlane Burnet (Burnet, 1964). At that time T cells had only recently been discovered, mainly through research on tissue transplantation, and they had no known function in defense against infection. A function for them was needed, and it made sense at the time to propose that their job might be to exercise surveillance over newly developed tumors. The idea was that most tumor cells would bear antigens resembling the alloantigens responsible for graft rejection; these tumor specific antigens would provoke a T-cell response which would cause rejection of most tumors at an early stage, before they had become visible. The proposal was prompted by the recent discovery of tumor specific transplantation antigens: cells of chemically induced tumors could evoke within an inbred mouse strain tumor specific immunity mediated by T cells (Prehn and Main, 1958), just as tissue allografts, or homografts as they were called at the time, were known to elicit transplantation immunity also mediated by T cells. The theory made two important predictions: one, that an individual with an impaired T-cell system would be more likely to develop cancer and the other, that cancer in a normal individual could be treated by nonspecifically enhancing the T-cell immune response. To these was later added a third prediction, that tumor incidence should show HLA associations.

The first prediction seemed to be born out by reports of a raised incidence of cancer in humans with congenital immunodeficiencies, and in congenitally athymic (nude) mice. Later studies failed to confirm these findings: an occasional outbreak of viral cancer might run through a colony of immunodeficient mice, but large colonies of virus free nude mice did not show heightened incidence of any cancer. "Approximately 6900 nudes were observed in the 7-year period 1969-1975... spontaneous malignant tumors were not observed in any animal" (Rygaard and Povlson, 1976). A large UK-Australian survey of immunocompromised individuals, mostly kidney transplant recipients, found a markedly heightened incidence of skin cancer (with exposure to sunlight as a major cofactor) and of B-cell lymphoma (Kinlen et al., 1979). These are precisely the tumors which would be expected to result from lowered resistance to viruses, for papilloma virus is strongly suspected of causing skin cancer, and EBV of causing B-cell lymphoma. In this connection, progress has been delayed by the difficulty of handling the antigens of papilloma virus, but the problem is now yielding to a potent combination of cellular immunology and molecular genetics (Stauss et al., 1992).

A heightened incidence of hepatoma was not recorded, presumably because the known oncogenic hepatitis virus B requires dietary cofactors which were lacking in the study population. An interesting parallel finding is that ultraviolet light induces highly immunogenic tumors in mice, perhaps again via papilloma virus (Mitchison and Kinlen, 1980). The findings in man have been extensively confirmed with the addition, still unexplained, of a heightened incidence of Kaposi's sarcoma following infection with human immunodeficiency virus (Penn, 1988). In summary, the particular pattern of heightened cancer incidence which is seen in the immunocompromised individuals can be better explained by immunosurveillance of viruses than of cancer per se.

The second prediction, that of cancer immunotherapy by nonspecific agents, has also had an up and down course. During the late 1970's there arose a vogue for treatment of cancer patients with such agents as live BCG (an attenuated form of *Mycobacterium tuberculosis*), killed *Corynebacterium parvum*, and lentinin (a fungal product). This stemmed from successful therapy in animal models with chemically induced tumors and gained impetus from success in treating bladder carcinoma with intravesicular BCG (Morales et al., 1976).

The tide has now receded, leaving over a thousand research publications, and BCG instilled into the bladder as the only immunotherapy of cancer is still widely practiced. In light of what has already been said, it is natural to suppose this form of treatment is holding its own because it is applied to a form of cancer which, like that of skin, may express papilloma virus antigens.

The third prediction, that of HLA associations with particular forms of cancer, can be dealt with even more briefly. A handful of reports of such associations can be found in the literature, but none appear to have been confirmed.

In at least one instance, for testicular cancer, the possibility has carefully been excluded (Forman et al., 1992). Various excuses have been offered, such as the variety of possible cancer antigens and the loss of HLA expression on tumors, but these negative findings remain a powerful argument against immunosurveillance. After all in other diseases, particularly those of a supposed autoimmune nature such as rheumatoid arthritis, it is the HLA associations which are accepted as providing the strongest evidence of immunological involvement.

Viral antigens

The survey above of immune surveillance has identified human papilloma virus and EBV, with the addition of hepatitis virus B, as the main cancer viruses; human lymphoma viruses types 1 and 2 should be added to this list. Together, these viruses are causal agents for some 8-10% of human cancer (with a lower figure in males and a higher one in females, attributable to papilloma virus causing cervical cancer). As such, their antigens are of great importance. It is likely that the first vaccine specifically for use against cancer in man will be developed against the strains 16 and 18 of human papilloma virus which are strongly implicated in cervical cancer (Munoz et al., 1988). The hepatitis B subunit vaccine currently in use is probably already providing some protection against hepatic carcinoma (Brown et al., 1990). Marek's disease, a lymphoma in chickens caused by avian poxvirus, has had an effective vaccine for many years.

These various forms of viral cancer can be dismissed as exceptions which confirm the rule that the immune system does not normally respond to cancers. They can equally well be regarded in a more favorable light as excellent targets for immune intervention of the type discussed in this volume, since it will be many years before any or all of them have been eliminated by vaccination. And yet again, they can be regarded as exquisite systems in which to analyse the interplay of viruses and the immune system. Nowhere is this more evident than in the maneuvers by which EBV encoded nuclear antigens (EBNA) manage to avoid provoking an effective immune response (Gavioli et al., 1992).

Mutations

It would now be generally accepted that cancer results from mutation, and that what matters are activating mutations in proto-oncogenes and inactivation of so-called tumor suppressor genes; indeed, the latter part of this statement is no more than a tautology. Most carcinogens are mutagens. Much of the history of tumor immunology can be understood in terms of the balance between mutations occurring in transformation associated genes and those occurring elsewhere in the genome.

As outlined above, modern ideas about tumor immunology started from the definition of tumor specific transplantation antigens in mice. In spite of much research effort, including some mounted by the luminaries of molecular biology (Sikora et al., 1979), these structures remain something of an enigma. They are unique in that each tumor expresses a different, noncross-reactive antigen. They are found most characteristically on tumors induced by powerfully mutagenic chemicals, such as methylcholanthrene and by the equally powerful mutagen ultraviolet light. These agents are powerful not so much in the sense that they are needed only in tiny doses (although that is generally true), but rather in that they induce cancer rapidly.

An important clue comes from the analysis of neoantigens expressed on tumor cells subjected to in vitro mutagenesis, the so called tum$^-$ antigens (Boon et al., 1989). Under the extreme conditions of mutagenesis used in this system, mutations encoding transplantation antigens (i.e.

antigens able to provoke tumor rejection in syngeneic mice, usually in cooperation with other tumor antigens, hence the name) seem to turn up at random in the expressed genome.

A second clue is the occurrence during chemical carcinogenesis of damage and repair, taking place in target cells which do not themselves go on to become tumors (Swann et al., 1976).

Taken together, these findings suggest a simple interpretation of tumor specific antigens. Mutations in somatic cells occur all the time, more or less at random. Whether they could ever be regarded as "spontaneous" is a matter of definition; certainly those caused by carcinogens from cigarettes would not. It is more helpful simply to regard the rate of mutation as variable, and at least partly under the control of environmental factors. Most mutations will occur in non-oncogenes and, when the rate of mutation is low, will eventually be eliminated either immunologically or through other mechanisms. Under these circumstances, when a critical proto-oncogene is finally hit by mutation and a cancer develops, the cell will harbor few other mutations and should be poorly immunogenic. On the other hand, when the mutation rate is high, cells will not have had time to be purged by immunoselection of neoantigens, and cancers should be more strongly immunogenic. In a sense this brings immunosurveillance back, but it is immunosurveillance of neoantigens occurring mainly in noncancer cells, rather than of neoantigens associated with cancer itself. It may be unfair to introduce a new theory into a chapter like this, but then again the idea may not be as new as the author believes, so vast is the literature of cancer immunology.

At least one example of chemical carcinogenesis is now known to result from mutation in a defined region within a defined proto-oncogene in a defined cell type at a defined point in development: ethylnitrosourea induced mutation in the *neu* gene of Schwann precursor cells producing schwannomas in the peripheral nervous system of the rat (Nikitin et al., 1991). Precise mutagenesis studies of this kind, linked to analysis of neoantigens, could test ideas of the sort outlined above. Eventually, they should help in choosing the optimal tumors for immunotherapy, in which the maneuvers described in subsequent chapters could be used. Meanwhile, they provide a reasonable explanation of why cancer in man has proved so much less immunogenic than strong carcinogen induced cancer in mice.

Oncogene encoded peptides

Tumor immunology has drawn new inspiration from the discovery, still quite recent, that CD8 T cells recognize peptides endogenous to the target cell. That seminal discovery has been warmly welcomed as making sense of cytotoxic T cells, which would otherwise be for ever in danger of killing innocent bystander cells, but can now be trusted to confine their attention to virus infected targets. This raises the possibility that these cells might also recognize peptides derived from mutated oncogenes, most of which are located inside tumor cells and are not, therefore, accessible to other immunological mechanisms.

This possibility has been much discussed and is being vigorously pursued (Boon et al., 1989; Melief et al., 1989; Smith et al., 1989). A notable recent development is the demonstration that human T-cell clones specific for p21 *ras* derived synthetic peptides can be generated (Gedde-Dahl et al., 1992); mutated *ras* genes are among the oncogenes most frequently identified in malignant tumors. This and other parallel work reinforces the hope that a direct path can now be taken towards raising truly cancer specific immune responses in patients.

Need immune responses have an evolutionary meaning?

In this chapter the arguments against natural immunosurveillance have been deployed in all their strength, leading to the conclusion that tumors, whether incipient or not, do not normally elicit an effective immune response. Yet we know that carcinogens can generate neoantigens and that the human T-cell repertoire contains receptors able to recognize mutant oncogenes. How much, then, does it matter that the immune response normally contributes so little? Should we be deterred from pursuing the maneuvers that promise to increase the immunogenicity of tumors, which form the main subject matter of this book?

An important part of the information which is pertinent to these questions is still under debate. We do not yet know with any confidence whether the immune system works as a random machine, endowed with the ability to make recognitions and discriminations within certain limits set by solely its chemistry; or alternatively, whether it is adapted through natural selection to performing a group of tasks which it has confronted continually during the long course of its evolution. If the former view is correct, we need not worry so much about the lack of normal immunosurveillance, but the implications of the latter view would be alarming for the present enterprise of immunotherapy. This divergence of views is central to the history of immunology. Our champions are, on the one hand, Karl Landsteiner, standing through the 1920's and 1930's for the chemical view, and Macfarlane Burnet, standing a little later for the evolutionary view. Their inheritors have been termed (Jerne, 1967) the trans-immunologists, whose aim is "to clarify the genetic mechanisms and cellular dynamics for maintaining a population of lymphocytes capable of synthesizing an enormous range of different antibody (or T-cell receptor) molecules", in contrast to the cis-immunologists who are "mainly concerned with the regulatory factors which permit or forbid an antigen to trigger the system successfully". Irun Cohen, a leading cis-immunologist of the present day, has formulated the view that the T-cell repertoire develops under the influence of highly conserved self macromolecules, such as heat shock proteins, so as to constitute a "homunculus" which enables the immune system to recognize similar proteins in microorganisms (Cohen and Young, 1991; Mitchison, 1993). If that view is valid, tumor immunology is in deep trouble.

Advances in immunological technology continue to make possible today what seemed impossible yesterday. Until recently, any improvement on a living cell as an immunogen seemed out of reach: a long history of grinding up extracted alloantigen with Freund's adjuvant or BCG led us nowhere. Yet all that has now changed, thanks to maneuvers to be described later in this volume, among which the technology of cytokine gene implantation looks particularly hopeful. It is for this reason that I would answer the questions above in the following way. Of course, it does matter that the immune response normally contributes so little, if only in teaching us to temper our hopes with caution; but by no means should we be deterred from pursuing the hope of increasing the immunogenicity of tumors in man.

References

Boon T, Van Peel A, De Plaen E, Chomez P, Lurquin C, Szikora JP, Sibille C, Mariame B, Van den Eynde B, Lethe B and Brichart V (1989). Genes coding for T-cell-defined tum transplantation antigens: Point mutations, antigenic peptides, and subgenic expression. Cold Spring Harb Symp Quant Biol **54**, 587-596.

Brown JL, Carman WF and Thomas HC (1990). The hepatitis B virus. Baillieres Clin Gastroenterol **4**, 721-747.

Burnet FM (1964). Immunological factors in the process of carcinogenesis. Brit Med Bull **20**, 154-158.

"Clinical Aspects of Immunology" (ed PJ Lachmann, DK Peters, FR Rosen and MJ Walport), Blackwell Scientific Publications, Boston.

Cohen IR and Young DB (1991). Autoimmunity, microbial immunity and the immunological homunculus. Immunol Today **12**, 105-110.

Forman D, Oliver RT, Brett AR, Marsh SG, Moses JH, Bodmer JG, Chilvers CE and Pike MC (1992). Familial testicular cancer: A report of the UK family register, estimation of risk and an HLA class I sib-pair analysis. Br J Cancer **65**, 255-262.

Gavioli R, De Campos Lima PO, Kurilla MG, Kieff E, Klein G and Masucci MG (1992). Recognition of the Epstein-Barr virus-encoded nuclear antigens EBNA-4 and EBNA-6 by HLA-A11-restricted cytotoxic T lymphocytes: Implications for down-regulation of HLA-A11 in Burkitt lymphoma. Proc Natl Acad Sci USA **89,** 5862-5866.

Gedde-Dahl T, Eriksen JA, Thorsby E and Gaudernack G (1992). T-cell responses against products of oncogenes: Generation and characterization of human T-cell clones specific for p21 ras-derived synthetic peptides. Hum Immunol **33,** 266-274.

Jerne NK (1967). Summary: Waiting for the end. Cold Spring Harb Symp Quant Biol **32,** 591-603.

Kinlen LJ, Sheil AGR, Peto J and Doll R (1979). Collaborative United-Kingdom-Australasian study of cancer in patients treated with immunosuppressive drugs. Brit Med J **2,** 1461-1466.

Melief CJM, Vasmel WRE, Offringa R, Sijts EJAM, Mathews EA, Peters PJ, Meloen RH, van der Eb AJ and Kast WM (1989). Immunosurveillance of virus-induced tumors. Cold Spring Harb Symp Quant Biol **54,** 597-603.

Mitchison NA (1993). A walk round the edges of self tolerance. Ann Rheum Dis **S2,** 3-5.

Mitchison NA and Kinlen LJ (1980). Present concepts in immune surveillance. Progr Immunol **8,** 641-650.

Morales A, Eidinger D and Bruce AW (1976). Intracavitary Bacillus Calmette-Guerin in the treatment of superficial bladder tumors. J Urol **116,** 180-183.

Munoz N, Basch X and Kaldor KM (1988). Does human papilloma virus cause cervical cancer? The state of the epidemiological evidence. Br J Cancer **57,** 1-5.

Nikitin AY, Ballering LAP, Lyons J and Rajewsky MF (1991). Early mutation of the neu (erbB-2) gene during ethylnitrosourea-induced oncogenesis in the rat Schwann cell lineage. Proc Natl Acad Sci USA **88,** 9939-9943.

Penn I (1988). Tumors of the immunocompromised patient. Ann Rev Med **39,** 63-73.

Prehn R and Main JM (1958). Immunity to methylcholanthrene-induced sarcomas. J Natl Cancer Inst **18,** 769-778.

Rygaard J and Povlson CO (1976). The nude mouse vs the hypothesis of immune surveillance. Transpl Rev **28**, 41-61.

Sikora K, Koch G, Brenner S and Lennox E (1979). Partial purification of tumor specific antigens from methylcholanthrene induced murine sarcomas by immobilised lectins. Br J Cancer **40**, 8-19.

Smith MEF, Bodmer JG, Kelly AP, Trowsdale J, Kirkland SC and Bodmer WF (1989). Variation in HLA expression on tumors: An escape from immune response. Cold Spring Harb Symp Quant Biol **54**, 581-586.

Stauss HJ, Davies H, Sadovnikova E, Chain B, Horowitz N and Sinclair C (1992). Induction of cytotoxic T lymphocytes with peptides in vitro: Identification of candidate T-cell epitopes in human papilloma virus. Proc Natl Acad Sci USA **89**, 7871-7875.

Swann PF, Magee PN, Mohr U, Reznik G, Green U and Kaufman DG (1976). Possible repair of carcinogenic damage caused by dimethylnitrosamine in rat kidney. Nature **263**, 134-136.

3

IDENTIFICATION OF TUMOR REJECTION ANTIGENS AND PERSPECTIVES FOR SPECIFIC IMMUNOTHERAPY

T. Boon, A. Van Pel, V. Brichard, B. Van den Eynde and T. Wolfel
Ludwig Institute for Cancer Research, Brussels Branch; Cellular Genetics Unit, Université Catholique de Louvain, Brussels, Belgium, and Medizinische Klinik der Johannes Gutenberg-Universitat, Mainz, Federal Republic of Germany

Introduction

Several decades have elapsed since the first evidence was obtained that some rodent tumors could elicit an immune rejection response (Gross, 1943; Prehn and Main, 1957; Klein et al., 1960). This response, which was specific enough to eliminate the tumor cells in the absence of significant adverse effects on normal tissues, led to the concept of tumor rejection antigens. Later, it became apparent that many rodent tumors did not elicit any immune rejection response in syngeneic animals (Hewitt et al., 1976; Middle and Embleton, 1981). However, we found that antigenic variants (tum^- variants) derived from such nonimmunogenic tumors, including spontaneous tumors, were able to generate a specific immune protection against the original tumor (Boon and Van Pel, 1978; Van Pel et al., 1983).

We therefore concluded that perhaps only a minority of tumors were immunogenic, i.e. capable of eliciting an immune rejection response, but that most rodent tumors were antigenic, i.e. they expressed structures that can be targets for specific immune rejection responses that are artificially induced. Much support for this notion came from later work carried out with tumor cells transfected with interleukin genes. Considerable increases in immunogenicity leading to tumor rejection were observed with several interleukins (Tepper et al., 1989; Fearon et al., 1990; Golumbek et al., 1991).

These observations suggest that the use of cytokines to improve the immunogenicity of tumor rejection antigens represents a major new approach in cancer immunotherapy. Of equal importance may be the identification of tumor rejection antigens and their encoding genes.

CYTOKINE-INDUCED TUMOR IMMUNOGENICITY
ISBN 0-12-261520-4

Mouse tumors

We have obtained evidence with mouse tumor P815 that antigens that are targets for CTL can elicit specific tumor rejection in vivo (Uyttenhove et al., 1983). Hence, we feel justified in naming as tumor rejection antigens not only those antigens that are shown to elicit tumor rejection in syngeneic animals but also those that are recognized by syngeneic CTL. It is not clear whether tumor reactive CTL exert their effect in vivo solely by causing target cell death through their lytic machinery. CTL are known to produce, for example, TNF and IFN-γ after contact with their target cells, and both mediators act synergistically in their cytotoxic and cytostatic effects on the tumor cells (Dealtry et al., 1987). Thus, when invading tumor tissue and recognizing antigen expressing cells, CTL can deliver lymphokines at the tumor site at concentrations not reached by systemic administration. This mechanism may contribute to tumor rejection as it probably does for the resolution of viral infection (Ramsay et al., 1993). In a sense, mounting a T-cell response against antigens expressed on tumor tissues would be a way of targeting lymphokines to tumor sites.

We have launched a systematic effort to identify the genes that code for antigens recognized on tumors by syngeneic CTL. In the mouse, we have identified an antigen recognized by CTL on mastocytoma tumor P815 (Van den Eynde et al., 1991). It results from the activation of a gene which is silent in normal mouse mastocytes and in mast cell lines that do not produce tumors. This gene is expressed in several but not in all tumorigenic mast cell lines. Recent evidence suggests that its activation is related to tumor transformation (Van den Eynde et al., in preparation).

Human tumors

For human tumors, autologous mixed lymphocyte-tumor cell cultures can generate CTL that lyse the autologous tumor but do not lyse NK cell targets or autologous fibroblasts (Hainaut et al., 1990). The antigens detected by these CTL on cultured tumor cells are also present on fresh metastatic tumor cells (Degiovanni et al., 1990). The gene coding for an antigen recognized on a melanoma tumor was recently identified. It was named MAGE-1 (Van der Bruggen et al., 1991; Traversari et al., 1992a). The sequence of this gene is identical to that of the equivalent gene found in normal cells. The gene is expressed at a high level in the tumor and is completely silent in all the normal tissues that have been tested, with the exception of testis. It is expressed in 40% of melanoma and in a significant fraction of breast and lung tumors (Brasseur et al., 1992).

The MAGE-1 gene family

Gene MAGE-1 belongs to a family of twelve closely related genes that are 85 to 99% identical. Four other genes of the MAGE family are expressed in many tumors. No member of the family is expressed in normal tissues other than testis. We feel that this specificity of the expression of the MAGE genes vindicates the approach of focusing on autologous CTL obtained by in vitro restimulation of blood cells as means for identifying target antigens for specific immunotherapy.

The MAGE-1 encoded antigen present on the original melanoma from which MAGE-1 was isolated was named MZ2-E. Most tumors that express gene MAGE-1 and HLA-A1 are recognized by the anti-MZ2-E CTL clone, suggesting that the MAGE-1 encoded peptide binds to HLA-A1 to constitute the MZ2-E antigen. This peptide has been identified. It is a nonapeptide (Traversari et al., 1992b).

Immunotherapy strategies with the MZ2-E peptide

Regarding strategies of active immunotherapy, important benefits can be expected from the identification of the gene that codes for a tumor rejection antigen. First, specific immunization can be applied selectively to those patients whose tumor expresses the antigen. Further, various forms of the antigen can be engineered in order to optimize its immunogenicity. Finally, it should be possible to monitor the effect of immunization by finding out whether specific CTL are generated in patients (Fig. 3.1). Here follows how these notions could be applied to the MZ2-E antigen. Because we know that the MAGE-1 encoded peptide has to combine with HLA-A1 to constitute antigen MZ2-E, we suggest that, before receiving surgery, patients bearing a melanoma, a breast tumor or a non-small cell lung carcinoma, should be HLA typed. If they carry HLA-A1, as is expected for 26% of Caucasian individuals, then small tumor samples could be frozen rapidly after surgery so as to ensure conservation of the RNA. This RNA could be tested by PCR amplification to identify those tumors that express gene MAGE-1. These tumors should express the MZ2-E antigen and may therefore be sensitive to anti-MZ2-E T lymphocytes.

Many different immunization modalities are contemplated. A first possibility involves the use of cells that express antigen MZ2-E after they have been killed by irradiation. This can be carried out with available melanoma cell lines. It is possible, however, that strong allogeneic responses may compete with the response directed against the tumor rejection antigen. This allogeneic reaction could be decreased by using human cells that normally do not express HLA genes and that would be transfected with the HLA-A1 and MAGE-1 genes. Such cells could also be transfected with various interleukin genes to augment their immunogenicity (Fearon et al., 1990; Golumbek et al., 1991).

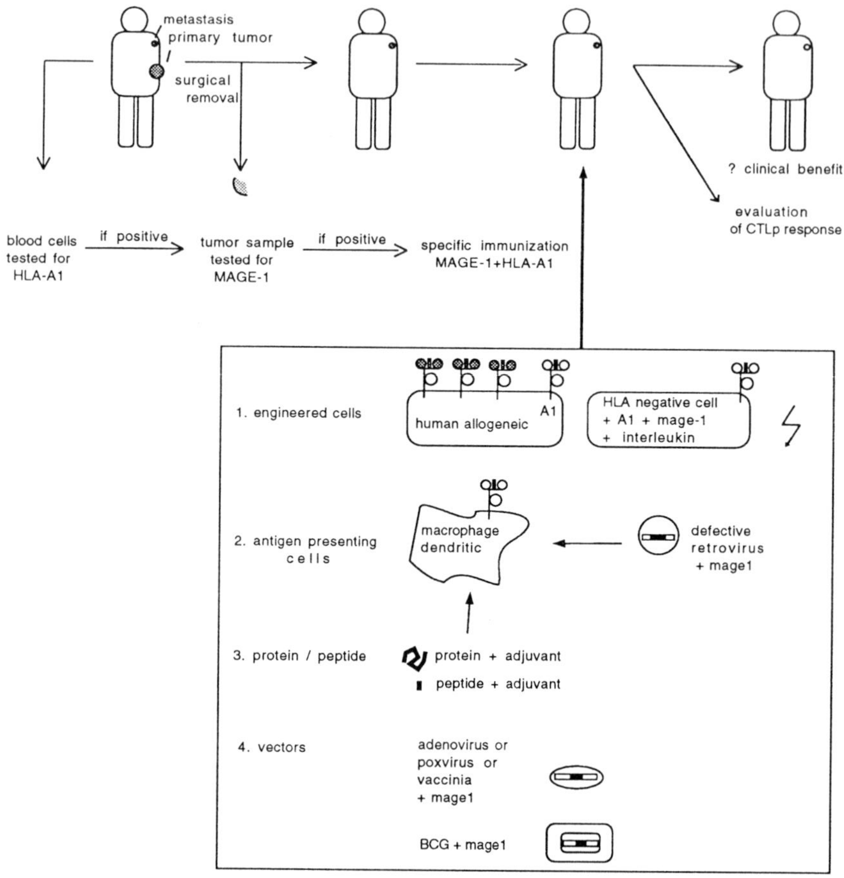

Fig. 3.1. *Strategy for immunotherapy directed against defined tumor antigens recognized by T lymphocytes.*

Another mode of immunization could be based on the ability of macrophages and dendritic cells to present peptides to CTL very efficiently (Aichele et al., 1990; Kast et al., 1991; Zheng et al., 1992). Macrophages or dendritic cells could be purified from the blood of patients, pulsed in vitro with peptide or protein and then reinfused into the patient. APC could also be infected in vitro with defective retroviruses carrying gene MAGE-1 (Cepko et al., 1984). Alternatively, patients could be immunized with the MZ2-E peptide combined with the appropriate adjuvant or with the entire MAGE-1 protein, in the expectation that it would be endocytosed by APC (Raychaudhuri et al., 1992). BCG bacteria carrying MAGE-1 also represent an interesting possibility (Stover et al., 1991). After phagocytosis of these bacteria, macrophages would present the tumor antigen on their class I MHC molecules. Finally, one should consider the use of poxviruses that do not multiply in human cells (Taylor et al., 1992). Cells infected with recombinant poxviruses containing gene MAGE-1 could transiently express the gene and present the MZ2-E peptide.

It would be very difficult to evaluate all these modes of immunization in clinical studies whose endpoint is represented by tumor regression or the absence of relapse. What may prove more feasible is to measure the frequency of CTL precursors directed against the MZ2-E antigen among blood lymphocytes of small groups of patients immunized with the various modalities. Such an approach is currently ongoing for patients from whom we have been able to derive a tumor cell line. We hope that it will be extended to situations in which autologous cell lines could not be established.

A successful immunization should generate a significant increase of these CTL precursors (Coulie et al., 1992). We intend to proceed as follows. First we will immunize eligible patients with the aim of finding a mode of immunization that reliably raises the frequency of CTL precursors directed against the antigen. If this can be achieved, a second wave of clinical trials will be attempted to find out if the raise in CTL frequency is associated with clinical benefit.

Conclusions

These are exciting prospects, but it is difficult to predict whether therapeutic success will be achieved, even if a significant increase in antitumor CTL is obtained by immunization.

Variants having lost the expression of MAGE-1 may arise and enable tumors to escape the immune response. Loss of HLA expression has been documented in many tumors, and this may render them refractory to immunotherapy (Lopez-Nevot et al., 1989). The prospects will undoubtedly improve if we can attack tumors through several antigens. This should improve the efficiency of the attack against antigenic cells. Moreover, it should decrease the probability of resistance due to antigen-loss variants. It is therefore of great interest to identify other genes coding for other tumor antigens. This should also increase the proportion of cancer patients that are eligible for specific immunization; at present this proportion is rather low: about 40% of melanoma tumors express MAGE-1, and about 26% of people in Caucasian populations express HLA-A1. Since expression of the MZ2-E antigen requires both MAGE-1 and HLA-A1 expression, this implies that at most 10% of patients can enter pilot studies. The identification of other antigens presented by different HLA class I molecules could raise the number of eligible patients. Among the other class I molecules, HLA-A2 appears to be a good candidate because it is present in almost 50% of the Caucasian population. Moreover, for melanomas and sarcomas, different groups reported the presence on HLA-A2 tumors of CTL-defined antigens that are shared by most tumors of the same histologic type. HLA-A2 was shown to be the presenting molecule for these common antigens (Slovin et al., 1986; Darrow et al., 1989; Wolfel et al., 1989). We recently found that one of these common antigens was encoded by the tyrosinase gene (Brichard et al., 1993). This gene is expressed in most melanomas but not in normal tissues, except for melanocytes. Because melanocytes also express class I MHC molecules, they presumably express the antigen as well. Whether they will be affected by immunization against tyrosinase and whether their destruction will generate important side effects remain to be assessed before considering clinical studies with tyrosinase. Meanwhile, the identification of other genes coding for tumor antigens is in progress.

Acknowledgments

V. Brichard is a Research Assistant with the Fonds National de la Recherche Scientifique, Brussels, Belgium.

References

Aichele P, Hengartner H, Zinkernagel RM and Schulz M (1990). Antiviral cytotoxic T cell response induced by in vivo priming with a free synthetic peptide. J Exp Med **171**, 1815-1820.

Boon T and Van Pel A (1978). Teratocarcinoma cell variants rejected by syngeneic mice: Protection of mice immunized with these variants against other variants and against the original malignant cell line. Proc Natl Acad Sci USA **75**, 1519-1523.

Brasseur F, Marchand M, Vanwijck R, Hérin M, Lethé B, Chomez P and Boon T (1992). Human gene MAGE-1, which codes for a tumor rejection antigen, is expressed by some breast tumors. Int J Cancer **52**, 839-841.

Brichard V, Van Pel A, Wolfel T, Wollfel C, De Plaen C, Lethé B, Coulie P and Boon T (1993). The tyrosinase gene codes for an antigen recognized by autologous cytolytic T lymphocytes on HLA-A2 melanomas. J Exp Med, **178**, 489-496.

Cepko C, Roberts BE and Mulligan RC (1984). Construction and applications of a highly transmissible murine retrovirus shuttle vector. Cell **37**, 1053-1062.

Coulie PG, Somville M, Lehmann F, Hainaut P, Brasseur F, Devos R and Boon T (1992). Precursor frequency analysis of human cytolytic T lymphocytes directed against autologous melanoma cells. Int J Cancer **50**, 289-297.

Darrow TL, Slingluff CL and Seigler HF (1989). The role of HLA class I antigens in recognition of melanoma cells by tumor-specific cytotoxic T lymphocytes. Evidence for shared tumor antigens. J Immunol **42**, 3329-3335.

Dealtry GB, Naylor MS, Fiers W and Balkwill FR (1987). DNA fragmentation and cytotoxicity caused by tumor necrosis factor is enhanced by interferon-γ. Eur J Immunol **17**, 689-693.

Degiovanni G, Hainaut P, Lahaye T, Weynants P and Boon T (1990). Antigens recognized on a melanoma cell line by autologous cytolytic T lymphocytes are also expressed on freshly collected tumor cells. Eur J Immunol **20**, 1865-1868.

Fearon ER, Pardoll DM, Itaya T, Golumbek P, Levitzky HI, Simons JW, Karasuyama H, Vogelstein B and Frost P (1990). Interleukin-2 production by tumor cells bypasses T helper function in the generation of an antitumor response. Cell **60**, 397-403.

Golumbek PT, Lazenby AJ, Levitsky HI, Jaffee LM, Karasuyama H, Baker M and Pardoll DM (1991). Treatment of established renal cancer by tumor cells engineered to secrete interleukin-4. Science **254**, 713-716.

Gross L (1943). Intradermal immunization of C3H mice against a sarcoma that originated in an animal of the same line. Cancer Res **3**, 326-333.

Hainaut P, Weynants P, Coulie PG and Boon T (1990). Anti-tumor T lymphocyte responses. Immunol Allergy Clin North Amer **10**, 639-662.

Hewitt H, Blake E and Walder A (1976). A critique of the evidence for active host defense against cancer based on personal studies of 27 murine tumors of spontaneous origin. Br J Cancer **33**, 241-259.

Kast WM, Roux L, Curren J, Blom HJJ, Voordouw AC, Meloen RH, Kolakofsky D and Melief CJM (1991). Protection against lethal Sendai virus infection by in vivo priming of virus-specific cytotoxic T lymphocytes with a free synthetic peptide. Proc Natl Acad Sci USA **88**, 2283-2287.

Klein G, Sjogren H, Klein E and Hellstrom KE (1960). Demonstration of resistance against methylcholanthrene-induced sarcomas in the primary autochtonous host. Cancer Res **20**, 1561-1572.

Lopez-Nevot MA, Esteban F, Ferron A, Gutierrez J, Oliva RM, Romero C, Huelin C, Ruiz-Cabello F and Garrido F (1989). HLA class I gene expression on human primary tumours and autologous metastases: Demonstration of selective losses of HLA antigens on colorectal, gastric and laryngeal carcinomas. Br J Cancer **59**, 221-226.

Middle J and Embleton M (1981). Naturally arising tumors of the inbred WAB/Not rat strain. II. Immunogenicity of transplanted tumors. J Natl Cancer Inst **67**, 637-643.

Prehn RT and Main JM (1957). Immunity to methylcholanthrene-induced sarcomas. J Natl Cancer Inst **18**, 769-778.

Ramsay AJ, Ruby J and Ramshaw IA (1993). A case for cytokines as effector molecules in the resolution of virus infection. Immunol Today **14**, 155-157.

Raychaudhuri S, Tonks M, Carbone F, Ryskamp T, Morrow WJW and Hanna N (1992). Induction of antigen-specific class I-restricted cytotoxic T cells by soluble proteins in vivo. Proc Natl Acad Sci USA **89**, 8308-8312.

Slovin SF, Lackman RD, Ferrone S, Kiely PE and Mastrangelo MJ (1986). Cellular immune response to human sarcomas: Cytotoxic T cell clones reactive with autologous sarcomas. I. Development, phenotype, and specificity. J Immunol **137**, 3042-3048.

Stover CK, de la Cruz VF, Fuerst TR, Burlein JE, Benson LA, Bennett LT, Bansal GP, Young LF, Lee MH, Hatfull GF, Snapper GB, Barletta RG, Jacobs WRJ and Bloom BR (1991). New use of BCG for recombinant vaccines. Nature **351**, 456-461.

Taylor J, Weinberg R, Tartaglia J, Richardson C, Alkhatib G, Briedis D, Appel M, Norton E and Paoletti E (1992). Nonreplicating viral vectors as potential vaccines: Recombinant canarypox virus expressing measles virus fusion (F) and hemagglutinin (HA) glycoproteins. Virology **187**, 321-328.

Tepper RI, Pattengale PK and Leder P (1989). Murine interleukin-4 displays potent anti-tumor activity in vivo. Cell **57**, 503-512.

Traversari C, Van der Bruggen P, Van den Eynde B, Hainaut P, Lemoine C, Ohta N, Old L and Boon T (1992a). Transfection and expression of a gene coding for a human melanoma antigen recognized by autologous cytolytic T lymphocytes. Immunogenetics **35**, 145-152.

Traversari C, Van der Bruggen P, Luescher IF, Lurquin C, Chomez P, Van Pel A, De Plaen E, Amar-Costesec A and Boon T (1992b). A nonapeptide encoded by human gene MAGE-1 is recognized on HLA-A1 by CTL directed against tumor antigen MZ2-E. J Exp Med **176**, 1453-1457.

Uyttenhove C, Maryanski J and Boon T (1983). Escape of mouse mastocytoma P815 after nearly complete rejection is due to antigen-loss variants rather than immunosuppression. J Exp Med **157**, 1040-1052.

Van den Eynde B, Lethé B, Van Pel A, De Plaen E and Boon T (1991). The gene coding for a major tumor rejection antigen of tumor P815 is identical to the normal gene of syngeneic DBA/2 mice. J Exp Med **173**, 1373-1384.

Van der Bruggen P, Traversari C, Chomez P, Lurquin C, De Plaen E, Van den Eynde B, Knuth A and Boon T (1991). A gene encoding an antigen recognized by cytolytic T lymphocytes on a human melanoma. Science **254**, 1643-1647.

Van Pel A, Vessière F and Boon T (1983). Protection against two spontaneous mouse leukemias conferred by immunogenic variants obtained by mutagenesis. J Exp Med **157**, 1992-2001.

Wolfel T, Klehmann E, Muller C, Schutt KH, Meyer zum Buschenfelde KH and Knuth A (1989). Lysis of human melanoma cells by autologous cytolytic T cell clones. Identification of human histocompatibility leukocyte antigen A2 as a restriction element for three different antigens. J Exp Med **170**, 797-810.

Zheng B, Brett SJ, Tite JP, Lifely MR, Brodie TA and Rhodes J (1992). Galactose oxidation in the design of immunogenic vaccines. Science **256**, 1560-1563.

4

IMMUNOLOGICAL RECOGNITION OF HUMAN MELANOMA: LESSONS FROM THE CLONAL ANALYSIS OF CYTOTOXIC T-CELL MEDIATED RESPONSE

A. Anichini, C. Maccalli, A. Mazzocchi, R. Mortarini, S. Salvi, M. L. Sensi and G. Parmiani
Division of Experimental Oncology D, Istituto Nazionale Tumori, Milan, Italy

Introduction

By the beginning of the past decade the development of techniques for the cloning and functional analysis of human T lymphocytes (Zagury et al., 1983) provided for the first time the experimental tools needed to address the issue of function and specificity of T cells reacting to human tumors.

During the last 10 years the application of these techniques, together with the general evolution in cellular and molecular procedures, has allowed to draw the first clear picture of the cellular and molecular basis for the host-tumor interaction as determined by analysis of antitumor CTL.

This has been accomplished by measuring the frequency, phenotype, specificity and mechanism of interaction of antitumor CTL clones. These cellular reagents have also been instrumental in the process leading to the identification of the first gene coding for a human CTL-defined tumor antigen (Van der Bruggen et al., 1991). By far the most informative results of this analysis have been obtained in the field of immune response to human melanoma.

This chapter outlines the current understanding of the host-tumor interaction as determined by studies performed with CTL clones isolated from melanoma patients. In addition, we will point out the challenges that still lie ahead before the full impact of the results of basic research can be translated in those clinical protocols aimed at the therapy of melanoma by exploiting immunological means.

CYTOKINE-INDUCED TUMOR IMMUNOGENICITY
ISBN 0-12-261520-4

Antigens recognized on human melanoma by autologous CTL clones: on breaking the tolerance to self

The analysis of specificity and mechanism of interaction of CTL clones directed against human melanomas has shown that some CTL clones, isolated from different patients and expressing the αβ form of the receptor, can interact with the autologous tumor in TCR dependent and HLA restricted fashion (reviewed in Anichini et al., 1987). This indicates that human melanoma associated antigens (MAA) are recognized through the same mechanism that controls the engagement of the TCR directed to conventional antigens, specifically the triggering of the receptor is induced by the interaction with the appropriate complex between MHC molecules and antigenic peptide (Davis and Bjorkman, 1988). In addition, the cellular analysis of the specificity of the CTL clones suggests that at least a fraction of clones reacts with autologous and allogeneic melanomas sharing the right MHC restricting element (Crowley et al., 1991; Anichini et al., 1993a).

Furthermore, some HLA class I alleles such as HLA-A2, the most frequent HLA-class I allele in the caucasian population, behave as dominant molecules for the restriction of MAA (Crowley et al., 1991). This means that the majority of CTL clones from a single HLA-A2$^+$ patient appear to be HLA-A2 restricted, even though the same patient has additional HLA class I alleles apparently not used to present MAA. These results led to the theory that most human melanomas are antigenic and express common antigens frequently seen in association with the same HLA allele (Kawakami et al., 1992). As expected from previous studies in animal models, HLA-A2 is not the only restricting element for the recognition of melanoma. In fact, the first melanoma antigen that has been characterized at the molecular level is a peptide originating from a family of genes named MAGE and is recognized in association with HLA-A1 (Van der Bruggen et al., 1991). In addition (Table 3.1), other HLA class I alleles, such as A3, A9 and DR-2, have been reported to restrict the recognition of MAA by CTL (Chen and Hersey, 1992; Anichini et al., 1993a). The molecular characterization of the MAGE family of genes has confirmed that these genes are expressed in transformed cells even outside the melanocyte lineage; normal tissues do not express MAGE 1 genes, with the exception of the testis (Boon, 1993). These results indicate that at least some of the antigens recognized on melanoma cells originate from the expression of normal genes and not from mutations, as occurs for the genesis of antigenic determinants encoded by mutated oncogenes of the *ras* family (Jung and Chluesener, 1991). Furthermore, the finding that the MAGE genes are expressed in at least one normal human tissue suggests that T cells potentially reactive to the corresponding peptides are not deleted by thymic selection. This hypothesis implies that the procedures employed to isolate MAGE 1 specific CTL clones have been able to break a state of tolerance to self antigens.

Table 3.1. *HLA class I and II alleles that restrict recognition of melanoma associated antigens by CTL clones*

HLA Allele	References
HLA-A1	Van der Bruggen et al., 1991
HLA-A2	Crowley et al., 1991
	Kawakami et al., 1992
	Hom et al., 1993
	Anichini et al., 1993b
HLA-A3	Chen and Hersey, 1992
HLA-A9	Anichini et al., 1993a
HLA-A24	Hom et al., 1993
HLA-B8	Hom et al., 1993
HLA-Cw7	Homet al., 1993
HLA-DR2	Chen and Hersey, 1992

According to results obtained with triple transgenic mice, a possible mechanism for breaking peripheral tolerance to self antigens is critically dependent on the chronic exposure of tolerant T cells to IL-2 and antigen (Heath et al., 1992). This situation is precisely what normally happens in vitro for the activation and cloning of antitumor T lymphocytes. In fact, most cloning procedures are based on a mixed lymphocyte-tumor culture prolonged for a few weeks in the presence of low dose IL-2, followed by cloning in the presence of tumor and further IL-2 (Anichini et al., 1989). Therefore, by the time T-cell clones are screened for function and specificity, these cells have already been exposed to IL-2 for more than a month. This period may be sufficient to break the tolerance status. Alternatively, tolerance could have been broken in vivo when a local increase of cytokines may have occurred, as a consequence either of a direct release by melanoma cells themselves (Colombo et al., 1992) or to inflammatory-like reactions mediated by infiltrating cells (Heath et al., 1992).

It is thus possible that even the common melanoma associated antigens detected by HLA-A2 restricted CTL are indeed self antigens frequently expressed by cells sharing their origin from the melanocyte lineage. If this hypothesis is true, then we argued that normal HLA-A2$^+$ melanocytes should be recognized by the same CTL that react with HLA-A2$^+$ melanoma cells. Evidence consistent with this hypothesis has been recently obtained in our laboratory (Anichini et al., 1993a). In fact, HLA-A2$^+$ CTL clones isolated from TIL and PBL of two patients lysed only HLA-A2$^+$ melanomas and HLA-A2$^+$ melanocytes, but not HLA-A2$^+$ or HLA-A2$^-$ normal and neoplastic cells of nonmelanocyte lineage in a panel of almost 50 targets (Anichini et al., 1993a). Furthermore, the recognition of melanocytes was not related to the proliferative status of these cells since both proliferating and growth arrested melanocytes were recognized. These results indicated that normal and neoplastic cells of the melanocyte lineage may express tissue related genes coding for normal proteins that, in turn, originate peptides recognized by autologous HLA-A2 restricted CTL clones.

Clearly, it is not yet known whether normal melanocytes express in vivo these tissue related antigens. This is a key question to be resolved to allow a safe use of tissue-antigen specific CTL clones in the therapy of melanoma. Furthermore, additional relevant evidence is lacking about the possibility that the potentially "autoreactive" T cells can be activated in vivo to become effective antitumor CTL. However, recent results from a sequential chemoimmunotherapy protocol, performed by combining chemotherapy with IFN-γ and IL-2, indicated that a significant subset of patients responding to therapy also developed a depigmentation syndrome most likely resulting from melanocyte destruction (Richards et al., 1992). Whether or not the effects observed by Richards and coworkers reflect the activation of autoreactive T cells capable of recognizing tissue-related antigens expressed only on normal and neoplastic cells of the melanocyte lineage requires further investigation.

It also remains to be defined whether T cells reacting to melanocyte lineage antigens can be exploited therapeutically or whether the potential destruction of normal melanocytes represents a clinically unacceptable risk. Alternatively, we should wait for the identification of truly tumor specific antigens. With the notable exception of antigens coded for by mutated oncogenes (Jung and Chluesener, 1991), evidence is still lacking for the existence of tumor associated antigens recognized by CTL clones that may satisfy the requirements of being frequently expressed on tumors from different patients and of being expressed only on the neoplastic cells but not on any normal tissue of each patient.

TCR repertoire against melanoma associated antigens: evidence for selective TCR expression at the tumor site

The results of the cellular analysis of the T-cell mediated response to melanoma have indicated that this tumor expresses an antigenic repertoire of unknown size. This repertoire of antigens might be recognized by a corresponding repertoire of TCR. Therefore, it is reasonable to speculate that the cellular basis of CTL mediated response to melanoma will be polyclonal.

The first evidence that this might be true came from experiments aimed at sequencing the TCR α and β chains of CTL clones expressing the same specificity (Sensi et al., 1991). That study indicated that the two clones studied differed at the TCR level even though their intratumor specificity was identical, as determined by differential lysability of melanoma clones isolated from the same metastatic tissue. Further studies have confirmed that even TIL represent an heterogeneous population composed of many different TCR (Ferradini et al., 1992). These results are not unexpected since an analysis of the TCR structure carried out without knowledge of the function of the TIL cannot discriminate between antitumor T cells and "innocent bystanders". A different approach has been that of characterizing the TCR structure of long term in vitro cultures of TIL (Morita et al., 1992). The results pointed towards an oligoclonal composition of the TIL population in terms of number of different TCR expressed. However, even in the study of Morita and coworkers it is not clear whether the selection for a few TCR occurs in vitro as a result of in vivo exposure to antigen(s) or as a result of differences among T cells in their growth ability in response to IL-2.

These studies should also be compared with findings on the TCR structure obtained in other human diseases that may have an autoimmune origin. In fact, the analysis of the TCR structure and relative frequency carried out in rheumatoid arthritis (Paliard et al., 1991), multiple sclerosis (Oksenberg et al., 1993), and autoimmune thyroid disease (Davies et al., 1992) has shown that there is a selective accumulation of T lymphocytes expressing a TCR that may be characterized either by a common V region segment or by a common V-D-J gene rearrangement. The finding of an amplification of a single V region segment is interpreted as evidence of the action of a superantigen (Paliard et al., 1991). Similar results, pointing towards the evidence of a predominant expression at the tumor site of T cells expressing a common V gene segment, have been obtained in ocular melanoma, where a predominant expression of $V\alpha$-7^+ TCR was found (Nitta et al., 1990). However, given the lack of functional analysis of these lymphocytes, one can not exclude that ocular melanoma cells may express a superantigen with affinity for the $V\alpha$-7^+ TCR. This would explain why the same population of TCR can be found overexpressed in different patients independently of the likely differences of MHC genotypes among unrelated individuals.

A second type of evidence, recently obtained in multiple sclerosis, is that the selective TCR amplification is limited to a specific V-D-J junctional sequence (Oksenberg et al., 1993). These results suggest that presentation of an antigenic peptide in the context of MHC may select/expand a single T-cell clone expressing a unique TCR. Evidence consistent with this possibility has been recently described in a case of regressive primary melanoma of the leg (Ferradini et al., 1993). The authors found evidence of selective amplification in the tumor of T cells defined by a unique V-D-J junctional sequence. However, attempts were not made to correlate the structure of the unique TCR with the function of the T cells bearing these receptors. Therefore, it is not possible to draw conclusions on the possibility that a melanoma associated antigen may be responsible for the observed bias in TCR distribution in comparison to the peripheral circulation.

More recently, we took a different approach aimed first at isolating and characterizing the antigenic specificity of CTL clones from TIL of melanoma patients and then at assessing whether the specific V-D-J junctional sequence of these CTL clones was preferentially expressed in the fresh tumor sample. The results (Sensi et al., submitted for publication) indicated that, indeed, the specific V-D-J junctional sequence expressed by an HLA-A2 restricted and melanocyte lineage specific TIL clone was selectively amplified in the fresh tumor specimen in comparison with the peripheral circulation, thus suggesting that a common melanocyte lineage specific antigen may drive in vivo a selective accumulation/expansion of an antigen specific T-cell clone.

These results raise the problem of understanding why the local response to the tumor, as detected by accumulation of a tumor reactive T-cell clone, does not succeed in eradicating the neoplastic lesion. A further relevant issue to be addressed is whether the same TCR can be found expressed and amplified in the tumor tissue of different patients whose tumors can be recognized by the same CTL clone. If this were the case, then it would be possible to identify the most relevant TCR associated with the CTL response to melanoma and to devise ways of using only antigen specific T cells for therapy of melanoma (Parmiani et al., 1990).

Impact of tumor heterogeneity on CTL mediated recognition of melanoma

Even though the cellular and molecular characterization of antigens recognized on melanoma cells by CTL clones can provide information essential to an understanding of the extent of the antigenic repertoire of melanoma, the clinical usefulness of this knowledge will be reduced unless means can be found to overcome the problem of tumor heterogeneity with regard to antigen expression. In fact, as shown by other laboratories as well as by our own, neoplastic populations isolated from primary and metastatic lesions of melanoma are heterogeneous in the expression of antigens recognized by autologous CTL (Anichini et al., 1989). This heterogeneity can be shown by cloning the tumor at an early in vitro passage and then by screening tumor clones for recognition by a single CTL clone. This analysis often reveals that some melanoma clones cannot be recognized by a given CTL clone (Anichini et al., 1989). An alternative approach consists of isolating antigen negative variants from the same tumor by multiple cycles of immunoselection with the same CTL clone (Van den Eynde et al., 1989). Both these approaches indicate that the same human melanoma can express multiple antigens, but not all neoplastic cells from the same tumor express all antigens.

A second mechanism leading to escape from T-cell mediated recognition is the down-modulation or loss of expression of HLA alleles acting as restricting elements for the recognition of MAA. This may occur in vivo, since an impairment in the expression of alleles as HLA-A2 on melanoma cells has been reported (Natali et al., 1989). The first consequence of the loss of peptides and of HLA alleles is that antigen negative or MHC negative variants will escape recognition by specific CTL clones. Every clinical approach to the immunological therapy of melanoma will have to be based on methods that can activate multiple subsets of T cells directed to distinct peptide/HLA complexes, otherwise a fraction of the neoplastic population will always survive.

In a recent study we addressed this issue by attempting to overcome the phenomenon of intratumor susceptibility to lysis by CTL clones using bispecific mAb (Nistico et al., 1992). The results of that study indicated that it is possible to induce CTL mediated lysis of otherwise resistant tumor clones by employing bispecific antibodies directed with one arm to the TCR associated CD3 complex and with the other arm to a commonly expressed MAA. This procedure can bypass the absence of TCR triggering due either to absence of antigen or to the absence of the MHC restricting element and is dependent uniquely on the expression on the tumor of the MAA recognized by the second arm of the bispecific antibody. It should, however, be noted that tumor heterogeneity affects also the MAA recognized by mAb. In light of a possible clinical use of bispecific antibodies, it will be necessary to produce pools of mAb made of multiple independent bispecific antibodies each capable of recognizing different MAA in addition to the CD3 complex.

Role of adhesion molecules in the CTL-tumor interaction

Heterogeneity is a hallmark of tumor biology and represents the manifestation of the evolutive processes that generate neoplastic cells with more aggressive behavior, thus providing a mechanism for tumor progression. This mechanism is effective also in contributing to the differential expression of adhesion molecules on melanoma cells from the same tumor. In a recent study we found that melanoma clones isolated from the same metastasis were heterogeneous for the expression of the adhesion molecule ICAM-1, and this heterogeneity correlated with the susceptibility of the neoplastic cells to lysability by specific and nonspecific T-cell clones (Anichini et al., 1990b). Even though all tumor clones expressed ICAM-1 molecules, only the clones with the highest expression, as determined by intensity of fluorescence by FACS analysis with anti-ICAM-1 antibodies, were efficiently recognized by all cytolytic effectors. More importantly, up-regulation of ICAM-1, either by treatment with IFN-γ or by gene transfer in low expressor clones, could boost tumor lysability only by specific, TCR dependent and HLA restricted effectors (Anichini et al., 1993b). These results were interpreted as indicating that efficient recognition of the neoplastic cells can be obtained only when T cells receive the appropriate activation signals from both the TCR and the ICAM-1 ligand, LFA-1. In antigen independent interactions, such as those between LAK cells and tumor cells, the mechanism of lysis is not triggered by TCR engagement. In these situations a signalling cooperation between TCR and LFA-1 is not possible. This may explain why LAK cells and nonspecific CTL clones are less sensitive than antigen specific CTL to the relative levels of ICAM-1 on the tumor.

Our results can also help to explain the apparent paradox due to the fact that expression of ICAM-1 is tightly regulated in association with tumor progression of melanoma (Johnson et al., 1989). Normal melanocytes and early primary melanomas do not express this molecule that appears during the conversion from the radial growth phase to the vertical growth phase and is always expressed in metastatic melanoma. The paradox originates from the notion that ICAM-1 is a fundamental adhesion molecule that regulates cellular interactions, and yet its expression does not appear to confer a better immunological "visibility" to metastatic melanoma cells in comparison to primary ICAM-1$^-$ tumors.

Our results with melanoma clones expressing various levels of this adhesion molecule indicate that the relative amount of ICAM-1 is a key factor for the outcome of the interaction with T cells. Therefore, it is possible that many ICAM-1$^+$ neoplastic cells, even within metastatic tumors, do not express this molecule to levels high enough for an efficient interaction with antigen specific T cells.

A further mechanism that can help melanoma cells to escape recognition by T cells is likely to be related to the serum levels of soluble ICAM-1 in melanoma patients. In fact, even in stage I patients, ICAM-1 serum levels are significantly elevated in comparison to normal controls and are inversely correlated with survival (Harning et al., 1991). Further studies, performed on human melanomas growing in nude mice, confirmed that the soluble, circulating form of ICAM-1 is indeed released by the neoplastic cells (Giavazzi et al., 1992). Therefore, the ability of melanoma to release ICAM-1 might provide a mechanism to block the ICAM-1 ligand on T cells, LFA-1, thus inhibiting the function of a key adhesion and signalling pathway for the function of T cells.

Other families of adhesion molecules, such as the extracellular matrix receptors known as integrins, can play a role in the lymphocyte-tumor interaction, and the differential expression of these molecules may have an impact on susceptibility to lysis similar to that produced by the differential expression of ICAM-1. By analyzing a panel of melanoma clones isolated from the same metastasis we found that VLA-1 and VLA-3 antigens were expressed on all tumor cells. However, only a subset of clones expressed VLA-2, VLA-5 and VLA-6 (Anichini et al., 1990b). Further studies indicated that clones with expression of VLA-2, VLA-5 and VLA-6 also expressed the highest susceptibility to lysis by LAK cells, by CD3$^+$ and CD3$^-$ IL-2 activated lymphocytes, as well as by specific and nonspecific CTL clones (Anichini et al., 1990a).

The potential role of integrins in this system was addressed in a recent study showing that indeed some VLA antigens can regulate the interaction of melanoma clones with specific and non specific T-cell clones (Anichini et al., 1992). In fact, blocking experiments with mAb confirmed that preincubation of VLA-2$^+$, -5$^+$ -6$^+$ melanoma clones with the corresponding anti-integrin antibodies could significantly reduce the lysability of the neoplastic cells by both specific and nonspecific CTL clones. No effect was seen by preincubating VLA-2⁻, VLA-5⁻ and VLA-6⁻ tumor clones with the same antibodies (Anichini et al., 1992). This indicated that integrins on melanoma cells do not act only as mediators of interaction with extracellular matrix components but also as mediators of interaction with T cells. Furthermore, determination of conjugate formation excluded that the clones expressing VLA-2, VLA-5 and VLA-6 could bind more efficiently to T-cell clones. This finding suggests that the increased lysability of some tumor clones expressing VLA-2, -5 and -6 may be dependent on activation signals delivered to the T cells. However, it remains unclear which receptors expressed on the T-cell clones can mediate the signals received upon interaction with tumor integrins.

Role of cytokines in the antitumor response mediated by CTL

The specific and HLA-restricted interaction of T lymphocytes with autologous tumors leads to the release of a set of cytokines and other soluble factors such as IL-2, IFN-γ and GM-CSF (Schwartzentruber et al., 1991; Salmeron et al., 1992; Hom et al., 1993). As described in other chapters of this book, these factors may contribute to the overall immune response to the tumor by a number of different mechanisms, such as the activation of distinct T-cell subsets (by IL-2), the up-regulation of MHC antigens and adhesion molecules on the tumor (by TNF-α and IFN-γ), and the recruitment/activation at the tumor site of macrophages and granulocytes (by GM-CSF).

A number of cytokines can also exert a direct antiproliferative or even cytotoxic effect on human melanoma. Results from our laboratory indicate that cytokines like IL-1β, IL-4, TNF-α and IFN-γ can inhibit the proliferation of human melanoma cells (Mortarini et al., 1990). The best results, in terms of the highest inhibition of proliferation on most melanomas, were observed by combining TNF-α and IFN-γ. The combination of these two cytokines could, in fact, bypass the heterogeneity among different tumors and among clones from the same tumor for susceptibility to a single cytokine.

Both TNF-α and IFN-γ can be released by T cells upon interaction with melanoma. This production may contribute to tumor cell destruction by a mechanism distinct from the direct lysis due to cell-cell contact between a CTL and the tumor cell. Evidence from our laboratory indicates that a single T-cell clone may mediate lysis of autologous melanoma through multiple mechanisms. The first mechanism depends on a direct effector-target interaction and is responsible for the tumor lysis seen within 18 h of the initiation of the interaction. Other mechanisms of tumor destruction are not strictly dependent on cell-cell contact. In fact, supernatants from a single CTL clone were shown to contain TNF-α, TNF-β and IFN-γ that can mediate further lysis of the tumor between 24 and 96 h after the initiation of the effector-target interaction (Mazzocchi et al., 1990). Experiments performed with supernatants from an antigen stimulated CTL clone indicate that the lytic activity on the tumor cells can be abrogated by antibodies directed against the individual cytokines released by the T cells. Furthermore, release of cytotoxic cytokines as TNF-α, triggered by TCR engagement, may be even more important to tumor destruction than direct lysis by cell-cell contact. This conclusion is corroborated by a more recent study indicating that some neoplastic clones from the same melanoma metastasis could not be lysed efficiently by a single CTL clone even in an 18 h assay (Mazzocchi et al., 1991). However, 96 h assays showed that susceptibility of the tumor cells to the cytotoxic effect of TNF-α, released by the CTL clone, was responsible for the lysis of all tumor clones independently of their susceptibility to the direct lytic activity of the same effector. These results were confirmed by in vivo Winn assays performed in nude mice which showed that anti-CD3 stimulated cells from a single clone could abrogate the growth of the tumor clones resistant to the direct cytotoxic activity of the effector (Mazzocchi et al., 1991). These data indicate that soluble factors released by CTL clones may contribute significantly to the local immune response to the tumor through a number of different mechanisms.

References

Anichini A, Fossati G and Parmiani G (1987). Clonal analysis of the cytolytic T cell response to human tumors. Immunol Today **8**, 385-388.

Anichini A, Mazzocchi A, Fossati G and Parmiani G (1989). Cytotoxic T lymphocyte clones from peripheral blood and from tumor site detect intratumor heterogeneity of melanoma cells. Analysis of specificity and mechanisms of interaction. J Immunol **142**, 3692-3701.

Anichini A, Mortarini R, Berti E and Parmiani G (1990a). Multiple VLA antigens on a subset of melanoma clones. Human Immunol **28**, 119-122.

Anichini A, Mortarini R, Supino R and Parmiani G (1990b). Human melanoma cells with high susceptibility to cell-mediated lysis can be identified on the basis of ICAM-1 phenotype, VLA profile and invasive ability. Int J Cancer **46**, 508-515.

Anichini A, Mortarini R and Parmiani G (1992). β-1 integrins on melanoma clones regulate the interaction with autologous cytotoxic T cell clones. J Immunother **12**, 183-186.

Anichini A, Maccalli C, Mortarini R, Salvi S, Mazzocchi A, Squarcina P,

Herlyn M and Parmiani G (1993a). Melanoma cells and normal melanocytes share antigens recognized by HLA-A2-restricted cytotoxic T cell clones from melanoma patients. J Exp Med **177**, 989-998.

Anichini A, Mortarini R, Alberti S, Mantovani A and Parmiani G (1993b). T-cell-receptor engagement and tumor ICAM-1 up-regulation are required to by-pass low susceptibility of melanoma cells to autologous CTL-mediated lysis. Int J Cancer **53**, 994-1001.

Boon T (1993). Genes encoding melanoma antigens. Melanoma Res **3**, supplement 1, 5-6.

Chen Q and Hersey P (1992). MHC-restricted responses of $CD8^+$ and $CD4^+$ T-cell clones from regional lymph nodes of melanoma patients. Int J Cancer **51**, 218-224.

Colombo MP, Maccalli C, Mattei S, Melani C, Radrizzani M and Parmiani G (1992). Expression of cytokine genes, including IL-6, in human malignant melanoma cell lines. Melanoma Res **2**, 181-189.

Crowley NJ, Darrow TL, Quinn-Allen MA and Seigler HF (1991). MHC-restricted recognition of autologous melanoma by tumor-specific cytotoxic T cells. Evidence for restriction by a dominant HLA-A allele. J Immunol **146**, 1692-1699.

Davies TF, Martin A, Concepcion ES, Graves P, Lahat N, Cohen WL and Ben-Nun A (1992). Evidence for selective accumulation of intrathyroidal T lymphocytes in human autoimmune thyroid disease based on T cell receptor V gene usage. J Clin Invest **89**, 157-162.

Davis MK and Bjorkman PJ (1988). T-cell antigen receptor genes and T-cell recognition. Nature **334**, 395-402.

Ferradini L, Roman-Roman S, Azocar J, Avril MF, Viel S, Triebel F and Hercend T (1992). Analysis of T cell receptor α/β variability in lymphocytes infiltrating a melanoma metastasis. Cancer Res **52**, 4649-4654.

Ferradini L, Mackensen A, Genevee C, Bosq J, Duvillard P, Avril MF and Hercend T (1993). Analysis of T cell receptor variability in tumor-infiltrating lymphocytes from a human regressive melanoma. J Clin Invest **91**, 1183-1190.

Giavazzi R, Chirivi RG, Garofalo A, Rambaldi A, Hemingway I, Pigott R and Gearing AJ (1992). Soluble intercellular adhesion molecule 1 is released by human melanoma cells and is associated with tumor growth in nude mice. Cancer Res **52**, 2628-2630.

Harning R, Mainolfi E, Bystryn JC, Henn M, Merluzzi VJ and Rothlein R

(1991). Serum levels of circulating intercellular adhesion molecule 1 in human malignant melanoma. Cancer Res **51**, 5003-5005.

Heath WR, Allison J, Hoffmann MW, Schonrich G, Hammerling G, Arnold B and Miller JF (1992). Autoimmune diabetes as a consequence of locally produced interleukin-2. Nature **359**, 547-549.

Hom SSH, Schwartzentruber DJ, Rosenberg SA and Topalian SL (1993). Specific release of cytokines by lymphocytes infiltrating human melanomas in response to shared melanoma antigens. J Immunother **13**, 18-30.

Johnson JP, Stade BG, Holzmann B, Schwanle W and Riethmuller G (1989). De novo expression of intercellular-adhesion molecule 1 in melanoma correlates with increased risk of metastasis. Proc Natl Acad Sci USA **86**, 641-644.

Jung S and Chluesener HJ (1991). Human T lymphocytes recognize a peptide of single point-mutated oncogenic ras proteins. J Exp Med **173**, 273-275.

Kawakami Y, Zakut R, Topalian SH, Stotter H and Rosenberg SA (1992). Shared melanoma antigens: Recognition by tumor-infiltrating lymphocytes in HLA-A2.1-transfected melanomas. J Immunol **148**, 638-643.

Mazzocchi A, Anichini A, Castelli A, Sensi ML, Poli F, Russo C and Parmiani G (1990). T lymphocytes can mediate lysis of autologous melanoma cells by multiple mechanisms: Evidence with a single T cell clone. Cancer Immunol Immunother **32**, 13-21.

Mazzocchi A, Rodolfo M, Parmiani G and Anichini A (1991). An autologous T cell clone overcomes intra-melanoma heterogeneity for susceptibility to cell-mediated lysis by using multiple lytic mechanisms: In vitro and in vivo analysis. Melanoma Res **1**, 169-176.

Morita T, Salmeron MA, Moser RP, Ross MI and Itoh K (1992). Oligoclonal expansion of Vβ8$^+$ cells in interleukin-2 activated T cells residing in subcutaneous metastatic melanoma. Clin Exp Metastasis **10**, 69-73.

Mortarini R, Belli F, Parmiani G and Anichini A (1990). Cytokine-mediated modulation of HLA-class II, ICAM-1, LFA-3 and tumor-associated antigen profile of melanoma cells. Comparison with anti-proliferative activity by rIL-1β, rTNF-α, rIFN-γ, rIL-4 and their combinations. Int J Cancer **45**, 334-341.

Natali PG, Nicotra MR, Bigotti A, Venturo I, Marcenaro L, Giacomini P and Russo C (1989). Selective changes in expression of HLA class I polymorphic determinants in human solid tumors. Proc Natl Acad Sci USA **86**, 6719-6723.

Nistico P, Mortarini R, De Monte LB, Mazzocchi A, Mariani M, Malavasi F, Parmiani G, Natali PG and Anichini A (1992). Cell retargeting by bispecific monoclonal antibodies: Evidence of bypass of intratumor susceptibility to cell lysis in human melanoma. J Clin Invest **90**, 1093-1099.

Nitta T, Oksenberg JR, Rao NA and Steinman L (1990). Predominant expression of T cell receptor Vα7 in tumor-infiltrating lymphocytes of uveal melanoma. Science **249**, 672-674.

Oksenberg JR, Panzara MA, Begovich AB, Mitchell D, Erlich HA, Murray RS, Shimonkevitz R, Sherritt M, Rothbard J, Bernard CCA and Steinman L (1993). Selection for T-cell receptor Vβ-Dβ-Jβ gene rearrangements with specificity for a myelin basic protein peptide in brain lesions of multiple sclerosis. Nature **362**, 68-70.

Paliard X, West SG, Lafferty JA, Clements JR, Kappler JW, Marrack P and Kotzin BL (1991). Evidence for the effects of a superantigen in rheumatoid arthritis. Science **253**, 325-329.

Parmiani G, Anichini A and Fossati G (1990). Cellular immune response against autologous human malignant melanoma: Are studies providing a framework for a more effective immunotherapy? J Natl Cancer Inst **82**, 361-370.

Richards JM, Mehta N, Ramming K and Skosey P (1992). Sequential chemoimmunotherapy in the treatment of metastatic melanoma. J Clin Oncol **8**, 1338-1343.

Salmeron MA, Morita T, Seki H, Platsoucas CD and Itoh K (1992). Lymphokine production by human melanoma tumor-infiltrating lymphocytes. Cancer Immunol Immunother **35**, 211-217.

Schwartzentruber DJ, Topalian SL, Mancini M and Rosenberg SA (1991). Specific release of granulocyte-macrophage colony-stimulating factor, tumor necrosis factor-α and IFN-γ by human tumor-infiltrating lymphocytes after autologous tumor stimulation. J Immunol **146**, 3674-3681.

Sensi ML, Anichini A., Grossberger D, Mazzocchi A, Mortarini R and Parmiani G (1991). Two autologous melanoma-specific and MHC-restricted human T cell clones with identical intra-tumor reactivity do not share the same TCR Vα and Vβ gene families. Melanoma Res **1**, 261-271.

Van den Eynde B, Hainaut P, Herin M, Knuth A, Lemoine C, Weynants P, Van der Bruggen P, Fauchet R and Boon T (1989). Presence on a human melanoma of multiple antigens recognized by autologous CTL. Int J Cancer **44**, 634-640.

Van der Bruggen P, Traversari C, Chomez P, Lurquin C, De Plaen E, Van Den Eynde B, Knuth A and Boon T (1991). A gene encoding an antigen recognized by cytolytic T lymphocytes on a human melanoma. Science **254**, 1643-1647.

Zagury A, Morgan D, Lenoir, G Fouchard M and Feldman M (1983). Human normal CTL clones: Generation and properties. Int J Cancer **31**, 427-432.

5

ACTIVE SPECIFIC IMMUNOTHERAPY IN HUMANS

A. N. Houghton and J. J. Lewis
Memorial Sloan-Kettering Cancer Center and Cornell University Medical College, New York, NY, USA

Introduction

Immunotherapy of cancer has traditionally been categorized as active (e.g. cancer vaccines), passive (e.g. adoptive cellular therapy or mAb therapy), and nonspecific (e.g. cytokine therapies). Perhaps the most absorbing approach has been active immunization, or vaccination, a strategy that dates back more than one hundred years. In some respects, the development of cancer vaccines parallels the evolution of viral and bacterial vaccines, although the state of cancer vaccines is quite primitive by comparison.

Despite similarities, there are important differences in vaccines against infectious agents versus tumor. Viral and bacterial antigens are clearly foreign to the host, while cancer antigens reflect either self or altered self components. Also, the viral gene structure is comparatively simple, encoding a finite set of defined antigens. On the other hand, immunogenic cancer antigens remain poorly defined, and the potential universe of antigens expressed by a single cancer cell is vast.

Thus, approaches to constructing cancer vaccines often have been torturous, unsophisticated and naive. This is a result of the substantial obstacles that have been, and continue to be, faced. Some have even argued that the whole concept of active vaccination against cancer is not valid. We will not review all the pitfalls in the field. Rather, for the purposes of this volume, we will give an overview of strategies for cancer vaccination in humans, based on the recent molecular identification of tumor antigens recognized by autoantibodies and, more recently, by autoreactive T lymphocytes.

In particular, we will focus on the two components of the immune system that carry specificity for an immune response - antibodies and T lymphocytes. We will address two fundamental questions for tumor immunology. First, is the immune system capable of rejecting tumors? Second, can the immune system specifically recognize antigens expressed by cancer cells?

CYTOKINE-INDUCED TUMOR IMMUNOGENICITY
ISBN 0-12-261520-4

Historical perspectives

Understanding of the immune response evolved with the germ theory of illness. Edward Jenner first reported inoculation against smallpox in 1798, almost a century before immunology became an established science (in the 19th century). The advent of modern clinical immunology arguably began with a report from Professor Louis Pasteur that immunization against chicken cholera was possible. These experiments led to the practical application of prophylactic immunization for infectious disease. Even at that time, it was recognized that a vaccine strategy might be applied to cancer. It was in 1893 that the cancer surgeon Dr William Coley first reported tumor regressions induced by immune system activation with bacterial toxins, called "Coley's vaccine". Although not strictly active immunization, Coley's studies opened the way for the field of cancer immunotherapy.

It soon became apparent that immunization with tumor cells was considerably more difficult than with infectious agents. Early studies were complicated by a lack of understanding of transplantation (histocompatibility) antigens. Experiments in mice and other species showed that immunization with allogeneic tumor cells could induce an immune response and tumor rejection, but this was due to recognition of transplantation antigens. Even in humans, transplanted allogeneic tumors (e.g. transplanted renal allografts containing donor tumor cells) can be readily rejected (Wilson et al., 1968). These phenomena do not predict whether a patient with cancer can reject an autologous tumor. However, these observations did establish one important fact: the immune system has the capacity to reject tumors, at least in the context of allograft rejection.

Over the next half century the role of transplantation antigens was unraveled, and syngeneic strains of mice with identical genetic backgrounds were established (Snell, 1963). By the 1950s, mouse experiments used syngeneic mice and transplantable, carcinogen induced tumors. These experiments established that immunization with syngeneic cancer cells can protect against rechallenge with the same tumor (Foley, 1953; Prehn and Main, 1957). The tumor rejection antigens defined on these carcinogen induced tumors were tumor specific, i.e. expressed by a specific tumor but not detected on independently derived tumors (even the same histologic tumor type induced by the same carcinogen in the same animal) nor on normal tissues.

These experiments established that carcinogen induced tumors can express unique antigens recognized by the immune system. However, experiments using spontaneous tumors elicited some pessimism (Hewitt et al., 1976). In transplantation experiments, it appeared that spontaneous tumors did not express tumor rejection antigens. The argument at the time was that the observations made in carcinogen induced models had little relevance to human tumors (Hewitt, 1978). However, it was subsequently shown that even spontaneous tumors can express weak tumor rejection antigens. For instance, when tumors were manipulated with mutagens, immunization with the mutagenized tumors led to rejection of both the mutagenized and the parental tumors. Thus, spontaneous tumors did express tumor rejection antigens, but these antigens were quite weak; immune responses against these weak tumor antigens could be unveiled by eliciting immunity against stronger, mutagen induced antigens. Thus, investigators have remained hopeful that immunotherapy based on targeting defined tumor rejection antigens might be possible in humans (Boon, 1993).

Tumor antigens defined by host immune reactions

One of the great mysteries of tumor immunology has been understanding the structures of tumor specific transplantation antigens (TSTA). Experiments using mutagenized mouse tumor strains have shown that single base point mutations within open reading frames of the genome can give rise to unique antigens that are targets for T-lymphocyte responses (Van den Eynde et al., 1991). The mutations can form new epitopes for recognition by the TCR or define new aggretopes (Lurquin et al., 1989) that can bind to MHC molecules for presentation to T cells.

These observations form two paradigms for TSTA defined for chemically induced tumors in mice. A third type of tumor rejection antigen that does not contain any mutation has been recognized in the mouse (Van den Eynde et al., 1991). This nonmutated gene product is similar to tumor autoantigens in humans (see below, differentiation autoantigens recognized by T lymphocytes).

Autoantigens defined in humans

Tumor rejection experiments are not possible in humans. Thus, evidence for immune recognition of tumors in humans has relied on a search for humoral or cellular responses against autologous tumors. Antibodies or lymphocytes that react with autologous tumor cells suggest immune responses, but it is important to recognize that active antitumor immune reactions cannot be distinguished readily from naturally occurring autoantibodies or lymphocytes that might be present in the immune repertoire.

Although humoral and cellular immune reaction have been extensively reported in the scientific literature, in only a handful of cases have potentially immunogenic cancer antigens been defined. Most immunogenic antigens have been defined by serologic techniques, mainly because biochemical methods are readily available to define antigens using antibodies (Old, 1981; Cote et al., 1986; Houghton, 1991). The

Table 5.1. *Serologically defined tumor autoantigens*

Class	Antigens	Tumors
Glycolipids		
Blood group	T, Tn, Sialosyl-Tn	Carcinomas
Gangliosides	GM2, GD2, 9-0-Ac GD3	Melanoma, glioblastoma, neuroblastoma, small cell lung cancer
Glycoproteins	M-TAA U-TT gp75	Melanoma Melanoma Melanoma

specificity and nature of antigens defined by cellular immune reactions have been much more difficult, but there have been recent advances in identifying cancer antigens recognized by T lymphocytes. A theme has begun to emerge in these initial studies: the tumor antigens recognized by the human immune system thus far have been antigens that are expressed by selected normal tissues.

These antigens are called differentiation antigens because they can distinguish one normal cell type from another or distinguish specific stages of development within a cell lineage. Serological techniques have been used to identify at least nine autoantigens expressed by human tumors (Table 5.1).

Antigens defined by antibodies

Of the nine serologically defined autoantigens expressed by human cancers, six are carbohydrate structures. Three of these are gangliosides, which are acidic glycoproteins expressed on the cell surface of melanomas and other neuroectodermal derived tumors, small cell lung cancers and some carcinomas. One antigen has been reproducibly shown to be immunogenic, i.e. capable of actively eliciting specific immune responses. This is the ganglioside GM2. Immunization of patients with metastatic melanoma using GM2 linked to a protein carrier elicits antibody responses against GM2 in most cases. Three carbohydrate antigens, T, Tn and sialosyl-Tn, are blood group antigens that are expressed on epithelial cancers. Immunization with carbohydrate antigens has given the first compelling data that antigens expressed by tumor cells are specifically immunogenic.

The best defined glycoprotein structure recognized by autologous antibodies is gp75, a member of the tyrosinase family of proteins. The tyrosinase related molecules are melanosomal membrane proteins that are specifically expressed by normal pigment cells, melanocytes. These autoantigens represent one of the best examples of differentiation antigens expressed by a specific cell lineage (melanocytes) and tumors derived from these cells.

Because these are intracellular proteins, it was predicted that the tyrosinase family of proteins would be recognized not only by antibodies but also by T lymphocytes.

Antigens defined by T lymphocytes

Until recently, the identity of antigens recognized by CTL against human cancers was elusive. CTL recognize peptide epitopes that are processed from endogenous cellular proteins, and the peptides are presented to the T cell by HLA antigens. CTL responses have been most intensively studied in human melanoma.

When lymphocytes of melanoma patients are stimulated in vitro by autologous tumor cells in the presence of T-cell growth factors (e.g. IL-2), CTL against autologous melanoma can frequently be isolated (Knuth et al., 1984; Herin et al., 1987). A major advance in human cancer biology has been the identification of a gene designated MAGE-1 (for Melanoma Antigen) that encodes an antigen recognized by autologous CTL in a patient with melanoma (Lurquin et al., 1989; Van der Bruggen et al., 1991). MAGE-1 belongs to a gene family that shows no similarity to any known gene sequence.

MAGE-1 is expressed by 40% of melanoma cell lines but not in normal adult tissue, with the exception of testes. The antigen encoded by MAGE-1 is class I MHC restricted and presented by HLA-A1. This system is the first formal identification of a HLA restricted human tumor autoantigen recognized by CTL.

More recently, a second gene encoding an antigen presented by HLA-A2 on melanoma has been identified. The antigen is encoded by the tyrosinase gene (Brichard et al., 1993). The tyrosinase gene is the second human gene discovered by Boon and his group that directs the expression of an antigen recognized on tumors by CTL (Table 5.2). Like the serologically defined gp75 antigen, tyrosinase is an example of a differentiation antigen that is expressed specifically by melanocytes, the normal cell counterparts of the tumor.

Table 5.2. *Genes encoding antigens on tumors recognized by CTL*

Antigen gene	Tumor cell	HLA presentation
P1A	Mouse mastocytoma	
MAGE-1	Human melanoma	HLA-A1 restricted
Tyrosinase	Human melanoma	HLA-A2 restricted

Other potential antigens: mutant oncogene products

To date, most autoantigens defined on human cancer cells have been normal cellular components. However, interest has been raised concerning recognition of mutated molecules within cancer cells. Several candidate mutant oncogenes have been identified in human cancers, including p53, *ras* and the unique fusion proteins generated in certain leukemias and lymphomas (e.g. *bcr-abl*).

Mutated versions of the p53 tumor suppressor gene are detected in a high proportion of human cancers. Approximately 10% of breast cancer patients have antibodies that recognize p53 (Crawford et al., 1982). Of note, these autoantibodies recognize both mutant and wild type p53. In a study of 60 patients with breast cancer, only patients who overexpressed p53 developed anti-p53 antibodies. In 46 patients with lung cancer, only patients expressing missense mutations developed autoantibodies. Thus, missense or overexpressed p53 may be recognized by B cells in the immune repertoire of cancer patients, but these autoantibodies are not specific for the mutated form of p53.

The *ras* family of protooncogenes encodes an ubiquitous highly conserved family of proteins that acquire transforming activity as a result of single point mutations, occurring most commonly around amino acid residues 12, 13 or 61. These activating mutations are present in a substantial number of human carcinomas. Human T cells are capable of recognizing *ras* mutations (Jung and Schluesener, 1991). In this study, $CD4^+$ T cells specifically recognizing mutated *ras* (valine for glycine at position 12) could be generated in vitro from peripheral blood in two of four healthy volunteers. This suggests that certain patients could generate autologous immune responses against aberrant *ras* proteins.

Viral gene products

Viruses have been implicated in a number of human cancers. Burkitt's lymphoma, nasopharyngeal carcinoma and possibly Hodgkin's lymphoma are associated with EBV (Young et al., 1988; Young et al., 1989; Herbst et al., 1991; Pallesen et al., 1991). The majority of cervical carcinomas are associated with human papilloma virus (HPV) (Durst et al., 1983; McCance et al., 1983). Viral proteins are expressed by these cancers, raising the possibility of developing both T-cell and antibody responses against viral gene products.

Indeed, vaccines against human cancers will probably be applied first to cancers associated with viruses. Hepatocellular carcinoma is closely associated with infection by the hepatitis B virus. It is likely that, at least in some parts of the world (particularly the Far East), this virus is involved in the pathogenesis of hepatocellular carcinoma. Widespread immunization with the hepatitis B virus vaccine is under way in China and other countries where infection is endemic. This epidemiological experiment could have a major influence on the incidence of hepatocellular carcinoma in these regions.

Issues of immune tolerance

Except for virus associated cancers, the potential targets for vaccination are emerging as either self or altered self antigens. Mechanisms of immune tolerance to self have begun to emerge, particularly from studies in transgenic mouse systems (Hanahan, 1990). Mechanisms of systemic immune tolerance include deletion of potentially autoreactive B or T cells (particularly during development), induction of anergy in B and T cells, and the poorly defined phenomenon of suppression of immune responses (e.g. by "suppressor" cells). While deletion of T and B cells largely occurs during normal development, a large array of potentially autoreactive T and B cells remains in the immune repertoire. Despite the presence of these potentially autoreactive cells, no autoimmunity is observed. This raises a fourth mechanism, called immune ignorance. It is thus possible to elicit an autoimmune response in animal models against tissue specific antigens by actively and effectively presenting the antigen.

Thus, presentation of antigen is crucial, but it is something that tumor cells generally do poorly. One strategy is to remove the tumor antigen from the tumor cell and express the antigen in more formal APC, e.g. B cells, macrophages or dendritic cells. Alternatively, the tumor cell could be engineered to present antigen more effectively.

The identification of crucial costimulatory molecules (e.g. B7/BB1) is permitting new strategies for engineering tumor cells (Linsley et al., 1991). Occupancy of TCR in the absence of these costimulatory signals could lead to T-cell anergy, one mechanism of tolerance induction (Jenkins et al., 1987a; Jenkins et al., 1987b; Harding et al., 1992).

Human vaccination trials

The results of 81 trials of tumor vaccine containing irradiated tumor cells or tumor cell fractions have been well summarized (Livingston, 1991). In seven of these trials, clinical responses were reported (but have not yet been confirmed at other centers), and two small randomized adjuvant trials have suggested prolongation of disease free interval and survival; however, these trials have pitfalls and need to be confirmed (Hoover et al., 1985; Hollinshead et al., 1987).

Although these trials suggest that tumor vaccines may occasionally (and infrequently) induce clinical responses, they fail to provide a solid foundation for future progress. Part of the difficulty in interpreting these data is a consequence of the diversity of the study designs. The antigen source was autologous or allogeneic tumor cells grown in culture or obtained from surgical specimens and injected as whole cells or lysates. A variety of routes of administration have been used, including intradermal, intralesional, intralymphatic and subcutaneous. Thus, there has been no clear evidence that patients immunized with undefined tumor cell preparations are specifically immunized against their tumors, and these trials have not established any principles about active immunization against cancer.

Clinical trials with defined antigenic preparations

Using defined antigen preparations for immunization allows host professional antigen presenting cells to present antigen. In a series of experiments immunizing procedures have been defined in the mouse that induce reproducible serologic responses to ganglioside antigens (Livingston et al., 1983a; Livingston et al., 1985; Livingston et al., 1987; Livingston et al., 1989). These include immunization with ganglioside attached to adjuvants or carriers such as keyhole limpet hemocyanin, BCG, liposomes containing endotoxin or monophosphoryl lipid A. These immunization procedures elicit higher titers of antibody against gangliosides. Using this approach, IgM antibodies were induced in 35 of 44 patients (80%) with metastatic melanoma receiving BCG-GM2 vaccine and in 35 of 39 patients (90%) receiving a full series of four or five immunizations (Livingston et al., 1983b). These antibodies were capable of lysing human tumor cells in the presence of human complement. Patients producing GM2 antibodies had a markedly prolonged disease-free interval and survival. Because GM2 is heterogenously expressed on melanoma cells in different patients (Real et al., 1985), and even within the same patient (Albino et al., 1981), it is likely that eventually a polyvalent vaccine will be required.

Anti-idiotype antibody vaccines

An alternate approach to immunization uses antibodies as immunogens. The idiotype defines the region of the antibody that is unique and includes the antigen binding site. The use of internal image anti-idiotypic antibodies may have advantages for immunization compared to certain weak immunogens. In addition, the anti-idiotypic antibody is available in purified form and in ready amounts for in vivo studies.

Anti-idiotypic mAb have been used effectively for immunization and protection against experimental infectious and tumor models (Reagan et al., 1983; Herlyn et al., 1986; Raychaudhuri et al., 1986; Dunn et al., 1987). Clinical trials with anti-idiotype antibody vaccines have been conducted in both colon carcinoma and melanoma (Herlyn et al., 1987; Kageshita et al., 1988).

It remains to be determined whether anti-idiotype vaccines against cancer are more or less effective at inducing immune responses than vaccines constructed with the corresponding antigen.

Genetic manipulation of tumors

Other chapters in this volume will deal in detail with genetic approaches to altering tumor cells. We will not review this area in any depth, but we will mention certain perspectives in our overview of this area.

Introduction of MHC genes

Antigen presentation and susceptibility to lysis by T cells require sufficient expression of class I or class II MHC antigens. Thus, inducing or enhancing class I MHC expression by gene transfection results in decreased tumorigenicity of some murine tumor models (Hui et al., 1984; Wallich et al., 1985). However, in other tumor systems induction of class I MHC expression actually leads to increased tumorigenicity, probably through loss of susceptibility to NK lymphocytes that prefer little or no class I MHC expression by target cells. These experiments point out the complex relationship between the host and tumor in the process of rejection and immunization.

There are no easy answers to issues of tumor rejection. Many cell types and other immune components can be involved in tumor rejection and active immunization.

Introduction of cytokine genes

Many cytokine genes have been introduced into tumor cells with varying effects on tumorigenicity and immunogenicity (Tepper et al., 1989; Watanabe et al., 1989; Fearon et al., 1990; Gansbacher et al., 1990a; Gansbacher et al., 1990b; Asher et al., 1991; Blankenstein et al., 1991; Colombo et al., 1991; Golumbek et al., 1991; Rollins and Sunday, 1991; Porgador et al., 1992; Restifo et al., 1992; Dranoff et al., 1993). There are two crucial issues in evaluating these experiments. First, they have given insights into immune or inflammatory cells that are involved with tumor regressions. Second, this approach has provided an alternate strategy for active immunization. Although introduction of cytokine genes was initially conceived as a method to induce costimulatory signals into tumor cells for antigen presentation, much of the data suggests that in most cases the cytokine elicits an inflammatory cell infiltration that can then present antigen, i.e. antigen presentation may not be performed by the tumor cell at all. In this view, cytokine gene therapy can be thought of as a sophisticated immune adjuvant.

One pivotal concept in utilizing cytokine transduced tumor vaccines was an attempt to model the paracrine physiology of local cytokine release and to avoid high systemic cytokine concentrations. This alteration in local immunologic environment of the tumor cell is targeted to enhance antigen presentation of the tumor antigens and activation of local immune or inflammatory responses. Variable results have been seen with cytokine transduced tumor cells when different tumor systems are analyzed.

Tumors transduced by IL-2 are characterized by massive infiltrates of either lymphocytes (T cells or NK cells), macrophages or neutrophils, depending on which tumor model is being evaluated. In some circumstances rejection of the IL-2 transduced tumor is dependent on $CD8^+$ and NK cells and not on $CD4^+$ cells (Fearon et al., 1990; Gansbacher et al., 1990b), but in other models either NK cells or macrophages (without any lymphocytic cell involvement) can lead to tumor regression. Expression of IL-4 in tumors results in infiltration of macrophages and eosinophils with regression of tumors. In one model, a systemic response was generated, and this was dependent on $CD8^+$ and $CD4^+$ T cells (Golumbek et al., 1991), but in other models no evidence of systemic immunization was observed. Thus, effective immunization using IL-2 and IL-4 transduced tumors may greatly depend on the underlying immunogenicity of the particular tumor cell. In summary, there are clearly many roads to tumor rejection.

IFN-γ up-regulates both class I and II MHC gene products. Expression of IFN-γ in some tumors results in tumor rejection and induction of systemic immunity. However, the effects of IFN-γ are dependent on the particular model, and cotransduction of IFN-γ with other cytokine genes may also inhibit the generation of a systemic immune response (Watanabe et al., 1989; Gansbacher et al., 1990a; Restifo et al., 1992). TNF-α transduced tumor cells, as compared with parental cells, grow more slowly in vitro and are occasionally rejected in syngeneic animals. Immunization with TNF-α transduced tumor cells does not enhance systemic immunity compared to nontransduced cells (Asher et al., 1991; Blankenstein et al., 1991). In fact, almost every cytokine gene introduced into tumors has led to some form of tumor rejection, depending on the particular tumor model. Active immunization appears to be more infrequent, and it must depend greatly on the inherent immunogenicity of the tumor.

Finally, in one model of a poorly immunogenic tumor, the mouse B16 melanoma, GM-CSF transduced cells were not rejected very efficiently but immunization with transduced irradiated tumor cells induced systemic antitumor immunity. This immunity was dependent on both $CD4^+$ and $CD8^+$ cells, even in MHC II^- tumors (Dranoff et al., 1993). It is hypothesized that part of the local effect may be the unique ability of GM-CSF to promote the differentiation of antigen presenting cells, for instance dendritic cells. Thus, antigen presentation might be performed by host cell responses induced by the cytokine, and it can lead to protection from even poorly immunogenic tumors.

Introduction of the costimulatory signals

Two reports have recently suggested that introduction of the costimulatory molecule B7 can enhance immunization by relatively nonimmunogenic tumors. B7, the ligand for CD28 and CTLA4 on T cells, is expressed normally on APC including B cells, macrophages and dendritic cells. Introduction of the B7 gene can result in rejection of tumors in syngeneic host and induce a systemic immune response that protects against challenge with the wild type tumor at distant sites (Chen et al., 1992; Baskar et al., 1993; Townsend and Allison, 1993). One scenario is that activation of $CD4^+$ T cells by B7 enhances lymphokine production for priming and generation of cytotoxic $CD8^+$ cells, but at this point it is not possible to discern whether tumor cells or other host immune or inflammatory cells actually present the tumor antigen (Lindstein et al., 1989; Fraser et al.,1991; Harding et al., 1992)

Conclusions

Innovations for construction of cancer vaccines are emerging from advances in molecular immunology and cancer biology. While vaccines against infectious agents are models for vaccine development, there are clearly distinct considerations and problems associated with cancer vaccines. Thus, despite the thousands of patients injected with various types of tumor cell vaccines during the last 30 years, we are little wiser for the experience. Because very few investigators were able to determine whether most of these vaccines contained relevant tumor antigens and whether the vaccines were consistently immunogenic, a foundation has not yet been laid on which to base future studies. The ganglioside antigens are, so far, one exception to this perception.

One pertinent issue in designing active specific immunotherapy is that cancer cells are derived from normal host cells. Thus, the antigenic profile of cancer cells closely mimics that of normal cells. How the immune system identifies and destroys cancer cells is therefore crucial. One method of solving this problem is the identification of target antigens that are unique to the cancer cell.

An alternate strategy is to induce autoimmunity against restricted normal cellular antigens present on cancer cells. Toxicity is a potential problem with this second approach, although such a strategy would still be highly specific compared to most present cancer therapies.

An ultimate goal of tumor vaccine design is the generation of antigen specific vaccines. The initial foray in identifying tumor antigens recognized by antibodies and T cells opens up potentially novel strategies. It is now feasible to develop vaccine strategies on a molecular basis. However, the role of vaccination in cancer therapy remains to be determined. Despite this necessary uncertainty, it is an exciting time for development in this area. Perhaps we can be cautiously optimistic as we enter an era where careful step-by-step vaccine development can be carried out.

References

Albino AP, Lloyd KO, Houghton AN, Oettgen HF and Old LJ (1981). Heterogeneity in surface antigen and glycoprotein expression of cell lines derived from different melanoma metastases of the same patient. Implications for the study of tumor antigens. J Exp Med, **154**, 1764-1778.

Asher AL, Mule JJ, Kasid A, Restifo NP, Salo JC, Reichert CM, Jaffe G, Fendly B, Kriegler M and Rosenberg SA (1991). Murine tumor cells transduced with the gene for tumor necrosis factor-alpha. Evidence for paracrine immune effects of tumor necrosis factor against tumors. J Immunol **146**, 3227-3234.

Baskar S, Ostrandrosenberg S, Nabavi N, Nadler LM, Freeman GJ and Glimcher LH (1993). Constitutive expression of B7 restores immunogenicity of tumor cells expressing truncated major histocompatibility complex class-II molecules. Proc Natl Acad Sci USA **90**, 5687-5817.

Blankenstein T, Qin ZH, Uberla K, Muller W, Rosen H, Volk HD and Diamantstein T (1991). Tumor suppression after tumor cell-targeted tumor necrosis factor alpha gene transfer. J Exp Med **173**, 1047-1052.

Boon T (1993). Teaching the immune system to fight cancer. Sci Am **266**, 82-89.

Brichard V, Van Pel A, Wolfel T, Wolfel C, De Plaen E, Lethe B, Coulie P and Boon T (1993). The tyrosinase gene codes for an antigen recognized by autologous cytolytic T lymphocytes on HLA-A2 melanomas. J Exp Med **178**, 489-495.

Chen L, Ashe S, Brady WA, Hellstrom I, Hellstrom KE, Ledbetter JA, McGowan P and Linsley PS (1992). Costimulation of antitumor immunity by the B7 counterreceptor for the T lymphocyte molecules CD28 and CTLA-4. Cell **71**, 1093-1102.

Colombo MP, Ferrari G, Stoppacciaro A, Parenza M, Rodolfo M, Mavilio F and Parmiani G (1991). Granulocyte colony-stimulating factor gene transfer suppresses tumorigenicity of a murine adenocarcinoma in vivo. J Exp Med **173**, 889-897.

Cote RJ, Morrissey DM, Houghton AN, Thomson TM, Daly ME, Oettgen HF and Old LJ (1986). Specificity analysis of human monoclonal antibodies reactive with cell surface and intracellular antigens. Proc Natl Acad Sci USA **83**, 2959-2963.

Crawford LV, Pim DC and Bulbrook RD (1982). Detection of antibodies against the cellular protein p53 in sera from patients with breast cancer. Int J Cancer **30**, 403-408.

Dranoff G, Jaffee E, Lazenby A, Golumbek P, Levitsky H, Brose K, Jackson V, Hamada H, Pardoll D and Mulligan RC (1993). Vaccination with irradiated tumor cells engineered to secrete murine granulocyte-macrophage colony-stimulating factor stimulates potent, specific, and long-lasting anti-tumor immunity. Proc Natl Acad Sci USA **90**, 3539-3543.

Dunn PL, Johnson CA, Styles JM, Pease SS and Dean CJ (1987). Vaccination with syngeneic monoclonal anti-idiotype protects against a tumour challenge. Immunology **60**, 181-186.

Durst M, Gissmann L, Ikenberg H and zur Hausen H (1983). A papillomavirus DNA from a cervical carcinoma and its prevalence in cancer biopsy samples from different geographic regions. Proc Natl Acad Sci USA **80**, 3812-3815.

Fearon ER, Pardoll DM, Itaya T, Golumbek P, Levitsky HI, Simons JW, Karasuyama H, Vogelstein B and Frost P (1990). Interleukin-2 production by tumor cells bypasses T helper function in the generation of an antitumor response. Cell **60**, 397-403.

Foley EJ (1953). Antigenic properties of methylcholanthrene-induced tumors in mice of the strain of origin. Cancer Res **13**, 835-837.

Fraser JD, Irving BA, Crabtree GR and Weiss A (1991). Regulation of interleukin-2 gene enhancer activity by the T cell accessory molecule CD28. Science **251**, 313-316.

Gansbacher B, Bannerji R, Daniels B, Zier K, Cronin K and Gilboa E (1990a). Retroviral vector-mediated gamma-interferon gene transfer into tumor cells generates potent and long lasting antitumor immunity. Cancer Res **50**, 7820-7825.

Gansbacher B, Zier K, Daniels B, Cronin K, Bannerji R and Gilboa E (1990b). Interleukin 2 gene transfer into tumor cells abrogates tumorigenicity and induces protective immunity. J Exp Med **172**, 1217-1224.

Golumbek PT, Lazenby AJ, Levitsky HI, Jaffee LM, Karasuyama H, Baker M and Pardoll DM (1991). Treatment of established renal cancer by tumor cells engineered to secrete interleukin-4. Science **254**, 713-716.

Hanahan D (1990). Transgenic mouse models of self-tolerance and autoreactivity by the immune system. Ann Rev Cell Biol **6**, 493-537.

Harding FA, McArthur JG, Gross JA, Raulet DH and Allison JP (1992). CD28-mediated signalling co-stimulates murine T cells and prevents induction of anergy in T-cell clones. Nature **356**, 607-609.

Herbst H, Dallenbach F, Hummel M, Niedobitek G, Pileri S, Muller-Lantzsch N and Stein H (1991). Epstein-Barr virus latent membrane protein expression in Hodgkin and Reed-Sternberg cells. Proc Natl Acad Sci USA **88**, 4766-4770.

Herin M, Lemoine C, Weynants P, Vessiere F, Van Pel A, Knuth A, Devos R and Boon T (1987). Production of stable cytolytic T-cell clones directed against autologous human melanoma. Int J Cancer **39**, 390-397.

Herlyn D, Ross AH and Koprowski H (1986). Anti-idiotypic antibodies bear the internal image of a human tumor antigen. Science **232**, 100-102.

Herlyn D, Wettendorff M, Schmoll E, Iliopoulos D, Schedel I, Dreikhausen U, Raab R, Ross AH, Jaksche H, Scriba M and Koprowski H (1987). Anti-idiotype immunization of cancer patients: Modulation of the immune response. Proc Natl Acad Sci USA **84**, 8055-8059.

Hewitt HB, Blake ER and Walder AS (1976). A critique of the evidence for active host defence against cancer, based on personal studies of 27 murine tumours of spontaneous origin. Br J Cancer **33**, 241-259.

Hewitt HB (1978). The choice of animal tumors for experimental studies of cancer therapy. Adv Cancer Res **27**, 149-200.

Hollinshead A, Stewart TH, Takita H, Dalbow M and Concannon J (1987). Adjuvant specific active lung cancer immunotherapy trials. Tumor-associated antigens. Cancer **60**, 1249-1262.

Hoover HC, Jr, Surdyke MG, Dangel RB, Peters LC and Hanna MG, Jr (1985). Prospectively randomized trial of adjuvant active-specific immunotherapy for human colorectal cancer. Cancer **55**, 1236-1243.

Houghton AN (1991). Cancer vaccines: Readdressing an old concept. In "Immunomodulation in Cancer: Critical Issues for Therapeutic Design" (ed EC Borden), pp 47-62. Mediscript, London.

Hui K, Grosveld F and Festenstein H (1984). Rejection of transplantable AKR leukaemia cells following MHC DNA-mediated cell transformation. Nature **311**, 750-752.

Jenkins MK, Pardoll DM, Mizuguchi J, Chused TM and Schwartz RH (1987a). Molecular events in the induction of a nonresponsive state in interleukin 2-producing helper T-lymphocyte clones. Proc Natl Acad Sci USA **84**, 5409-5413.

Jenkins MK, Pardoll DM, Mizuguchi J, Quill H and Schwartz RH (1987b). T-cell unresponsiveness in vivo and in vitro: Fine specificity of induction and molecular characterization of the unresponsive state. Immunol Rev **95**, 113-135.

Jung S and Schluesener H (1991). Human T lymphocytes recognize a peptide of single point-mutated oncogenic ras proteins. J Exp Med **173**, 273-276.

Kageshita T, Chen ZJ and Kim J-W (1988). Murine anti-idiotypic monoclonal antibodies to syngeneic antihuman high molecular weight melanoma associated antigen monoclonal antibodies: Development, characterization and clinical applications. Pigment Cell Res **1**, 185-190.

Knuth A, Danowski B, Oettgen HF and Old L (1984). T-cell mediated cytotoxicity against autologous malignant melanoma: Analysis with interleukin-2-dependent T-cell cultures. Proc Natl Acad Sci USA **81**, 3511-3515.

Lindstein T, June CH, Ledbetter JA, Stella G and Thompson CB (1989). Regulation of lymphokine messenger RNA stability by a surface-mediated T cell activation pathway. Science **244**, 339-343.

Linsley PS, Brady W, Grosmaire L, Aruffo A, Damle NK and Ledbetter JA (1991). Binding of the B cell activation antigen B7 to CD28 costimulates T cell proliferation and interleukin 2 mRNA accumulation. J Exp Med **173**, 721-730.

Livingston PO, DeLeo AB, Jones M and Oettgen HF (1983a). Comparison of approaches for augmenting the serologic response to the individually specific methylcholanthrene-induced sarcoma-Meth A: Pretreatment with cyclophosphamide is most effective. J Immunol **131**, 2601-2605.

Livingston PO, Takeyama H, Pollack MS, Houghton AN, Albino A, Pinsky CM, Oettgen HF and Old LJ (1983b). Serological responses of melanoma patients to vaccines derived from allogeneic cultured melanoma cells. Int J Cancer **31**, 567-575.

Livingston PO, Jones M, DeLeo AB, Oettgen HF and Old LJ (1985). The serologic response to Meth A sarcoma vaccines after cyclophosphamide treatment is additionally increased by various adjuvants. J Immunol **135**, 1505-1509.

Livingston PO, Calves MJ and Natoli EJ, Jr (1987). Approaches to augmenting the immunogenicity of the ganglioside GM2 in mice: Purified GM2 is superior to whole cells. J Immunol **138**, 1524-1529.

Livingston PO, Ritter G and Calves MJ (1989). Antibody response after immunization with the gangliosides GM1, GM2, GM3, GD2 and GD3 in the mouse. Cancer Immunol Immunother **29**, 179-184.

Livingston P (1991). Active specific immunotherapy in the treatment of patients with cancer. Immunol Allergy Clin North Am **11**, 401-423.

Lurquin C, Van Pel A, Mariame B, De Plaen E, Szikora JP, Janssens C, Reddehase MJ, Lejeune J and Boon T (1989). Structure of the gene of tum-

6

CYTOKINE DEPENDENT TUMOR RECOGNITION

A. Modesti, P. Musiani, G. Cortesina and G. Forni
Institute of Experimental and Social Medicine, University of Chieti, Chieti, Italy; Otorhinolaryngology Clinic and CNR Center for Immunogenetics and Experimental Oncology, University of Turin, Turin, Italy

Introduction

The immune system does not appear to have evolved in response to a pressing need to ward off tumors (Mitchison and Kinlen, 1980). Even so, immune mechanisms may affect tumor establishment and growth.

This tumor-immune relationship depends on the expression of mutated peptides (Van der Bruggen et al., 1991) and other membrane traits (Forni and Santoni, 1984) associated with the process whereby cells that become neoplastic are recognized by both "nonspecific" and "specific" leukocytes.

This relationship, however, is ambiguous and may result in either the inhibition or the enhancement of tumor growth (Forni and Comoglio, 1973; Giovarelli et al., 1976). The fact that a tumor does not exercise any function means that it can readily submit to attacks on the part of the immune system and suffer extensive necrosis without thereby halting its progression.

Moreover, during their long period of latency spontaneous tumors acquire mechanisms that enable them to sneak through the meshes of the immune control (Fig. 6.1). Experiments in vitro have shown that these escape mechanisms are potent. A balanced estimate of their weight in vivo is hard to achieve.

Despite their simultaneous presence in vivo, indeed, most tumors that are lethal in syngeneic hosts are recognized and rejected by the immune system of allogeneic hosts even when differences of only a few minor antigens are involved (Forni et al., 1982). Furthermore, it is possible to preimmunize syngeneic hosts towards what are regarded as nonimmunogenic tumors and bring about the rapid rejection of the neoplastic cells (Boon, 1985). It is clear that despite their escape mechanisms tumors can be both recognized and destroyed once an intense immune reactivity is activated.

CYTOKINE-INDUCED TUMOR IMMUNOGENICITY
ISBN 0-12-261520-4

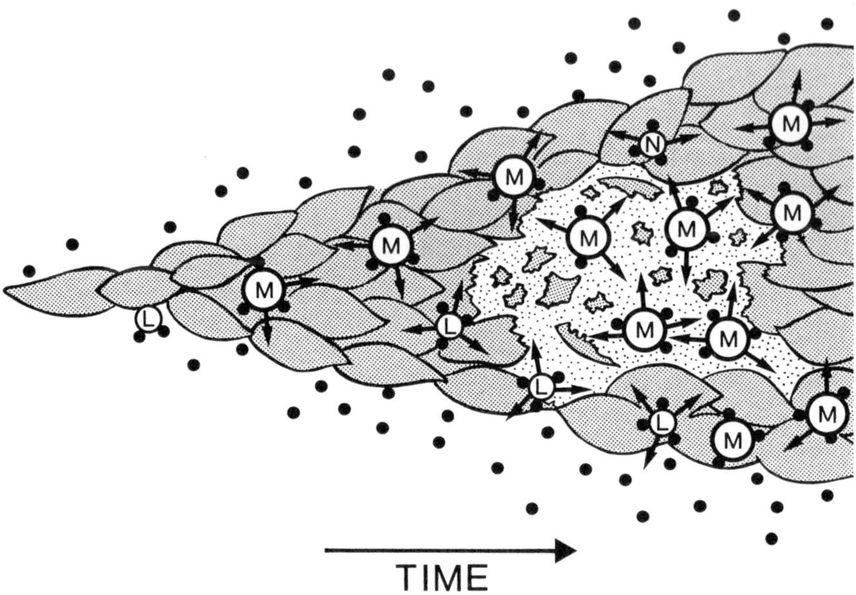

Fig. 6.1. *A few tumor mechanisms used to fool the immune system.* Tumors release inhibitory catabolites, cytokines and proteases (●) that subvert the functionality of the immune system. They also produce a particular extracellular matrix so as to attract various kinds of leukocytes (**L**, lymphocytes; **N**, neutrophils; **M**, macrophages) and then paralyze their reactivity. Many tumors even exploit the ability of trapped leukocytes to release cytokines and other factors (black arrows) that favor angiogenesis and enhance cell proliferation. The absence or rarefaction of class I and II MHC antigens on tumor cell membrane, or their expression in the absence of costimulatory signals or adhesion molecules, favor the onset of specific T-cell tolerance. Functional data show that infiltrating T lymphocytes and those in tumor draining lymph nodes are unable to react against the tumor and polyclonal activators. Lengthy culture in the presence of IL-2 is needed to rescue their reactivity. T lymphocytes often display abnormal TCR composition and signal transduction. During the progression of a tumor, those of its clones that are more resistant to particular reaction mechanisms can be selected. Lastly, a tumor is a particular kind of antigen whose size increases with time (for a review of these data see Colombo et al., 1992).

Cytokine induced tumor recognition

An ambiguity that may exist in the relationship between a tumor and the host immune system is resolved when particular "inducer" cytokines are present at the tumor site, since they elicit a strong reaction which can completely tear down a tumor and possibly lead to the establishment of an immune memory (Bubenik et al., 1983; Forni et al., 1988). Both reactions are of major interest, especially since their clinical exploitation can be envisaged. Several contributions to this volume describe the features of immune reactions induced by a variety of cytokines administered locally, though by means of different techniques.

Inducer cytokines, however, do not impair tumor proliferation directly, but rather through their activation of an immune reaction on the part of the host. Interestingly, the mechanisms involved in these in vivo reactions differ from those predictable on the strength of the in vitro data. Attention has thus been focused on hitherto neglected aspects of the action of cytokines in vivo.

The reaction following cytokine induced tumor recognition consists of two successive stages, clearly distinct but closely interlinked. The first stage is substantially nonspecific and resembles an inflammatory reaction (this term will be used hereafter). The second stage takes the form of a specific T-cell dependent immune response. T lymphocytes often make the inflammatory reaction more effective, and this reaction itself is needed to induce an immune memory and determines its characteristics.

The inflammatory reaction

Similar inflammatory reactions are induced by:

- repeated local injections of exogenous cytokines
- cytokines released by cytokine gene engineered (hereafter called "engineered") tumor cells
- cytokines released by various kinds of engineered cells antigenically not related to the tumor.

Repeated injection of exogenous cytokines at the site of a tumor or around its draining lymph nodes quickly induces an inflammatory reaction that is often powerful enough to lead to the rejection of incipient experimental tumors (reviewed by Colombo et al., 1992). Its potential has also been illustrated in a large series of clinical trials (Pericle et al., 1992).

In mice engineered cells activate a similar reaction that inhibits their oncogenic capacity. Its efficacy will depend on the features of the inducer cytokine, as well as on the tumor histotype and extracellular matrix (Colombo et al., 1992). In this case, the reaction is governed in a feedback fashion by its own intensity. The quantity of cytokine increases at first as the engineered cells expand; it then decreases in function of the efficacy of the reaction that is being activated and results in the killing of the engineered cells.

These characteristics mean that engineered cells can be regarded as true biological minipumps whose regulated action can be exploited in vivo against unrelated tumors (Bubenik et al., 1988; Roth et al., 1992; Tepper et al., 1992; Cavallo et al., 1993). Instead of transducing cytokine genes directly into neoplastic cells derived from every single patient, inoculation of already engineered syngeneic or allogeneic cell lines around a tumor is an easier, but still very effective, procedure.

Repertoires of inflammatory leukocytes

Strong reactions are elicited by numerous locally injected cytokines (IL-1, IL-2, IL-4, IFN-γ, G-CSF) and by cytokines released by engineered tumor cells (IL-1, IL-2, IL-3, IL-4, IL-5, IL-6, IL-7, IL-8, IL-10, IL-12, G-CSF, GM-CSF, IFN-α, IFN-γ, etc.) (reviewed by Colombo et al., 1992; Colombo and Forni, 1994). However, differences are evident in the efficacy of the reaction and repertoire of cells involved. Since these data were obtained with tumors of dissimilar origin, immunogenicity and histotype engineered to release different amounts of inducer cytokines and did not always coincide or were apparently contradictory, we engineered the parental cells of a spontaneous mammary adenocarcinoma of BALB/c mice (TS/A^{par}) with a set of distinct cytokines.

The reaction activated by TS/A cells releasing IL-2 ($TS/A^{IL\text{-}2}$) is mainly sustained by neutrophils that efficiently kill them and fully inhibit their oncogenicity (Cavallo et al., 1992). The one activated by TS/A cells releasing IL-4 ($TS/A^{IL\text{-}4}$), too, efficiently inhibits their growth through the early intervention of eosinophils (Pericle et al., in preparation). Necrosis always proceeds from the edge to the center of the tumor. Interestingly, TS/A cells engineered to release IL-5 ($TS/A^{IL\text{-}5}$) are deeply infiltrated by eosinophils (Kruger-Krasagakes et al., 1993) that neither inhibit tumor formation nor kill $TS/A^{IL\text{-}5}$ cells, suggesting that more complex signals are required to guide eosinophil killer activity (Modesti et al., manuscript in preparation). By contrast, macrophages activated by TS/A cells engineered to release IFN-γ ($TS/A^{IFN\text{-}\gamma}$) lead to prolonged TS/A dormancy, but tumor takes were eventually observed in the majority of mice.

The death of $TS/A^{IFN\text{-}\gamma}$ cells depends on direct lysis by macrophages and blood vessel alteration. Necrosis is associated with ischemic areas and an epicenter within the tumor mass rather than at the border. Each of these reaction patterns is close to those observed following the repeated local injection of recombinant IL-2, IL-4, and IFN-γ (Musiani et al., 1994), showing that the characteristics of the reaction are not influenced by the way in which local cytokine concentration is obtained. However, engineered cells appear to induce a more efficacious reaction than recombinant cytokines, probably because their secretion is both continuous and regulated.

Tumor rejection mechanisms

In the presence of local concentrations of IL-2 and IL-4, incipient TS/A tumors become infiltrated by mononuclear cells and granulocytes that kill tumor cells. Destroyed tumor cells and degenerated inflammatory leukocytes releasing their lytic enzymes are often in the vicinity of fully intact tumor cells, showing that inhibition rests on direct cell lysis.

Ultrastructural data show that cell death is the outcome of membrane and cytoplasmic damage, and only sometimes of apoptosis. Activated lymphocytes are also detected in the process of killing tumor cells. However, they are late arrivals and too few to take any significant part in tumor rejection.

Killer activity of inflammatory cells appears to result from an intense cell-cell cross-talk, since contacts between granulocytes, macrophages, fibroblasts and lymphocytes form the dominant and recurrent ultrastructural hallmark. These cell interactions may be of significance in tuning up the effectiveness of a reaction. By interacting, leukocytes receive or deliver regulatory messages, and build up something similar to a synaptic chamber, where polarized secretion of cytokines and other regulatory factors may take place (Forni et al., 1988; Colombo et al., 1992). In situ hybridization experiments showed that during these reactions there is an intense local release of secondary cytokines (Stoppacciaro et al., 1994).

On the other hand, treatment with cyclosporin A, which impairs the release of certain cytokines, blocks these reactions (Forni et al., 1988). Even though the lymphocytes are few in number, their continuous contacts with the inflammatory cells suggest that they act as guides. The importance of this particular guidance is confirmed by the fact that in mice selectively deprived of T lymphocytes, the antitumor reaction of inflammatory cells elicited by IL-2, IL-4 and G-CSF becomes marginal (Bosco et al., 1990; Cavallo et al., 1992; Colombo et al., 1992). Thus, the efficacy of the reaction elicited by several cytokines is significantly improved by T-lymphocyte guidance, even if it is not T-lymphocyte mediated.

Unsolved problems

The inflammatory reaction activated by the local presence of inducer cytokines has been contrasted with those induced by their systemic inoculation.

It is possible, however, that local reactions depend on the same mechanisms, independently from the dramatic modulation of blood leukocyte populations and onset of LAK cells elicited by systemic administration of huge doses of cytokines, particularly IL-2.

Morphologic evidence shows that local tumor cell destruction by cytokines is not associated with damage to adjacent normal tissues. Even though a cytokine activated reaction is nonspecific, its mechanisms recognize, therefore, some characteristic traits of neoplastic cells. This selectivity is evident both when engineered mouse cells and recombinant cytokines are used (Bubenik et al., 1988; Roth et al., 1992; Tepper et al., 1992; Cavallo et al., 1993).

Therapeutic prospects

In cancer patients, the nonspecific reactions elicited by the local presence of cytokines can be impressive and as effective as surgery, and this approach could perhaps be developed as a new procedure for the management of tumors in situ or localized in a few body cavities. Cancer must primarily be regarded as a systemic disease, and local effects, however impressive, are devoid of long term therapeutic significance.

This situation is well illustrated by the results of our trials with patients with recurrent head and neck squamous cell carcinomas. Activation of local reactivity acquires particular importance in these tumors, since they display local growth, tendency to form local recurrences, and regional lymph node metastatization. Moreover, there is a direct correlation between the degree of immune depression and the severity of the prognosis. Patients no longer responsive to conventional forms of treatment received, irrespective of the site of their tumors, low dose IL-2 injections 1.5 cm from the insertion of the sternocleidomastoid muscle on the mastoid, no matter where the recurrent tumor had appeared. A significant number experienced a dramatic and immediate functional benefit associated with complete or partial remission in the absence of toxic effects. Unfortunately, the responses lasted only 3-5 months and additional IL-2 courses had no further effect (Cortesina et al., 1988; Cortesina et al., 1991).

It should be noted that these temporary responses were obtained in tumors that seemed unlikely to respond as these patients display marked immunosuppression that may impair the establishment of systemic reactivity (Cortesina et al., 1982).

Cytokine induced immune memory: towards cytokine based vaccines

During the inflammatory phase T cells seem to guide the activity of effector leukocytes. Later on the dialogue between T lymphocytes and leukocytes is reversed. The inflammatory setting and the presence of nonspecific leukocytes is instrumental for the induction of a specific and long lasting T-lymphocyte response even towards tumors previously classed as nonimmunogenic. Inflammatory leukocytes first help the presentation of tumor antigens to $CD4^+$ and $CD8^+$ T lymphocytes, and they then assist their activation and expansion.

The rise of these T-lymphocyte reactivities against poorly or nonimmunogenic tumors was unexpected. It relies on the unique conditions occurring during the inflammation reaction guided by the inducer cytokine. Intense tumor cell destruction, release of dead tumor cell remnants, and their phagocytosis by cytokine activated macrophages and granulocytes, give rise to a massive antigen load and build up the conditions for effective indirect tumor antigen presentation to T lymphocytes. The inducer cytokine released by the engineered tumor cells, or repeatedly injected at the tumor site, up-modulates the expression of adhesion molecules and MHC glycoproteins on leukocytes, fibroblasts and endothelia, thus facilitating cell-to-cell interactions and presentation of tumor antigens.

This up-modulation of membrane antigens may be the result of the direct action of the inducer cytokine and/or may depend on its ability to induce the release of secondary cytokines. It can also come from the synergism between the cytokine and the many proinflammatory chemokines released by the granulocytes and macrophages recruited in the reaction. Moreover, the cytokine may also both directly and indirectly provide the accessory signals required for the activation and expansion of T lymphocytes (Fig. 6.2).

This passage from an inflammation like reaction to specific immunity is of particular interest since it enables apparently nonimmunogenic tumors to generate an effective immunogenic signal.

The role of the inducer cytokine in memory induction

The primary role of the inducer cytokine in memory induction is not yet clearly established, and three different, though not mutually exclusive, possibilities can be considered.

1. One possibility is that the inducer cytokine has no direct role at all. What it is doing is simply regulating the intensity and efficacy of the inflammatory reaction. In several cases this is directly proportional to the local concentration of the inducer cytokine. However, when a reaction is very intense and efficacious, quick disappearance of the incipient tumor does not load APC cells with a sufficient quantity of tumor antigen, and no effective memory is elicited. In less effective reactions tumor cells grow at first and are then rejected. Inducer cytokines that elicit a sound immune memory against a nonimmunogenic tumor are merely bringing about, albeit rather slowly, the regression of tumors that have already achieved a certain degree of growth. In this respect, therefore, they provide an elegant, repeated equivalent of what was obtained at the beginning of the century by ligating a tumor's blood vessels so as provoke its necrosis.

If this, then, is the main role of inducer cytokines, some problems arise with regard to the use of irradiated or mitomycin-C treated engineered cells, which are much less effective than proliferating engineered cells in the induction of an immune memory. The reason for this weakness is probably not an outcome of the fact that they release the inducer cytokine for a shorter period, but stems from their inability to proliferate and give rise to small tumors. A failure to initiate tumor growth, in other words, is what prevents them from generating a substantial load of tumor antigen.

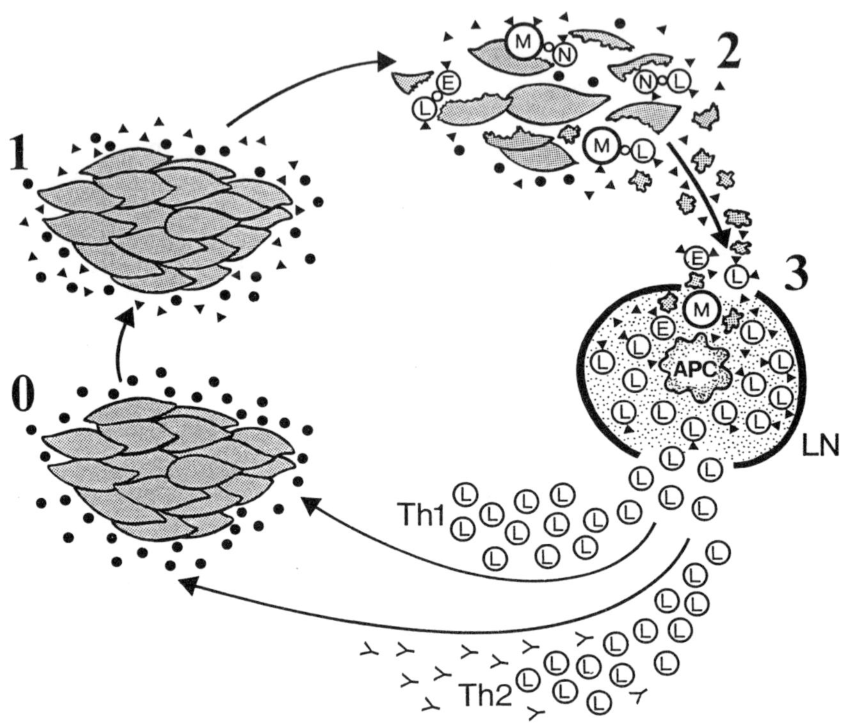

Fig. 6.2. *Interactions between the inflammatory and specific reactions elicited by engineered cells.* **(0)** The scenario dominated by immunosuppressive factors (●) released by the incipient parental tumor (see Fig. 6.1) is radically altered **(1)** by the presence of exogenous cytokines or those released by engineered cells (▲). The eventual size of the incipient tumor mass, in fact, will be determined by the intensity and efficacy of the inflammatory reaction induced. **(2)** This reaction will be characterized by the intervention of different cells (**L**, lymphocytes, **M**, macrophages, **N**, neutrophils, **E**, eosinophils), depending on the type of cytokine involved, and these cells often interact with each other (**o**). The efficacy of this nonspecific inflammatory reaction can be augmented by T-cell guidance. **(3)** Specific recognition of tumor antigens is favoured by their indirect presentation by MHC class II$^+$ inflammatory cells and by the massive load of tumor cell debris. Activation of tumor draining lymph nodes (**LN**) is also evident. Induction of tumor specific immune response and its direction towards **Th1** or **Th2** type of reactivity and antibody (**-<**) response will be determined by the inducer cytokine released by the engineered cells (see Forni et al., 1988, and Colombo and Forni, 1994 for a review of these data).

2. An alternative view is that inducer cytokines do not influence memory T lymphocytes directly, but orchestrate the repertoire of cells involved in the inflammatory reaction. This in itself is sufficient to determine the effectiveness of the memory thus induced, since inflammatory cells differ in their ability to express MHC glycoproteins, provide accessory signals and release distinct chemokine repertoires. An inducer cytokine may thus regulate in this way the efficacy of the indirect presentation of tumor antigen and the activation of T lymphocytes.

3. A further possibility is that the inducer cytokine delivers crucial costimulatory signals during both the inflammatory stage and induction of the immune memory. This is the assumption underlying several experiments in which lymphokines were injected locally to activate TIL immunodepressed by tumor environment and soluble factors. Data from nontumor systems, in fact, indicate that the presence of IL-2 during some of the critical phases of antigen recognition may have important consequences. Four consecutive i.p. injections of as little as 150 U of IL-2 reverse T-lymphocyte unresponsiveness in *Mycobacterium bovis* BCG infected mice (Colizzi, 1984). The solid, long lasting tolerance induced by i.v. injection of newborn mice with semiallogenic lymphoid cells can be nullified by injection of IL-2. When neonatal mice carrying tolerated healthy skin allografts received 1200 U i.p. of IL-2 on days 58 and 59 after graft, there was a significant acceleration of graft rejection (Malkowsky et al., 1984).

The possibility that exogenous IL-2 converts a nonimmunogenic stimulus into an immunogenic one is further supported by experiments showing that injections of 200 U convert a stimulus causing unresponsiveness (picrylated spleen cells injected i.v.) into one causing contact hypersensitivity (Colizzi et al., 1985). Moreover, immunization of low-responder mice with the antigen mixed with 100000 U of IL-2 in complete Freund's adjuvant augments the antibody response to levels similar to those in high responder mice. IL-2 has no effect in congenic, high-responder mice, suggesting that it enhances suboptimal T-lymphocyte help in low responders, whereas T-lymphocyte help is not a limiting factor in high responders (Kawamura et al., 1985).

The potential of IL-1 administered locally or systemically as an adjuvant for a tumor cell vaccine was also explored in mice. Improved vaccine effectiveness was found when human IL-1α or IL-1β were administered in combination with irradiated, weakly immunogenic syngeneic lung tumor cells. The effect is dose dependent, and equally evident when IL-1 is administered at the vaccine site or remotely (McCune and Marquis, 1990).

The distinct Th1 like and Th2 like responses elicited by TS/A^{IL-2} and TS/A^{IL-4} cells against TS/A^{par} cells underline this direct role of the inducer cytokine (Cavallo et al., 1994).

Choice of a Th1 or a Th2 type response in the presence of cells engineered to release IL-2 or IL-4 could be also exploited to get the most efficient response against different microorganisms in situations where its type could be a question of life or death (Finkelman and Urban, 1992).

In a tumor system, too, the type of reaction elicited is of importance. The cure of an established parental tumor with repeated injections of TS/A^{IL-2} (Cavallo et al., 1994) eliciting a Th1 type of response contrasts with the total ineffectiveness of TS/A^{IL-4} cells which elicit a Th2 type response. Since both types of tumors grow to a similar size before being rejected by the reaction they elicit, these contrasting findings suggest that the inducer cytokine is of greater importance in determining the mechanisms and efficacy of memory induction. Nevertheless, experiments are under way to explore this point directly.

Therapeutic prospects

Several critical questions must be answered before the new and attractive prospects offered by vaccination through the use of engineered cells can be considered as a possible form of therapy.

Definition of the predominant role of inducer cytokines in induction of an immune memory is an essential preliminary step, since it opens the way to rational delineation of the strategies to be employed. It must also be conclusively demonstrated that engineered cells provide a more immunogenic signal than plain, living tumor cells either inhibited in proliferation or admixed with conventional adjuvants. One is indeed encouraged to suppose that they are more immunogenic in light of a number of theoretical considerations and the impressive consequences of the inflammatory reaction, while a certain weight must undoubtedly be attached to the positivistic pleasure held out by the prospect of working with fancy technologies.

The results of Hock et al. (1993), however, have sprinkled not a little cold water on this enthusiasm. Preliminary but none the less provocative, they suggest that the immunogenicity of tumor cells that have been engineered is not greater than that of those which have been simply mixed with *Corynebacterium parvum,* that familiar member of the large armamentarium of adjuvants which kept immunologists busy for so long but have apparently led to no results of any consequence. Hock et al.'s data were obtained with irradiated cells from a highly immunogenic tumor. A broad series of experiments designed to provide a straightforward evaluation of the vaccination potential of living TS/A^{IL-2} tumor cells in a variety of settings has shown that they never enhanced the takes of the TS/A^{par} tumor, but constantly and specifically protected against its subsequent challenges. They also cured a significant number of mice bearing both solid TS/A^{par} tumors and established TS/A^{par} metastases, though their efficacy decreased as the tumor mass increased (Cavallo et al., 1993). In tumor bearing mice, vaccinations with mitomycin-C treated TS/A^{par} cells alone or admixed with *Corynebacterium parvum,* complete or incomplete Freund's adjuvants, were totally ineffective (Consalvo et al., submitted).

These findings fit in well with and complete reports of the preimmunizing and curative potential of IL-2 releasing tumor cells (reviewed by Colombo and Forni, 1994). Their value is also enhanced by the fact that they were obtained with a spontaneous and rapidly invasive adenocarcinoma against which it is difficult to induce a significant cell reactivity.

Direct comparison with the human scenario is obviously out of the question. Even so, a similar degree of success in the experimental animal does at least hold out the hope that this new kind of vaccine may be endowed with a certain measure of efficacy, and perhaps will be open to employment in the management of minimum residual disease and small incipient metastases.

Acknowledgments

We thank Dr John Iliffe for careful review of the manuscript and Paola Feraiorni for skillful technical help. This work was supported by grants from the Istituto Superiore di Sanita Italy-USA special projects on "Therapy of Tumors" and AIDS, the Italian Association for Cancer Research (AIRC), the National Research Council (CNR PF-ACRO) and the Italian Ministry of Education (MPI 40% and 60%).

References

Boon T (1985). Tum$^-$ variants: Immunogenic variants obtained by mutagen treatment of tumor cells. Immunol Today **6**, 307-310.

Bosco MC, Giovarelli M, Forni M, Modesti A, Scarpa S, Masuelli L and Forni G (1990). Low doses of IL-4 injected perilymphatically in tumor-bearing mice inhibit the growth of poorly and apparently nonimmunogenic tumors and induce a tumor-specific immune memory. J Immunol **145**, 3136-3143.

Bubenik J, Perlmann P, Indrova M, Simova J, Jandlova T and Neuwirt J (1983). Growth inhibition of an MC-induced mouse sarcoma by TCGF (IL 2)-containing preparations. Cancer Immunol Immunother **14**, 205-206.

Bubenik J, Voitenok NN, Kieler J, Prassolov VS, Chumakov PM, Bubenikova D, Simova J and Jandlova T (1988). Local administration of cells containing an inserted IL-2 gene and producing IL 2 inhibits growth of human tumours in nu/nu mice. Immunol Lett **19**, 279-282.

Cavallo F, Giovarelli M, Gulino A, Vacca A, Stoppacciaro A, Modesti A and Forni G (1992). Role of neutrophils and CD4$^+$ T lymphocytes in the primary and memory response to nonimmunogenic murine mammary adenocarcinoma made immunogenic by IL-2 gene transfection. J Immunol **149**, 3627-3635.

Cavallo F, Di Pietro F, Giovarelli M, Gulino A, Vacca A, Stoppacciaro A, Forni M, Modesti A and Forni G (1993). Protective and curative potential of vaccination with IL-2-gene-transfected cells fron a spontaneous mouse mammary adenocarcinoma. Cancer Res **53**, 5067-5970.

Cavallo F, Consalvo M, Allione A, Novelli F and Giovarelli M (1994). Antitumor respons elicited by mouse adenocarcinoma cells engineered to release IL-2 and IL-4. In "Cytokine Induced Tumor Immunogenicity" (eds G Forni, R Foa, A Santoni and L Frati), pp 203-216. Academic Press, London.

Colizzi V (1984). In vivo and in vitro administration of interleukin-2 containing preparations reverses T-cell unresponsiveness in Mycobacterium bovis BCG-infected mice. Infect Immunity **45**, 25-28.

Colizzi V, Malkowsky M, Lang G and Asherson GL (1985). In vivo activity of interleukin-2. Conversion of a stimulus causing unresponsiveness to a stimulus causing contact hypersensistivity by injection of interleukin-2. Immunology **56**, 653-659.

Colombo MP, Modesti A, Parmiani G and Forni G (1992). Local cytokine availability elicits tumor rejection and systemic immunity through granulocyte-T-lymphocyte cross-talk. Cancer Res **52**, 4853-4857.

Colombo MP and Forni G (1994). Cytokine gene transfer in tumor inhibition and tentative tumor therapy. Where are we now? Immunol Today **15**, 48-51.

Cortesina G, Cavallo GP, Beatrice F, Sartoris A, Bussi M, Morra B, Di Fortunato V, Poggio E and Rendine S (1982). Production of leukocyte migration inhibition factor by lymphocytes of larynx cancer patients stimulated by laryngeal carcinoma solubilized membrane antigens. Tumori **68**, 39-44.

Cortesina G, De Stefani A, Giovarelli M, Barioglio MG, Cavallo GP, Jemma C and Forni G (1988). Treatment of recurrent squamous cell carcinoma of the head and neck with low doses of interleukin-2 injected perilymphatically. Cancer **62**, 2482-2485.

Cortesina G, De Stefani A, Galeazzi E, Cavallo GP, Jemma C, Giovarelli M, Vai S and Forni G (1991). Interleukin-2 injected around tumor draining lymph nodes in head and neck cancer. Head and Neck **13**, 125-131.

Finkelman FD and Urban JF Jr (1992). Cytokines: Making the right choice. Parassitol Today **8**, 311-314.

Forni G and Comoglio PM (1973). Growth of syngeneic tumors in unimmunized newborn and adult hosts. Brit J Cancer **27**, 120-127.

Forni G, Giovarelli M, Lanfrancone L and Varesio L (1982). Suppressor macrophages in tumor bearing mice. Inconsistency between in vivo and in vitro findings? Int J Cancer **29**, 695-698.

Forni G and Santoni A (1984). Immunogenicity of non immunogenic tumors. J Biol Resp Modif **13**, 128-133.

Forni G, Fujiwara H, Martino F, Hamaoka H, Jemma C, Caretto P and Giovarelli M (1988). Helper strategy in tumor immunology. Expansion of helper lymphocytes and utilization of helper limphokines for experimental and clinical immunotherapy. Cancer Metastasis Rev **7**, 289-309.

Giovarelli M, Comoglio MP and Forni G (1976). Induction of resistance or enhancement to a transplantable murine plasmacytoma by transfer of non-immune leukocytes. Br J Cancer **34**, 233-238.

Hock H, Dorsch M, Kunzendorf U, Uberla K, Qin Z, Diamantstein T and Blankenstein T (1993). Vaccinations with tumor cells genetically engineered to produce different cytokines. Effectivity not superior to a classical adjuvant. Cancer Res **53**, 714-716.

Kawamura H, Rosenberg SA and Berzofsky JA (1985). Immunization with antigen and interleukin 2 in vivo overcomes Ir gene low responsiveness. J Exp Med **162**, 381-385.

Kruger-Krasagakes S, Li W, Richter G, Diamantstein T and Blankenstein T (1993). Eosinophils infiltrating interlelukin-5 gene-transfected tumors do not suppress tumor growth. Eur J Immunol **23**, 992-995.

Malkowsky M, Medawar PB, Thatcher DR, Toy J, Hunt R, Rayfield LS and Dore C (1984). Acquired immunological tolerance of foreign cells is impaired by recombinant interleukin 2 or vitamin A acetate. Proc Natl Acad Sci USA **83**, 536-538.

McCune CS and Marquis DM (1990). Interleukin 1 as adjuvant for active specific immunotherapy in a murine tumor model. Cancer Res **50**, 1212-1215.

Mitchison NA and Kinlen LJ (1980). Present concepts in immune surveillance. Progr Immunol **8**, 641-650.

Musiani P, Modesti A, Brunetti M, Modica A, Gulino A, Bosco MC, Colombo MP, Nanni P, Cavallo F, Pericle F, Giovarelli M, Cavallo R and Forni G (1994). The nature and potential of the reactive response to mouse mammary adenocarcinoma engineered with IL-2, IL-4 of IFN-γ gene. Nat Immun **13**, 93-101.

Pericle F, Di Pierro F and Forni G. (1992). Clinical trials with local administration of lymphopoietic growth factors. In "Lympho-hematopoietic Growth Factors in Cancer Therapy II" (ed R Mertelsmann), pp 87-96. Springer-Verlag, Berlin.

Roth C, Mir LM, Cressent M, Quintin-Colonna F, Ley V, Fradelizi D and Kourlinsky P (1992). Inhibition of tumor growth by histoincompatible cells expressing interleukin-2. Int Immunol **4**, 1439-1436.

Stoppacciaro A, Forni G and Colombo MP (1994). Different tumors transduced with different cytokine genes as G-CSF and IL-2 show inhibition of tumor take through neutrophil activation but differ in T cell functions. Folia Biol, in press.

Tepper RI, Coffman RL and Leder P (1992). An eosinophil-dependent mechanism for the anti-tumor effect of interleukin 4. Science **257**, 548-551.

Van der Bruggen P, Traversari C, Chomez P, Lurquin C, De Plaen E, Van den Eynde B, Knuth A and Boon T (1991). A gene encoding an antigen recognized by cytolytic T lymphocytes on a human melanoma. Science **254**, 1643-1647.

EXPERIMENTAL MODELS

7

GENE MODIFIED TUMOR VACCINES

D. M. Pardoll
Departments of Oncology, Medicine and Molecular Biology and Genetics, Johns Hopkins University School of Medicine, Baltimore, MD, USA

Introduction

The classic concept of a vaccine derives from the area of viral vaccines. For viral vaccines, individuals are immunized against viral antigens prior to encountering the pathogenic virus. Such a strategy is feasible with viruses because viral genomes are relatively simple, possessing a limited number of defined antigens. Such is not the case for most tumors, whose potential universe of antigens can be virtually limitless (see below). Therefore, when we talk about tumor vaccines, the most common clinical setting is one in which induction of a systemic immune response by the vaccine must occur subsequent to, rather than prior to, the antigenic insult. In the past, and for the near future, the major tumor specific antigens upon which we can focus the immune system have not been molecularly defined. We must therefore produce the vaccine by using the patient's own tumor cells as the source of antigen. In considering how to most effectively enhance immune responses against tumor antigens, it is important to appreciate the emerging principle that T-cell responses, rather than antibody responses, are the primary target of effective antitumor immunization strategies.

The importance of T cells in vaccine-based tumor immunotherapy has a number of critical implications. First, since T cells can recognize peptide antigens brought to the cell surface in conjunction with MHC proteins, the universe of potential tumor antigens recognizable by T cells involves any cellular peptide sequence, whether membrane bound or intracellular. Second, the three dimensional conformation of a tumor antigen becomes far less important than the primary amino acid sequence characterizing that antigen. Finally, since different MHC haplotypes bind largely nonoverlapping sets of peptides, it is likely that, with the exception of common shared MHC haplotypes (such as HLA-A2.1), the specific epitopes recognized by the immune system will differ from individual to individual.

CYTOKINE-INDUCED TUMOR IMMUNOGENICITY
ISBN 0-12-261520-4

Tumors possess neoantigens potentially recognizable by T cells

Eventually, the spectrum of tumor specific antigens capable of being recognized by T cells may become identified. Over the past five years preliminary headway has indeed been made in identifying examples of tumor specific antigens capable of being recognized by T cells. Without question, one of the largest categories of potential antigens is represented by mutated or rearranged oncogenes or mutated antioncogenes known to be a hallmark of essentially all malignancies (Hollstein et al., 1991; Hunter, 1991; Marshall, 1991). Most of these altered proteins are intracellular in location, so that antibodies recognizing them would likely not be of therapeutic value. The problem with these particular gene products as targets for vaccination/immunization strategies is that in most cases the site of the mutation or rearrangement tends to vary among tumors. Nonetheless, T cells specific for *bcr-abl,* mutated *ras* and p53 have indeed been identified in various murine and human systems (Jung and Schluesener, 1991; Peace et al., 1991).

Another important category of commonly expressed potential tumor antigens is represented by viral gene products. It is now known that a number of common human malignancies have viral associations and are indeed characterized by the expression of specific viral gene products. Thus, a proportion of Hodgkin's as well as Burkitt's lymphomas and nasopharingeal carcinomas is associated with EBV (Young et al., 1988; Young et al., 1989; Herbst et al., 1991; Pallesen et al., 1991), and the majority of cervical carcinoma is associated with human papilloma virus (Durst et al., 1983; McCance et al., 1983). In the case of Hodgkin's disease the EBV gene products EBNA 1 and LMP appear to be expressed, whereas in cervical carcinoma the HPV E6 and E7 gene products are expressed. Under certain circumstances, both T-cell and antibody responses against these gene products can indeed be identified.

Among the most exciting recent advances in tumor biology has been the identification of a melanoma specific gene product termed MAGE-1 by Boon and colleagues (Van der Bruggen et al., 1991). This gene was identified as encoding the antigen recognized by a human melanoma specific T-cell clone. The MAGE-1 gene expressed in the melanoma was found to have a sequence identical to the germ line gene. This gene was found to be expressed in roughly 50% of melanomas but was not expressed in any normal adult tissue (with the exception of the testes). While MAGE-1 itself may not be a dominant tumor antigen, it represents an important category of tumor antigens, namely embryonic gene products that are reactivated in tumors. While examples of expression of embryonic gene products in tumors such as CEA and AFP have been recognized for quite a while, MAGE-1 indeed represents the first such example identified as a tumor antigen specifically recognizable by T cells from the patient.

For purposes of discussion, it is worth separating tumor vaccine strategies into two categories:

- strategies that use the tumor cell itself as a source of tumor antigens and therefore do not require knowledge of the specific molecular identity of the antigen
- antigen specific vaccine strategies which are designed to generate immune responses against specific gene products known to be expressed by a particular tumor.

Immune tolerance in the antitumor immune response

Recognition of a tumor specific peptide presented on class I MHC by the $CD8^+$ T cell, as well as the signals generated from that recognition event, are in and of themselves insufficient to activate the CTL precursor to discharge its lytic machinery. At least one set of additional signals - represented by lymphokines - are necessary for complete activation of the CTL. Lymphokines are not produced constitutively but, rather, by Th cells, which tend to be $CD4^+$. To produce lymphokines, $CD4^+$ Th cells must recognize separate sets of antigenic peptides presented on class II MHC molecules by dedicated antigen presenting cells such as dendritic cells or macrophages, which must endocytose these antigens from the exterior and place them into the endosomal/lysosomal compartment. Protein antigens are degraded in that compartment, and the resultant peptides associate with class II MHC molecules. We now know that one of the features which distinguishes a "professional" antigen presenting cell is the expression of costimulatory molecules (such as B7) which engage receptors on Th cells (such as CD28) concomitant with antigen recognition by the TCR (Linsley et al., 1991). TCR occupancy in the absence of the stimulatory signals can lead to T-cell anergy, a mechanism potentially responsible for tolerance induction to self antigens expressed in the periphery (Jenkins et al., 1987a; Jenkins et al., 1987b; Gaspari et al., 1988; Lo et al., 1988; Markmann et al., 1988; Harding et al., 1992).

A failed immune response against a tumor cell can be due to reasons other than absence of a tumor specific antigen or absence of $CD8^+$ T cells capable of recognizing that antigen. A defect in the helper arm of the immune response would result in failure to produce appropriate lymphokines as a second signal to activate CTL. A helper arm deficiency could result from either an absence of class II MHC restricted tumor antigens, the failure of antigen presenting cells to endocytose sufficient quantities of tumor specific antigens, or the generation of tolerance among Th cells specific for class II MHC restricted tumor antigens. Indeed, an intriguing notion is that the same mechanisms which generate T-cell tolerance to peripheral self antigens may cause tolerance among Th cells specific for newly arising class II MHC restricted tumor antigens. For example, expression of class II MHC by epithelial tumors (which typically do not express costimulatory molecules such as B7) could result in presentation of tumor specific antigens to Th cells in a manner that results in clonal anergy. As will be discussed below, many of the more recent genetically modified tumor vaccine strategies were developed specifically to either bypass a defective helper arm of the immune response or modify the way in which tumor antigens are presented to T cells, such that the resulting responses lead to activation rather than tolerance.

Tumor vaccines with adjuvants: the past

Whole cell tumor vaccines have been used to treat cancer patients with certain solid tumors for the past few decades (Livingston et al., 1985; Berd et al., 1986; McCune et al., 1990; Berd et al., 1991). In the past, these vaccines have consisted of either irradiated autologous or allogeneic tumor cells, tumor cell lysates, or in some cases irradiated virus infected tumor cells. In general, most of these vaccine strategies have involved coinjection of cells with adjuvants such as BCG. Indeed, the concept of inducing antitumor immune responses by coinjection of tumor cells with bacterially derived products as adjuvants derives from the experiments of William Coley first reported exactly 100 years ago. Most tumor vaccine studies to date have not compared parameters in a randomized fashion; this, together with the uncharacterized nature of many adjuvants, has made it difficult to determine which approach generates maximal immune responses. Nonetheless, either stabilization, partial remission, or occasionally complete remission have been reported infrequently for renal cancer and melanoma.

In man, melanoma and renal cancer are notable for occasional spontaneous remissions which appear to be immunologically mediated. They may therefore represent cases in which the tumor and immune system are somehow balanced, and even relatively inefficient strategies for activation of immune responses may tip the balance in favor of the immune system over the tumor. Indeed, in some studies that reported antitumor responses, the responses were shown to correlate with the return of DTH responses to recall antigens and the development of a DTH response to autologous tumor cells.

Genetic alteration of tumors to enhance immunogenicity of tumor cells: the present

Introduction of MHC genes into tumor cells

It has long been known that increased expression of class I MHC proteins on tumor cells enhances their ability to be lysed by tumor specific CTL in vitro. Studies in the 1980's demonstrated that enhancing class I MHC expression by gene transfection resulted in decreased tumorigenic capacity and/or metastatic capacity of murine tumor models (Hui et al., 1984; Wallich et al., 1985). The decreased tumorigenicity has been interpreted as due to enhanced presentation of tumor specific peptides allowing the enhanced rejection of tumor cells by $CD8^+$ CTL in vivo. Despite these positive studies, the relationship between levels of MHC expression and tumorigenicity is, in fact, quite inconsistent when multiple tumor models are studied. Furthermore, enhanced expression of self-class I MHC molecules does not always increase the immunologic potency of a tumor when used in a vaccine capacity (either with or without adjuvants).

While most epithelial tumors express class I MHC products constitutively, the vast majority are class II^- MHC. Class II MHC is inducible in most epithelioid tumors by certain cytokines such as IFN-γ. In certain murine tumors, introduction of class II MHC genes into tumor cells decreases their immunogenicity and can result in the generation of enhanced systemic immune responses against the parental class II^- MHC tumor. It was postulated that the expression of class II MHC molecules on tumor cells allowed for the presentation of class II restricted tumor specific antigens to Th cells which ultimately could provide enhanced in vivo help for CTL activation, thereby resulting in protection from challenges with the parental tumor expressing only class I MHC products. The major problem with this hypothesis is that most epithelioid tumor cells do not express costimulatory signals such as B7 and are therefore typically not considered to be effective antigen presenting cells for Th cells. Indeed, as mentioned earlier, expression of class II MHC molecules on epithelioid tumor cells represents a potential mechanism for generating Th cell tolerance that could allow tumors to escape immune recognition. Recently, the coexpression of B7 together with class II MHC in tumors has been shown to enhance their immunogenicity (Baskar et al., 1993).

While the introduction of autologous class I MHC and MHC class II molecules into tumor cells represents an important experimental model, it is important to recognize that because of the tremendous polymorphisms in the outbred population, as well as the number of different loci in an individual that could potentially be critical in the presentation of specific tumor antigens, such a genetic approach in humans would necessarily be highly individualized and therefore impractical for treating large numbers of patients. Nonetheless, many of these experimental studies have provided a basis for approaches that introduce the IFN-γ gene into tumors as a means of regulating endogenous MHC gene products (see below).

Introduction of the B7 gene into tumor cells

Another cell surface molecule that has recently shown promise in enhancing the vaccination potential of tumor cells is B7. B7 is normally expressed as an activation antigen on APC such as B cells, macrophages and dendritic cells. B7 is the ligand for two receptors expressed on T cells - CD28 and CTLA4. While the function of CTLA4 (expressed exclusively on $CD8^+$ T cells) is as yet unknown, much evidence has accumulated that CD28 (expressed on all $CD4^+$ T cells and the majority of $CD8^+$ T cells) is a critical receptor for costimulatory signals in T-cell activation. Crosslinking of CD28 has been shown to enhance the level of lymphokine production by $CD4^+$ T cells subsequent to TCR engagement. The mechanism for enhanced lymphokine production appears to be at the level of transcription and of enhanced RNA stability (Lindsten et al., 1989; Fraser et al., 1991). In some in vitro systems, CD28 engagement not only enhances T-cell activation but is necessary for driving T cells toward activation rather than anergy (Harding et al., 1992). More recently, CD28 engagement on $CD8^+$ cells has also been suggested to be important for CTL activation and possibly CTL priming. With this background, the B7 gene was a logical choice for introduction into tumor cells as an approach to enhanced immunization capacity.

A number of laboratories have indeed shown that, subsequent to transfection with a B7 gene, autologous tumors were rejected in their syngeneic host, and systemic immune responses capable of generating protection against challenges with the wild type tumor at a distant site were achieved (Chen et al., 1992; Baskar et al., 1993; Townsend and Allison, 1993). In general, the potency of B7 transduction remains to be established. In one case, the transduction of B7 into a tumor by itself was insufficient to produce rejection or systemic immune responses; it was necessary to introduce an additional "strong" tumor antigen into the tumors as well in order to see a B7 effect. In the study mentioned above (Baskar et al., 1993), introduction of B7 into a class I^+ MHC, class II^- MHC tumor did not enhance immunogenicity, but cointroduction of B7 and class II MHC did enhance immunogenicity. While B7 engineered tumors are not the final answer to tumor vaccine design, these studies have focused important attention on the role of costimulatory molecules in T-cell activation as part of the generation of a systemic antitumor immune response.

Introduction of cytokine genes into tumor cells

Most recently, there has been intense activity in generating tumor cell vaccines engineered to secrete various cytokines (Tepper et al., 1989; Watanabe et al., 1989; Fearon et al., 1990; Gansbacher et al., 1990a; Gansbacher et al., 1990b; Asher et al., 1991; Blankenstein et al., 1991; Colombo et al., 1991; Golumbek et al., 1991; Hock et al., 1991; Rollins and Sunday, 1991; Colombo et al., 1992; Pordagor et al., 1992; Restifo et al., 1992; Dranoff et al., 1993). This strategy seeks to locally alter the immunologic environment of the tumor cell so as to either enhance antigen presentation of tumor specific antigens to the immune system or to enhance the activation of tumor specific lymphocytes. One of the most important concepts underlying the use of cytokine gene transduced tumor vaccines is that the cytokine is produced at very high concentrations local to the tumor. Systemic concentrations are generally quite low. This paracrine physiology much more closely mimics the natural biology of cytokine action than does the systemic (i.v.) administration of recombinant cytokines. Many cytokine genes have been introduced into tumor cells with varying effects on both tumorigenicity and immunogenicity. Some of these cytokines, when produced by tumors, induce a local inflammatory response which results in elimination of the injected tumor. This local inflammatory response is most often predominantly dependent on leukocytes other than classical T cells.

These systems have been used to uncover in vivo effects of cytokines that result in activation of tumoricidal potential by various types of leukocytes. In a subset of the studies, systemic immune responses have been demonstrated to be generated against challenge with the wild type parental tumor. In all cases in which systemic immunity against wild type tumor challenge has been analyzed, it is mediated by T cells. Given the number of studies done to date with cytokine transduced tumor cells, it is not surprising that variable results have been seen when different tumor systems are analyzed. Additional variables such as cell dose, level of cytokine expression, and location of immunization and challenge site are crucial parameters affecting vaccine efficacy for cytokine gene engineered tumor cells. Given the number of potential cytokine genes in the armamentarium and the technical difficulties in transducing human tumor cells to make vaccines, it is critical that they be compared head to head for efficacy. Also, given that most mouse tumors show significant immunogenicity when simply irradiated, identification of genes which truly enhance a tumor's "immunogenicity" significantly above that of irradiated wild type cells is important. The first study which directly compared multiple cytokine and other genes in murine tumor models used a highly transmissible defective retroviral vector. This study demonstrated that, in a number of poorly and moderately immunogenic tumors, immunization with GM-CSF transduced tumors produced the greatest degree of systemic immunity, which was enhanced relative to irradiated nontransduced tumors (Dranoff et al., 1993). Immunity was dependent on both $CD4^+$ and $CD8^+$ T cells, despite the fact that the tumors were MHC class II^-. The potency of GM-CSF's effect locally may relate to its unique ability in promoting the differentiation of hematopoietic precursors to dendritic cells, which are the most potent APC for Th cells (Inaba et al., 1992a; Inaba et al., 1992b).

Development of genetically modified human tumor vaccines

While the recent preclinical studies performed in mouse tumor models are indeed paving the way for clinical trials with genetically engineered tumor vaccines, a number of critical considerations remain that will impact significantly on whether or not this new immunotherapy approach will translate into true clinical benefit for human cancer patients.

High efficiency gene transfer systems for primary human tumor explants must be developed and optimized. There are currently a number of barriers to efficient gene transfer into primary human tumors. Most human tumors are difficult to transfect using standard techniques such as calcium phosphate or lipofection. Furthermore, most human solid tumors are difficult to establish as long-term lines. Even in cases where subselection of continuously growing tumor lines can be achieved, there is a high likelihood that after extended passage the antigenic composition will change significantly relative to the original primary tumor from which a patient's metastases have originated.

All of these issues emphasize the importance of high efficiency vector systems capable of achieving gene transfer into the majority of explanted cells. Ideally, such high efficiency gene transfer would obviate the need for cointroduction of a selection marker, thereby minimizing the level of in vitro passage and maximally preserving the original antigenic composition of the explanted primary tumor. Defective retroviral vector systems are unquestionably the gold standard for high efficiency gene transfer in the initial stages of human gene therapy. These vectors have the potential for being helper virus free and, therefore, extremely safe. Nonetheless, they require at least one cell cycle in order for integration and gene expression to occur. Early generations of retroviral vectors have been notoriously inconsistent in their gene transfer efficiency as well as consistency of gene expression. Recently, newer generations of retroviral vectors that have taken advantage of modifications in the intronic *gag* sequences as well as positioning of the start sites of the inserted gene appear to have successfully circumvented a number of these problems and allowed efficient gene transfer into primary human tumor explants in the absence of selection (Jaffee et al., 1993). Other gene transfer systems, such as adenovirus based systems, as well as nonviral gene transfer systems (such as liposomes), are currently under evaluation though their ultimate usefulness in clinical settings remains to be determined.

Antigen specific tumor vaccines: the future

The ultimate goal of tumor vaccine design is the generation of antigen specific vaccines. If we knew the specific antigens expressed by tumor cells, it would be possible to produce generic vaccines using vectors, such as recombinant vaccinia virus, thereby obviating the need for the individual's own tumor cells as the source of antigen. The initial success of Boon and colleagues in identifying T-cell recognized tumor specific antigens, as well as the demonstration of specific examples of T-cell recognition of mutated oncogenes, tumor suppressor genes and tumor associated viral associated gene products, has rekindled excitement about these approaches. It is important to note that certain categories of tumor specific antigen will be more useful as generic therapeutic targets than others.

For example, a tumor antigen such as MAGE-1, representing an embryonic gene product reactivated in the tumor cell, represents a whole protein's worth of potential epitopes to be chosen by an individual's MHC molecules for presentation to T cells. On the other hand, a mutated oncogene or antioncogene will produce a tumor specific peptide with much more limited possibilities in terms of binding to a particular individual's MHC molecules.

Currently, the gold standard for T-cell based vaccine design utilizes recombinant vaccinia virus (Mackett et al., 1982). This vaccine has the advantage of having a large capacity (25 kDa) for genetic insertion as well as the feature of strong promoters for high level gene expression. Currently, modified versions of vaccinia vectors with reduced risk of potential virulence are under active investigation. While infection with recombinant vaccinia viruses has been shown to protect animals from subsequent challenge with tumor cells containing the gene product cloned into the recombinant vaccinia, it has so far been impossible to demonstrate cure of animals with even small doses of established tumors by recombinant vaccinia injection (Lathe et al., 1987; Meneguzzi et al., 1991). It will thus be important to enhance the immunization potency of antigen specific vaccines. This may eventually be accomplished through increased understanding of pathways for antigen presentation.

Other antigen specific vaccination strategies such as peptide vaccines may also be applied in the case of tumors. One approach demonstrating significant promise in both animal models and humans has been idiotype vaccination with B-cell lymphoma idiotypes (Campbell et al., 1990; Kwak et al., 1992).

Idiotype vaccination has been shown to induce idiotope-specific CD4 responses resulting in anti-idiotype humoral responses. Recently, the engineering of chimeric idiotype-GM-CSF molecules has resulted in enhanced immunization in the absence of additional adjuvants (Tao and Levy, 1993). It is intriguing to speculate that the same mechanisms responsible for the enhanced immunization capacity of GM-CSF gene transduced tumor cells described above are operative in the case of the idiotype-GM-CSF chimeras.

While most tumor vaccines must be employed subsequent to the development of cancer, in certain cases, such as regions with high endemic rates of HPV infection and consequently high rates of cervical carcinoma, it may be feasible to vaccinate children prophylactically. Ultimately, the development of multiple strategies that can be utilized in synergy will most likely be responsible for the major advances against cancer.

It still remains to be seen what place immunotherapy in general will have in the armamentarium. Without question, though, it is now feasible to design vaccine strategies on a molecular basis rather than "with eye of newt and toe of frog".

References

Asher AL, Mule JJ, Kasid A, Restifo NP, Salo JC, Reichert CM, Jaffe G, Fendly B, Kriegler M and Rosenberg SA (1991). Murine tumor cells transduced with the gene for tumor necrosis factor-α. Evidence for paracrine immune effects of tumor necrosis factor-α against tumors. J Immunol **146**, 3227-3234.

Baskar S, Ostrand-Rosenberg S, Nabavi N, Nadler LM, Freeman GJ and Glimcher L (1993). Constitutive expression of B7 restores immunogenicity of tumor cells expressing truncated major histocompatibility complex class II molecules. Proc Nat Acad Sci USA **90**, 5687-5690.

Berd D, Maguire HC and Mastrangelo MJ (1986). Induction of cell-mediated immunity to autologous melanoma cells and regression of metastases after treatment with a melanoma cell vaccine preceded by cyclophosphamide. Cancer Res **46**, 2572-2577.

Berd D, Maguire HC, McCue P and Mastrangelo MJ (1991). Treatment of metastatic melanoma with an autologous tumor-cell vaccine: Clinical and immunologic results in 64 patients. J Clin Oncol **8**, 1858-1862.

Blankenstein T, Qin ZH, Uberla K, Muller W, Rosen H, Volk HD and Diamantstein T (1991). Tumor suppression after tumor cell-targeted tumor necrosis factor α gene transfer. J Exp Med **173**, 1047-1052.

Campbell MJ, Esserman L, Byars NE, Allison AC and Levy R (1990). Idiotype vaccination against murine B cell lymphoma. J Immunol **145**, 1029-1036.

Chen L, Ashe S, Brady W, Hellstrom I, Hellstrom KE, Ledbetter J, McGowan P and Linsley P (1992). Costimulation of antitumor immunity by the B7 counter receptor for the T lymphocyte molecules CD28 and CTLA-4. Cell **71**, 1093-1102.

Colombo M, Ferrari G, Stoppacciaro A, Parenza M, Rodolfo M, Mavilio F and Parmiani G (1991). Granulocyte colony-stimulating factor gene transfer suppresses tumorigenicity of a murine adenocarcinoma in vivo. J Exp Med **173**, 889-897.

Colombo M, Modesti A, Parmiani G and Forni G (1992). Local cytokine availability elicits tumor rejection and systemic immunity through granulocyte-T-lymphocyte cross-talk. Cancer Res **52**, 4853-4857.

Dranoff G, Jaffee E, Lazenby A, Golumbek P, Levitsky H, Brose K, Jackson V, Hirofumi H, Pardoll D and Mulligan RC (1993). Vaccination with irradiated tumor cells engineered to secrete murine granulocyte-macrophage colony stimulating factor stimulates potent, specific and long-lasting anti-tumor immunity. Proc Natl Acad Sci USA **90**, 3539-3543.

Durst M, Gissman L, Ikenberg H and Zur Hausen H (1983). A papilloma virus DNA from a cervical carcinoma and its prevalence in cancer biopsy samples from different geographic regions. Proc Natl Acad Sci USA **80**, 3812-3815.

Fearon ER, Pardoll DM, Itaya T, Golumbek P, Levitsky HI, Simons JW, Karasuyama H, Vogelstein B and Frost P (1990). Interleukin-2 production by tumor cells bypasses T helper function in the generation of an antitumor response. Cell **60**, 397-403.

Fraser JD, Irving BA, Crabtree GR and Weiss A (1991). Regulation of interleukin-2 gene enhancer activity by the T cell accessory molecule CD28. Science **251**, 313-316.

Gansbacher B, Bannerji R, Daniels B, Zier K, Cronin K and Gilboa E. (1990a). Retroviral vector-mediated γ-interferon gene transfer into tumor cells generates potent and long lasting antitumor immunity. Cancer Res **50**, 7820-7825.

Gansbacher B, Zier K, Daniels B, Cronin K, Bannerji R and Gilboa E (1990b). Interleukin-2 gene transfer into tumor cells abrogates tumorigenicity and induces protective immunity. J Exp Med **172**, 1217-1224.

Gaspari AA, Jenkins MK and Katz SI (1988). Class II MHC-bearing keratinocytes induce antigen-specific unresponsiveness in hapten-specific TH1 clones. J Immunol **141**, 2216-2220.

Golumbek PT, Lazenby AJ, Levitsky HI, Jaffee LM, Karasuyama H, Baker M and Pardoll DM (1991). Treatment of established renal cancer by tumor cells engineered to secrete interleukin-4. Science **254**, 713-716.

Harding FA, McArthur JG, Gross JA, Raulet DH and Allison JP (1992). CD28-mediated signalling co-stimulates murine T cells and prevents induction of anergy in T-cell clones. Nature **356**, 607-609.

Herbst H, Dallenbach F, Hummel M, Niedobitek G, Pileri S, Muller-Lantzsch N and Stein H (1991) Epstein-Barr virus latent membrane protein expression in Hodgkin's and Reed-Sternberg cells. Proc Natl Acad Sci USA **88**, 4766-4770.

Hock H, Dorsch M, Diamantstein T and Blankenstein T (1991). Interleukin 7 induces $CD4^+$ T cell-dependent tumor rejection. J Exp Med **174**, 1291-1298.

Hollstein M, Sidransky D, Vogelstein B and Harris CC (1991). p53 mutations in human cancers. Science **253**, 49-53.

Hui K, Grosveld F and Festenstein H (1984). Rejection of transplantable AKR leukaemia cells following MHC DNA-mediated cell transformation. Nature **311**, 750-752.

Hunter T (1991). Cooperation between oncogenes. Cell **64**, 249-270.

Inaba K, Inaba M, Romani N, Aya H, Deguchi M, Ikehara S, Muramatsu S and Steinman RM (1992a). Generation of large numbers of dendritic cells from mouse bone marrow cultures supplemented with granulocyte/macrophage colony-stimulating factor. J Exp Med **176**,1693-1702.

Inaba K, Steinman RM, Pack MW, Aya H, Inaba M, Sudo T, Wolpe S and Schuler G. (1992b). Identification of proliferating dendritic cell precursors in mouse blood. J Exp Med **175**, 1157-1167.

Jaffee E, Dranoff G, Cohen L, Hauda K, Clift S, Marshall F, Mulligan RC and Pardoll D (1993). High efficiency gene transfer into primary human tumor explants without cell selection. Cancer Res **53**, 2221-2226.

Jenkins MK, Pardoll DM, Mizuguchi J, Chused TM and Schwartz RH (1987a). Molecular events in the induction of a nonresponsive state in interleukin 2-producing helper T-lymphocyte clones. Proc Natl Acad Sci USA **84**, 5409-5413.

Jenkins MK, Pardoll DM, Mizuguchi J, Quill H and Schwartz RH (1987b). T-cell unresponsiveness in vivo and in vitro: Fine specificity of induction and molecular characterization of the unresponsive state. Immunol Rev **95**, 113-135.

Jung S and Schluesener HJ (1991). Human T lymphocytes recognize a peptide of single point-mutated, oncogenic ras proteins. J Exp Med **173**, 273-276.

Kwak LW, Campbell MJ, Czerwinski DK, Hart S, Miller RA and Levy R (1992). Induction of immune responses in patients with B-cell lymphoma against the surface-imunoglobulin idiotype expressed by their tumors. N Engl J Med **327**, 1209-1215.

Lathe R, Kieny MP, Gerlinger P, Clertant P, Guizani I, Cuzin F and Chambon P (1987). Tumour prevention and rejection with recombinant vaccinia. Nature **326**, 878-880.

Lindsten T, June CH, Ledbetter JA, Stella G and Thompson CB (1989). Regulation of lymphokine messanger RNA stability by a surface-mediated T cell activation pathway. Science **244,** 339-343.

Linsley PS, Brady W, Grosmaire L, Aruffo A, Damle NK and Ledbetter JA (1991). Binding of the B cell activation antigen B7 to CD28 costimulates T cell proliferation and interleukin 2 mRNA accumulation. J Exp Med **173,** 721-730.

Livingston PO, Albino AP, Cheng TJC, Real FX, Houghton AN, Oettgen HF and Old LJ (1985). Serological response of melanoma patients to vaccines prepared from VSV lysates of autologous and allogeneic cultured melanoma cells. Cancer **55,** 713-716.

Lo D, Burkly LC, Widera G, Cowing C, Flavell RA, Palmiter RO and Brinster RL (1988). Diabetes and tolerance in transgenic mice expressing class II MHC molecules in pancreatic beta cells. Cell **53,** 159-168.

Mackett M, Smith GL and Moss B (1982). Vaccinia virus: A selectable eukaryotic cloning and expression vector. Proc Natl Acad Sci USA **79,** 7415-7419.

Markmann J, Lo D, Naji A, Palmiter RD, Brinster RL and Heber-Katz E (1988). Antigen presenting function of class II MHC expressing pancreatic beta cells. Nature **336,** 476-479.

Marshall CJ (1991). Tumor suppressor genes. Cell **64,** 313-326.

McCance D, Walker P, Dyson J, Coleman D and Singer A (1983). Presence of human papillomavirus DNA sequences in cervical intraepithelial neoplasia. Br Med J **287,** 784-788.

McCune CS, O'Donnell RW, Marquis DM and Sahasrabudhe DM (1990). Renal cell carcinoma treated by vaccines for active specific immunotherapy: Correlation of survival with skin testing by autologous tumor cells. Cancer Immunol Immunother **32,** 62-66.

Meneguzzi G, Cerni C, Kieny MP and Lathe R (1991). Immunization against human papillomavirus type 16 tumor cells with recombinant vaccinia viruses expressing E6 and E7. Virology **181,** 62-69.

Pallesen G, Hamilton-Dutoit SJ, Rowe M and Young LS (1991). Expression of Epstein-Barr virus latent gene products in tumour cells of Hodgkin's disease. Lancet **337,** 320-322.

Peace DJ, Chen W, Nelson H and Cheever MA (1991). T cell recognition of transforming proteins encoded by mutated ras proto-oncogenes. J Immunol **146,** 2059-2065.

Porgador A, Tzehoval E, Katz A, Vadai E, Revel M, Feldman M and Eisenbach L (1992). Interleukin 6 gene transfection into Lewis lung carcinoma tumor cells suppresses the malignant phenotype and confers immunotherapeutic competence against parental metastatic cells. Cancer Res **53**, 3679-3686.

Restifo NP, Spiess PJ, Karp SE, Mule JJ and Rosenberg SA (1992). A nonimmunogenic sarcoma transduced with the cDNA for interferon-γ elicits CD8+ T cells against the wild-type tumor: Correlation with antigen presentation capability. J Exp Med **175**, 1423-1431.

Rollins BJ and Sunday ME (1991). Suppression of tumor formation in vivo by expression of the JE gene in malignant cells. Mol Cell Biol **11**, 3125-3131.

Tao M and Levy R (1993). A novel vaccine for B cell lymphoma: Idiotype/granulocyte-macrophage colony stimulating factor fusion protein. Nature **362**, 755-758.

Tepper RI, Pattengale PK and Leder P (1989). Murine interleukin-4 displays potent anti-tumor activity in vivo. Cell **57**, 503-512.

Townsend SE and Allison JP (1993). Tumor rejection after direct costimulation of $CD8^+$ T cells by B7-transfected melanoma cells. Science **259**, 368-370.

Van der Bruggen P, Traversari C, Chomez P, Lurquin C, De Plaen E, Van den Eynde B, Knuth A and Boon T (1991). A gene encoding an antigen recognized by cytolytic T lymphocytes on a human melanoma. Science **254**, 1643-1647.

Wallich R, Bulbuc N, Hammerling GJ, Katzav S, Segal S and Feldman M (1985). Abrogation of metastatic properties of tumor cells by de novo expression of H-2K antigens following H-2 gene transfection. Nature **315**, 301-305.

Watanabe Y, Kuribayashi K, Miyatake S, Nishihara K, Nakayama E, Taniyama T and Sakata T (1989). Exogenous expression of mouse interferon-γ cDNA in mouse neuroblastoma C1300 cells results in reduced tumorigenicity by augmented anti-tumor immunity. Proc Natl Acad Sci USA **86**, 9456-9460.

Young LS, Dawson CW, Clark D, Rupani H, Busson P, Tursz T, Johnson A and Rickinson AB (1988). Epstein-Barr virus gene expression in nasopharyngeal carcinoma. J Gen Virol **69**, 1051-1065.

Young LS, Alfieri C, Hennessy K, Evans H, O'Hara C, Anderson KC, Ritz J, Shapiro RS, Rickinson A and Kieff E (1989). Expression of Epstein-Barr virus transformation-associated genes in tissues of patients with EBV lymphoproliferative disease. N Engl J Med **321**, 1080-1085.

8

EXOGENOUS CYTOKINES ENHANCE TUMOR SPECIFIC RESPONSES TO VACCINES

C. S. McCune
University of Montana and Missoula Medical Oncology, Missoula, MO, USA

Introduction

Those cytokines which cause the activation or proliferation of immune cells are of potential clinical interest as agents to amplify the response to vaccines. As these agents become available, they can be tested in settings where they are administered with vaccines. Conceptually, they could be used as adjuvants in immunizing procedures by either local or systemic administration. If given locally, a cytokine is mixed with an immunogen and administered at one site, such as an intradermal injection. We can hypothesize that a high level of a lymphocyte activating cytokine, such as IL-1 or IL-2, given with a vaccine at a localized site, might facilitate the initial interaction of the immunogen with the immune cells. If a cytokine is given systemically concurrent with an intradermal vaccine, we can hypothesize that newly activated lymphocytes, responding to the antigens of the vaccine, are arriving in the lymphatic and vascular circulation. The presence in the circulating fluids of high levels of lymphocyte activating cytokine molecules would potentially expand this new population of antigen activated lymphocytes. Experiments testing the use of systemically administered cytokines as adjuvants to vaccines have been conducted in particular with IL-1 and IL-2, and the resulting data will be summarized in this chapter.

Animal studies with IL-2

IL-2 has been observed to have a strong adjuvant effect on experimental vaccines of bacterial, viral and tumor derivation. Examples of the data from these publications will be given in this section. One of the earliest studies was conducted in pigs by Anderson et al. (1987) using a formalin inactivated bacterin to protect the pigs from challenge with

CYTOKINE-INDUCED TUMOR IMMUNOGENICITY
ISBN 0-12-261520-4

Table 8.1. *IL-2 given with a bacterin vaccine*[a]

	% of lung affected by HPP lesions at necroscopy	Average increase in body weight after HPP challenge
Challenge control	36	8
Bacterin immunized control	32	32
Bacterin plus low dose IL-2 (1 dose)	41	13
Bacterin plus low dose IL-2 (5 daily doses)	12	42
Bacterin plus high dose IL-2 (1 dose)	16	22
Bacterin plus high dose IL-2 (5 daily doses)	<1	44

[a]Modified from Anderson et al. (1987)

Haemophilus pleuropneumoniae (HPP), a bacterium that causes severe pneumonia in pigs. Treated animals received immunization with the bacterin on days 0 and 21. Recombinant human IL-2 was given by i.m. injection for 1 day or 5 days following each vaccine. The results are shown in Table 8.1. Two endpoints were used: first the percentage of the lungs involved with pneumonia as determined at necropsy, and second the ability of the animals to gain weight as an indication of freedom from illness. The results indicate that even one injection of high dose IL-2 could improve the vaccine's effectiveness, while five daily injections could almost completely prevent disease.

IL-2 was given with *Herpes simplex* vaccines by Weinberg and Merigan (1988). They used a guinea pig model of genital herpes infection. Two herpes vaccines were studied: one was a crude extract (HSV) and the other a recombinant glycoprotein D (rgD). Human recombinant IL-2 was given as the adjuvant. The vaccines were administered s.c. every 2 weeks for either two or three doses. The IL-2 was given s.c. as 17 consecutive daily doses beginning on the day the first vaccine was given. The virus challenge was given intravaginally 9 days after the last vaccine. The results are shown in Table 8.2. The endpoint used was the appearance of localized clinical infection or virus shedding from challenged animals. They found a marked reduction in the incidence of infection in all groups given IL-2, including complete prevention of infection when the crude vaccine was given three times along with rIL-2.

Table 8.2. *IL-2 given with a herpes simplex vaccine*[a]

Vaccine	No. of doses	Incidence of infections	
		with IL-2	with placebo
None	-	8/10	9/10
rgD	2	4/10	10/10
rgD	3	1/ 8	5/ 8
HSV	2	2/10	8/10
HSV	3	0/ 8	6/ 8

[a]Modified from Weinberg and Merigan (1988)

The effect of IL-2 on a tumor vaccine was evaluated by Nomi et al. (1986) and Naito et al. (1988). They used a murine fibrosarcoma tumor model for which they had previously developed methods to obtain a biochemically extracted tumor antigen. The treatment regimen had three components which were used in varying combinations: tumor antigen, cyclophosphamide and IL-2. Of particular interest were the variations in IL-2 administration which they tested. They compared bolus versus continuous infusion (c.i.) and they compared i.v., i.p. and intrasplenic (i.s.) routes of administration. The unique i.s. route was achieved by splenic translocation to a s.c. site. Their findings are exemplified by the data from one experiment shown in Table 8.3. The tumor challenge was given s.c. on day 0. Treatment began on day 7 with the tumor antigen injected s.c. and repeated weekly. A single dose of cyclophosphamide was also given on day 7. IL-2 was given by the method and route indicated on days 10 through 19. Using the endpoint of median survival time, IL-2 was found to be effective for every group except one: those which received IL-2 by i.v. bolus. Continuous infusion was consistently superior to bolus IL-2, although not to statistical significance. The administration of IL-2 i.s. by c.i. had the best result in this experiment and in another in which treatment was begun on day 14 rather than on day 7. The success of the i.s. route suggests the importance of getting sufficient amounts of IL-2 to the lymphoid organs of the host where activated lymphocytes are concentrated. Using a murine model of colon cancer, Arroyo et al. (1990) evaluated the contribution of IL-2 in an active immunotherapy regimen. The mice were challenged on day 0 with an i.s. injection of tumor cells to

Table 8.3. *Comparison of different methods and routes of administration of IL-2*[a]

Group	Method	Infusion route	Dose (U/day)	Survival	MST[b] (days)	*p*
A			0	1/10	50.9	-
B	bolus	i.v.	120	1/10	52.1	NS
C	c.i.	i.v.	120	3/10	67.3	<0.005
D	bolus	i.p.	120	2/10	60.4	<0.05
E	c.i.	i.p.	120	4/ 9	71.4	<0.005
F	bolus	i.s.	120	3/10	66.9	<0.01
G	c.i.	i.s.	120	6/ 9	78.9	<0.001

[a]Modified from Nomi et al. (1986) and Naito et al. (1988)
[b]Mean survival time

Table 8.4. *Comparison of the effects of IL-2 and IFN-α when given with VCO type of vaccine*[a]

Group	Mean liver weight (g)	Mean liver nodule count
Control	> 2.0	>100
IFNα	> 2.0	>100
VCO	> 2.0	>100
VCO + IFNα	1.8	90
IL-2 only	1.5	55
IL-2 + IFNα	1.5	55
VCO + IL-2	1.5	20
VCO + IL-2 + IFNα	1.2	<1

[a]Modified from Arroyo et al. (1990)

Table 8.5. *IL-2 given with a hepatitis vaccine (HBVax) for renal dialysis patients*[a]

Patient groups	Unresponsive to immunization	Developed protective anti-HBs antibody titers
HBVax alone	7/ 8	1/ 8
HBVax + IL-2	3/10	7/10

[a]Modified from Meuer et al. (1989)

produce artificial liver metastases. On days 4 and 10 a viral oncolysate (VCO), consisting of vaccinia virus infected and sonicated murine colon tumor cells, was given s.c.

Two cytokines were studied as adjuvants. IL-2 and IFN-α were given i.p. twice daily, on days 11, 12 and 13. The endpoints used were liver weight and tumor nodule counts. The results they obtained are adapted to a tabular form in Table 8.4. When the vaccine (VCO) or IFN-α were given alone, they had no therapeutic effect. IL-2 alone had a moderate effect. IL-2 was more effective than IFN-α when given as an adjuvant with the vaccine. Note that treatment with the vaccine alone still yielded tumor nodule counts >100, but the addition of IL-2 to the vaccine reduced the nodule counts to 20. Interestingly, when both IFN-α and IL-2 were given with VCO, 3 animals had no nodules in the liver, and the 1 remaining animal had only 3 nodules.

IL-2 has also been evaluated when given in preparations for local administration with the immunogen rather than by systemic administration, as in the previous data. Good et al. (1988) emulsified IL-2 with several adjuvants for incorporation in malaria peptide vaccines. They consistently achieved high titer antibody responses to peptides which otherwise were weak immunogens. IL-2 liposomes were tested with an influenza A virus vaccine by Mbawuike et al. (1990). Low dose vaccines given with the IL-2 liposomes became equal to high dose vaccines.

Clinical studies with IL-2

In 1989, Meuer et al. reported an evaluation of a hepatitis vaccine (HBVax) given with or without IL-2. The 18 participating patients had renal failure and were on dialysis. All patients had previously received at least three immunizations with a hepatitis vaccine, and none had

Table 8.6. *IL-2 given with autologous tumor vaccines for renal carcinoma and melanoma*

IL-2 dose (Cetus U/24 hr)	Partial response	Mixed response	Progressive disease
1×10^5 U/m^2	--	1[a]	1[a] 1[b]
3×10^5 U/m^2	--	1[a] 1[b]	2[b]
1×10^6 U/m^2	1[b]	--	3[b]
3×10^6 U/m^2	3[a]	--	7[a] 2[b]
Total	4	3	16

[a]Renal cell carcinoma
[b]Melanoma

produced a detectable antibody titer which is typical of the immune impairment of such patients. The study treatment involved giving HBVax i.m. followed 4 h later by the injection of IL-2 at the same site. The control group received HBVax only. The results are summarized in Table 8.5 and indicate that the addition of a single dose of IL-2 at the vaccine site converted nonresponders to responders most of the time. We conducted a phase I study of IL-2 given with active immunotherapy vaccines for patients with advanced renal carcinoma and melanoma (manuscript in preparation). Four doses of IL-2 were used. The vaccines consisted of irradiated autologous tumor cells plus the adjuvant Detox, given every 3 weeks for three doses. The IL-2 was given as i.v. c.i. for 5 days, beginning on the same day the vaccine was given s.c. As seen in Table 8.6, some partial responses and mixed responses were noted. Mixed responses refer to findings where some metastases regressed more than 50%, but other lesions in the same patient failed to respond.

Both vaccines and IL-2 have independently produced regressions in these two tumor types, and we cannot determine whether the combination

Table 8.7. *Adjuvant effect of IL-1 on the antibody response to BSA immunization*[a]

Primary immunization	Dose of IL-1	Anti-BSA antibody
BSA	--	63
BSA + IL-1	200 U	320
BSA + IL-1	1000 U	491

[a]Modified from Staruch and Wood (1993)

is superior since the study was only designed to determine a feasible dose of IL-2.

Animal studies with IL-1

An early study evaluating the use of IL-1 as a systemically administered adjuvant was performed by Staruch and Wood (1983). They immunized mice with bovine serum albumin (BSA) given s.c. IL-1 was prepared from human and murine cell lines known to be high secretors. The IL-1 was given i.p. at various times after BSA, and 2 h was found to be most effective. Only a single dose of IL-1 was given. The endpoint was the secondary antibody levels to a boost of BSA given 11 days following the primary dose. The results from one experiment are shown in Table 8.7 and indicate a strong adjuvant effect from this single dose of IL-1. In other experiments they compared IL-1 given i.p. with two other adjuvants, muramyldipeptide (MDP) and complete Freund's adjuvant (CFA), which were given s.c. mixed with BSA. Both MDP and CFA, given locally, produced higher antibody levels than IL-1 given systemically. However, our experiments (reported below) indicated that several daily doses of systemic IL-1, rather than a single dose, are required to give a maximum effect.

Nencioni et al. (1987) demonstrated that both the recombinant human IL-1β molecule and a peptide fragment of the molecule can produce a systemic adjuvant effect on the antibody response to immunization with sheep red blood cells.

We evaluated three types of IL-1 molecules as systemic adjuvants given with a vaccine of irradiated tumor cells in a model of active specific

Table 8.8. *IL-1β given as a systemic adjuvant with a murine lung carcinoma tumor vaccine*

Group	Vaccine	IL-1 (ng)	Mean survival (days)	Tumor free/ Total
1	No	--	14.9	0/10
2	Yes	--	17.0	0/10
3	Yes	120	33.8	7/10
4	Yes	360	40.0	10/10

immunotherapy (McCune and Marquis, 1990). Recombinant human IL-1α, IL-1β and a peptide fragment of IL-1β were studied, and all were found to be effective as adjuvants. The essential observations were that the IL-1 should be given for a period of approximately 8 days following immunization, that the effect was systemic rather than localized to the vaccine site, and that weak vaccines could become highly effective. One of the experiments is shown in Table 8.8. Mice were immunized on day 1 with a s.c. injection of irradiated syngeneic lung carcinoma tumor cells. IL-1β was injected s.c. once daily for 8 days beginning on day 1. A s.c. challenge of live tumor cells was given on day 15. Neither the vaccine nor IL-1 (data not shown) when given alone had significant effects, but IL-1 was a powerful adjuvant when given in combination with the vaccine, rendering 100% of the mice tumor free.

Clinical study with IL-1

We are presently conducting an investigation for colon cancer patients where the treatment regimen consists of a vaccine given with IL-1. The vaccine consists of 4 human colon cancer cell lines which are irradiated, prepared with the adjuvant Detox, and injected intradermally every 3 weeks. Human recombinant IL-1α is administered daily by s.c. injection, beginning on the day of vaccine injection and continued for 8 days. Four sequential patient groups are being studied. The first group receives only the vaccine, while the second, third and fourth groups receive increasingly greater doses of IL-1 with the vaccine. A second vaccine for hepatitis B is given on the same days as the colon vaccine as a means of evaluating the effect of IL-1. Antibody responses to the hepatitis vaccine are being followed to determine if the addition of varying doses of IL-1 achieves higher antibody responses than the vaccine alone. Among the first 10 evaluable patients, two events of clinical interest have been observed. Two patients with serially documented rising CEA (carcinoembronic antigen) levels have had serially documented distinct drops in this tumor marker. These patients had minimal metastatic disease, but the reality of an antitumor effect seems highly likely. The higher doses of IL-1 have not yet been evaluated.

References

Anderson G, Urban O, Fedorka-Cray P, Newell A, Nunberg J and Doyle M (1987). Interleukin-2 and protective immunity in haemophilus pleuropneumoniae: Preliminary studies. In: "Vaccines 87" (eds RM Chanock, RA Lerner, F Brown and H Ginsberg), pp 22-25. Cold Spring Harbor Laboratory, New York.

Arroyo PJ, Bash JA and Wallack MK (1990). Active specific immunotherapy with vaccinia colon oncolysate enhances the immunomodulatory and antitumor effects of interleukin-2 and interferon alpha in a murine hepatic metastasis model. Cancer Immunol Immunother **31**, 305-311.

Good MF, Pombo D, Lunde MN, Maloy WL, Halenbeck R, Koths K, Miller LH and Berzofsky JA (1988). Recombinant human IL-2 overcomes genetic nonresponsiveness to malaria sporozoite peptides. J Immunol **141**, 972-977.

Mbawuike IN, Wyde PR and Anderson PM (1990). Enhancement of the protective efficacy of inactivated influenza A virus vaccine in aged mice by IL-2 liposomes. Vaccine **8**, 347-352.

McCune CS and Marquis DM (1990). Interleukin 1 as an adjuvant for active specific immunotherapy in a murine tumor model. Cancer Res **50**, 1212-1215.

Meuer SC, Dumann H, Buschenfelde KMZ and Kohler H (1989). Low-dose interleukin-2 induces systemic immune responses against HBsAg in immunodeficient non-responders to hepatitis B vaccination. Lancet i, 15-18.

Naito K, Pellis NR and Kahan BD (1988). Effect of continuous administration of interleukin 2 on active specific chemoimmunotherapy with extracted tumor-specific transplantation antigen and cyclophosphamide. Cancer Res **48**, 101-108.

Nencioni L, Villa L, Tagliabue A, Antoni G, Presentini R, Perin F, Silvestri S and Boraschi D (1987). In vivo immunostimulating activity of the 163-171 peptide of human IL-1 beta. J Immunol **139**, 800-804.

Nomi S, Naito K, Kahan BD and Pellis NR (1986). Intrasplenic administration of interleukin-2 to potentiate specific chemo-immunotherapy in tumor-bearing mice. Cancer Res **46**, 5606-5610.

Staruch MJ and Wood DD (1983). The adjuvanticity of interleukin 1 in vivo. J Immunol **130**, 2191-2194.

Weinberg A and Merigan TC (1988). Recombinant interleukin 2 as an adjuvant for vaccine-induced protection. J Immunol **140**, 294-299.

9

IMMUNE RECOGNITION AND REJECTION OF IL-1α GENE TRANSDUCED TUMOR CELLS

R. N. Apte, A. Douvdevani, M. Zoller, R. M. White, T. Dvorkin, N. Shimoni, E. Fima, M. Hacham, E. Voronov and S. Segal
Department of Microbiology and Immunology, Faculty of Health Sciences, Ben-Gurion University of the Negev, Beer-Sheva, Israel, and Institute for Radiology and Pathophysiology, German Cancer Research Center, Heidelberg, Federal Republic of Germany

Introduction

The local activity of cytokines expressed at the site of tumor development is capable of mounting an efficient antitumor immune response (Rosenberg, 1988; Talmadge 1989; Bubenik, 1990; Foon, 1992). Immunotherapeutical approaches based on the local administration of recombinant cytokines (intratumorally or perilymphatically) or the transfer of cytokine genes into transformed cells were thus established. Murine tumor cells expressing genes for IL-1α, IL-2, IL-4, IL-6, IL-7, IFN-γ, TNF-α and G-CSF have been found to manifest decreased tumorigenicity or slower growth in vivo (Gilboa, 1990; Russel, 1990; Blankenstein et al., 1991; Colombo et al., 1992; Gutierrez et al., 1992).

Some of the cytokines were shown to mediate their effects through T-cell dependent mechanisms and to induce long-lasting memory, while others potentiate non T-cell dependent, nonadaptive antitumor effector mechanisms. This diversity in the immunosurveillance mechanisms elicited by the transfected cytokine genes possibly stems from the differential effects of cytokines on various immune responses and also from the susceptibility of the target tumor cells to these effector mechanisms. Because of the efficiency of using cytokine gene transfer immunotherapeutical approaches in experimental tumor systems, attempts are underway to adapt them for future use in humans (Kasid et al., 1990).

IL-1 is one of the most pleiotropic cytokines capable of potentiating both specific (humoral and cellular) as well as nonadaptive immune responses (Oppenheim et al., 1986; Dinarello, 1989; Apte and Huliehel,

CYTOKINE-INDUCED TUMOR IMMUNOGENICITY
ISBN 0–12–261520–4

1992). The antitumor activities of IL-1 are multiple, and they include effects on the proliferation, differentiation and function of the diverse immunosurveillance cells and also direct cytostatic/cytotoxic effects on some tumor cells. Furthermore, IL-1 is one of the strongest inducers of cytokine production by T cells, macrophages and mesenchymal cells. Indeed, successful antitumor effects of IL-1 have been described in animal tumor models (Castelli et al., 1988; Nakata et al., 1988; North et al., 1988; Belardelli et al., 1989). However, in most of the studies the i.v. route of IL-1 administration was used, and multiple injections of high doses of the cytokine were needed in order to achieve immunotherapeutical effects; these were often accompanied by severe deleterious effects (such as fever, anorexia, neutrophilia, etc) on the host.

The discovery of the efficiency of local activity of cytokines at the site of tumor development triggered the initiation of immunotherapeutical protocols using IL-1. In studies performed in our laboratory we focused on assessing the effects of IL-1 expression by fibrosarcoma cells on tumor development. Here, we report on the reduced tumorigenicity pattern of malignant fibroblasts which express IL-1 constitutively (either as a result of the transformation event or following transfer of the IL-1α gene) or transiently (following activation of the tumor cells in vitro by cytokines/immunomodulators).

Results and discussion

IL-1α expression in transformed fibrosarcoma cell lines

In our previous studies on cytokine production by primary fibroblasts we have shown that the induction of IL-1 requires a sequential two step activation protocol, which consists of a preactivation of fibroblasts with each of a variety of individual recombinant cytokines (IL-2, GM-CSF, TNF, IL-1β and IFNs α, β and γ) (for 72 h), followed by a challenge with lipopolysaccharide (LPS) (for 24 h) (Huleihel et al., 1990; Apte and Huleihel, 1992; Huleihel et al., 1993). IL-1 like activity on fibroblasts was detected in cell lysates or associated with the producing cell membrane, but not in culture fluids, and it was mainly neutralized with anti-IL-1α antibodies. Such immune activated fibroblasts were shown to actively transcribe mRNA of both IL-1 genes (α and β). For the expression of IL-1 specific mRNA, a single stimulus, provided by either a cytokine or LPS, was sufficient (Huleihel et al., 1990; Apte and Huleihel, 1992; Huleihel et al., 1993).

We next assessed the pattern of IL-1α expression in transformed fibroblastoid cell lines and its influence on fibroblastoid cell tumorigenic potential. For this purpose three distinct experimental tumor systems were employed.

1. First, we used oncogene transformed fibroblastoid cell lines which constitutively express IL-1 activity, possibly as an aberration resulting from the in vitro transformation process (Apte et al., 1992; Douvdevani et al., 1992; Zoller et al., 1992a). These cell lines were derived from NIH-3T3 cells which were transfected in culture with defined oncogene sequences. It appears that constitutive IL-1α membrane associated activity correlates with the transformation induced by protein kinase encoding oncogenes (TPR-*met*, murine *c-met*, *c-abl*, *v-src*, *v-raf* and *v-mos*). While intracellular activity was also detected in cell lines transformed with the above mentioned oncogenes or *v-myc*, the *v-myc* nuclear oncogene induced only intracellular, but not membrane bound, activity. In *v-H-ras*-transformed cells no constitutive IL-1α mRNA or IL-1α activity were detected. No DNA rearrangement, amplification of the IL-1α gene or constitutive expression of IL-1β mRNA was detected in any of the cell lines. These aberrations in IL-1α production by transformed fibroblasts may result from direct effects of the oncogene product on the pathway(s) of IL-1α generation, or it may be indirect, as a consequence of alterations in some other signal transduction pathway(s). Further experiments are being conducted in our laboratory to elucidate the general role of the specific oncogenes in the regulation of IL-1α production.

2. IL-1α nonexpressing fibroblastoid tumor cell lines were induced to express constitutively IL-1α after gene transfer (Apte et al., 1992; Douvdevani et al., 1992). For this purpose we introduced into an IL-1 negative cell line an expression vector encoding for the IL-1α gene, under the control of the LTR of RSV, cotransfected with an expression vector (pMAMneo) bearing the gene for resistance to G418 (Gibco, Grand Island, NY). In a successful transfection experiment we obtained 40 colonies resistant to G418, and after initial screening 5 colonies constitutively expressing IL-1α were recovered and further characterized. The IL-1α producing clones generated moderate constitutive levels of the cytokine, in the range of 0.1-1 U/ml (about 5-20 pg/ml). We are currently using retroviral constructs bearing the IL-1α gene in an attempt to more precisely correlate the amounts of IL-1 released and the patterns of tumorigenicity.

3. IL-1α nonexpressing transformed fibroblastoid cell lines were induced to express IL-1 transiently, upon activation with cytokines and LPS, as required for IL-1 induction in primary fibroblasts (Huleihel et al., 1990; Apte and Huleihel, 1992; Zoller et al., 1992a; Zoller et al., 1992b; Huleihel et al., 1993).

Such immune activated transformed fibroblasts express IL-1α activity in culture for about 5-7 days after washing out the stimulators and adding fresh medium to the monolayers. In all these experimental systems IL-1α (and not IL-1β) detection was based on the levels of mRNA as well as on the biological activity of the cytokine, documented in both cell lysates and cell membrane.

The compartmentalization patterns of IL-1 in normal and transformed fibroblasts suggest that cells of this lineage are possibly devoid of the enzymatic transport mechanisms needed for IL-1β secretion.

It should also be noted that the process of IL-1 secretion or its insertion into the membrane are still obscure, as the molecule lacks a signal peptide or a hydrophobic region essential for active secretion or insertion into the membrane (Oppenheim et al., 1986; Dinarello, 1989; Apte and Huleihel, 1992).

IL-1α expression in the oncogene transformed cell lines and in the IL-1α gene transfectants are shown in Fig. 9.1. In Fig. 9.1A the constitutive activity of IL-1 in cell lysates from the TPR-*met* cell line is demonstrated, compared to the activity induced by the cytokine following stimulation of the cells with cytokines/LPS (CK/LPS), detected in lysates from murine peritoneal macrophages (Mac), murine embryo fibroblasts (MEF), parental NIH/3T3 cells and a line transfected with the *H-ras* oncogene.

Fig. 9.1B demonstrates the constitutive IL-1α activity in cell lysates obtained from the parental cell line used for transfection (met-1), the IL-1α positive transfectants (met-IL1-1 and met-IL1-5) and two lines which were transfected only with the selection marker (met-1p2 and met-1p6).

The role of endogenous IL-1α expression by transformed fibroblasts on tumor development

We next focused on the role of endogenous IL-1α production by malignant cells on tumor development by assessing the in vivo growth patterns of constitutive IL-1α producing vs nonproducing fibroblastoid cell lines (Apte et al., 1992; Douvdevani et al., 1992; Zoller et al., 1992a; Zoller et al., 1992b).

No differences in the in vitro cell growth rate were observed between IL-1α producing or nonproducing cell lines, and no autocrine growth effects of IL-1 could be demonstrated. The IL-1α producing tumors, from the various fibroblastoid series of cell lines, either did not grow at all or started to develop but within 14-20 days completely regressed, in contrast to IL-1 nonproducing cell lines which grew progressively until the tumor bearing animals died. The same correlation between IL-1 expression and reduced tumorigenicity was also valid for fibroblastoid cell lines which acquired the ability to constitutively express IL-1α following gene transfer, or for fibroblastoid cell lines which express this cytokine transiently, following immune activation with cytokine/LPS.

Fig. 9.2 demonstrates the in vivo growth characteristics of the oncogene transformed cell lines and the IL-1α transfectants. In Fig. 9.2A the growth characteristics of the following cell lines is demonstrated: TPR-*met* and *v-mos* which constitutively produce IL-1α and the *H-ras* line (IL-1 negative), untreated or activated by cytokine/LPS to express IL-1α (*ras*/A).

The kinetics of tumor development of the IL-1α transfectants is demonstrated in Fig. 9.2B; the parental cell line (met-1), the IL-1α positive clones (met-IL1-1 and met-IL1-5) and the control clone met-1p6.

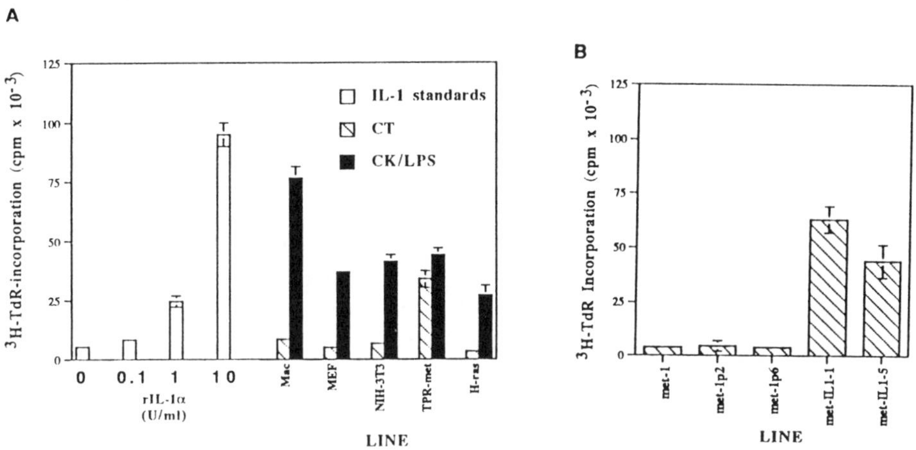

Fig. 9.1. *IL-1 activity in cell lysates from oncogene-transformed cell lines and IL-1α transfectants.* Panel A demonstrates the constitutive (CT), or induced (CK/LPS), IL-1 activity in control, and oncogene transformed fibroblastoid cells. Panel B demonstrates the constitutive IL-1 activity in cell lysates from the various IL-1α transfectants. Cell lysates (final concentration 10% v/v) were tested for IL-1 activity on the D10.G4.1 T-cell line in a proliferation assay, as previously described in detail (Huleihel et al., 1990; Douvdevani et al., 1991).

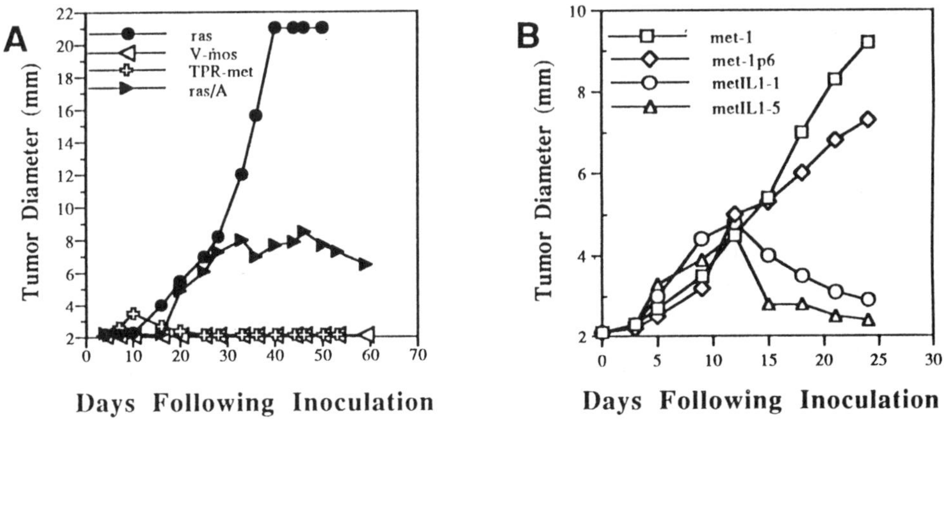

Fig. 9.2. *In vivo growth patterns of constitutive IL-1α producing and nonproducing fibroblastoid cell lines.* 10^5 cells of each cell line, untreated or immune activated, were injected intrafootpad into syngeneic NFS/N mice and tumor growth was assessed at different intervals thereafter.

The nature of immune responses leading to reduced tumorigenicity of IL-1α producing fibrosarcomas

1. *Role of specific antitumor immune response.* Initial observations suggesting that specific immunity may play a dominant role in the eradication of IL-1α producing fibrosarcomas came from experiments in which the various fibrosarcomas were injected into sublethally irradiated mice, where progressive tumor growth occurred independently of the patterns of IL-1α production (Zoller et al., 1992a). The development of an immune response against the IL-1α producing tumor lines was further documented by a dense lymphocytic infiltrate observed around or in regressing tumors and replacing their mass.

By contrast, no infiltrate was detected in histological sections from progressively growing IL-1 nonproducing fibrosarcomas. Experiments are being performed to further define the nature of T cells which characterize the infiltrate; it is, however, evident, as will be discussed below, that both Th cells and CTL contribute to the reduced tumorigenicity of IL-1α producing fibrosarcomas. Both cell lines, nonexpressing vs constitutive/transient IL-1α producers, bear approximately equal amounts of MHC class I and II surface antigens, i.e. high to moderate levels of MHC class I antigens and undetectable levels of MHC class II encoded antigens (Zoller et al., 1992b). This indicates that the fibrosarcoma cell lines have the potential to be recognized by CTL and to stimulate their generation, provided that the auxillary signals needed for their expansion/maturation are provided. Indeed, specific CTL were detected in the spleen of mice with regressing IL-1α producing tumors, in contrast to spleen of IL-1 nonproducing tumors, where such effector cells could not be demonstrated. CTL could also not be detected in the spleen of IL-1α producing fibrosarcomas which did not form tumors in mice.

Further studies were performed in order to analyze the mechanisms of CTL development in lymphoid organs of IL-1α producing fibrosarcomas, in comparison to nonproducing lines. This was performed by estimating the rate and activity of CTL, CTL precursors and proliferating Th cells, using mice immunized with fibrosarcoma cell lines which constitutively/transiently express IL-1α (Zoller et al., 1992a; Zoller et al., 1992b). It was shown, using limiting dilution analyses, that the frequencies of CTL precursors in the spleen are in the same range for IL-1α expressing (constitutive/transient) and nonproducing fibrosarcomas. Thus, differences in the potential and frequencies of CTL for IL-1α expressing vs nonexpressing fibroblasts were only detected after in vivo stimulation, where activation of CTL depended on the presence of stimulatory cytokines of Th cell origin. CTL recovered from the spleen of mice immunized with fibrosarcoma cell lines, immune activated to express IL-1α, could kill equally well the parental IL-1 nonexpressing line. This indicates that IL-1α expression does not alter MHC nor does it create new antigenic epitopes on the cell surface.

The in vivo expansion/maturation of CTL was further shown to be the consequence of activation of Th cells by transformed fibroblast associated IL-1α, that delivers the costimulatory signal. Analyses of the frequencies and proliferation rates of tumor specific Th cells confirmed this assumption. Unlike CTL, differences in the proliferative response of spleen cells depending on IL-1α expression by the stimulatory fibrosarcoma cells were observed also in primary cultures. Thus, both during the in vitro primary cultures of spleen cell, and after in vivo stimulation, the constitutive IL-1α producing cell lines stimulated higher proliferative responses as compared to negative lines. On the other hand, the proliferative rate of the IL-1 negative fibrosarcomas was increased after immune activation of the fibrosarcoma stimulator cells to express the cytokine. A similar picture emerged when the frequencies of splenic proliferating cells were determined: the frequency of proliferating T cells was higher when the stimulator fibrosarcoma cells expressed IL-1α constitutively/transiently in comparison to negative fibroblastoid lines.

The regression of IL-1α producing fibrosarcomas induced a specific immune memory, which is the ultimate goal of immunotherapy. Thus, mice in which IL-1α producing tumors, expressing the cytokine following immune activation with cytokine/LPS, regressed developed a specific memory and were resistant to a challenge with the parental IL-1 nonproducing tumor cell line. Specific immunity was detected at least three months after tumor rejection, and the challenge with the aggressive tumor cell line could be extended to any anatomical site, irrespective of that of the primary tumor. In Fig. 9.3, CTL activity in the spleen of mice challenged with the various fibroblastoid cell lines, differing in their patterns of IL-1α expression, is shown. The CTL activity of different oncogene transformed cell lines, as well as that of the IL-1α transfectants, is demonstrated: the nontumorigenic NIH-3T3 parental cell line, the IL-1α positive cell line which is a regressor (TPR-met), the IL-1α negative met-1 line which was used for transfection and which develops progressive tumors in mice, the IL-1α positive regressor transfectants (met-IL1-1 and met-IL1-5) and the clones transfected only with the selection marker (met-1p2 and met-1p6) which are highly tumorigenic.

It was next assessed whether the fibrosarcoma cell lines can serve as APC, and possibly present their own antigens to Th cells. This assumption was based on our previous studies showing that primary fibroblasts are able to present antigen to proliferating T cells, provided that they are treated with IFN-γ to express MHC class II antigens. It was demonstrated, in an in vitro antibody production assay, that fibrosarcoma cells can present soluble antigens to lymphocytes provided that they express IL-1α as well as MHC class II antigens. Thus, constitutive IL-1α expressing fibrosarcomas are able to present antigen if they are treated with IFN-γ, while IL-1 nonexpressing fibrosarcomas have to be immuneactivated (cytokine/LPS) and treated with IFN-γ to express IL-1.

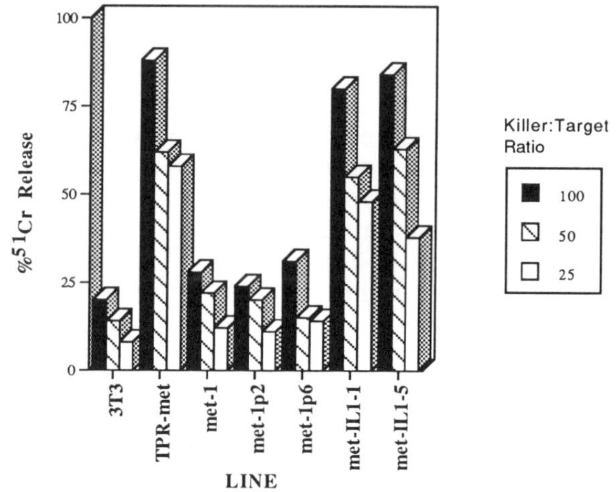

Fig. 9.3. *CTL activity in the spleen of IL-1α producing fibrosarcoma cell lines.* Spleen cells, obtained from mice 3 weeks after tumor cell inoculation, were cultured for 5 days in a mixed lymphocyte-tumor cell reaction (MLTR) and afterwards assessed for lytic activity in a standard 4 h chromium release assay. Mitomycin-C inactivated cells from the same line which were injected into the mice served as targets.

We next assessed whether IL-1α expressing fibrosarcomas are able to present their own antigens to proliferating T cells. For this purpose highly purified T cells, devoid of MHC class II positive cells, were mixed with fibrosarcoma cells, and the proliferation rate was assessed. Indeed, IL-1α expressing fibrosarcomas, induced to express MHC class II antigens by IFN-γ, stimulated efficient T-cell proliferation. However, fibrosarcoma cell lines expressing only IL-1α, but not MHC class II antigens, also stimulated T-cell proliferation. The latter observation suggests that IL-1α, and possibly other cytokines concomitantly induced by IL-1α (i.e. IL-6 and CSF), may locally stimulate the maturation or activation of APC. Thus, IL-1α producing fibrosarcomas have the potential of presenting their own antigens or alternatively of activating other tissue resident APC cells, and by this characteristics antitumor immune responses are potentiated.

2. *Effects of nonadaptive immunity on the reduced tumorigenicity of IL-1α producing fibrosarcoma cell lines.* From the experiments described above it is evident that specific immunity plays a major role in IL-1α induced tumor rejection. It was, however, of importance to assess whether nonadaptive mechanisms also contribute to such reduced tumorigenicity (Zoller et al., 1992a). We suspected the NK cells and activated macrophages might also be involved in IL-1α mediated reduced tumorigenicity, since some IL-1α producing fibrosarcoma cell lines failed completely to grow in vivo, and in the spleen of such mice CTL could not be detected (Apte et al., 1992; Douvdevani et al., 1992).

No enhanced NK or activated macrophage activity could be observed in the spleen of mice bearing IL-1α positive tumors at regression. However, locally, at the site of tumor cell injection, the IL-1α producing fibrosarcoma cell lines stimulated these immunosurveillance effector cells. Thus, when the various mitomycin-C treated fibrosarcoma cell lines which differ in their ability to generate IL-1α were injected i.p., enhanced NK and cytocidal macrophage activity could be demonstrated in cells recovered from the peritoneum.

3. *Cytokines involved in the reduced tumorigenicity of IL-1α producing fibrosarcomas.* IL-1 is a strong inducer of proinflammatory cytokine production; it was, therefore, of interest to characterize the cytokine network which mediates IL-1α induced tumor regression. For this purpose the above mentioned oncogene transformed cell lines were tested for their ability to generate other proinflammatory cytokines/mediators, such as IL-6, CSF, TNF and prostaglandin E_2 (PGE_2) (Apte et al., 1992). All transformed fibroblastoid cell lines, irrespective of their capacity to express IL-1α, produced IL-6, CSF and PGE_2, whereas we could not detect the production of TNF. This correlation was also found when using cell lines which differ in their ability to generate IL-1α positive cells, and that during propagation in culture shifted from a regressor to a progressor tumor phenotype. The IL-1α generating clones failed to develop tumors, whereas the negative clones generated progressive tumors. The majority of IL-1α expressing fibrosarcomas are rejected possibly because of its broad immunomodulatory activities, which affects most of the immunosurveillance cellular mechanisms. Other proinflammatory cytokines/mediators, induced by endogenous IL-1α, may amplify and sustain antitumor immune responses and contribute to the reduced malignancy patterns of fibrosarcomas.

Immunotherapeutic effects induced by IL-1α expressing fibrosarcoma cell lines

Attempts were made to establish immunotherapeutic approaches to eradicate IL-1 nonproducing fibrosarcomas with IL-1α producing fibrosarcoma cells (Apte et al., 1992). Initial results indicated that it is possible to reduce the tumorigenicity of aggressive IL-1 nonexpressing cells by injecting fibrosarcoma cells which express IL-1α either constitutively or transiently following activation with cytokine/LPS. The latter was of special interest to us, as it is assumed that most tumors isolated from human subjects or experimental animals are IL-1 negative. Endogenous expression of IL-1 is possibly effective against the tumor because it elicits an immune response against it. However, it is easy to

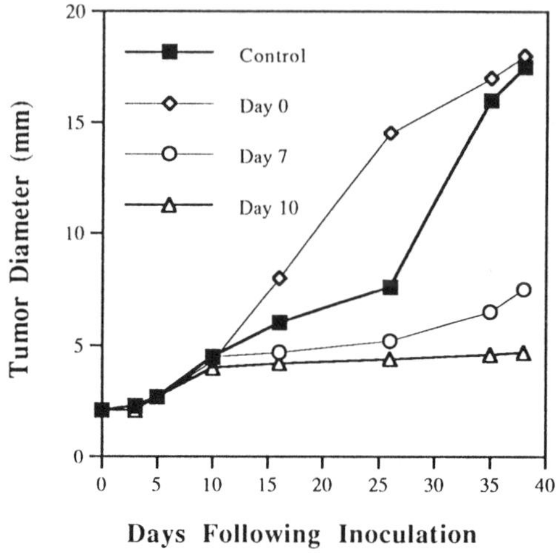

Fig. 9.4. *Effects of transient IL-1α expressing cells on the tumorigenicity of growing fibrosarcomas.* NFS/N mice were injected with 10^5 *ras*-transformed cells (IL-1 negative), and at different intervals 2 x 10^5 cells from the same line (mitomycin-C inactivated), stimulated with cytokine/LPS to express IL-1α, were injected into the same footpad and tumor development was assessed. The methods to activate fibrosarcoma cells and to monitor tumor development have been previously described in detail (Douvdevani et al., 1991; Apte et al., 1992; Douvdevani et al., 1992).

induce the production of IL-1α in transformed fibroblastoid cells, and this can be used for the development of immunotherapy protocols.

It is noteworthy that induced IL-1α activity persists in fibrosarcoma cells for 5-7 days following stimulation. We tried this type of manipulation in our tumor experimental systems. Thus, we treated IL-1 nonproducing developing fibrosarcomas with cells from the same line activated in culture to generate IL-1α. It was found that the timing of the injection of the IL-1α producing cells is crucial for achieving positive immunotherapeutic effects on existing tumors. When the IL-1α expressing cells were injected 5-10 days after the aggressive fibrosarcoma cells, tumor regression was observed. When the IL-1α expressing fibrosarcoma cells were inoculated concomitantly with the aggressive tumor cells or at intervals later than day 14, the treatment was ineffective and no antitumor effects could be observed.

Fig. 9.4 shows the effects of IL-1α expression by *ras*-transformed cells, following immune activation, on the development of an aggressive tumor from the same line. When the activated cells were injected concomitantly with the IL-1 negative wild type cells no inhibition of tumor development occurred. In this specific experiment tumor growth potentiation was in fact recorded. However, when the immune activated cells were injected on days 7 or 10 after inoculation of the malignant *ras*-transformed cells, a significant inhibition in tumor development was observed.

The crucial timing for the application of the IL-1α producing cells as an immunotherapeutic device corresponds to that of the spontaneous regressions of constitutive IL-1α producing tumors, occurring concomitantly with the development of antitumor Th cell function. Dependence on time of immmunotherapy protocols for solid tumors was also reported for treatments with recombinant cytokines or cytokine engineered tumor cells (Maas et al., 1989; Bosco et al., 1990; Golumbek et al., 1991). Thus, when planning cytokine therapeutic approaches for existing tumors one has to carefully consider their timing.

Concluding remarks

Using different tumor experimental systems we have shown that fibrosarcomas which express constitutive or transient IL-1α activity exhibit reduced tumorigenic patterns, manifested either by the inability to grow in vivo or by the regression of existing tumors (Douvdevani et al., 1991; Apte et al., 1992; Douvdevani et al., 1992; Zoller et al., 1992a; Zoller et al., 1992b).

In both normal and transformed cells of the fibroblastoid lineage, intracellular or membrane associated IL-1α is expressed, whereas the secreted form of the cytokine (IL-1β) could not be detected (Douvdevani et al., 1991; Apte and Huleihel, 1992; Apte et al., 1992; Douvdevani et al., 1992; Zoller et al., 1992a; Zoller et al., 1992b; Huleihel et al., 1993). The unique expression of IL-1α on the cell surface could facilitate cell to cell interactions necessary for the induction of local immune responses. In IL-1α producing fibrosarcomas the cytokine is expressed on the surface of the malignant cells, possibly in proximity to MHC antigens, and these joint entities may efficiently stimulate specific Th cells which subsequently potentiate the development of the various antitumor effector mechanisms (Apte et al., 1992; Douvdevani et al., 1992; Zoller et al., 1992a; Zoller et al., 1992b).

In addition, membrane associated IL-1α can be slowly released from the tumor cells and mediate immunoregulatory activities. Another source of the cytokine is the IL-1α precursor, located in the cytosol, which is possibly released from necrotizing fibrosarcoma cells. Antitumor immune responses may be further potentiated by improved antigen presentation mediated by accessory cells. It is brought about also by the transformed cells themselves, induced to express MHC class II antigens by IFN-γ as part of the local immune/inflammatory response against the tumor (Zoller et al., 1992b). IL-1 is a good candidate for immunotherapy since it is a pleiotropic cytokine which potentiates most of the specific and nonspecific antitumor effector mechanisms, and it is a strong inducer of a proinflammatory cytokine cascade (Oppenheim et al., 1986; Dinarello, 1989; Apte and Huleihel, 1992).

In accordance with this, we have demonstrated that IL-1α expressed by fibrosarcoma cells potentiates specific cellular immune response to weakly immunogenic epitopes on the surface of the malignant cells (Apte et al., 1992; Douvdevani et al., 1992; Zoller et al., 1992b). IL-1α in such a form supplies auxiliary signals to tumor specific Th cells which subsequently potentiate the development of CTL, the latter being important effector cells for tumor eradication (Apte et al., 1992; Douvdevani et al., 1992; Zoller et al., 1992b). The involvement of local nonadaptive immunosurveillance mechanisms was also shown to contribute, to some extent, to the reduced tumorigenicity patterns of IL-1α expressing fibrosarcomas (Zoller et al., 1992a). Tumor rejection of IL-1α producing fibrosarcomas is accompanied by the establishment of an immune memory, a fact which is of utmost importance for establishing efficient immunotherapeutic approaches.

We assume that most naturally occurring fibrosarcomas are possibly IL-1α negative and that during tumor progression IL-1α negative variants are selected, as this cytokine potentiates the immunogenicity of the malignant cells.

However, cells from IL-1α nonproducing lines, induced in vitro to transiently express IL-1α upon immune activation by cytokine/LPS, shifted from a progressor to a regressor tumor phenotype. IL-1α expression by immune activated fibrosarcoma cells persists for a few days following treatment with cytokine/LPS, thus enabling the cells to potentiate antitumor immune responses following their injection into mice.

Positive immunotherapeutic effects were achieved when existing tumors of IL-1α nonproducing fibrosarcoma cells were treated with cells of the same line immune activated to transiently express the cytokine. Thus, the use of tumor cells induced to express IL-1α may be a realistic approach to the management of fibrosarcomas, at least of the superficial ones.

This is of particular importance in view of the present limitations on using cytokine gene transfer approaches in humans and of the feasibility of inducing the transient production of IL-1α by fibrosarcoma cells. The use of tumor cells induced to express IL-1α may be particularly significant in a clinical setting following reduction of the tumor burden.

In conclusion, the present study emphasizes the effectiveness of local IL-1α expression by fibrosarcoma cells for tumor rejection. Cell surface expression of IL-1α is possibly advantageous for immunotherapy, in comparison to secreted IL-1β, because it allows the cellular interaction required for efficient antitumor immune responses. The local expression of IL-1α at low levels may also circumvent the toxic side effects which occur following the systemic application of this cytokine. Attempts are focused on elucidating the mechanisms through which fibrosarcoma associated IL-1α exerts its antitumor effects and also on the development of novel immunotherapy modalities based on constitutive/transient induction of IL-1α on the surface of tumor cells.

Acknowledgments

R. N. Apte is supported by The Charles H. Revson Endowment Fund for Basic Research in Life Sciences of the Israel Academy of Sciences and Humanities, and the National Council for Research and Development, Israel in a joint project with The Deutsches Krebsforschungszentrum Heidelberg, Germany, The Israel Cancer Association, The Israel Cancer Research Fund (ICRF) and The Chief Scientist's Office, The Israel Ministry of Health.

References

Apte RN and Huleihel M (1992). Immunoregulatory activities of fibroblasts. Implications for rheumatoid arthritis. Isr J Med Sci **28**, 133-136.

Apte RN, Douvdevani A, Huleihel M, Fima E, Hacham M, Shimoni N, Dvorkin T and Segal S (1992). IL-1 and pro-inflammatory cytokines produced by primary and transformed fibroblasts abrogate the tumorigenic potential of fibrosarcomas. Folia Biol **38**, 240-257.

Belardelli F, Proietti E, Ciolli V, Sestili P, Carpinelli G, De Vito M, Ferretti A, Woodrow D, Boraschi D and Podo F (1989). Interleukin-1β induces tumor necrosis and early morphologic and metabolic changes in transplantable mouse tumors. Similarities with the anti-tumor effects of tumor necrosis factor α and β. Int J Cancer **44**, 116-123.

Blankenstein T, Rowley DA and Schreiber H (1991). Cytokines and cancer: Experimental systems. Current Opinion Immunol **3**, 694-698.

Bosco M, Giovarelli M, Forni M, Modesti A, Scarpa S, Masuelli L and Forni G (1990). Low doses of IL-4 injected perilymphatically in tumor-bearing mice inhibit the growth of poorly and apparently non-immunogenic tumors and induce a tumor-specific immune memory. J Immunol **145**, 3136-3143.

Bubenik J (1990). Local and regional immunotherapy of cancer with interleukin-2. J Cancer Res Clin Oncol **116**, 1-7.

Castelli MP, Black PL, Schneider M, Pennington R, Furminori A and Talmadge JE (1988). Protective, restorative and therapeutic properties of recombinant human IL-1 in rodent models. J Immunol **140**, 3830-3837.

Colombo MP, Mattei S and Parmiani G (1992). Cytokine gene transfer in tumor cells as an approach to antitumor therapy. Int J Clin Lab Res **21**, 278-282.

Dinarello CA (1989). Interleukin-1 and its biologically related cytokines. Adv Immunol **44**, 153-205.

Douvdevani A, Huleihel M, Segal S and Apte RN (1991). Aberrations in interleukin-1α expression in oncogene-transformed fibrosarcoma lines: Constitutive interleukin-1α transcription and manifestation of biological activity. Eur Cytokine Net **2**, 257-264.

Douvdevani A, Huleihel M, Zoller M, Segal S and Apte RN (1992). Reduced tumorigenicity of fibrosarcomas which constitutively generate IL-1α either spontaneously or following IL-1α gene transfer. Int J Cancer **51**, 822-830.

Foon KA (1992). Biological response modifiers: The new immunotherapy. Cancer Res **49**, 1621-1639.

Gilboa E (1990). Retroviral gene transfer: Applications to human therapy. Prog Clin Biol Res **352**, 301-311.

Golumbek PT, Lazenby AJ, Levitzky HI, Jaffe LM, Karasuyama H, Baker M and Pardoll DM (1991). Treatment of established renal cancer by tumor cells engineered to secrete interleukin-4. Science **254**, 713-716.

Gutierrez AA, Lemoine NR and Sikora K (1992). Gene therapy for cancer. Lancet **339**, 715-721.

Huleihel M, Douvdevani A, Segal S and Apte RN (1990). Regulation of interleukin-1 generation in immune-activated fibroblasts. Eur J Immunol **20**, 731-738.

Huleihel M, Douvdevani A, Segal S and Apte RN (1993). Different regulatory levels involved in the generation of hemopoietic cytokines (CSFs and IL-6) in fibroblasts stimulated by inflammatory products. Cytokine **5**, 47-56.

Kasid A, Morecki S, Abersold P, Cornetta K, Culver K, Freeman S, Director E, Lotze MT, Blaese RM, Anderson WF and Rosenberg SA (1990). Human gene transfer: Characterization of human tumor-infiltrating lymphocytes as vehicles for retroviral-mediated gene transfer in man. Proc Natl Acad Sci USA **87**, 473-477.

Maas RA, Dullens HFJ, De Jong WH and Den Otter W (1989). Immunotherapy of mice with a large burden of disseminated lymphoma with low-dose interleukin-2. Cancer Res **49**, 7037-7040.

Nakata K, Kishimoto S, Yoshida J, Oku T and Nakamura S (1988). Augmented antitumor effect of recombinant human interleukin-1α by indomethacin. Cancer Res **48**, 584-588.

North RJ, Neubauer RH, Huang JJ, Newton RC and Loveless SE (1988). Interleukin-1-induced T cell-mediated regression of immunogenic murine tumors. Requirement for an adequate level of already acquired concomitant immunity. J Exp Med **168**, 2031-2043.

Oppenheim JJ, Kovacs EJ, Matsushima K and Durum SK (1986). There is more than one interleukin-1. Immunol Today **7**, 45-56.

Rosenberg SA (1988). The development of new immunotherapies for the treatment of cancer using interleukin-2. Ann Surg **208**, 121-135.

Russell SJ (1990). Lymphokine gene therapy for cancer. Immunol. Today **11**, 196-200.

Talmadge JM (1989). Development of immune therapeutic strategies for the treatment of neoplasms. Biotherapy **4**, 215-236.

Zoller M, Douvdevani A, Segal S and Apte RN (1992a). Interleukin-1 produced by tumorigenic fibroblasts influences tumor rejection. Int J Cancer **50**, 443-449.

Zoller M, Douvdevani A, Segal S and Apte RN (1992b). Interleukin-1 production by transformed fibroblasts. II. Influence on antigen presentation and T-cell-mediated anti-tumor response. Int J Cancer **50**, 450-457.

10

UTILIZATION OF IL-2 AND IL-2 GENE TRANSFER FOR REGIONAL IMMUNOTHERAPY OF CANCER

J. Bubenik
Institute of Molecular Genetics, Academy of Sciences of the Czech Republic, Prague, Czech Republic

Introduction

The aim of this chapter is to review how local and regional administration of IL-2 in preclinical models has contributed to our understanding of the tumor restricted immune recognition and rejection mechanisms, as well as to show how it has led to the development of the experimental IL-2 gene therapy and to the first clinical trials (Table 10.1). We would also like to discuss how future tumor therapy based on IL-2 administration is likely to evolve, what are its prospects and limitations and, particularly, how IL-2 gene transfer could be used to decrease the high toxicity of the IL-2 doses required for cancer therapy. The chapter is primarily focused on IL-2; other interleukins, cytokines and factors are beyond the scope of this review. However, when relevant, the results obtained with biological response modifiers other than IL-2 are briefly mentioned for comparison.

Tumor inhibitory effects of local IL-2 administration

In 1983 it was demonstrated that peritumoral injection of rat lymphoid IL-2 could suppress or markedly inhibit the growth of methylcholanthrene-induced sarcomas in syngeneic mice (Bubenik et al., 1983). Equally effective inhibition of the sarcoma growth as that obtained with rat IL-2 was subsequently described also with purified IL-2 of murine lymphoid origin and with molecularly homogeneous human recombinant IL-2 (Bubenik et al., 1985; Bubenik et al., 1986). The tumor inhibitory effects of local IL-2 administration were confirmed in various experimental tumor systems; IL-2 was found to be particularly effective

CYTOKINE-INDUCED TUMOR IMMUNOGENICITY
ISBN 0–12–261520–4

Table 10.1. *How one finding has led to another: IL-2 and gene therapy of cancer*

Discovery of IL-2	Morgan et al. (1976)
Mitogenic IL-2 activity and long-term cultures of tumor specific CTL	Gillis and Smith (1977)
Finding of activating IL-2 signal (LAK cells)	Grimm et al. (1982)
Local administration of IL-2 inhibits tumor growth in preclinical models	Bubenik et al. (1983)
Local administration of IL-2 inhibits growth of human tumors	Pizza et al. (1984)
Systemic administration of IL-2 inhibits growth of metastatic experimental tumors	Rosenberg et al. (1985a)
Utilization of recombinant IL-2 for successful clinical trials in patients with generalized tumors	Rosenberg et al. (1985b)
Regional IL-2 gene therapy inhibits tumor growth in preclinical models	Bubenik et al. (1988b)
Utilization of TIL or tumor cells carrying an inserted IL-2 gene for the first clinical trials	for reviews see Anderson (1992b); Foa et al. (1992)

when injected repeatedly during the early period of tumor growth in murine lung and mammary carcinoma (Kedar et al., 1984; Vaage, 1987; Vaage et al., 1987; Vaage, 1988), myeloma (Maekawa et al., 1986; Bubenik et al., 1990; Bubenik et al., 1991a; Bubenik et al., 1991b), melanoma (Silagi and Schaefer, 1986; Silagi et al., 1988) and murine lymphomas (Liu et al., 1990; Maas et al.,1991). Local therapy with IL-2 was also found to be efficient in rat hepatoma (Konno et al., 1991) and bovine ocular squamous cell carcinoma (Rutten et al., 1989).

Adoptive transfer of locally IL-2 activated but not nonactivated lymphocytes from tumor bearing mice could inhibit methylcholanthrene-induced sarcoma growth in syngeneic recipients (Forni et al., 1985). Interestingly, in this latter model mice injected with methylcholanthrene-induced sarcoma cells and only IL-2 did not display a significant inhibition of tumor growth (Forni et al., 1985), and prolonged daily local injections of IL-2 at the tumor challenge site induced only a limited inhibition of tumor growth (Forni et al., 1987).

The use of a panel of murine methylcholanthrene-induced sarcomas, carrying individual tumor associated transplantation antigens, showed that the growth of approximately 80% of such antigenic tumors in syngeneic mice could be substantially inhibited or prevented by local administration of human recombinant IL-2 (Bubenik and Indrova, 1987).

Similar results were obtained by Vaage (1988) using a panel of mouse mammary tumors both immunogenic and nonimmunogenic. Of the immunogenic tumors 80% were sensitive to local IL-2 immunotherapy, whereas only approximately 30% of the nonimmunogenic tumors were affected. These results indicate that the local immunotherapeutic effect of exogenous IL-2 is associated with immunogenicity of the responding tumor cells.

The therapeutic effectiveness of IL-2 was dependent on the ability of a tumor to attract monocytes; it was reduced by fast tumor growth (Vaage, 1988), and the effect was probably the result of IFN-γ induction at the IL-2 injection site. The injection of an anti-IFN-γ mAb abrogated the systemic therapeutic effect of peritumoral IL-2 administration (Vaage, 1991). Daily injections of the antibody to murine IFN-γ were also capable of abrogating the effect of local IL-2 administration in the experimental tumor system used by Forni et al. (1987). The effects of IFN-γ were different in various experimental models.

While in murine mammary carcinoma peritumoral injections of IL-2 and IFN-γ had no additive effect when used in combination (Vaage, 1991), in mouse melanomas the results with this combination were additive in s.c. tumors and synergistic in i.p. melanomas (Silagi et al., 1988). It has, in fact, been demonstrated both in vitro and in vivo that IL-2 stimulates the production of lymphokines like IFN-γ and TNF-α (Kasahara et al., 1983; Kasid et al., 1989; Kasid et al., 1990; Boccoli et al., 1990), and that these lymphokines may either mediate, or cooperate with, the effect of IL-2 in various experimental models.

The importance of the recruitment of systemic effector cells at the tumor site was reported by Forni et al. (1985 & 1987), who demonstrated that sublethal whole-body irradiation can completely prevent destruction of s.c. implants of methylcholanthrene-induced sarcoma cells mixed with immune lymphocytes and treated with IL-2. The induction of systemic therapeutic effects by local (peritumoral) administration of IL-2 has been observed in mice carrying transplants of syngeneic mammary carcinomas (Vaage, 1987; Vaage et al., 1987; Vaage, 1988) or syngeneic lymphomas

(Maas et al., 1991). Similar induction of both local and systemic therapeutic effects by peritumoral IL-4 injection has been described in mice carrying syngeneic transplants of melanoma (Gallagher and Zaloom, 1992).

The first positive results of clinical trials using regional IL-2 administration have been reported in bladder carcinoma patients. Pizza et al. (1984) found that repeated intratumoral and peritumoral injections of gibbon IL-2 performed under cystoscopic control led to complete regression of 3 bladder carcinomas and to partial regression of an additional 3 of 10 urinary bladder carcinomas. Huland and Huland (1989) used continuous IL-2 perfusion of the urinary bladder after incomplete transurethral resection of the tumor. In 1 of the 5 treated patients a complete histologically confirmed remission lasting for more than 6 months was observed. In both reports the percentage of complete responses was rather low. The high number of IL-2 nonresponders among urinary bladder carcinoma patients could be due to the fact that patients with invasive, high grade urinary bladder carcinoma are deficient in endogenous IL-2 production (Bubenik et al., 1988a) and, therefore, may not respond properly to exogenous IL-2 administration.

Interestingly, in the majority of patients with urinary bladder carcinoma treated by continuous IL-2 perfusion of the urinary bladder, tumor associated local eosinophilia (constituting up to 65% of leukocytes) was observed in the urinary sediment (Huland and Huland, 1992). Activated eosinophils were attached to bladder tumor cells and active degranulation of eosinophils took place on the tumor cells.

Adler et al. (1984) reported positive response in recurrent cutaneous melanoma and mammary carcinoma nodules after intralesional injection of IL-2 cultured autologous lymphocytes. Arrest of tumor growth and partial or complete regressions were recorded (Yasumoto et al., 1987) with intrapleural instillations of recombinant human IL-2 in 11 patients with malignant pleurisy due to lung cancer. Pleural effusions and cancer cells in the effusions disappeared in 9 of the 11 patients 4-10 days after the start of the treatment.

Yoshida et al. (1988) observed that 6 of 23 (26%) patients with recurrent malignant glioma treated by repeated injections of autologous LAK cells plus IL-2 directly into the cavities of the brain tumors responded with tumor regression. The patients had not responded to previous standard therapy, including surgery and radiochemotherapy; in 3 of the 6 patients, continuous remission of the brain tumor lasted for more than 16 months.

Cortesina et al. (1988) have treated 10 patients with recurrent squamous-cell carcinoma of the head and neck by repeated injections of human lymphoid IL-2 given perilymphatically near the insertion of the sternocleidomastoid muscle on the mastoid, ipsilaterally to the site of the recurrent tumors. Three of the patients responded with a complete regression, and 3 others with a partial tumor regression. Clinical responses were observed only in patients who did not undergo radical cervical lymph node dissection, i.e. in those whose contralateral or bilateral cervical lymph nodes remained intact. Where radical excision of cervical lymph nodes had previously been done, local injections into the affected region were ineffective. The complete responses were followed by a 5 to 8 month tumor free interval. In agreement with preclinical studies (Maas et al., 1993), the local administration of repeated, extremely low IL-2 doses was effective. These findings were extended by other groups. Ishikawa et al. (1989) activated lymphocytes from head and neck cancer patients by cultivation with tumor cells and recombinant IL-2 and infused the activated lymphocytes intraarterially into the lymphocyte donors. Significant reduction of primary tumor size was reported. Regional injections of recombinant IL-2 plus LAK cells have been performed in 14 patients with recurrent head and neck cancer (Squadrelli-Saraceno et al., 1990). Three partial and 3 minor responses were seen.

In another study 15 patients with locally advanced, non pretreated head and neck squamous cell carcinoma were treated with 10 daily perilymphatic injections of low dose recombinant IL-2 (Mattijssen et al., 1991). In contrast to the previously reported data, no tumor regressions were observed and no histological changes were found after treatment as compared to pretreatment biopsies. To access whether regional injections of IL-2 can induce an in vivo lymphocyte activation in cervical lymph nodes of patients with head and neck carcinoma, patients were treated for 1 week prior to surgery by repeated daily IL-2 injections in the perimastoid region. It was found that regional IL-2 injections are capable of activating $CD16^+$ NK cells both locally and in contralateral lymph nodes and that the injections can also expand $CD5^+$ T cells expressing autologous tumor specific cytolysis (Rivoltini et al., 1990).

Local inhalatory administration of human lymphoid IL-2 has been described by Huland et al. (1992) in patients with metastatic renal cell carcinoma. Pulmonary metastases of the renal cell carcinomas responded in 5 of 5 patients and the response lasted for 2-9 months. The low toxicity of the inhalatory IL-2 administration has even allowed a long term outpatient treatment.

IL-2 gene transfer

In 1988 a novel method of long-lasting therapeutic IL-2 administration was proposed, later designated IL-2 gene therapy of cancer (Bubenik et al., 1988b). The therapeutic procedure was based on insertion of a cloned IL-2 gene into the genome of somatic cells, followed by transfer of the genetically manipulated cells, constitutively producing IL-2, to the vicinity of the growing tumor.

Peritumoral administration of rat fibroblastoid (RAT-1) cells, transformed by IL-2 cDNA and constitutively producing IL-2, inhibited the growth of human tumor xenografts in congenitally athymic (nu/nu) mice (Bubenik et al., 1988b), as well as growth of RSV-induced sarcomas transplanted in syngeneic rats (Bubenik, 1989). The tumor inhibitory effects of the IL-2 gene therapy were confirmed in various experimental tumor systems, including murine plasmacytoma (Bubenik et al., 1990; Bubenik et al., 1991a; Bubenik et al., 1991b), fibrosarcoma (Gansbacher et al., 1990b), colon carcinoma (Fearon et al., 1990), renal cell carcinoma (Golumbek et al., 1991) and mastocytoma (Ley et al., 1991). Interestingly, the therapy using other cloned cytokine genes, such as IL-4 (Tepper et al., 1989), IFN-γ (Watanabe et al., 1989; Gansbacher et al., 1990a), TNF-α (Asher et al., 1991), IL-6 (Porgador et al., 1992) and IL-7 (Aoki et al., 1992) was also shown to display a tumor inhibitory effect. The tumor inhibitory effect of cells carrying an inserted lymphokine gene could be demonstrated by vaccination with the genetically modified cells prior to tumor challenge (Table 10.2), or by coinjection of the genetically manipulated tumor cells into a syngeneic recipient (Table 10.3), or by local administration of the genetically manipulated cells in the vicinity of the growing tumor (Table 10.4).

Although the vaccination with genetically manipulated somatic cells prior to tumor challenge (Table 10.2) and coinjection of the genetically engineered cells and tumor cells (Table 10.3) has substantially contributed to our knowledge of the mechanism underlying the effect and specificity of the genetically engineered cell vaccines, only the administration of the genetically modified cells performed after a temporary growth of the tumor (Table 10.4) represents a truly therapeutic approach which can prospectively be utilized in the clinic.

In most of the experimental studies (Tables 10.2-10.4) the cloned genes encoding immunoregulatory lymphokines were inserted into the genome of the same tumor cells which were used for challenge. Therefore, the tumorigenicity of the genetically manipulated cells could be compared to that of the original parental tumor cells. It has been found that the insertion of foreign genes can dramatically decrease the tumorigenicity of the genetically engineered tumor cells. However, the reduction or abrogation of tumorigenicity of the transfected cells was not specific for genes encoding immunoregulatory lymphokines; it has not been observed exclusively after insertion of the genes encoding for IL-2, IL-4, IL-6, IL-7, TNF-α and IFN-γ (Tables 10.2-10.4), but also after transfection with DNA coding for the influenza hemagglutination antigen (Fearon et al., 1988).

It has been reported that the tumor inhibitory effect of IL-2 gene therapy is critically time and dose dependent. The percentage of cured/tumor challenged mice increased proportionally with the dose of transferred IL-2 producing cells, and decreased with an increasing dose of the inoculated tumorigenic parental cells; the percentage of cured/tumor challenged mice has also decreased with an increasing time interval between tumor inoculation and peritumoral administration of the genetically manipulated IL-2 producing cells (Bubenik et al., 1990; Bubenik et al., 1991b; Bubenik et al., 1992a). In the experimental systems which have utilized transfer of somatic cells carrying an inserted gene for

Table 10.2. *Tumor inhibitory effect of vaccines containing genetically manipulated cells producing immunoregulatory lymphokines and administered prior to tumor challenge*

Murine tumor	Lymphokine	Reference
Neuroblastoma	IFN-γ	Watanabe et al. (1989)
Colon carcinoma	IL-2	Fearon et al. (1990)
Fibrosarcoma	IFN-γ	Gansbacher et al. (1990a)
Mastocytoma	IL-2	Ley et al. (1991)
Renal cell carcinoma	IL-4	Golumbek et al. (1991)
Glioma	IL-7	Aoki et al. (1992)

immunoregulatory lymphokines (IL-2, IL-4, IL-6, IL-7, TNF-α, IFN-γ) in the gene therapy of cancer, T-cell immune responses and the existence of an intact host immunity were required for the effectiveness of the genetically manipulated cells in euthymic mice (Watanabe et al., 1989; Fearon et al., 1990; Gansbacher et al., 1990a; Gansbacher et al., 1990b; Asher et al., 1991; Golumbek et al., 1991; Aoki et al., 1992; Porgador et al., 1992). In all the experimental models examined (murine fibrosarcoma plus IL-2 or IFN-γ, murine neuroblastoma plus IFN-γ, murine glioma plus IL-7), transplantation resistance directed specifically against rejected tumors was observed (Watanabe et al., 1989; Gansbacher et al., 1990a; Gansbacher et al., 1990b; Aoki et al., 1992).

It has been shown that murine colon carcinoma (CT-26) cells carrying an inserted IL-2 gene and secreting constitutively IL-2 did not grow s.c. in syngeneic mice; rejection of the genetically manipulated cells completely protected these mice against a challenge with the parental CT-26 cells (Fearon et al., 1990). Vaccination with genetically manipulated CT-26 cells led to CTL activation in the spleen of the vaccinated mice and the majority of the in vitro CTL activity could be blocked by antibodies to CD8; mice depleted of $CD8^+$ cells were incapable of rejecting the genetically manipulated CT-26 cells, whereas mice depleted of $CD4^+$ cells were fully capable of rejecting the CT-26 transfectants. Since CT-26 cells secreting IL-2 produced an antitumor response even in the absence of $CD4^+$ cells, it has been concluded that IL-2 production by the genetically engineered cells bypasses T helper function during the generation of an antitumor immune response. Similarly, mice immunized with syngeneic renal cell car-

Table 10.3. *Tumor inhibitory effect of vaccines containing genetically manipulated cells producing immunoregulatory lymphokines and coinjected with tumor cells*

Murine tumor	Lymphokine	Reference
Plasmacytoma	IL-4	Tepper et al. (1989)
Mammary carcinoma	IL-4	Tepper et al. (1989)
Fibrosarcoma	IL-2	Gansbacher et al. (1990b)
Mastocytoma	IL-2	Ley et al. (1991)
Sarcoma	TNF-α	Asher et al. (1991)

cinoma engineered to secrete IL-4 could reject parental (Renca) tumor cells challenges at a distant site (Golumbek et al., 1991). When $CD8^+$ cells were eliminated before immunization, the parental renal cell carcinoma challenges could not be rejected. Likewise, murine glioma cells engineered to secrete IL-7 were rejected in syngeneic mice even when the recipients were depleted of NK 1.1^+ and $CD4^+$ cells; however, the glioma cells were not rejected in mice depleted of $CD8^+$ cells (Aoki et al., 1992). In the neuroblastoma model the growth of murine neuroblastoma cells engineered to secrete IFN-γ was suppressed in syngeneic mice; the tumor suppression was abolished by depletion of Lyt 2.2^+ cells but not by depletion of activated ($AcM1^+$) macrophages (Watanabe et al., 1989). Methylcholanthrene induced sarcomas genetically engineered to produce TNF-α regressed in a significant number of syngeneic hosts in contrast to the progressively growing parental (MCA-205, WP-4) cells. The ability of the genetically manipulated tumor cells to regress was abrogated by in vivo depletion of $CD4^+$ or $CD8^+$ T cells (Asher et al., 1991).

The first clinical trials based on the results of the preclinical model experiments have recently been started (see reviews by Rosenberg 1991; Rosenberg, 1992; Foa et al., 1992). In pilot experiments it has been found that human renal cell carcinoma and melanoma carrying an inserted IL-2 gene can produce IL-2 and that after inoculation in nu/nu mice the IL-2 production can inhibit tumor growth (Gansbacher et al., 1992; Gastl et al., 1992), in accordance with previous findings (Bubenik et al., 1987; Bubenik et al., 1988b).

The constitutive production of IL-2 by the genetically manipulated cells could be maintained for several weeks after irradiation of the cells

Table 10.4. *Tumor inhibitory effect of vaccines containing genetically manipulated cells producing immunoregulatory lymphokines and administered after the tumor challenge*

Tumor	Lymphokine	Reference
Human cervical carcinoma xenograft	IL-2	Bubenik et al. (1988b)
Rat sarcoma	IL-2	Bubenik (1989)
Murine plasmacytoma	IL-2	Bubenik et al. (1990) Bubenik et al. (1991b)
Murine renal cell carcinoma	IL-4	Golumbek et al. (1991)
Murine lung carcinoma	IL-6	Porgador et al. (1992)

with the doses capable of inhibiting cell proliferation (Gansbacher et al., 1992).

The cloned IL-2 gene can thus be inserted into human tumor cells and human fibroblasts, and repeated injections of the genetically manipulated and irradiated cell vaccines, which no longer proliferate, but secrete IL-2, are being performed (Abbott, 1992; Foa et al., 1992; Rosenberg, 1992).

Since foreign genes have not, until recently, been inserted into humans, safety studies had to be performed prior to the insertion of the IL-2 or similar genes into human cells followed by injections of the genetically engineered cells into cancer patients. Using retroviral vectors, a successful insertion of the *neo* gene has been accomplished in human TIL $CD4^{+}$ and $CD8^{+}$ subsets (Morecki et al., 1991). It has been demonstrated that after i.v. infusion of the genetically modified TIL carrying an inserted *neo* gene, the genetically engineered cells can be found in the circulation of tumor bearing patients for 3 weeks up to 6 months (Rosenberg et al., 1990; Rosenberg, 1992). It is still unclear whether the insertion of foreign genes, such as IL-2 or TNF, can or can not alter the ability of TIL to home into the tumor. The unaltered "homing" ability of the genetically manipulated TIL is a necessary prerequisite for targeting the constitutive lymphokine production to the site of the growing tumor and, therefore, also for successful utilization of the genetically manipulated lymphocytes for gene therapy of cancer in clinical trials (Anderson, 1992a).

Prospects and limitations

The toxicity of the extremely high doses required for systemic IL-2 treatment represents the main obstacle towards a large scale use of IL-2 in the management of cancer patients (see reviews by Bubenik 1990; Maas et al., 1993). Both local IL-2 administration and local IL-2 gene therapy can be used to decrease the dosage and, consequently, also the toxicity of the IL-2 treatment. Unfortunately, cancer is usually a systemic rather than a local disease, and this fact considerably limits the usefulness of local IL-2 immunotherapy. As shown in the section of the book dedicated to Clinical Perspectives, local and regional IL-2 therapy has been successfully used for the treatment of some patients with nongeneralized, unresectable tumors resistant to irradiation and chemotherapy. Prospectively, the local IL-2 administration could also be considered for treatment of minimum residual disease in patients with resectable, nongeneralized tumors, particularly in combination with conventional therapy or other cytokines. For both purposes, a local source of long-lasting exogenous IL-2 production is required. The source could be provided by mini osmotic pumps (Nishimura et al., 1986); however, this approach still has substantial technical difficulties. An alternative approache is based on the insertion of the cloned IL-2 gene into the genome of somatic cells, followed by transfer of the genetically modified cells constitutively producing IL-2 to the vicinity of the tumor (Bubenik et al., 1988b).

Theoretically, the IL-2 gene can be inserted into tumor cells, and the genetically engineered cells can be irradiated in order to prevent their tumorigenicity. Another possibility is to insert the cloned IL-2 gene into the genome of autochthonous fibroblasts and to use the genetically modified, nonmalignant cells as vaccines. Still another possibility is to insert the IL-2 gene into the genome of leukocytes which can, after systemic administration of the genetically modified cells, provide the targeting of IL-2 production by a selective leukocyte-tumor cell adherence.

The amounts of exogenous IL-2 released from the genetically manipulated locally administered cells into the circulation are negligible and, therefore, no concentration dependent adverse effects of IL-2 administration are to be expected. Even if only a small proportion of the genetically engineered cells will localize in the tumor, the cells may release sufficient amounts of IL-2 for extended periods. In other words, it can be expected that the level of IL-2 released by the genetically manipulated cells in the vicinity of the tumor will greatly exceed the local IL-2 concentration achievable by systemic administration of exogenous IL-2, without causing a systemic toxicity. The provision of IL-2 source at the tumor site may help to generate local effector cells which can then recirculate and reach the other sites of metastatic tumor process.

The construction and utilization of the genetically engineered and irradiated tumor vaccines or fibroblast vaccines carrying an inserted IL-2 gene is presently in progress in patients with solid tumors (Abbott, 1992). In patients with generalized forms of the solid tumors and in patients suffering from leukemia (Foa et al., 1992) the systemic administration of irradiated tumor cell vaccines is being considered. The alternative approach, i.e. the insertion of the lymphokine gene into the leukocytes followed by infusion of the genetically engineered leukocytes into cancer patients, is at present not uniformly accepted. The first clinical trials pioneered by Rosenberg's group (Rosenberg et al., 1990; Rosenberg, 1991; Rosenberg, 1992) have so far not provided the expected results, and some technical problems have still to be solved before definitive conclusions can be drawn (see review by Anderson, 1992b). In addition to the problem of the homing of the genetically manipulated leukocytes (Anderson, 1992b), a possible autocrine growth and tumorigenicity of some leukocyte subsets transfected with the IL-2 gene should also be considered (Karasuyama et al., 1989).

Another interesting problem is the persistence of genetically modified lymphokine producing cells after their administration. Genetically modified tumor cells have a substantially reduced tumorigenicity. It has been shown in vitro that killer cells generated by cocultivation of their precursors with genetically modified IL-2 producing cells are cytolytic also for the genetically modified tumor cells (Bubenik et al., 1992b). Genetically modified IL-2 producing cells, therefore, may also be expected to activate in vivo effector killer cells which can terminate IL-2 production by killing the IL-2 producers locally. However, despite these findings, the only ethically acceptable approach in human trials is to prevent both the tumorigenicity and the long lasting IL-2 production by irradiation of the genetically modified tumor cells prior to their in vivo administration (Gansbacher et al., 1992). The problem of tumorigenicity does not arise when genetically manipulated autochthonous fibroblasts are considered for vaccination. Although it has been reported that genetically engineered fibroblasts tend to gradually inactivate the introduced genes (Palmer et al., 1991), the long lasting IL-2 production in the fibroblast derived vaccines should probably also be prevented by irradiation of the genetically modified cells.

References

Abbott A (1992). German state unexpectedly approves first gene trials. Nature **360**, 702-706.

Adler A, Stein JA, Kedar E, Naor D and Weiss DW (1984). Intralesional injection of interleukin-2-expanded autologous lymphocytes in melanoma and breast cancer patients: A pilot study. J Biol Resp Modif **3**, 491-500.

Anderson C (1992a). Gene therapy researcher under fire over controversial cancer trials. Nature **360**, 399-400.

Anderson WF (1992b). Human gene therapy. Science **256**, 808-813.

Aoki T, Tashiro K, Miyatake SI, Kinashi T, Nakano T, Oda Y, Kikuchi H and Honjo T (1992). Expression of murine interleukin 7 in a murine glioma cell line results in reduced tumorigenicity in vivo. Proc Natl Acad Sci USA **89**, 3850-3854.

Asher AI, Mule JJ, Kasid A, Restifo NP, Salo JC, Reichert CM, Jaffe G, Fendly B, Kriegler M and Rosenberg SA (1991). Murine tumor cells transduced with the gene for tumor necrosis factor. J Immunol **146**, 3227-3234.

Boccoli G, Masciulli R, Ruggeri EM, Carlini P, Gianella G, Montesoro E, Mastroberardino G, Isacchi G, Testa U, Calabresi F and Peschle C (1990). Adoptive immunotherapy of human cancer: The cytokine cascade and monocyte activation following high-dose interleukin 2 bolus treatment. Cancer Res **50**, 5795-5799.

Bubenik J, Perlmann P, Indrova M, Simova J, Jandlova T and Neuwirt J (1983). Growth inhibition of a MC-induced mouse sarcoma by TCGF (IL-2)-containing preparations. Cancer Immunol Immunother **14**, 205-206.

Bubenik J, Indrova M, Perlmann P, Berzins K, Mach O, Kraml J and Toulcova A (1985). Tumour-inhibitory effects of TCGF (IL-2)-containing preparations. Cancer Immunol Immunother **19**, 57-61.

Bubenik J, Kieler J and Indrova M (1986). Local treatment with human recombinant interleukin 2 inhibits growth of MC-induced sarcomas in syngeneic mice. Folia Biol **32**, 209-211.

Bubenik J and Indrova M (1987). The anti-tumor efficacy of human recombinant interleukin 2. Correlation between sensitivity of tumours to the cytolytic effect of LAK cells in vitro and their susceptibility to interleukin 2 immunotherapy in vivo. Cancer Immunol Immunother **24**, 269-271.

Bubenik J, Kieler J, Tromholt V, Jandlova T and Lotzova E (1987). Recombinant interleukin 2 inhibits growth of human tumour xenografts in congenitally athymic mice. Immunol Lett **14**, 325-330.

Bubenik J, Kieler J, Tromholt V, Hermann G and Jandlova T (1988a). Defect in lectin-induced interleukin 2 production by peripheral blood lymphocytes of patients with invasive urinary bladder carcinoma. Immunol Lett **18**, 115-118.

Bubenik J, Voitenok NN, Kieler J, Prassolov VS, Chumakov PM, Bubenikova D, SimovaJ and Jandlova T (1988b). Local administration of cells containing an inserted IL-2 gene and producing IL-2 inhibits growth of human tumours in nu/nu mice. Immunol Lett **19**, 279-282.

Bubenik J (1989). Local immunotherapy of cancer with interleukin 2. Immunol Lett **21**, 267-274.

Bubenik J (1990). Local and regional immunotherapy of cancer with interleukin 2. J Cancer Res Clin Oncol **116**, 1-7.

Bubenik J, Simova J and Jandlova T (1990). Immunotherapy of cancer using local administration of lymphoid cells transformed by IL-2 cDNA and constitutively producing IL-2. Immunol Lett **23**, 287-292.

Bubenik J, Lotzova E, Indrova M, Simova J, Jandlova T and Savary CA (1991a). IL 2 gene transfer in immunotherapy of cancer: Local administration of LAK and X63-m-IL 2 cells transformed by IL 2 cDNA and constitutively producing IL 2 inhibits growth of plasmacytomas in syngeneic mice. Nat Immun Cell Growth Regulation **10**, 247-255.

Bubenik J, Lotzova E, Zeuthen J, Bubenikova D, Simova J and Indrova M (1991b). Gene therapy of cancer: Use of interleukin-2 (IL-2) gene transfer in local immunotherapy. J Exp Clin Cancer Res **10**, 213-220.

Bubenik J, Lotzova E, Indrova M, Simova J, Jandlova T and Bubenikova D (1992a). Use of IL-2 gene transfer in local immunotherapy of cancer. Cancer Lett **62**, 257-262.

Bubenik J, Zeuthen J, Simova J, Jandlova T, Bubenikova D and Radzikowski C (1992b). Activation and phenotyping of LAK generated by exposure to product of cells transformed by interleukin 2 cDNA. Anticancer Res **12**, 965-968.

Cortesina G, De Stefani A, Giovarelli M, Barioglio MG, Cavallo GP, Jemma C and Forni G (1988). Treatment of recurrent squamous cell carcinoma of the head and neck with low doses of interleukin-2 injected perilymphatically. Cancer **62**, 2482-2485.

Fearon ER, Itaya T, Hunt B, Vogelstein B and Frost P (1988). Induction in a murine tumor of immunogenic tumor variants by transfection with a foreign gene. Cancer Res **48**, 2975-2980.

Fearon ER, Pardoll DM, Itaya T, Golumbek P, Levitsky HJ, Simons JW, Karasuyama H, Vogelstein B and Frost P (1990). Interleukin-2 production by tumor cells bypasses T helper function in the generation of an antitumor response. Cell **60**, 397-403.

Foa R, Guarini A and Gansbacher B (1992). IL 2 treatment for cancer: From biology to gene therapy. Br J Cancer **66**, 992-998.

Forni G, Giovarelli M and Santoni A (1985). Lymphokine activated tumor inhibition in vivo. I. The local administration of interleukin 2 triggers nonreactive lymphocytes from tumor bearing mice to inhibit tumor growth. J Immunol **134**, 1305-1311.

Forni G, Giovarelli M, Santoni A, Modesti A and Forni M (1987). Interleukin-2-activated tumor inhibition in vivo depends on the systemic involvement of host immunoreactivity. J Immunol **138**, 4033-4041.

Gallagher G and Zaloom Y (1992). Peritumoural IL-4 treatment induces systemic inhibition of tumour growth in experimental melanoma. Anticancer Res **12**, 1019-1024.

Gansbacher B, Bannerji R, Daniels B, Zier K, Cronin K and Gilboa E (1990a). Retroviral vector-mediated γ-interferon gene transfer into tumor cells generates potent and long lasting antitumor immunity. Cancer Res **50**, 7820-7825.

Gansbacher B, Zier K, Daniels B, Cronin K, Bannerji R and Gilboa E (1990b). Interleukin 2 gene transfer into tumor cells abrogates tumorigenicity and induces protective immunity. J Exp Med **172**, 1217-1224.

Gansbacher B, Zier K, Cronin K, Hantzopoulos PA, Bouchard B, Houghton A, Gilboa E and Golde D (1992). Retroviral gene transfer induced constitutive expression of interleukin-2 or interferon-gamma in irradiated human melanoma cells. Blood **80**, 2817-2825.

Gastl G, Finstad CL, Guarini A, Bosl G, Gilboa E, Bander NH and Gansbacher B (1992). Retroviral vector-mediated lymphokine gene transfer into human renal cancer cells. Cancer Res **52**, 6229-6236.

Gillis S and Smith K (1977). Long term culture of tumour-specific cytotoxic T cells. Nature **268**, 154-156.

Golumbek PT, Lazenby AJ, Levitsky HJ, Jaffee LM, Karasuyama H, Baker M and Pardoll DM (1991). Treatment of established renal cancer by tumor cells engineered to secrete interleukin-4. Science **245**, 713-716.

Grimm EA, Mazuneder A, Zhang HZ and Rosenberg SA (1982). Lymphokine activated killer cell phenomenon: Lysis of natural killer-resistant fresh solid tumor cells by interleukin 2-activated autologous human peripheral blood lymphocytes. J Exp Med **155**, 1823-1841.

Huland E and Huland H (1989). Local continuous high dose interleukin 2: A new therapeutic model for the treatment of advanced bladder carcinoma. Cancer Res **49**, 5469-5474.

Huland E and Huland H (1992). Tumor-associated eosinophilia in interleukin-2-treated patients: Evidence of toxic eosinophil degranulation on bladder cancer cells. J Cancer Clin Oncol **118**, 463-467.

Huland E, Huland H and Heinzer H (1992). Interleukin 2 by inhalation: Local therapy for metastatic renal cell carcinoma. J Urol **147**, 344-348.

Ishikawa T, Ikawa T, Eura M, Fukiage T and Masuyama K (1989). Adoptive immunotherapy for head and neck cancer with killer cells induced by stimulation with autologous or allogeneic tumour cells and recombinant interleukin 2. Acta Otolaryngol **107**, 346-351.

Karasuyama H, Tohyama N and Tada T (1989). Autocrine growth and tumorigenicity of interleukin 2-dependent helper T cells transfected with IL-2 gene. J Exp Med **169**, 13-25.

Kasahara T, Hooks JJ, Dougherty SF and Oppenheim JJ (1983). Interleukin 2-mediated immune interferon (IFN-gamma) production by human T cells and T cell subsets. J Immunol **130**, 1784-1788.

Kasid A, Director EP and Rosenberg SA (1989). Induction of endogenous cytokine-mRNA in circulating peripheral blood mononuclear cells by IL-2 administration to cancer patients. J Immunol **143**, 736-740.

Kasid A, Director EP, Stovroff MC, Lotze MT and Rosenberg SA (1990). Cytokine regulation of tumor necrosis factor-alpha and -beta (lymphotoxin)-messenger RNA expression in human peripheral blood mononuclear cells. Cancer Res **50**, 5072-5076.

Kedar E, Chriqui-Zeira E and Kyriazis AP (1984). Immunotherapy of murine and human tumours in mice with lymphokines and interleukin-2-propagated lymphocytes. J Biol Resp Modif **3**, 517-526.

Konno H, Yamashita A, Tadakuma T and Sakaguchi S (1991). Inhibition of growth of rat hepatoma by local injection of liposomes containing recombinant interleukin-2. Biotherapy **3**, 211-218.

Ley V, Langlade-Demoyen P, Kourilsky P and Larsson Sciard EL (1991). Interleukin 2 dependent activation of tumor-specific cytotoxic T lymphocytes in vivo. Eur J Immunol **21**, 851-854.

Liu CM, Suzuki Y, Chen L, Okayasu T, Calkins CE and Wheelock EF (1990). Maintenance and cure of the L5178Y murine tumor-dormant state by interleukin 2: In vivo and in vitro effects. Cancer Res **50**, 1361-1367.

Maas RA, Van Weering DHJ, Dullens HFJ and Den Otter W (1991). Intratumoral low-dose interleukin-2 induces rejection of distant solid tumour. Cancer Immunol Immunother **33**, 389-394.

Maas RA, Dullens HFJ and Den Otter W (1993). Interleukin 2 in cancer treatment: Disappointing or (still) promising? A review. Cancer Immunol Immunother **36**, 141-148.

Maekawa R, Matsumoto M, Kitagawa T, Harada M and Sato K (1986). Effect of recombinant interleukin 2 (R-IL 2) on in vivo growth of murine myeloma X5563. Cancer Immunol Immunother **23**, 25-30.

Mattijssen V, De Mulder PH, Schornagel JH, Verweij J, Van den Broek P, Galazka A, Roy S and Ruiter DJ (1991). Clinical and immunopathological results of a phase II study of perilymphatically injected recombinant interleukin-2 in locally far advanced, nonpretreated head and neck squamous cell carcinoma. J Immunother **10**, 63-68.

Morecki S, Karson E, Cornetta K, Kasid A, Aebersold P, Blaese RM, Anderson WF and Rosenberg SA (1991). Retrovirus-mediated gene transfer into CD4+ and CD8+ human T cell subsets derived from tumor-infiltrating lymphocytes and peripheral blood mononuclear cells. Cancer Immunnol Immunother **32**, 342-352.

Morgan DA, Ruscetti FW and Gallo R (1976). Selective in vitro growth of T-lymphocytes from normal human bone marrow. Science **193**, 1007-1008.

Nishimura T, Uchiyama Y, Yagi H and Hashimoto Y (1986). Administration of slowly released recombinant interleukin 2. J Immunol Meth **91**, 21-30.

Palmer TD, Rosman GJ, Osborne WRA and Miller D (1991). Genetically modified skin fibroblasts persist long after transplantation but gradually inactivate introduced genes. Proc Natl Acad Sci USA **88**, 1330-1334.

Pizza G, Severini G, Menniti D, De Vinci C and Corrado F (1984). Tumour regression after intralesional injection of interleukin 2 (IL-2) in bladder cancer. Preliminary report. Int J Cancer **34**, 359-367.

Porgador A, Tzehoval E, Katz A, Vadai E, Revel M, Feldman M and Eisenbach L (1992). Interleukin 6 gene transfection into Lewis lung carcinoma tumor cells suppresses the malignant phenotype and confers immunotherapeutic competence against parental metastatic cells. Cancer Res **52**, 3679-3686.

Rivoltini L, Gambacorti-Passerini C, Squadrelli-Saraceno M, Grosso MI, Cantu G, Molinari R, Orazi A and Parmiani G (1990). In vivo interleukin 2-induced activation of lymphokine-activated killer cells and tumor cytotoxic T cells in cervical lymph nodes of patients with head and neck tumors. Cancer Res **50**, 5551-5557.

Rosenberg SA, Mule JJ, Spiess PJ, Reichert CM and Schwarz SL (1985a). Regression of established pulmonary metastases and subcutaneous tumor mediated by the systemic administration of high-dose recombinant interleukin 2. J Exp Med **161**, 1169-1188.

Rosenberg SA, Lotze MT, Muul LM, Leitman S, Chang AE, Ettinghauser SE, Matory YL, Skibber JM, Shiloni E, Vetto JT, Seipp CA, Simpson C and Reichert CM (1985b). Observations on the systemic administration of autologous lymphokine-activated killer cells and recombinant interleukin-2 to patients with metastatic cancer. N Engl J Med **313**, 1485-1492.

Rosenberg SA, Aebersold P, Cornetta K, Kasid A, Morgan RA, Moen R, Karson EM, Lotze MT, Yang JC, Topalian SL, Merino MJ, Culver K, Miller AD, Blaese RM and Anderson WF (1990). Gene transfer into humans-Immunotherapy of patients with advanced melanoma, using tumor-infiltrating lymphocytes modified by retroviral gene transduction. N Engl J Med **323**, 570-578.

Rosenberg SA (1991). Immunotherapy and gene therapy of cancer. Cancer Res (Suppl) **51**, 5074s-5079s.

Rosenberg SA (1992). The immunotherapy and gene therapy of cancer. J Clin Oncol **10**, 180-199.

Rutten VPMG, Klein WR, De Jong WAC, Misdorp W, Den Otter W, Steerenberg PA, DeJong WH and Ruitenberg EJ (1989). Local interleukin-2 therapy in bovine ocular squamous cell carcinoma. Cancer Immunol Immunother **30**, 165-169.

Silagi S and Schaefer AE (1986). Successful immunotherapy of mouse melanoma and sarcoma with recombinant interleukin-2 and cyclophosphamide. J Biol Resp Modif **5**, 411-422.

Silagi S, Dutkowski R and Schaefer A (1988). Eradication of mouse melanoma by combined treatment with recombinant human interleukin 2 and recombinant murine interferon-gamma. Int J Cancer **41**, 315-322.

Squadrelli-Saraceno M, Rivoltini L, Cantu G, Ravagnani F, Parmiani G and Molinari R (1990). Local adoptive immunotherapy of advanced head and neck tumours with LAK cells and interleukin-2. Tumori **76**, 566-571.

Tepper RI, Pattengale PK and Leder P (1989). Murine interleukin-4 displays potent anti-tumor activity in vivo. Cell **57**, 503-512.

Vaage J (1987). Local and systemic effects during interleukin-2 therapy of mouse mammary tumors. Cancer Res **47**, 4296-4298.

Vaage J, Pauly JL and Harlos JP (1987). Influence of the administration schedule on the therapeutic effect of interleukin-2. Int J Cancer **39**, 530-533.

Vaage J (1988). Local interleukin 2 therapy of mouse mammary tumors of various immunogenicities. Cancer Res **48**, 2193-2197.

Vaage J (1991). Peritumor interleukin-2 causes systemic therapeutic effect via interferon gamma induction. Int J Cancer **49**, 598-600.

Watanabe Y, Kuribayashi K, Miyatake S, Nishihara K, Nakayama E, Taniyama T and Sakata T (1989). Exogenous expression of mouse interferon gamma cDNA in mouse neuroblastoma C1300 cells results in reduced tumorigenicity by augmented anti-tumor immunity. Proc Natl Acad Sci USA **86**, 9456-9460.

Yasumoto K, Miyazaki K, Nagoshima A, Ishida T, Kuda T, Yano T, Sugimachi K and Nomoto K (1987). Induction of lymphokine-activated killer cells by intrapleural instillations of recombinant interleukin-2 in patients with malignant pleurisy due to lung cancer. Cancer Res **47**, 2184-2187.

Yoshida S, Tanaka R, Takai N and Ono K (1988). Local administration of autologous lymphokine-activated killer cells and recombinant interleukin 2 to patients with malignant brain tumors. Cancer Res **48**, 5011-5016.

11

MURINE MODELS FOR CANCER IMMUNOTHERAPY USING CYTOKINE GENE MODIFIED TUMOR VACCINES

E. Gilboa
Department of Experimental Surgery, Duke University Medical Center, Durham, NC, USA

Introduction

It is now well recognized that the major cause for therapeutic failures in clinical oncology is not the initial tumor presented during diagnosis but rather the consequences of local recurrence and distant metastases originating from minimal residual disease (MRD) that was not eliminated during initial treatment of the tumor. It is, therefore, difficult to overstate the importance of devising means to target MRD and prevent recurrence of disease. One such approach is active immunotherapy, the purpose of which is to induce a state of immunity in the patient with low tumor burden that would eliminate the remaining cancer cells or, at least, provide constant and long term protection against recurrence of disease. Active immunotherapy of cancer patients is not a new idea, but it is fair to say that, in spite of considerable efforts over the past 20-30 years, the development of effective tumor vaccines has not been achieved (Livingston, 1991).

Disenchantment with immunological intervention modalities in the treatment of cancer has been further fueled by the observations of Hewitt et al. (1976) and Middle and Embleton (1981) that spontaneously occurring tumors in animals do not induce protective immunity and that the incidence of cancer is not elevated in T-cell deficient mice (Rygaard and Povlsen, 1976). If, as these studies imply, naturally occurring tumors do not encode tumor specific antigens capable of inducing an effective immune response, active immunotherapy of cancer would be futile. In a landmark study, Van Pel and Boon (1982) have shown that a nonimmunogenic murine tumor could be converted to an immunogenic form which was capable of inducing systemic immunity against the parental, apparently nonimmunogenic, tumor. This finding was first in a series of studies to demonstrate that "strictly defined" nonimmunogenic tumors can encode tumor antigens.

CYTOKINE-INDUCED TUMOR IMMUNOGENICITY
ISBN 0-12-261520-4

The defect is that in their original form the tumors are incapable of presenting the tumor antigen(s) to the immune system. Furthermore, experimental manipulation of nonimmunogenic tumors was capable of exposing the tumor antigen(s) to the immune system and of generating a measurable immune response directed against the tumor antigens present in the apparently nonimmunogenic tumors.

Rationale for using cytokine gene modified tumor vaccines

What were then the reasons for the failure to develop successful tumor vaccines? One possibility that we considered was the fact that most attempts to develop tumor vaccines have focused, consciously or unconsciously, on the humoral arm of the immune system, while it became increasingly clear that the cellular immune system, in particular CTL, are best suited to recognize tumor cells and initiate a cascade of events leading to their eventual destruction. Indeed, animal studies support the notion that, with few exceptions, tumor specific CTL constitute an important effector arm in the antitumor immune response (see reviews by Hamaoka and Fujiwara, 1987; Greenberg, 1991, and references therein).

It was, therefore, the initial premise of our studies to try to induce and enlist CTL as the primary antitumor immune effector arm, and the rationale for doing so is shown in Fig. 11.1. $CD8^+$ CTL recognize antigens in the context of MHC class I molecules expressed on the surface of the target cell (Zinkernagel and Doherty, 1979). CTL epitopes consisting of 9-10 amino acid peptides are derived from endogenously synthesized proteins regardless of whether the proteins are destined to the cell surface or to other cellular compartments (Townsend and Bodmer, 1989; Monaco, 1992). Thus, CTL should be capable of recognizing tumor cells expressing a wide range of gene products to which the immune system has not been tolerized. Interaction between the tumor cell and its cognate T cell is sufficient for tumor cell lysis by an activated CTL and may be also sufficient to activate a memory CTL. However, induction of CTL responses in the naive animal, i.e. activation of an unprimed CTL, was shown to require the participation of antigen specific MHC class II restricted $CD4^+$ T cells (Th cells) (Zinkernagel et al. 1978; Von Boehmer and Haas, 1979; Keene and Forman, 1982). $CD4^+$ T cells are believed to mediate their accessory role via the secretion of IL-2 which is required for the full activation and maturation of CTL as shown in Fig. 11.1, left panel. If activation of tumor specific Th cells is limiting in this process, then providing IL-2 directly to the unprimed CTL would augment the antitumor CTL response. Indeed, the systemic administration of IL-2 had profound inhibitory effects on tumor progression in the experimental animal, but exhibited only limited therapeutic benefit when administered

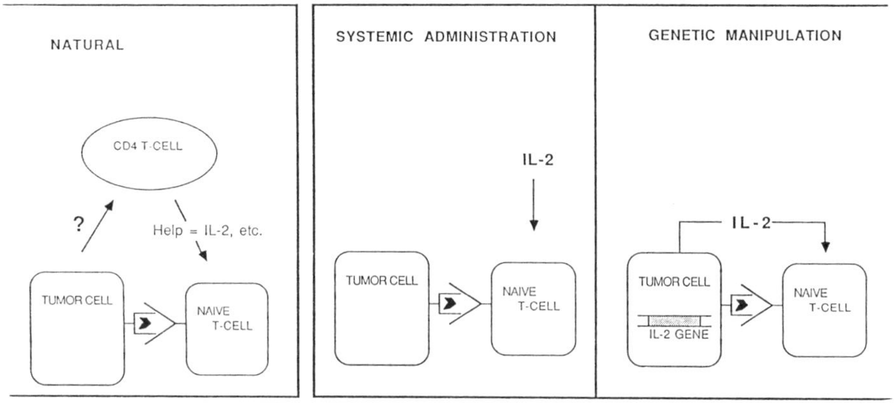

Fig. 11.1. *Experimental induction of CD8+ T-cell responses in vivo.* The prevailing, though not uncontested, view is that induction of a $CD8^+$ T cell involves not only presentation of antigen, tumor antigen or viral antigen, in association with MHC class I molecules to a T-cell precursor, but also requires activation of a $CD4^+$ T cell, known as Th cell, whose function is to provide cytokines, such as IL-2 which is required for the full activation of the $CD8^+$ T-cell precursor (left panel). To the extent that activation of the $CD4^+$ T cell is a limiting factor in the induction of effective antitumor CTL responses, the systemic administration of IL-2 would eliminate the need for CD4 T-cell activation (center panel). An alternative approach aimed at providing IL-2 to the T-cell precursor, the subject of this chapter, is to insert the corresponding gene into the tumor cells; this would result in localized and continuous secretion of IL-2 in the vicinity of the T cells (right panel).

to cancer patients (Rosenberg, 1988) (Fig. 11.1, center panel). The limited efficacy of IL-2 in cancer immunotherapy is explained at least in part by the toxicity resulting from the necessity of administering high doses of the cytokine to the patients (Lee et al., 1989).

Since most immunomodulators act as local hormones whose accumulation in the serum is prevented by virtue of their short half-life, it was argued that local delivery of cytokines would have a profound therapeutic benefit (Bubenik et al., 1983; Forni et al., 1986). Several studies have attempted to induce the regression of established tumors or to induce an immunological memory by repeated injections of cytokines at the site of the tumor.

Overall, these studies have failed to demonstrate a distinct advantage for this mode of cytokine delivery, perhaps because it did not approximate closely enough the physiological rate of cytokine secretion required to elicit an optimal immune response. Another approach that would result in the highly localized secretion of cytokines at the site of the tumor would consist of inserting cytokine genes into tumor cells (Bubenik et al., 1988; Bubenik et al., 1989).

We hypothesized that this approach would mimick more closely the physiological mode of lymphokine delivery to its target and that the locally secreted cytokine would lead to the activation of the CTL in the absence of Th cells (Fig. 11.1, right panel).

The CMS-5 tumor model system

The CMS-5 tumor is a methylcholanthrene induced fibrosarcoma which originated in a BALB/c mouse (DeLeo et al., 1977). The two cytokines that we chose for these studies were IL-2 and IFN-γ, because each cytokine was shown to play a role in the activation and/or function of CTL. The two main observations made in these studies were that:

- CMS-5 tumor cells engineered to secrete low levels of either IL-2 or IFN-γ lost their ability to form tumors in mice
- mice which rejected the cytokine secreting tumor cells became resistant to a subsequent challenge with highly tumorigenic doses of unmodified CMS-5 cells, but not of an unrelated tumor such as CMS-13 or METH A (Gansbacher et al., 1990a; Gansbacher et al., 1990b).

These observations, together with similar ones reported using IFN-γ secreting C1300 neuroblastoma tumor cells (Watanabe et al., 1989) or IL-2 secreting CT26 coloncarcinoma tumor cells (Fearon et al., 1990), have apparently vindicated the working hypothesis depicted in Fig. 11.1. They show in fact that the localized secretion of cytokines mediated via cytokine gene transfer is capable of inducing effective and tumor specific immunity in the animal. It was these and subsequent observations using other tumor models and cytokines that indicated the generality of these conclusions and fueled the interest in considering the use of cytokine secreting tumor cells as cellular vaccines in the treatment of cancer.

It did not take very long, however, to recognize that these initial observations were largely misinterpreted, and this because several features of the experimental systems used in these studies have severely limited their relevance to human cancer. First, tumor free animals were treated with the cytokine secreting tumor cells and shown to develop resistance to a subsequent tumor challenge, while attempts were not made to cure mice with a preestablished tumor, as would be the case in the cancer patient.

Second, and this was a critical flaw of these experimental systems, live (cytokine secreting) tumor cells were used to immunize the animals. The use of live tumor cells as cellular vaccines in cancer patients is, of course, undesirable. Live tumor cells could be used in these studies because, fortuitously, the cytokine secreting tumor cells also lost their ability to grow in the animals, i.e. to form tumors. The problem is that since live tumor cells were used as immunogens, it was unclear whether the cytokine contributed to the induction of an antitumor immunity or whether its only function was to inhibit tumor growth in the mouse, thereby allowing the manifestation of the immunogenicity intrinsic to the tumor. In view of the fact that spontaneously arising tumors are apparently nonimmunogenic, it was argued that the use of even weakly immunogenic tumor cells in animal models is undesirable (Hewitt et al., 1976). Dranoff et al. (1993) have recently shown that the tumors used in these earlier studies, including the CMS-5 tumor used in our studies, were highly immunogenic, as judged by the fact that irradiated, non-cytokine secreting tumor cell preparations were also capable of inducing a potent antitumor immunity; thus, cytokine secretion from the tumor cells contributed little if any to the development of antitumor immunity.

Clearly, the need to develop more relevant animal models that simulate as closely as possible the conditions prevailing in the cancer patient is essential in order to determine whether cytokine secreting tumor vaccines can be of use in cancer patients. In addition to using growth inactivated tumor vaccines and treating tumor bearing animals, it is also important to consider the following issues:

- the choice of a relevant animal tumor model that corresponds in its origin and resembles in its biology a particular form of human cancer
- the use of orthotopic rather than ectotopic (i.e. s.c.) tumor implantation to assess effectiveness of a particular treatment, in view of a large body of evidence showing that the site of tumor implantation can greatly influence its biological properties and its susceptibility to treatment (Fidler, 1990)
- the use of a meaningful endpoint such as metastasis or survival, and not local tumor growth
- the use of cytokine gene modified tumor vaccines as adjuvant therapy and not as primary treatment
- the side-by-side comparison of a treatment protocol using cytokine secreting tumor cells and an established vaccination protocol such as treatment with BCG or *Corynebacterium parvum*.

Drs Leah Eisenbach and Michael Feldman were the first to recognize the pitfalls of using animal models in immunotherapy studies. In a series of elegant studies, these investigators and their colleagues have shown that nonimmunogenic tumor cells engineered to secrete either IL-2, IL-6 or IFN-γ were capable of inhibiting the growth of preexisting lung micrometastases, and even of succeeding to cure a fraction of animals from their disease (Porgador et al., 1992; Porgador et al., 1993a; Porgador et al., 1993b).

A mouse model for immunotherapy of bladder cancer

Bladder cancer is the second most common cancer among men. There are approximately 50000 new cases of bladder carcinoma leading to 11000 deaths in the United States annually. Transitional cell carcinoma (TCC) is the most common form of bladder cancer in the United States. The incidence of TCC is increased in industrialized countries; it is associated with cigarette smoking and can be considered the result of carcinogen induction.

Although transurethral resection followed by intravesical BCG treatment will result in prolonged remission, a significant fraction of patients will eventually relapse locally and/or progress to metastatic disease (Herr et al., 1989). Some patients with metastatic disease respond to cytotoxic chemotherapy such as M-VAC or cisplatin plus radiotherapy. However, most complete responders will eventually relapse and die (Raghavan et al., 1990). Because of the serious limitations of available therapies for bladder cancer, new approaches are clearly needed. Bladder cancer, especially presenting as superficial disease, is responsive to immunotherapeutic agents such as BCG and may represent a good candidate for immunological intervention using tumor vaccines.

The murine MBT-2 tumor constitutes an excellent model to evaluate new approaches to the treatment of bladder cancer. The MBT-2 tumor was induced by the oral administration of FANFT, a potent carcinogen inducing bladder neoplasms in 80-90% of mice, and is highly specific for the urothelium. This highly malignant tumor retains the histological appearance of a poorly differentiated TCC and resembles, both grossly and histologically, its human counterpart. Most importantly, the MBT-2 model system has allowed the testing and development of a number of therapeutic strategies that now form part of the standard treatment for human superficial bladder cancer. The MBT-2 model system is unique in this regard since treatments that have shown promise in this murine model have been similarly effective in human bladder cancer (Soloway and Murphy, 1979). Clearly, the MBT-2 tumor model resembles in its origin, biology, and response to treatment the metastatic form of human bladder cancer; it therefore constitutes a very good, if not excellent, model to explore treatment strategies to bladder cancer at this incurable stage. We have, therefore, used the MBT-2 tumor model to reexamine the utility of cytokine gene modified tumor vaccines in the treatment of bladder cancer.

Retroviral vectors were used to deliver and express the genes for the human IL-2 and mouse IFN-γ in MBT-2 cells. Utilizing a selectable marker present in the retroviral vector, the bacterial neomycin resistance gene, individual clones were isolated and cytokine expression determined. Clones secreting the highest level of each cytokine were chosen for further studies. Secretion of IL-2 or IFN-γ had no discernable effects on cell morphology or on the growth rate of the MBT-2 cells in culture when compared to parental MBT-2 cells or to MBT-2 cells transduced with another retroviral vector encoding the human ADA minigene. Parental MBT-2 cells or clones transduced with a control vector did not secrete detectable levels of either IL-2 or IFN-γ.

As previously noted in the CMS-5 tumor system, neither IL-2 nor IFN-γ secreting MBT-2 cells grew in the animals, and animals which rejected the cytokine secreting tumor cells became fully resistant to a subsequent challenge with parental MBT-2 cells, but not to a challenge with 38C13 B lymphoma cells, a tumor of the same genetic background. To further enhance the relevance of these studies to human bladder cancer, we used experimental designs that simulate as closely as possible the conditions prevailing in the patient (Connor et al., 1993).

First, we tested the ability of growth inactivated cytokine secreting tumor cell preparations to affect the course of disease in mice carrying an established tumor load. Second, since the microenvironment in which the tumor is growing can greatly influence its biology and growth pattern, we developed an orthotopic implantation protocol in which the MBT-2 tumor cells are injected into the bladder wall of the mouse. As shown in Fig. 11.2, parental MBT-2 cells were first implanted into the bladder wall of the mouse, and treatment with tumor cell preparations, inactivated by exposure to 7000 rads, was initiated 7 days later when the implanted tumor was well established. Untreated mice developed a visible tumor within 10-20 days of implantation and succumbed to disease within 6-7 weeks. Treatment of the tumor bearing animals with irradiated non-cytokine secreting tumor cell preparations had no beneficial effect on their survival, which means that under these experimental conditions unmodified MBT-2 cells do not exhibit measurable intrinsic immunogenicity and that any therapeutic benefit seen can be attributed to the consequences of genetic modification, i.e. secretion of cytokine. Indeed, treatment of the tumor bearing animals with IL-2 secreting tumor cells not only resulted in significant prolongation of survival of all treated mice, but was also capable of curing a significant fraction of mice from the implanted tumor, usually 60% as shown in Fig. 11.2, and sometimes 80%. IFN-γ secreting tumor cell preparations, on the other hand, were considerably less effective than IL-2 secreting preparations.

Since mice could be cured from an existing tumor burden using IL-2 secreting tumor vaccines, it was of interest to test whether such mice developed an immunologic memory. As shown in Table 11.1, no tumor growth was observed in cured mice that were rechallenged with a highly tumorigenic dose of MBT-2 cells, while all control mice implanted with 5-fold less tumor cells developed a tumor. Although specificity controls have not been used in this experiment, this result indicates that an effective immunological memory was established in the tumor bearing mice cured by treatment with inactivated IL-2 secreting tumor cell preparations.

We have recently begun to test additional cytokines, and preliminary studies have indicated that GM-CSF secreting tumor vaccines are consistently less effective than IL-2 secreting vaccines, while IL-1α or IL-1β

Table 11.1. *Challenge of mice cured from their original tumor with parental MBT-2 cells*

Mice	No. of parental MBT-2 cells instilled intravesically	Mice with tumors/ mice injected
"Cured" mice[a]	1×10^5	0/9
Age-matched control	2×10^4	5/5

[a]Mice which exhibited complete tumor regression following treatment with IL-2

secreting tumor vaccines have only a marginal effect (Saito et al., unpublished observations).

In studies using a rat model for prostate cancer we also found that IL-2 secreting tumor vaccines were superior to GM-CSF and IFN-γ secreting vaccines (Vieweg et al., submitted for publication). Due to the complexity of the experimental systems employed in these studies, it would be highly premature, if not irresponsible, to suggest a general conclusion as to the comparative values of cytokines used in this setting.

Having used irradiated tumor cell preparations, it was possible to conclude that the IL-2 secreted from the genetically engineered tumor cells contributed in a significant manner to the induction of an immune response which had considerable therapeutic benefit even in animals with a preestablished tumor. Nevertheless, the animal model in which treatment with IL-2 secreting MBT-2 cells is initiated shortly after establishment of the parental tumor in the bladder of the experimental mouse fails to take into account the fact that cancer patients generally become candidates for immunotherapy after a long history of disease, including treatments designed to reduce the tumor burden present at the time of diagnosis. Clearly, immunotherapy will not be provided as a primary treatment to patients with extensive tumor burden, but rather as adjuvant therapy to protect the patient from recurrence of disease. In bladder cancer, candidates for immunotherapy with cytokine secreting tumor vaccines may be individuals whose initial cancer was treated by BCG and went into remission, but these patients have a high risk of relapse and progression to metastatic disease.

A second concern has been raised that individuals with a history of cancer, even when in remission, may be in a state of generalized or tumor specific immunosuppression; the question would, therefore, be whether treatment with cytokine secreting tumor vaccines could also reverse a possible state of immune unresponsiveness.

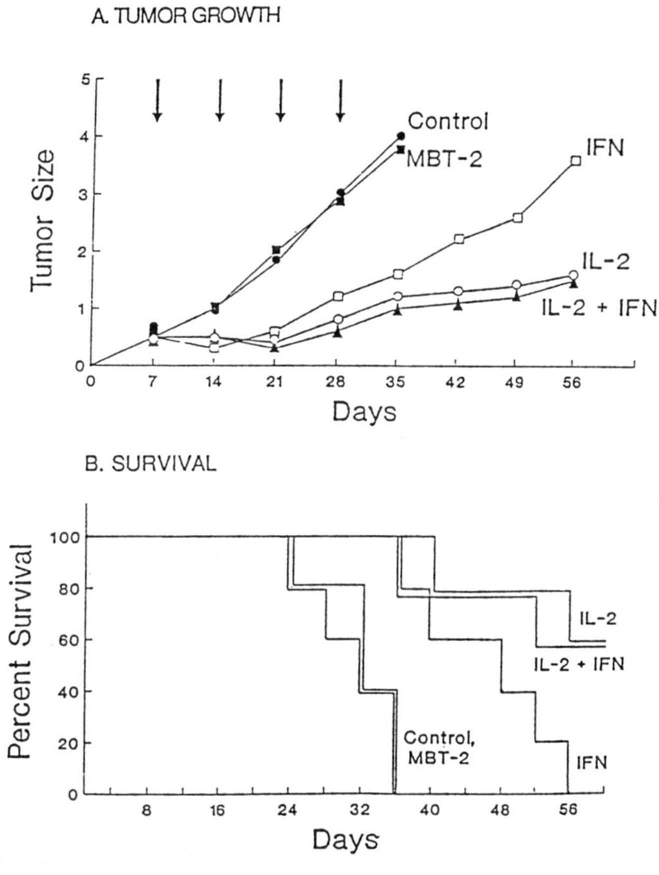

Fig. 11.2. *Immunotherapy of tumor bearing mice with inactivated IL-2 or IFN-γ gene modified MBT-2 cells.* Tumors were established in the bladder of C3H mice by intravesical instillation of 2×10^3 MBT-2 cells. Seven days later, animals (five mice per group) were injected i.p. with 5×10^6 X-irradiated cells that consisted of either unmodified MBT-2 cells, MBT/IL-2 cells, MBT/IFN-γ cells, or a combination of both MBT/IL-2 and MBT/IFN-γ cells. Injections were repeated weekly for a total of four times, as indicated by the arrows. Control mice were injected with PBS. (**A**) The origin and irregular shape of the bladder induced tumors prevented an accurate measure of tumor size. Tumors were, therefore, graded on a relative scale of 0-5: 0, no tumor; 0.5, palpable but not visible tumor; 1.0, small visible tumor; 2-4, increasingly larger tumors; and 5, abdomen full of tumor with erosion through the skin. Mice were killed when moribund. Average tumor size was calculated for each group of five or remaining mice. (**B**) Survival of mice >60 days correlated with complete tumor regression.

From animal models to cancer patients

The use of animal models that involve orthotopic tumor implantation is labour intensive, and treatment of animals with preestablished tumors generates results which are less dramatic. Nevertheless, the use of more stringent, though increasingly realistic, experimental systems adds to the confidence that the observations made in such animal studies would also apply to the human situation. A number of hurdles have to be overcome in clinical studies using cytokine gene modified tumor vaccines in cancer patients. One such hurdle is the availability of sufficient tumor tissue for vaccine preparations. In cancers in which the amount of tumor available post surgery or biopsy is limited, procedures will have to be developed, involving either ex vivo culture protocols or use of immunocompromised animals to generate sufficient material for vaccine preparations.

A second hurdle in the generation of human tumor vaccines is the gene transfer step, i.e., efficient introduction of the cytokine genes into the patient's tumor cells. So far, retroviral vectors have been the method of choice to introduce and express cytokine genes in tumor cells. The main feature that recommended their use was the high efficiency of gene transfer and stable insertion and expression of the cytokine in the target cells. Nevertheless, retroviral vectors constitute a rather complex experimental system, and their use is limited to actively dividing cells (which was not a problem in animal tumor models since tissue culture established tumor cell lines were used in these studies). Retroviral vectors may not be the method of choice in a clinical setting, primarily because of difficulties in establishing even short term cultures from primary tumor tissue. Moreover, since stable and long term expression of cytokines may not be a prerequisite in this instance (because the irradiated tumor vaccines are expected to persist in vivo for a limited period of time), alternative and simpler methods of gene transfer that would result in efficient, though transient, expression of the cytokine in the tumor cell may also be useful. Examples of alternative gene transfer approaches consist of physical methods employing liposomes, lipophilic agents, or a "gene gun", or viral vectors derived from adenoviruses or poxviruses. Since, unlike retroviruses, such gene transfer methods are not limited to actively dividing cells, primary tumor tissue obtained directly from the cancer patient without further ex vivo culture could also be used.

Animal models, realistic as they may be, can only provide general, though critically important, guidance as to what may work in man. There is no set of objective rules one can use to determine when preclinical data justify the initiation of clinical studies. A decision always remains, and properly so, a judgement call. Clearly, either delaying or prematurely engaging in costly clinical studies is detrimental to real progress (it is the latter practice which prevails at present).

References

Bubenik J, Perlmann P, Indrova M, Simova J, Jandlova T and Neuwirt J (1983). Growth inhibition of an MC-induced mouse sarcoma by TCGF (IL-2)-containing preparations. Cancer Immunol Immunother **14**, 205-206.

Bubenik J, Voitenok NN, Kieler J, Prassolov VS, Chumakov PM, Bubenikova D, Simova J and Jandlova T (1988). Local administration of cells containing an inserted IL-2 gene and producing IL-2 inhibits growth of human tumours in nu/nu mice. Immunol Lett **19**, 279-282.

Bubenik J, Simova J and Jandlova T (1989). Immunotherapy of cancer using local administration of lymphoid cells transformed by IL-2 cDNA and constitutively producing IL-2. Immunol Lett **23**, 287-292.

Connor J, Bannerji R, Saito S, Heston W, Fair W and Gilboa E (1993). Regression of bladder tumors in mice treated with interleukin 2 gene-modified tumor cells. J Exp Med **177**, 1127-1134.

DeLeo A, Shiku H, Takahashi T, John M and Old L (1977). Cell surface antigens of chemically induced sarcomas of the mouse. I. Murine leukemia virus related antigens and alloantigens on cultured fibroblasts and sarcoma cells: Description of a unique antigen on BALB/c Meth A sarcoma. J Exp Med **146**, 720-734.

Dranoff G, Jaffee E, Lazenby A, Golumbek P, Levitsky H, Brose K, Jackson V, Hamada H, Pardoll D and Mulligan R (1993). Vaccination with irradiated tumor cells engineered to secrete murine granulocyte-macrophage colony-stimulating factor stimulates potent, specific, and long-lasting anti-tumor immunity. Proc Natl Acad Sci USA **90**, 3539-3543.

Fearon ER, Pardoll DM, Itaya T, Golumbek P, Levitsky HI, Simons JW, Karasuyama H, Vogelstein B and Frost P (1990). Interleukin-2 production by tumor cells bypasses T helper function in the generation of an antitumor response. Cell **60**, 397-403.

Fidler IJ (1990). Critical factors in the biology of human cancer metastasis: Twenty-eighth G.H.A. Clowes Memorial Award Lecture. Cancer Res **50**, 6130-6138.

Forni G, Giovarelli M, Santoni A, Modesti A and Forni M (1986). Tumor inhibition by interleukin-2 at the tumour/host interface. Biochim Biophys Acta **865**, 307-327.

Gansbacher B, Zier K, Daniels B, Cronin K, Bannerji R and Gilboa E (1990a). Interleukin 2 gene transfer into tumor cells abrogates tumorigenicity and induces protective immunity. J Exp Med **172**, 1217-1224.

Gansbacher B, Bannerji R, Daniels B, Zier B, Cronin K and Gilboa E (1990b). Retroviral vector-mediated gamma-interferon gene transfer into tumor cells generates potent and long lasting antitumor immunity. Cancer Res **50**, 7820-7825.

Greenberg PD (1991). Adoptive T cell therapy of tumors: Mechanisms operative in the recognition and elimination of tumor cells. Adv Immunol **49**, 281-355.

Hamaoka T and Fujiwara H (1987). Phenotypically and functionally distinct T-cell subsets in anti-tumor responses. Immunol Today **8**, 267-269.

Herr HW, Badalament RA, Amato DA, Laudone VP, Fair WR and Whitmore WF, Jr (1989). Superficial bladder cancer treated with bacillus Calmette-Guerin: A multivariate analysis of factors affecting tumor progression. J. Urol **141**, 22-28.

Hewitt HB, Blake ER and Walder AS (1976). A critique of the evidence for active host defense against cancer, based on personal studies of 27 murine tumors of spontaneous origin. Br J Cancer **33**, 241-259.

Keene J and Forman J (1982). Helper activity is required for the in vivo generation of cytotoxic T lymphocytes. J Exp Med **155**, 768-782.

Lee RE, Lotze MT, Skibber JM, Tucker E, Bonow RO, Ognibene FP, Carrasquillo JA, Shelhamer JH, Parrillo JE and Rosenberg SA (1989). Cardiorespiratory effects of immunotherapy with interleukin-2. J Clin Oncol **7**, 7-20.

Livingston PO (1991). Active specific immunotherapy in the treatment of patients with cancer. In "Immunology and Allergy Clinics of North America: Human Cancer Immunology II" (ed HF Oettgen), pp 401-403. WB Saunders Company Ltd, London.

Middle JG and Embleton MJ (1981). Naturally arising tumors of the inbred WAB/Not rat strain. II. Immunogenicity of transplanted tumors. J Natl Cancer Inst **67**, 637-643.

Monaco JJ (1992). A molecular model of MHC class-I-restricted antigen processing. Immunol Today **13**, 173-179.

Porgador A, Tzehoval E, Katz A, Vadai E , Revel M, Feldman M and Eisenbach L (1992). Interleukin 6 gene transfection into Lewis lung carcinoma tumor cells suppresses the malignant phenotype and confers immunotherapeutic competence against parental metastatic cells. Cancer Res **52**, 3679-3686.

Porgador A, Bannerji R, Watanabe Y, Feldman M, Gilboa E and Eisenbach L (1993a). Anti-metastatic vaccination of tumor-bearing mice with two types of IFN-gamma gene-inserted tumor cells. J Immunol **150**, 1458-1470.

Porgador A, Gansbacher B, Bannerji R, Tzehoval E, Gilboa E, Feldman M and Eisenbach L (1993b). Anti-metastatic vaccination of tumor-bearing mice with IL-2-gene-inserted tumor cells. Int J Cancer **53**, 471-477.

Raghavan D, Shipley WU, Garnick MB, Russell PJ and Ritchie JP (1990). Biology and management of bladder cancer. N Engl J Med **322**, 1129-1138.

Rosenberg SA (1988). Immunotherapy of cancer using interleukin-2: Current status and future prospects. Immunol Today **9**, 58-61.

Rygaard J and Povlsen CO (1976). The nude mouse vs the hypothesis of immunological surveillance. Transplant Rev **28**, 43-61.

Soloway MS and Murphy WM (1979). Experimental chemotherapy of bladder cancer: Systemic and intravesical. Semin Oncol **6**, 166-183.

Townsend A and Bodmer H (1989). Antigen recognition by class I-restricted T lymphocytes. Ann Rev Immunol **7**, 601-624.

Van Pel A and Boon T (1982). Protection against a nonimmunogenic mouse leukemia by an immunogenic variant obtained by mutagenesis. Proc Natl Acad Sci USA **79**, 4718-4722.

Von Boehmer H and Haas W (1979). Distinct Ir genes for helper and killer cells in the cytotoxic response to H-Y antigen. J Exp Med **150**, 1134-1142.

Watanabe Y, Kuribayashi K, Miyatake S, Nishihara K, Nakayama E, Taniyama T and Sakata T (1989). Exogenous expression of mouse interferon gamma cDNA in mouse neuroblastoma C1300 cells results in reduced tumorigenicity by augmented anti-tumor immunity. Proc Natl Acad Sci USA **86**, 9456-9460.

Zinkernagel RM, Callahan GN, Althage A, Cooper S, Streilein JW and Klein J (1978). The lymphoreticular system in triggering virus plus self-specific cytotoxic T cells: Evidence for T help. J Exp Med **147**, 897-911.

Zinkernagel RM and Doherty PC (1979). MHC-restricted cytotoxic T cell: Studies on the biological role of polymorphic major transplantation antigens determining T cell restriction-specificity, function and responsiveness. Adv Immunol **27**, 51-106.

12

IL-2 IN THE INDUCTION AND EXPANSION OF TUMOR SPECIFIC IMMUNITY

D. J. Peace, W. Chen, M. L. Disis and M. A. Cheever
Division of Hematology/Oncology, Loyola University Chicago, Maywood, IL, USA and Division of Medical Oncology, University of Washington, Seattle, WA, USA

Introduction

The cardinal function of the T-cell immune system is the preservation of self integrity. It has been postulated that one of the most important aspects of T-cell function is the selective recognition and elimination of host cells that become altered as a consequence of malignant transformation (Burnet, 1970). The theory of immunological surveillance implies that, to become fully established, tumors by necessity employ strategies to evade or suppress host T-cell immunity. The primary goal of tumor immunology is to develop effective counter strategies in order to elicit or augment specific immune responses capable of eradicating established malignancies. The impetus for efforts to identify and utilize tumor specific T-cell immune responses to treat human malignancies derives from:

- the demonstration in preclinical models that tumor specific T cells exist and can be used to mediate tumor therapy
- the development through recombinant DNA technology of an expanding pharmacopoeia of T-cell growth factors that demonstrably potentiate T-cell mediated immunotherapy
- the recent advent of successful methods for the isolation and characterization of tumor specific antigens, recognizable by human T cells.

The first convincing evidence for the existence of tumor specific antigens capable of mediating tumor rejection in vivo emerged with the establishment of syngeneic tumor transplant rejection models (Foley, 1953; Baldwin, 1955; Prehn and Main, 1957). Subsequent studies performed in a number of animal tumor models documented the presence of a wide assortment of immunogenic tumor antigens (Klein et al., 1960; Old et al., 1962; Prehn, 1962; Globerson and Feldman, 1964). However, for many years intensive efforts to characterize the nature of antigens responsible for mediating tumor rejection largely proved unsuccessful. Notwithstan-

CYTOKINE-INDUCED TUMOR IMMUNOGENICITY
ISBN 0-12-261520-4

ding the paucity of precisely characterized tumor antigens, it was possible for investigators to demonstrate in various models that thymic derived lymphocytes (T cells) are responsible for specific recognition of tumor antigens. In addition, T cells have been shown to have important direct and indirect effector functions during tumor elimination (Cerottini and Brunner, 1974; Greenberg, 1991). The preeminent role of the T cell in tumor specific immunity has prompted the exploration of the potential use of tumor specific T cells as reagents for the treatment of established syngeneic tumors in vivo.

The development of treatment strategies based on tumor specific T cells has been facilitated by the discovery and production in recombinant form of T-cell growth factors, particularly IL-2. To date, the principles for generating and utilizing antigen specific T cells in conjunction with IL-2 to augment specific antitumor immunity have been developed to the extent that attempts to treat human malignancies with "antigen specific" T cells are both warranted and feasible. However, the clinical application of T-cell immunotherapy has been constrained by slow progress in identifying and characterizing antigenic determinants on human tumors that are appropriate for directing T-cell immune responses. Many human studies have demonstrated that T cells specifically reactive to autologous tumors can be isolated and grown from the autologous tumor, regional lymph nodes and/or peripheral blood, indicating that many different types of human tumors express distinguishing antigenic determinants (Vose and Bonnard, 1982; Mitsuya et al., 1983; Slovin et al., 1986; Cozzolino et al., 1987; Itoh et al., 1988; Anichini et al., 1989). However, few successful efforts have been made in isolating and defining the molecular nature of putative tumor specific determinants on human tumors (Knuth et al., 1989; Boon, 1993).

Recent conceptual advances concerning the nature of antigens which are recognized by immune T cells have illuminated the search for potential tumor specific determinants. It is now understood that antigens recognized by T cells are not proteins but derivative peptide fragments bound to class I and class II MHC gene molecules (Germain, 1986; Berzofsky et al., 1988; Grey et al., 1989; Townsend and Bodmer, 1989). Peptide antigens may be derived either from cellular proteins that are processed internally without exposure of the intact protein to the extracellular milieu or from secreted proteins that are endocytosed and processed by specialized antigen presenting cells. Thus, virtually any protein which is synthesized within an organism is potentially subject to the scrutiny of the T-cell immune system.

Many tumors express abnormal oncogenic proteins that result from the mutation of protooncogenes and that contribute directly to the establishment and maintenance of the malignant phenotype. We have examined the hypothesis that mutated oncogenic proteins might be immunogenic to the T-cell immune system and, thus, might serve as "ideal" tumor specific antigens that are shared amongst many different tumors. The current article will first review animal studies from our laboratory that have established a number of basic tenets for the use of tumor specific T cells and IL-2 for cancer therapy, and will subsequently review recent studies to elucidate the conditions wherein mutated oncogenic proteins might serve as targets for T-cell mediated therapy.

Murine studies evaluating the treatment of disseminated tumors with specifically immune T cells and/or IL-2

Adoptive immunotherapy with tumor specific T cells

The demonstrated capacity of antigenic tumors to become established in immunocompetent hosts attests to the ability of tumors to defeat host immune responses. A number of ploys have been discovered whereby tumors avoid immunological destruction (North, 1985; Mizoguchi et al., 1992; Restifo et al., 1993). Tumors frequently prevent or limit the generation of a quantitatively sufficient immune response. The principal determinant of the magnitude of an immune response is the number of antigen specific T cells that can be recruited to the site of antigen. Thus, one potential strategy to strengthen inadequate host immune responses to intrinsically antigenic tumors is to increase the number of relevant tumor specific T cells within a tumor bearing host by adoptive cellular therapy.

The cotransfer of tumor specific immune T cells with tumor cells in Winn type assays effectively prevents the establishment of tumor in many different models. However, it has been much more difficult to show that adoptive transfer of immune T cells can eliminate or prevent the growth of tumors already well established in vivo prior to adoptive T-cell therapy. Initial demonstrations that transferred immune T cells could mediate the elimination of established malignancies were achieved by the use of chemotherapeutic "conditioning" regimens to prepare tumor bearing hosts for adoptive cell therapy (Fefer, 1969). Our laboratory has extensively studied the requirements for effective adoptive chemoimmunotherapy (ACIT). These studies have focused upon the transplantable Friend virus induced leukemia line, FBL-3, as a prototypical tumor model (Fefer et al., 1976).

FBL-3 is a highly virulent, antigenic murine tumor line. As few as 100 FBL-3 cells injected i.p. will rapidly disseminate and kill syngeneic C57BL/6 mice. Mice inoculated with a lethal dose of FBL-3 leukemia cells on day 0 can be cured by the subsequent administration of cyclophosphamide (180 mg/kg) on day 5 followed in 6 h by specifically immune syngeneic lymphocytes (Fefer et al., 1976). The interval between the administration of chemotherapy and immune T cells ensures adequate clearance of the drug and, thus, viability of the donor T cells. The administration of cyclophosphamide prior to cell transfer is essential for effective ACIT, even though the chemotherapy doses are noncurative with only limited direct effects upon the tumor. More importantly, the chemotherapy appears to condition the host for successful engraftment of the transferred T cells. The principal mechanism by which chemotherapy potentiates adoptive T cell therapy is unknown, but may involve inhibition of suppressor cells, creation of "space" for the transferred cells and/or the induction of host derived growth factors that potentiate engraftment.

The therapeutic efficacy of ACIT against FBL-3 is exquisitely tumor specific (Cheever et al., 1980) and occurs after the tumor has become widely disseminated within the host. The antitumor effect is proportionate to the number of transferred cells with increasing numbers of donor T cells achieving greater percentages of cures. Both class I MHC restricted and class II MHC restricted T-cell subsets have the capacity to specifically recognize FBL-3 tumor cells and to independently mediate tumor eradication in ACIT (Greenberg, 1991). The donor T cells must be viable and capable of prolonged survival in vivo to effectively eradicate established FBL-3 (Fefer et al., 1976). Abrogation of transferred immunity by selective T-cell depletion as long as 1 month after adoptive cell transfer can lead to recurrent, fatal disease (Greenberg et al., 1980). These findings imply that complete eradication of FBL-3 is a protracted phenomenon reflecting the temporal requirements for T cells to migrate to and mediate the destruction of every vestige of tumor.

As humans rarely have access to syngeneic donors who can provide T cells sensitized to nominal tumor antigens, consideration has been given to the development of alternative sources of immune tumor specific T cells for clinical application. In vitro culture techniques have evolved to the point that it is feasible to sensitize autologous T cells to a nominal antigen and to expand them to sufficient numbers for therapy. In preclinical studies long term cultured T cells expanded in vitro by periodic stimulation with tumor antigen retain the ability to specifically recognize antigen (Cheever et al., 1986). In vivo transfer of antigen driven cultured T cells can augment antigen specific T-cell immunity to a greater extent than is possible by active immunization of the host. Moreover, such cells retain normal T-cell functions in vivo including the ability to migrate widely, to respond specifically to antigen, to mediate tumor specific therapy and to provide long term immunological memory (Cheever et al., 1986).

Cloned T cells with specificity to FBL-3 have been successfully expanded and used in ACIT to cure mice with disseminated disease (Klarnet et al., 1987). Thus, the progeny of a single tumor reactive T cell - if expanded to large numbers in vitro under conditions that allow retention of the ability to respond subsequently to antigen - can eradicate tumor and survive in vivo in sufficient numbers to mediate sustained tumor specific immunological memory.

The use of IL-2 to potentiate the therapeutic efficacy of tumor specific T cells

Though technically possible, the in vitro expansion of antigen specific T cells by periodic stimulation with antigen alone is so laborious and time consuming as to be impractical. However, the use of recombinant T-cell growth factors, particularly IL-2, during T-cell culture routinely enables the enormous numbers of T cells required for adoptive therapy to be generated in a timely fashion. The elucidation of the normal physiological role of IL-2 has contributed greatly to the optimal utilization of IL-2 to enhance T-cell proliferation and expansion in vitro, as well as in vivo.

Mature, resting T cells are activated to proliferate by two sequential signals. The primary signal results from the interaction of antigen with clonotypic TCR in the context of appropriate accessory molecular interactions.

The second signal emanates from the interaction of IL-2 with specialized high affinity receptors, designated IL-2R. The expression of IL-2R on T cells is strictly dependent on the engagement of TCR by specific antigen (Cantrell and Smith, 1983). Once activated by antigen, T cells remain sensitive to the trophic effects of IL-2 for a limited period of time, approximately 1 to 2 weeks (Cantrell and Smith, 1983). TCR are internalized following antigenic stimulation, promptly terminating the inductive signal for IL-2R expression. Eventually, IL-2R expression wanes and T cells stop proliferating in response to IL-2. The cells become quiescent and remain refractory to further stimulation by either antigen or IL-2 until functional TCR are reexpressed (Fathman and Frelinger, 1983; Wilde and Fitch, 1984).

Although antigen engagement of TCR initiates a specific T-cell immune response, the magnitude of the response ultimately is determined by the amount of IL-2 available to drive T-cell proliferation during the interlude of IL-2R expression (Smith, 1993). IL-2 is transiently produced by Th cells following antigen stimulation (Gillis et al., 1978). The amount of IL-2 produced in response to antigen is limiting. Consequently, the amplitude of a given T-cell immune response can be augmented, in terms of increased proliferation and cell number, by the administration of supplemental IL-2 at appropriate intervals following antigen stimulation.

The principles governing the physiological effects of IL-2 have been applied to the development of IL-2 as a reagent to augment tumor specific T-cell immunotherapy. The growth of tumor specific T cells in vitro can be greatly expedited by the addition of exogenous IL-2 following antigen activation. The timing and concentration of supplemental IL-2 are critical for optimal expansion of tumor reactive T cells. Excessive or prolonged exposure of T cells to IL-2 can disrupt T-cell function and foster dependency on IL-2 for continued survival (Cheever et al., 1984; Lenardo, 1991).

However, judicious application of exogenous IL-2 can markedly increase the yield of functional T cells that can persist in vitro for extended periods in the absence of either tumor antigen or IL-2, yet proliferate in response to specific stimulation by tumor.

Importantly, T cells grown in vitro with intermittent antigen and brief courses of low dose IL-2 function normally in vivo as evidenced by the retention of the ability to migrate widely, to proliferate in response to tumor, to mediate immunotherapy and to survive long term, thus providing effective immunologic memory (Cheever et al., 1981).

IL-2 has also been developed as a pharmacologic agent for adjuvant use with adoptive immunotherapy. The administration of IL-2 to induce the growth and augment the overall activity of donor T cells in vivo has become a well established principle. Cultured tumor specific T cells adoptively transferred to tumor bearing animals are therapeutically more effective if given in conjunction with exogenous IL-2 (Cheever et al., 1982). Short courses of IL-2 (up to 7 days) have been shown to increase the number and antitumor activity of donor tumor specific T cells in vivo, proportionate to the level of IL-2 administered and sustained (Cheever et al., 1984; Cheever et al., 1985). Moreover, sufficient doses of IL-2 can obviate the need for conditioning chemotherapy prior to adoptive cellular therapy, presumably by providing an adaptive growth advantage to the donor T cells (unpublished observation).

The rules that govern the actions of exogenous IL-2 on cultured T cells in vitro can be used to expand T cells in vivo. Thus, optimal growth of donor T cells in vivo with exogenous IL-2 is achieved by intermittent injection of specific antigen in vivo followed by short courses of IL-2 (Chen et al., 1990).

Periodic stimulation with antigen and IL-2 can support donor tumor specific T cells at expanded levels with retention of specific therapeutic activity for extended periods of time (Chen et al., 1990). The regimen of exogenous IL-2 employed has been shown to be critical to the outcome of therapy. Prolonged or excessive courses of IL-2 do not increase the efficacy of transferred cultured T cells and may actually impair their antitumor efficacy.

The double edged nature of IL-2 as a pharmacological reagent is well known. IL-2 can potentiate antitumor T-cell immune responses, but it can also induce myriad toxic effects which most likely do not contribute to the antitumor effects of IL-2.

The most problematical toxicities observed in animal models and clinical trials have been the development of hypotensive shock and a pronounced vascular leak syndrome (Lotze et al., 1985; Rosenstein et al., 1986; Rosenberg et al., 1987). The NK cell subpopulation appears to be the principal cellular mediator of IL-2 mediated toxicity (Peace and Cheever, 1989). In contrast to T cells, NK cells constitutively express functional IL-2R, mostly of intermediate affinity (Siegel et al., 1987; Kehrl et al., 1988; Caligiuri et al., 1990). High dose IL-2 administered in vivo can activate intermediate IL-2R bearing NK cells, thereby recruiting approximately 10% of all peripheral blood mononuclear cells (Smith, 1993).

IL-2 induces NK cells to release various cytokines, notably TNF-α, which have been implicated as proximate mediators of IL-2 induced toxicity (Mier et al., 1988; Fraker et al., 1989). Fortunately, the doses of IL-2 that optimally stimulate antigen reactive T cells in vivo are far below threshold levels needed to activate the majority of host NK cells. Thus, marked adjuvant stimulation of antigen specific T cells can be procured without triggering substantial systemic toxicity.

Studies evaluating the antigenicity of mutated oncogenic proteins

Characteristics of mutated oncogenic proteins as potential tumor associated antigens

The principles and methods for generating and utilizing tumor specific T cells have been defined. The major remaining issue for prospectively testing adoptive T-cell therapy in humans is the determination of antigens that are not only within the host T-cell repertoire but also appropriate targets for selective tumor cell destruction. Theoretically, any protein that is expressed within an organism might be processed and presented to host T cells. In general, a host is tolerant to peptides derived from normal self proteins; by contrast, foreign peptides or self peptides altered by mutation may be recognized as alien antigens.

Mutations of somatic cell genes are considered to be common primary events in the transformation of normal cells to malignant cells. A small class of cancer related genes, termed protooncogenes, has been identified which is frequently altered by mutation in transformed cells. The mutations, characterized by point mutations, translocations, truncations and inversions, often result in the expression of structurally altered proteins that function inappropriately and result in malignant transformation. Potentially, mutated proteins responsible for malignant transformation are antigenic to the host T-cell immune system.

There are a number of attributes which commend mutated oncogenic proteins as candidate tumor antigens. First, by definition, the proteins are specifically associated with transformed cells, i.e. the mutated proteins are tumor specific. Second, the proteins are intimately linked to the transforming process itself and may be indispensable for the maintenance of the malignant phenotype. Thus, potential antigenic determinants created by activating mutations may be especially vulnerable sites for immunological attack. Third, activating mutations of protooncogenes are often highly characteristic and predictable. The same aberrant protein containing the same mutation related determinant may be "shared" by many different tumors.

Finally, particular oncogenic mutations are highly prevalent across a wide spectrum of tumor types. Epitopes engendered by the most prevalent oncogenic proteins might be suitable for the induction of prophylactic immune responses, i.e. used as tumor vaccines. Recent investigations in our laboratory have examined whether mutated oncogenic proteins are immunogenic to the T-cell immune system, i.e. whether transforming proteins can serve as tumor specific antigens. Initial studies have focused on two prototypical oncogenes which become activated by characteristic mutations to encode abnormal oncogenic proteins, namely *ras* and *bcr-abl*.

T-cell recognition of oncogenic p21ras proteins

The mammalian genome contains at least three *ras* protooncogene homologs, denoted as *H-ras-1, K-ras-2,* and *N-ras,* which are located on chromosomes 11, 12 and 1, respectively, and which encode a highly conserved family of 21 kDa proteins collectively designated as p21*ras* (Barbacid, 1987).

The p21*ras* proteins are G-binding proteins with signal transduction functions that regulate cell growth and differentiation (Bourne et al., 1990). A wide variety of malignancies are associated with characteristic point mutations of the *ras* protooncogenes, usually within codon 12, 13 or 61 (Barbacid, 1987; Guerrero and Pellicer, 1987; Bos, 1989). The mutations result in single amino acid substitutions at the corresponding residues of p21*ras* and confer transforming activity. Approximately 20% of all human tumors harbor *ras* genes activated by specific mutations (Barbacid, 1987; Bos, 1989). The same characteristic mutations of *ras* are detected amongst different patients with different kinds of tumors.

Studies performed in our laboratory to determine whether activated p21*ras* proteins bearing single amino acid substitutions are potentially immunogenic have established that it is possible to elicit specific T-cell immunity to mutated p21*ras* by immunization with synthetic peptides corresponding to the mutated portion of the p21*ras* protein (Peace et al., 1991). For example, T cells from C57BL/6 mice immunized with a synthetic peptide corresponding to the amino acid sequence of p21*ras* from residue 5-16 with the aberrant substitution of arginine for glycine normally resident at "residue 12" (termed Arg-12) proliferated in response to the Arg-12 *ras* peptide but not to the normal homologous peptide with glycine at residue 12 or other homologous peptides with alternative residue 12 substitutions.

Multiple strains of mice representing various inbred H-2 haplotypes were able to mount highly specific T-cell proliferative responses to various mutated synthetic *ras* peptides, including peptides encompassing single amino acid substitutions at the other two common activation sites of p21*ras*, residue 13 and residue 61. Not all strains of mice responded specifically to all of the substituted peptides; however, most strains responded specifically to the majority of aberrant peptides tested. Similar specific T-cell responses to mutated *ras* peptides have also been detected in humans (Jung and Schluesener, 1991; Gedde-Dahl et al., 1992; and unpublished observations). Together, the results indicate that the mammalian T-cell repertoire against the common activating mutations of p21*ras* is quite extensive and exquisitely specific.

From a therapeutic standpoint, the critical question is whether or not mutated oncogenic p21*ras* proteins containing defined T-cell epitopes and expressed by malignant cells can be recognized by T cells and can stimulate effective immunologic responses. In general, T cells recognize peptide fragments excised from intact proteins and bound to MHC molecules.

The ability to generate T cells specific for a mutated *ras* peptide need not necessarily mean that the parental oncogenic p21*ras* protein is similarly immunogenic. Inherently antigenic peptides embedded within a larger protein may not be recognized by T cells due to inadequate processing or the presence of competing or inhibitory domains within the same protein (Gammon et al., 1987; Adorini and Nagy, 1990; Perkins et al., 1991).

In general, processed antigenic determinants are presented in the context of either class I or class II MHC molecules. In our studies, all responder T cells elicited by in vivo immunization with synthetic *ras* peptides were classical CD4+ class II MHC restricted T cells. More recently, we have been able to elicit CD8+ class I MHC restricted T cells with specificity for single amino acid substituted *ras* peptides by immunizing unprimed murine lymphocytes in vitro with synthetic peptides constructed to incorporate defined class I MHC binding motifs (Falk et al., 1991; Peace et al., 1992).

The processing of antigenic proteins for presentation by class I versus class II MHC molecules is quite different (Grey et al., 1989; Townsend and Bodmer, 1989). However, we have observed that an intact, mutated p21*ras* protein (189 amino acids in length) can be appropriately processed and presented to both class I MHC restricted and class II MHC restricted T cells specific for the mutated segment of p21*ras* (Peace et al., 1991; Peace et al., 1992).

An important difference between class I and class II MHC restricted determinants is that the former are generally derived from endogenous cellular proteins, whereas the latter are derived from exogenous secreted proteins (Germain, 1986; Grey et al., 1989; Townsend and Bodmer, 1989). Normally the p21*ras* protein is bound to the inner aspect of the cell membrane and, therefore, may be sequestered from the class II MHC restricted T-cell immune system.

However, in tumor bearing hosts considerable amounts of soluble p21*ras* protein are released into the extracellular environment in vivo (Niman et al., 1985; Hamer et al., 1991). Thus, oncogenic p21ras proteins expressed by tumors in vivo are likely to be available to both class I and class II MHC restricted antigen processing pathways. To date, T-cell immune responses to mutated p21ras proteins have been examined in the context of a limited number of MHC haplotypes. It is highly improbable that all oncogenic p21*ras* proteins will be immunogenic in all circumstances. However, the results predict that, at least in some circumstances, mutated p21*ras* proteins expressed by tumor cells in vivo might be appropriately processed and presented in a fashion capable of directing specific CD4+ and/or CD8+ T-cell immune responses. It remains to be determined whether an endogenously expressed oncogenic p21*ras* protein can effectively target specific immunological attack. Ongoing studies are examining the conditions whereby class I and class II MHC restricted T cells specific for mutated oncogenic p21*ras* protein can mediate tumor therapy in vivo.

T-cell recognition of the p210 fusion protein associated with chronic myelogenous leukemia

Major genetic dislocations including translocations, intragenic inversions and truncations can lead to oncogene activation. In many instances the dislocation results in the expression of hybrid or scrambled proteins with aberrant and, thus, potentially immunogenic amino acid sequences. For example, in greater than 95% of patients with chronic myelogenous leukemia the c-*abl* gene on chromosome 9 is translocated to the break point cluster region of chromosome 22 creating a fusion gene designated *bcr-abl* (reviewed in Deisseroth and Arlinghaus, 1991).

The translocation site is intronic and typically aligns the c-*abl* gene in frame with either the second or third exon of the *bcr* gene (Kurzrock et al., 1987; Canaani et al., 1991). The resulting fusion genes encode one of two major 210 kDa proteins consisting of 1004 amino acids encoded by c-*abl*, a unique amino acid at the fusion site and either 902 or 927 amino acids encoded by *bcr*, respectively designated p210*bcr*3-*abl*2 and p210*bcr*2-*abl*2.

We have adapted the same strategy used to characterize the immunogenicity of p21*ras* proteins to assess the immunogenicity of p210*bcr-abl*. Initial studies demonstrated that synthetic peptides corresponding to the fusion region of one of the two common p210*bcr-abl* proteins can elicit highly specific T-cell immune responses (Chen et al., 1992).

BALB/c mice immunized with a synthetic peptide corresponding to the junction region of p210*bcr*3-*abl*2 developed T cells that proliferated specifically in response to the unique amino acid sequence created by the fusion. T cells elicited by in vivo immunization with synthetic *bcr*3-*abl*2 peptide exhibited the CD4+ cell surface phenotype and were class II MHC restricted. Importantly, the peptide induced T cells specifically recognized the corresponding p210*bcr*3-*abl*2 protein presented by syngeneic antigen presenting cells in vitro (Chen et al., 1992).

P210*bcr-abl* is normally found in the cytosol and is not considered to be a secreted protein. However, recent studies indicate that p210*bcr-abl* protein can be detected in spent culture supernatants (Chen et al., 1992), implying that the protein may be released to the external cell environment as the result of cell damage or death (Chen et al., 1992). Therefore, it is plausible that the p210*bcr-abl* protein might be available for class II MHC restricted presentation by antigen presenting cells in the proximate tumor microenvironment in vivo as the result of normal cell turnover or interventions such as cytotoxic chemotherapy.

An alternative approach to focus specific immune responses against malignant cells expressing the p210*bcr-abl* protein would be to develop CD8+ class I MHC restricted cytolytic responses. It has been possible to generate murine CD8+ T-cell clones with specificity for the junction region of p210*bcr-abl* by priming BALB/c mice with a corresponding synthetic peptide, then periodically restimulating primed lymphocytes with the sensitizing peptide and syngeneic antigen presenting cells in vitro (Chen et al., 1993).

More recently, we have observed that appropriately constructed peptides corresponding to the *bcr-abl* fusion region can bind to the human class I MHC molecule, HLA-A2. Therefore, it is possible that similar *bcr-abl* specific CTL can be generated in humans. As in the case of *ras* specific T cells, the therapeutic potential of both class I MHC restricted and class II MHC restricted T cells with specificity for the fusion domain of p210bcr-*abl* is a focus of current investigations.

Conclusions

The principles and methods for successful adoptive T-cell immunotherapy of established malignancies have been established in extensive preclinical studies. The judicious use of recombinant T-cell growth factors, most notably IL-2, has played a central role in the development of adoptive therapy techniques that can be applied at the clinical level.

A major obstacle to the widespread use of tumor specific adoptive immunotherapy has been the lack of defined tumor associated antigens appropriate for directing therapy. Increasingly it appears that in many circumstances the mutated proteins responsible for malignant transformation may themselves represent ideal tumor specific antigens for recognition by the T-cell immune system. Analytical techniques are rapidly emerging which should enable the accurate prediction of putative antigenic determinants that may arise as the result of a particular oncogenic mutation within a particular individual (Falk et al., 1991). Accordingly, the routine generation of "customized" tumor specific T cells for the treatment of human malignancies may be imminent.

Acknowledgments

This work was supported by U.S. Public Health Service grants R37CA43081, R01CA5456102 and K08CA01299 from the National Cancer Institute.

References

Adorini L and Nagy ZA (1990). Peptide competition for antigen presentation. Immunol Today **11**, 21-24.

Anichini A, Mazzocchi A, Fossati G and Parmiani G (1989). Cytotoxic T lymphocyte clones from peripheral blood and from tumor site detect intratumor heterogeneity of melanoma cells: Analysis of specificity and mechanisms of interaction. J Immunol **142**, 3692-3701.

Baldwin RW (1955). Immunity to methylcholanthrene-induced tumours in inbred rats following atrophy and regression of the implanted tumours. Br J Cancer **9**, 652-657.

Barbacid M (1987). Ras genes. Ann Rev Biochem **56**, 779-827.

Berzofsky JA, Brett SJ, Streicher HZ and Takahashi H (1988). Antigen processing for presentation to T lymphocytes: Function, mechanisms and implications for the T cell repertoire. Immunol Rev **106**, 5-31.

Boon T (1993). Teaching the immune system to fight cancer. Sci Am **266**, 3, 32-39.

Bos JL (1989). Ras oncogenes in human cancer: A review. Cancer Res **49**, 4682-4689.

Bourne HR, Sanders DA and McCormick F (1990). The GTPase superfamily: A conserved switch for diverse cell functions. Nature **348**, 125-132.

Burnet FM (1970). The concept of immunological surveillance. Progr Exp Tumor Res **13**, 1-27.

Caligiuri MA, Zmuidzinas A, Manley TJ, Levin H, Smith KA and Ritz J (1990). Functional consequences of IL-2 receptor expression on resting human lymphocytes: Identification of a novel NK cell subset with high affinity receptors. J Exp Med **171**, 1509-1526.

Canaani E, Marcelle C, Fainstein E, Gale RP, Croce C, Prokocimer M and Berrebi A (1991). In "Chronic Myelogenous Leukemia: Molecular Approaches to Research and Therapy" (eds A Deisseroth and RB Arlinghaus), pp 217-240. Marcel Dekker, New York.

Cantrell DA and Smith KA (1983). Transient expression of interleukin-2 receptors: Consequences for T cell growth. J Exp Med **158**, 1895-1911.

Cerottini JC and Brunner KT (1974). Cell-mediated cytotoxicity, allograft rejection and tumor immunity. Adv Immunol **18**, 67-129.

Cheever MA, Greenberg PD and Fefer A (1980). Specificity of adoptive chemoimmunotherapy of established syngeneic tumors. J Immunol **125**, 711-714.

Cheever MA, Greenberg PD and Fefer A (1981). Specific adoptive therapy of established leukemia with syngeneic lymphocytes sequentially immunized in vivo and in vitro and nonspecifically expanded by culture with interleukin-2. J Immunol **126**, 1318-1322.

Cheever MA, Greenberg PD, Fefer A and Gillis S (1982). Augmentation of the anti-tumor therapeutic efficacy of long-term cultured T lymphocytes by in vivo administration of purified interleukin 2. J Exp Med **155**, 968-980.

Cheever MA, Greenberg PD, Irle C, Thompson JA, Urdal DL, Mochizuki DY, Henney CS and Gillis S (1984). Interleukin 2 administered in vivo induces the growth of cultured T cells in vivo. J Immunol **132**, 2259-2265.

Cheever MA, Thompson JA, Kern DE and Greenberg PD (1985). Interleukin 2 (IL 2) administered in vivo: Influence of IL 2 route and timing on T cell growth. J Immunol **134**, 3895-3900.

Cheever MA, Thompson DB, Klarnet JP and Greenberg PD (1986). Antigen-driven long-term cultured T cells proliferate in vivo, distribute widely, mediate specific tumor therapy and persist long-term as functional memory T cells. J Exp Med **163**, 1100-1112.

Chen W, Reese VA and Cheever MA (1990). Adoptively transferred antigen-specific T cells can be grown and maintained in large numbers in vivo for extended periods of time by intermittent restimulation with specific antigen plus IL-2. J Immunol **144**, 3659-3666.

Chen W, Peace DJ, Rovira DK, You SG and Cheever MA (1992). T cell immunity to the joining region of p210BCR-ABL protein. Proc Natl Acad Sci USA **89**, 1468-1472.

Chen W, You SG, Disis ML and Cheever MA (1993). Evaluation of the joining region segment of p210BCR-ABL chimeric protein as a potential leukemia-specific antigen to elicit class I MHC-restricted CTL responses. J Cell Biochem Suppl **17D**, 107.

Cozzolino F, Torcia M, Carossino AM, Giordani R, Selli C, Talini C, Reali E, Novelli A, Pistoia V and Ferrarini M (1987). Characterization of cells from invaded lymph nodes in patients with solid tumors: Lymphokine requirement for tumor-specific lymphoproliferative response. J Exp Med **166**, 303-318.

Deisseroth A and Arlinghaus RB (1991). "Chronic Myelogenous Leukemia: Molecular Approaches to Research and Therapy". Marcel Dekker, New York.

Falk K, Rötzschke O, Stevanovic S, Jung G and Rammensee HG (1991). Allele-specific motifs revealed by sequencing of self-peptides eluted from MHC molecules. Nature **351**, 290-296.

Fathman CG and Frelinger JG (1983). T-lymphocyte clones. Ann Rev Immunol **1**, 633-655.

Fefer A (1969). Immunotherapy and chemotherapy of Moloney sarcoma virus-induced tumors in mice. Cancer Res **29**, 2177-2183.

Fefer A, Einstein AB Jr, Cheever MA and Berenson JR (1976). Models for syngeneic adoptive chemoimmunotherapy of murine leukemias. Ann NY Acad Sci **276**, 573-583.

Foley EJ (1953). Antigenic properties of methylcholanthrene-induced tumors in mice of the strain of origin. Cancer Res **13**, 835-837.

Fraker DL, Langstein HN and Norton JA (1989). Passive immunization against tumor necrosis factor partially abrogates interleukin 2 toxicity. J Exp Med **170**, 1015-1020.

Gammon G, Shastri N, Cogswell J, Wilbur S, Sadegh-Nasseri S, Krzych U, Miller A and Sercarz E (1987). The choice of T-cell epitopes utilized on a protein antigen depends on multiple factors distant from, as well as at, the determinant site. Immunol Rev **98**, 53-73.

Gedde-Dahl T III, Eriksen JA, Thorsby E and Gaudernack G (1992). T cell responses against products of oncogenes: Generation and characterization of human T cell clones specific for p21 ras-derived synthetic peptides. Hum Immunol **33**, 266-274.

Germain RN (1986). The ins and outs of antigen processing and presentation. Nature **322**, 687-689.

Gillis S, Ferm MM, Ou W and Smith KA (1978). T cell growth factor: Parameters of production and a quantitative microassay for activity. J Immunol **120**, 2027-2032 .

Globerson A and Feldman M (1964). Antigenic specificity of benzo[a]pyrene-induced sarcomas. J Natl Cancer Inst **32**, 1229-1243.

Greenberg PD, Cheever MA and Fefer A (1980). Detection of early and delayed anti-tumor effects following curative adoptive chemoimmunotherapy of established leukemias. Cancer Res **40**, 4428-4432.

Greenberg PD (1991). Adoptive T cell therapy of tumors: Mechanisms operative in the recognition and elimination of tumor cells. Adv Immunol **49**, 281-355.

Grey HM, Sette A and Buus S (1989). How T cells see antigen. Sci Am **261**, 11, 56-64.

Guerrero I and Pellicer A (1987). Mutational activation of oncogenes in animal model systems of carcinogenesis. Mutation Res **185**, 293-308.

Hamer PJ, LaVecchio J, Ng S, DeLellis R, Wolfe H and Carney WP (1991). Activated V12 ras p21 in cell culture fluids and mouse plasma. Oncogene **6**, 1609-1615.

Itoh K, Platsoucas CD and Balch CM (1988). Autologous tumor-specific cytotoxic T lymphocytes in the infiltrate of human metastatic melanomas: Activation by interleukin-2 and autologous tumor cells and involvement of the T cell receptor. J Exp Med **168**, 1419-1441.

Jung S and Schluesener HJ (1991). Human T lymphocytes recognize a peptide of single point-mutated, oncogenic ras proteins. J Exp Med **173**, 273-276.

Kehrl JH, Dukovich M, Whalen G, Katz P, Fauci AS and Greene WC (1988). Novel interleukin 2 (IL-2) receptor appears to mediate IL-2-induced activation of natural killer cells. J Clin Invest **81**, 200-205.

Klarnet JP, Matis LA, Kern DE, Mizuno MT, Peace DJ, Thompson JA, Greenberg PD and Cheever MA (1987). Antigen-driven T cell clones can proliferate in vivo, eradicate disseminated leukemia and provide specific immunologic memory. J Immunol **138**, 4012-4017.

Klein G, Sjogren HO, Klein E and Hellström KE (1960). Demonstration of resistance against methycholanthrene-induced sarcomas in the primary autochthonous host. Cancer Res **20**, 1561-1572.

Knuth A, Wolfel T, Klehmann E, Boon T and Meyer zum Buschenfelde KH (1989). Cytolytic T cell clones against an autologous human melanoma: Specificity study and definition of three antigens by immunoselection. Proc Natl Acad Sci USA **86**, 2804-2808.

Kurzrock R, Kloetzer WS, Talpaz M, Block M, Walters R, Arlinghaus RB and Gutterman JU (1987). Identification of molecular variants of p210bcr-abl in chronic myelogenous leukemia. Blood **70**, 233-236.

Lenardo MJ (1991). Interleukin-2 programs mouse T lymphocytes for apoptosis. Nature **353**, 858-861.

Lotze MT, Matory YL, Ettinghausen SE, Rayner AA, Sharrow SO, Seipp CAY, Custer MC and Rosenberg SA (1985). In vivo administration of purified human interleukin 2. J Immunol **135**, 2865-2875.

Mier JW, Vachino G, Vander Meer JWM, Numerof RP, Adams S, Cannon JG, Bernheim HA, Atkins MB, Parkinson DR and Dinarello CA (1988). Induction of circulating tumor necrosis factor (TNF-α) as the mechanism for the febrile response to interleukin 2 (IL-2) in cancer patients. J Clin Immunol **8**, 426-436 .

Mitsuya H, Matis L, Megson M, Bunn P, Murray C, Mann D, Gallo R and Broder S (1983). Generation of an HLA-restricted cytotoxic T cell line reactive against cultured tumor cells from a patient infected with human T cell leukemia/lymphoma virus. J Exp Med **158**, 994-999.

Mizoguchi H, O'Shea JJ, Longo DL, Loeffler CM, McVicar DW and Ochoa AC (1992). Alterations in signal transduction molecules in T lymphocytes from tumor-bearing mice. Science **258**, 1795-1798.

Morgan DA, Ruscetti FW and Gallo RC (1976). Selective in vitro growth of T lymphocytes from normal human bone marrows. Science **193**, 1007-1008.

Niman HL, Thompson AMH, Yu A, Markman M, Willems JJ, Herwig KR, Habib NA, Wood CB, Houghten RA and Lerner RA (1985). Anti-peptide antibodies detect oncogene-related proteins in urine. Proc Natl Acad Sci USA **82**, 7924-7928.

North RJ (1985). Down-regulation of the antitumor immune response. Adv Cancer Res **45**, 1-43.

Old LJ, Boyse EA, Clarke DA and Carswell EA (1962). Antigenic properties of chemically induced tumors. II. Antigens of tumor cells. Ann NY Acad Sci **101**, 80-106.

Peace DJ and Cheever MA (1989). Toxicity and therapeutic efficacy of high-dose interleukin-2. In vivo infusion of antibody to NK-1.1 attenuates toxicity without compromising efficacy against murine leukemia. J Exp Med **169**, 161-173.

Peace DJ, Chen W, Nelson H and Cheever MA (1991). T cell recognition of transforming proteins encoded by mutated ras proto-oncogenes. J Immunol **146**, 2059-2065.

Peace DJ, Smith J, Disis N, Cosand W, Blake J, You S-G, Chen W and Cheever MA (1992). Class I MHC-restricted T cell recognition of the mutated segment of activated p21ras protein. FASEB J **6**, A1403.

Perkins DL, Berriz G, Kamradt T, Smith JA and Gefter ML (1991). Immunodominance: Intramolecular competition between T cell epitopes. J Immunol **146**, 2137-2144.

Prehn RT and Main JM (1957). Immunity to methycholanthrene-induced sarcomas. J Natl Cancer Inst **18**, 769-778.

Prehn RT (1962). Specific isoantigenicities among chemically-induced tumors. Ann NY Acad Sci **101**, 107-113.

Restifo NP, Esquivel F, Kawakami Y, Yewdell JW, Mule JJ, Rosenberg SA and Bennink JR (1993). Identification of human cancers deficient in antigen processing. J Exp Med **177**, 265-272.

Rosenberg SA, Lotze MT, Muul LM, Chang AE, Avis FP, Leitman S, Linehan WM, Robertson CN, Lee RN, Rubin JT, Seipp CA, Simpson CG and White DE (1987). A progress report on the treatment of 157 patients with advanced cancer using lymphokine-activated killer cells and interleukin-2 or high-dose interleukin-2 alone. N Engl J Med **316**, 889-897.

Rosenstein M, Ettinghausen SE and Rosenberg SA (1986). Extravasation of intravascular fluid mediated by systemic administration of recombinant interleukin 2. J Immunol **137**, 1735-1742.

Siegel JP, Sharon M, Smith PL and Leonard WJ (1987). The interleukin 2 receptor chain (p70): Role in mediating signals for LAK, NK, and proliferative activities. Science **238**, 75-78.

Slovin SF, Lackman RD, Ferrone S, Kiely PE and Mastrangelo MI (1986). Cellular immune response to human sarcomas: Cytotoxic T cell clones reactive with autologous sarcomas. J Immunol **137**, 3042-3048.

Smith KA (1993). Lowest dose interleukin-2 immunotherapy. Blood **81**, 1414-1423.

Townsend A and Bodmer H (1989). Antigen recognition by class I-restricted T lymphocytes. Ann Rev Immunol **7**, 601-624.

Vose BM and Bonnard GD (1982). Human tumor antigens defined by cytotoxic and proliferating responses of cultured lymphoid cells. Nature **296**, 359-361.

Wilde DB and Fitch FW (1984). Antigen-reactive cloned helper T cells. I. Unresponsiveness to antigenic restimulation develops after stimulation of cloned helper T cells. J Immunol **132**, 1632-1638.

13

INHIBITION OF TUMOR GROWTH BY HISTOINCOMPATIBLE CELLS EXPRESSING IL-2

C. Roth, S. Delassus, J. Even and P. Kourilsky
U.277 - Biologie Moleculaire du Gène, Institut Pasteur,
Paris, France

Introduction

Since the first demonstration that the immune system plays a key role in the recognition and rejection of tumor cells, considerable efforts have been made to stimulate a specific host immune response against cancer cells.

Considering the weak immune response against tumor cells in syngeneic hosts, the development of various strategies to overcome the poor immunogenicity of tumors has been the focus of active investigation. These strategies of cell based immunotherapy can be classified into three different categories. Nonspecific immunotherapy aims at potentiating the host immune response with microbial products, such as BCG and *Corynebacterium parvum,* or with cytokines that are normally elicited during infections, such as IFN-γ, IL-2 and TNF. Adoptive immunotherapy involves the expansion of tumor specific lymphocytes in vitro followed by their reinfusion into the host. Although in some cases these approaches have been shown to be efficient, as for instance BCG in the treatment of superficial bladder cancer and IFN-α in the treatment of hairy cell leukemia, these nonspecific and adoptive immunotherapies have led to complete tumor regressions only in a small proportion of patients. Reconstitution with TIL that have been stimulated ex vivo with IL-2 can also induce some antitumor effect (Rosenberg et al., 1988), but most responses are only partial, and complete tumor elimination is rare. Active immunotherapy was developed in an attempt to elicit a specific immune response against tumor antigens. Initial studies involved immunization with irradiated cancer cells or cancer cell lysates, in the hope of provoking protective immunity against emerging or existing tumors. However, even in conjunction with adjuvants such as BCG, the immune response elicited is relatively weak, and there has been little evidence that this approach is effective against established tumors.

CYTOKINE-INDUCED TUMOR IMMUNOGENICITY
ISBN 0-12-261520-4

Different reasons could account for the low antitumor response. Some tumors do not present antigens to T cells because of their reduced expression of MHC class I molecules (Forni et al., 1988; Elliot et al., 1989; Melief et al., 1989). Alternatively, the absence of response to tumor antigens could be due to a failure of T-cell help, either because of low presentation of immunogenic peptides in the context of class II MHC molecules (Ostrand-Rosenberg et al., 1990), or because cancer cells may lack the costimulatory molecule B7 which is needed to fully activate the $CD4^+$ T-cell population (Linsley et al., 1991; Harding et al., 1992; Schwartz, 1992). For these reasons, several groups have investigated the possibility of promoting the activation of truly tumor specific lymphocytes in different tumor models in mice by enhancing the level of tumor antigen presentation and by locally altering the immunological environment of the tumor cell. We report here a summary of our attempts to elicit a systemic immune response that could affect the growth of emerging or already existing tumor cells, as well as an outline of novel analytical methods which we devised to better understand and follow the corresponding immune responses.

Self peptides as potential tumor specific T-cell epitopes

Major conceptual changes have taken place in the last decade as the function of the MHC has become better understood. Several major findings have unified the diverse world of MHC molecules. The crystallographic studies first published in 1987 for HLA-A2 (Bjorkman et al., 1987) have revealed that MHC molecules share very similar structures. Functional studies, on the other hand, have demonstrated that both class II (Rudensky et al., 1991) and class I (Townsend et al., 1986; Falk et al., 1991) MHC molecules are peptide receptors of relatively loose specificity. Finally, it is now established that MHC molecules are permanently loaded with mixtures of peptides derived from self proteins. In the initial formulation (Kourilsky and Claverie, 1986) of the so called "peptidic self" model, we reasoned that certain tumor associated antigens could be peptides derived from self proteins. Because class I MHC molecules were in particular expected to present a large variety of self peptides derived from internal proteins, the concept of immune surveillance could be broadened to the display of mutant self peptides, or peptides derived from nonmutated up-regulated proteins, or peptides derived from nonmutated self proteins unappropriately expressed in certain cells (Kourilsky and Claverie, 1986; Kourilsky et al., 1989; Kourilsky et al., 1991).

The underlying hypothesis that the immune system is tolerant to only a subset of all the potentially "presentable" self peptides (while normally ignoring the others) was approached in the following way. Transgenic mice which expressed hen egg white lysozyme (HEL) in all their tissues were prepared. The promoter which we used to drive the transcription of the transgene had the property of yielding very different levels of expression. It was thus possible to test the tolerance to HEL as a function of the concentration of HEL in the serum. We found that animals expressing modest amounts of HEL were tolerant to the protein, to the immunodominant peptide and not to the subdominant or cryptic peptides (Cibotti et al., 1992). This is in agreement with other reports showing that T cells may evade tolerance induction for minor determinants on the antigen (Gammon and Sercarz, 1989; Milich et al., 1991), suggesting the existence of a reservoir of presentable self peptides, not normally visible to the host immune system, which harbours potentially autoreactive T cells. The dose dependence strongly suggested that peptides from up-regulated proteins could be a source of epitopes involved in tumor rejection or, perhaps, autoimmune attack.

T. Boon and his coworkers have characterized several tumor associated epitopes which are targets for antitumor CTL. Quite remarkably, both in the murine mastocytoma P815 and in human melanoma the epitopes were derived from deregulated, up-regulated self proteins. It can thus be reasonably hoped that tumor specific immunogens will be identified in larger numbers and will become usable to manipulate the immune system in order to eradicate cancer cells. It will be important, however, to evaluate possible adverse autoimmune reactions triggered by such manipulations.

Immunization with engineered cells to activate tumor specific effector cell populations

It is now well established that most rodent tumors induced by oncogenic viruses express tumor associated antigens that constitute the targets for immune response. Tumors caused by polyoma, SV40 and murine leukemia viruses (MuLV) are highly immunogenic and induce an effective cell mediated immune response. This immune reaction involves both helper T lymphocytes and cytotoxic T cells that participate in rejection of these antigenic tumors (Leclerc and Cantor, 1980; Greenberg et al., 1981; Klein, 1991). In a previous report, the nature of retroviral antigens was identified at the molecular level using transfectants that express *env* or *gag* gene products and which are recognized by H-2 restricted specific CTL (Plata et al., 1987). It appears that L cells transfected with the H-2D^d gene are killed very efficiently if they only express the *env* gene, while the *gag* gene products are preferentially recognized in association with the H-2K^b antigen, thus demonstrating that the antigenicity of the viral proteins varies according to the H-2 antigen with which the viral protein is associated. More strikingly, mice immunized with L cells expressing D^d or K^b molecules and transfected with the MuLV *env* or *gag* gene develop a strong antitumor immunity and a resistance to subsequent challenge with virus induced tumor cells (Plata et al., 1987). Since these two virus encoded proteins are truly tumor associated antigens, these experiments were amongst the first to show that effective T-cell mediated antitumor immunity can be generated in vivo with cells expressing cloned tumor specific antigens.

In other virus induced tumors, such as Friend leukemia virus or human adenovirus, whose early region 1 can transform murine cells, or EBV, immunogenic antigens recognized by H-2 restricted specific CTL have been identified. The extent to which these epitopes expressed on transfected target cells are capable of inducing an in vivo antitumor response still remains to be investigated (Melief, 1992).

As for viral epitopes presented in the context of class I or class II MHC molecules, the same prediction can be made for tumor associated epitopes which are derived from processed mutated or nonmutated up-regulated proteins. This was first shown for tumor epitopes derived from mutagenized tumors. In this case, mutational events create a novel sequence or a neo-epitope that is a tumor specific antigen. T. Boon and colleagues elegantly demonstrated that such neo-antigens on the P815 tum$^-$ variants could constitute tumor rejection antigens presented by class I MHC molecules to CTL (Lurquin et al., 1989; Sibille et al., 1990). In other situations, the tumor peptides derive from a normal but infrequently expressed sequence (Van den Eynde et al., 1991; Boon, 1992) which can be considered as a rare epitope and which, thereby, behave as foreign to the immune system.

It seems clear, then, that the failure of an antitumor immune response is not due to the absence of tumor specific antigens, but rather to defects in immune regulation. As a consequence, it was reasoned that local secretion of cytokines at the site of the tumor might restore the missing signal required for mounting an efficient immune response against the tumor cells.

The constitutive release of cytokines at the site of the tumor was first attempted in experimental models, using tumor cells engineered with the IL-2 gene (Bubenik et al., 1988; Bubenik et al., 1990; Fearon et al., 1990; Gansbacher et al., 1990a; Gansbacher et al., 1990b; Ley et al., 1990; Ley et al., 1991; Russell et al., 1991). The objective was to bypass a defective helper arm of the immune response by stimulating various cell types, including CTL, helper T cells, NK and LAK cells that can participate in an antitumor response. This local cytokine secretion is more suitable than systemic administration, probably because of the dose limiting toxicity and the short half life of the cytokine in vivo (Lotze et al., 1985). In all the experimental systems reported, the production of IL-2 by tumor cells that are weakly immunogenic resulted in growth inhibition of the inoculated, modified tumor cells. Histological analysis and experiments performed in immunodeficient mice revealed the participation of T lymphocytes in tumor growth arrest (Bubenik et al., 1988; Russell et al., 1991). Furthermore, in vivo depletion studies with anti-CD4 and anti-CD8 mAb indicated that CD8 cells are involved in the rejection of tumor cells producing IL-2, thus rendering T helper cells irrelevant in this rejection response (Fearon et al., 1990). In addition, in most of the tumor models a systemic immune response against the parental tumor can be observed. This parental tumor growth inhibition is correlated with the stimulation of specific CTL that recognize both parental and IL-2 gene transduced tumor cells, as shown by the increase of CTL precursors and effector frequencies in spleen cells of mice immunized with IL-2 gene transfected P815 tumor cells (Ley et al., 1991).

Finally, based on these experiments dealing with genetically modified tumor cells, it appears that the local production of cytokines such as IL-2 by the tumor cell itself is capable of activating an immune reaction responsible for tumor growth suppression and, ultimately, a systemic long lasting antitumor immunity mediated by $CD8^+$ CTL.

Use of allogeneic or xenogeneic cells releasing IL-2 to control antitumor immune responses

Most tumors cannot fully stimulate T cells by themselves, either because they express very low levels of class II molecules (Ostrand-Rosenberg et al., 1990) or because they lack the B7 costimulatory signal (Chen et al., 1992; Schwartz, 1992). As a consequence, CTL precursors cannot proliferate or differentiate into cytolytic effector cells. This anergy appears to be heavily dependent on the total amount of IL-2 provided by the immediate environment since it can be reversed by the addition of exogenous IL-2.

Using transfectants that express variable levels of IL-2, it was shown that tumors releasing higher levels of IL-2 are more efficient in promoting tumor rejection than those producing low levels of IL-2 (Cavallo et al., 1992). This effect could result from the direct activation of effector cells other than CTL, such as NK cells and monocytes, that contribute to tumor eradication but not to long lasting immune protection. Indeed, in the P815 mastocytoma model there is a direct correlation between the level of IL-2 locally secreted and the fraction of protected animals, even though the P815 cells are insensitive to lysis by NK cells (Roth et al., 1992). To determine whether the beneficial effect of IL-2 correlates with the activation of CTL in a dose dependent manner, the analysis of the immune memory generated in animals inoculated with variable amounts of locally secreted IL-2 would be informative. In this tumor system two points should be borne in mind:

- a small percentage of transfected tumor cells is sufficient to induce tumor growth inhibition
- IL-2 is required only for a few days after inoculation, as shown by the protection generated by irradiated transfected cells that secrete normal IL-2 levels for 5 days or less in vitro (Ley et al., 1991).

Thus it appears that a small percentage of transfectants secreting high amounts of IL-2, even over a short period of time, can efficiently induce tumor growth inhibition. Furthermore, IL-2 does not have to be secreted by the tumor cells themselves, since its beneficial effect can be induced by vector cells that are allogeneic with respect to the tumor and the host animal (Roth et al., 1992). In two experimental models, a low number of histoincompatible cells expressing IL-2 was capable of inducing tumor rejection. In DBA/2 mice allogeneic fibroblasts caused the rejection of P815 cells, and in C57BL/6 mice allogeneic P815 cells induced growth inhibition of Lewis lung tumors (Table 13.1, part A). Even xenogeneic cells, which release IL-2 over a shorter period of time since they are more rapidly eliminated, can protect mice against Lewis tumor growth. Similarly, in another experimental system a certain degree of protection was shown in nude mice inoculated with a human tumor xenograft mixed with a large number of cells from a rat fibroblast line secreting IL-2 (Bubenik et al., 1988).

Interestingly, previous reports indicated that inoculation of tumor cells expressing allogeneic determinants induces a strong immune reaction that results in reduced tumorigenicity of the modified tumor. Moreover, mice immunized with these transfectants developed systemic immunity directed against the parental tumor cells, since the latter were rejected upon further inoculation (Itaya et al., 1987). This is in agreement with the protection observed when histoincompatible cells are co-inoculated with the original tumor, although allogeneic and syngeneic determinants are not carried by the same cell line; in this setting, IL-2 is required to restore the full activation of CTL against the parental tumor.

Identification of the effector cells that are stimulated by the allogeneic cells secreting IL-2 and that lyse a given tumor has revealed that their nature depends on the tumor model. In the Lewis tumor system IL-4 secreting allogeneic cells conferred only limited protection against tumor growth and, even more, they inhibited the beneficial effect of IL-2 locally delivered by allogeneic cells. This is in sharp contrast to the protection observed when IL-2 and/or IL-4 are expressed by the P815 tumor cells, suggesting major differences in the mechanisms involved in tumor rejection. Indeed, in protection of nude mice, and in cell depletion experiments with anti-NK antibodies, NK cells were shown to be a major component of the protective effect against Lewis tumor growth (Roth et al., 1992). More interestingly, in animals treated with an anti-NK antibody, a delay in tumor growth was observed when a combination of the two interleukins - IL-2 and IL-4 - that synergize in vivo to induce CTL activation, was used (Table 13.1, part B). It thus seems likely that the immune reaction that leads to tumor rejection involves (in addition to NK or other nonspecific killer cells) CTL against foreign or allogeneic determinants, as well as CTL recognizing self tumor associated antigens on the parental tumor.

These results are consistent with other reports, showing a "two wave" immune reaction against tumor cells. In the early phase (within 18 h), nonspecific effector cells, such as macrophages, activated eosinophils or neutrophils and NK cells, are associated with the local inflammatory reaction observed, playing a crucial role in the tumoricidal effect initiated by the cytokine (Cavallo et al., 1992; Hock et al., 1993a; Hock et al., 1993b). This first reaction enables the secondary specific immune reaction to take place by slowing tumor growth and by increasing the level of tumor antigen presentation by class I and class II MHC molecules on specialized APC that surround the tumor cells. This gives specific T cells the time to proliferate and to acquire cytolytic activity. In the later phase, generation of tumor specific CD8+ T cells can then arise as a result of the powerful tumor epitope presentation that occurs during these cell-to-cell interactions (Colombo et al., 1992); this can potentially lead to a specific and long term immune protection against the parental tumor growth.

The fact that animals inoculated with a mixture of Lewis tumor cells and IL-2 secreting cells do not display a long term immunity against the same tumor could be due either to a local secretion of IL-2 insufficient to fully activate antitumor CTL, or to a weak antigen presentation by class I MHC molecules which determines a poor susceptibility to CTL (Chadwick and Miller, 1992; Glas et al., 1992). Experiments are currently in progress to verify this aspect.

Analytical tools

As new approaches for possible cancer therapy develop, there appears to be an increasing need for improved analytical tools. The latter should help to evaluate, both in animal models and in human cases, immune parameters which may correlate with the evolution of the disease. With this objective in mind, we have designed new molecular methods which allow: (1) a more detailed definition of T-cell repertoires, and (2) a relatively precise quantification of mRNA transcripts in biological samples. Both techniques have been recently described (Cochet, et al., 1992; Pannetier et al., 1993a; Pannetier et al., 1993b). Following is a discussion of a few issues related to their potential uses.

The general approach is based on the PCR technique, thanks to which a sample of RNA, once reverse transcribed into cDNA, is amplified with two oligonucleotide primers. After amplification, a third oligonucleotide, labelled with a fluorescent tag, is used as a primer in a one step run-off reaction.

Assumption is made (as will be the case below) that the amplified molecules are distributed into several size classes. The fluorescent run-off products can be separated by electrophoresis. Since the run-off reaction is quantitative, the ratios of the size peaks accurately reflect the ratios of the number of molecules in the amplified population. This readout has the advantage that the use of a third primer adds to the specificity of the PCR reaction. Occasional amplified contaminants, therefore, will not be detected and will not interfere with the measurements, and subclasses of the amplified molecules can be specifically detected with the appropriate primer.

Table 13.1. *Tumor growth inhibition induced by IL-2 or IL-4 secreting allogeneic tumor cells*

	Strains	Cells injected		Mean tumor volume (cm^3) (days post injection)		Fraction of protected mice
		Parental tumor	Allogeneic cells (5×10^5)	D_1[a]	D_2[b]	
A	DBA2	P815	--	2.8	21.7	0/5[c]
		P815	LM1 (10^6)	1.6	9.6	0/5
		P815	LM1 (5×10^6)	1.0	12.3	0/5
		P815	LM1-(IL-2) (10^6)	0.2	3.2	4/5
		P815	LM1-(IL-2) (5×10^6)	0.2	0.7	4/5
	C57BL/6	Lewis	--	0.46	1.7	0/6
		Lewis	P815 (5×10^5)	0.45	2.3	0/6
		Lewis	P815 (10^6)	0.49	1.5	0/6
		Lewis	P815 (2×10^6)	0.44	2.3	0/6
		Lewis	P815-(IL-2) (5×10^5)	0	0 (n=5), 3.48 (n=1)	5/6
		Lewis	P815-(IL-2) (10^6)	0	0 (n=5), 3.63 (n=1)[d]	5/6
		Lewis	P815-(IL-2) (2×10^6)	0	0 (n=5), 5.54 (n=1)[d]	5/6
		Lewis	P815-(IL-4) (10^6)	0.064	0.39	0/6
		Lewis	P815-(IL-2) (5×10^5) + P815-(IL-4) (5×10^5)	0.058	0.71	1/6
B	NK depleted C57BL/6[e]	Lewis	--	0.25	2.2	0/5
		Lewis	P815 (5×10^5)	0.42	2.3	0/5
		Lewis	P815-(IL-2) (5×10^5)	0.12	2.0	0/5
		Lewis	P815-(IL-4) (5×10^5)	0.14	2.8	0/5
		Lewis	P815-(IL-2) (5×10^5) + P815-(IL-4) (5×10^5)	0	1.3	0/5

A second operational principle deals with the amplification reaction itself (rather than with the readout step). When a mixture of molecules is co-amplified to saturation, an internal competition takes place; provided that the molecules are near identical in sequence and length, the composition of the amplified mixture closely resembles the original mixture. We have observed, on the contrary, that if the sequences between the two amplification primers are largely different, when the reaction reaches saturation the ratio of the amplification products can diverge dramatically from the ratio of initial molecules (Pannetier et al., 1993a).

Using these two operational principles, we have developed two series of tools which should help to provide a picture of the state of the immune system in biological samples. First, in order to measure the mRNA content for any given gene, our approach is to clone the corresponding cDNA and slightly modify its size (usually by four base pairs). The modified cDNA is used as an internal standard in the PCR amplification, and the runoff reaction provides two peaks of distinct sizes, out of which the amount of mRNA converted into cDNA in the sample can be estimated. It is appropriate to use two or three concentrations of an internal DNA standard as shown in Fig. 13.1 and the result of the quantification is then precise to the extent of $\pm 25\%$ under optimal conditions. This has been applied to a series of mouse cytokines and cell surface receptors. For the first time, we could also quantify a T-cell clone in vivo. We have shown that in the mouse spleen and blood samples can be readily analyzed (unpublished data) and work is in progress to adapt these tools to human samples.

Secondly, we have developed a way of analyzing in greater detail the T-cell repertoires. We have focused on the diversity of the mouse β chains of the TCR and devised the following scheme. Starting from any biological sample (blood, nodes, spleen, thymus, etc.), we extract RNA and reverse transcribe it. We then amplify it with a Cβ primer and with

---

(*Table 13.1, notes*)

[a] 15 days after cell inoculation in DBA/2 mice and 11 days after cell inoculation in C57BL/6 mice (part A) or 9 days after cell inoculation in NK depleted C57BL/6 mice (part B)

[b] 20 days after cell inoculation in DBA/2 mice and 17 days after cell inoculation in C57BL/6 mice or 16 days after cell inoculation in NK depleted C57BL/6 mice (part B)

[c] Fraction of animals without detectable tumors at day 90 (part A) or day 30 (part B)

[d] By day 15, one mouse from each group of six mice had developed a deep tumor. None of the other five mice showed any detectable tumor

[e] In vivo antibody depletion was performed by i.p. injection of 100 μg of purified NK specific mAb for 3 days, starting 1 day prior to injection of the tumor. Efficient elimination of NK cells was assessed on endogenous and poly-IC induced NK activities in a ^{51}Cr release cytotoxicity assay using YAC-1 target cells

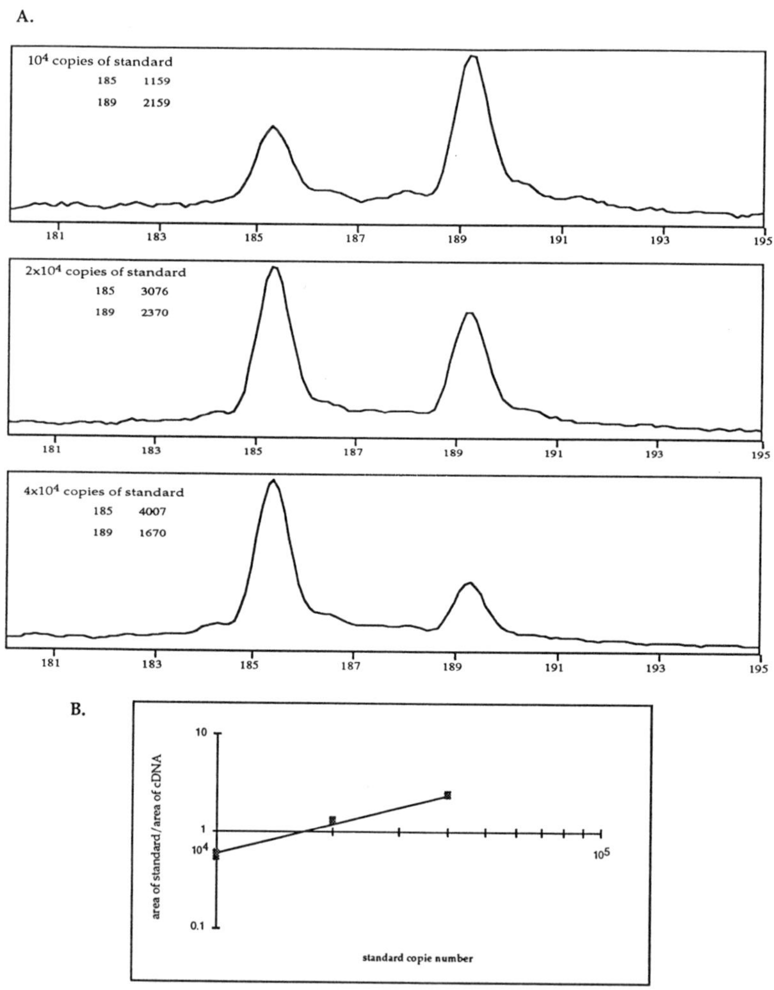

Fig. 13.1. *Quantitative titration of IL-10 mRNA in mouse blood.*

A. Fluorescence peaks obtained upon electrophoresis of the runoff products generated by the co-amplification of mRNA corresponding to 1.5×10^3 cells with 10^4, 2×10^4 and 4×10^4 copies of pIL-10. The lengths (in nucleotides) of the products are 185 bp for pIL-10 and 189 bp for IL-10 mRNA.

B. Quantification of the IL-10 mRNA obtained from the fluorescence peaks above.

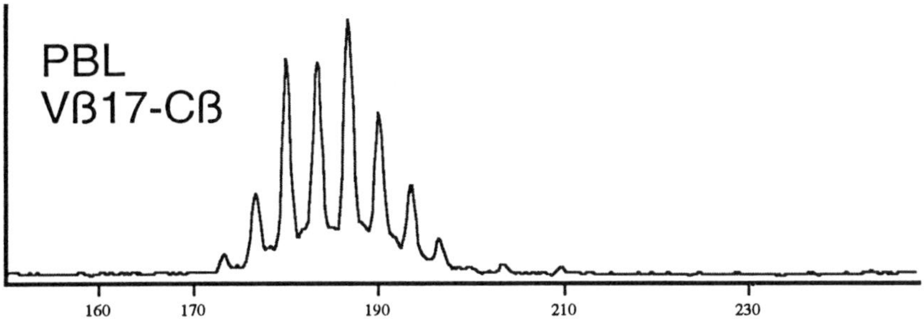

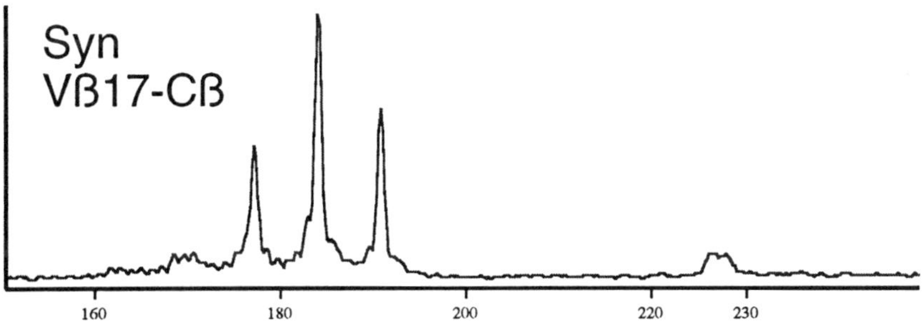

Fig.13.2. *Size distribution of Vβ17-Cβ transcripts present in PBL and in a synovial biopsy of a rheumatoid arthritis patient.* cDNA made from total RNA extracted from PBL (top) and a synovial biopsy (bottom) was amplified by PCR with a couple of Vβ17 and Cβ specific primers. The PCR products were copied by a linear Taq polymerase reaction initiated by an internal fluorescent Cβ primer; these runoff products were subjected to electrophoresis on a 373A DNA sequencer (Applied Biosystems) and analyzed with a specially devised software. The size differences are due to the variable lengths of the CDR3 region reflecting the great diversity of the T-cell response generated by the VDJ joining mechanism.

each of the 20 or so specific Vβ primers. It is worth noting that the mixture of molecules thus amplified to saturation satisfies the second operational principle detailed above. Finally, each amplification reaction is divided into 12 and subjected to a runoff step using each of the 12 specific Jβ fluorescent primers. In each runoff reaction the number of molecules which have distinct lengths in their CDR3 like regions is recorded. Upon electrophoresis, a series of size peaks spaced by 3 nucleotides which correspond to the in frame rearranged β chain encoding transcripts can be observed. Since there are 20-25 Vβ fragments, 12 Jβ segments and 7-8 CDR3 size classes, about 2000 measurements are obtained. We have shown that, upon immunization of mice with a protein antigen, dominant T-cell responses could be detected in vivo (Cochet et al., 1992). We hoped, therefore, that the analysis of human samples infiltrated by T cells might provide information on the complexity of the infiltrates. The above technology was adapted to the analysis of human samples, and the results are quite similar to those expected on the basis of the observations made in mice. First, in most healthy individuals we observed the regular, bell shaped, near gaussian distributions of size peaks illustrated in Figure 13.2. Second, when analyzing TIL or lymphocytes infiltrating organs undergoing an autoimmune attack, we observed distributions with clear indications of oligoclonality (Fig. 13.2). In certain cases, the suspicion of oligoclonality could be confirmed by direct sequencing of the amplified peak. These results suggest that the increased resolution provided by this type of analysis will allow us to describe the state and evolution of TIL. Work is in progress to analyze a series of melanoma samples.

Concluding remarks

The idea of a vaccine to activate a specific immune response against malignant cells has been a dream for many years. The recent progress in active immunotherapy brought about by engineering techniques capable of inducing tumor cells to produce cytokines has made this objective more realistic. The goal is to bypass the need for T-cell help and to overcome the state of immune suppression mediated by the tumor cells by providing the signals that are required for cytotoxic effector function. Firstly, it appeared that a number of cytokines - including IL-2, IL-4, TNF and IFN-γ - when released by a poorly immunogenic tumor were capable of mediating a powerful inhibition of tumor growth in the absence of any toxic side effects. In that sense, the use of allogeneic or xenogeneic vector cells for the in vivo delivery of various molecules, including syngeneic MHC antigens and tumor epitopes, could offer an alternative to individual gene therapy to promote an efficient and specific tumor rejection by the host immune system. Secondly, the development of new means to follow the immune response in vivo should help in analyzing dominant T-cell responses among the tumor infiltrate and in evaluating the efficacy of these recent immunotherapeutic procedures. Finally, the molecular characterization of an increasing number of tumor epitopes, selected according to their binding motifs that interact with MHC antigens, will allow the identification of immunogenic tumor specific peptides against which beneficial responses may be generated.

Acknowledgments

We thank Drs L.M. Mir, F. Quintin-Colonna, D. Fradelizi, J. Kanellopoulos, C. Rochlitz, as well as J.P. Levraud and C. Pannetier for helpful suggestions and discussions. We acknowledge A. Lim for technical assistance, A. Toubert for providing human samples, and Dr D. Ojcius for careful review of the article. We are particularly indebted to Ms. V. Caput for preparing the manuscript.

This work was supported by grants from Fondation de France and M.R.E. (Ministere de la Recherche et de l'Espace).

References

Bjorkman PJ, Saper MA, Samraoui B, Bennett WS, Strominger JL and Wiley DC (1987). Structure of the human class I histocompatibility antigen, HLA-A2. Nature **329**, 506-512.

Boon T (1992). Toward a genetic analysis of tumor rejection antigens. Adv Cancer Res **58**, 177-210.

Bubenik J, Voitenok NN, Kieler J, Prassolov VS, Chumakov PM, Bubenikova D, Simova J and Jandlova T (1988). Local administration of cells containing an inserted IL-2 gene and producing IL-2 inhibits growth of human tumours in nu/nu mice. Immunol Lett **19**, 279-282.

Bubenik J, Simova J and Jandlova T (1990). Immunotherapy of cancer using local administration of lymphoid cells transformed by IL-2 cDNA and constitutively producing IL-2. Immunol Lett **23**, 287-292.

Cavallo F, Giovarelli M, Gulino A, Vacca A, Stoppacciaro A, Modesti A and Forni G (1992). Role of neutrophils and $CD4^+$ T lymphocytes in the primary and memory response to non immunogenic murine mammary adenocarcinoma made immunogenic by IL-2 gene. J Immunol **149**, 3627-3635.

Chadwick BS and Miller RG (1992). Hybrid resistance in vitro. Possible role of both class I MHC and self-peptides in determining the level of target cell sensitivity. J Immunol **148**, 2307-2313.

Chen L, Ashe S, Brady WA, Hellstrem I, Hellstrem KE, Ledbetter JA, McGowan P and Linsley PS (1992). Costimulation of antitumor immunity by the B7 counter receptor for the T lymphocyte molecules CD28 and CTLA-4. Cell **71**, 1093-1102.

Cibotti R, Kanellopoulos JM, Cabaniols JP, Halle-Panenko O, Kosmatopoulos K, Sercarz E and Kourilsky P (1992). Tolerance to a self-protein involves its immunodominant but does not involve subdominant determinants. Proc Natl Acad Sci USA **89**, 416-420.

Cochet M, Pannetier C, Regnault A, Darche S, Leclerc C and Kourilsky P (1992). Molecular detection and in vivo analysis of the specific T cell response to a protein antigen. Eur J Immunol **22**, 2639-2647.

Colombo MP, Modesti A, Parmiani G and Forni G (1992). Local cytokine availability elicits tumor rejection and systemic immunity through granulocyte-T-lymphocyte cross-talk. Cancer Res **52**, 4853-4857.

Elliot BE, Carlow DA, Rodrick AM and Wade A (1989). Perspectives on the role of MHC antigens in normal and malignant cell development. Adv Cancer Res **53**, 181-244.

Falk K, Retzschke O, Stevanovic S, Jung G and Rammensee HG (1991) Allele-specific motifs revealed by sequencing of self-peptides eluted from MHC molecules. Nature **351**, 290-296.

Fearon ER, Pardoll DM, Itaya T, Golumbek P, Levitsky HI, Simons JW, Karasuyama H, Vogelstein B and Frost P (1990). Interleukin-2 production by tumor cells bypasses T helper function in the generation of an antitumor response. Cell **60**, 397-403.

Forni G, Fujiwara H, Martino F, Hamaoka T, Jemma C, Caretto P and Giovarelli M (1988). Helper strategy in tumor immunology: Expansion of helper lymphocytes and utilization of helper lymphokines for experimental and clinical immunotherapy. Cancer Met Rev **7**, 289-309.

Gammon G and Sercarz E (1989). How some T cells escape tolerance induction. Nature **342**, 183-185.

Gansbacher B, Bannerji R, Daniels B, Zier K, Cronin K and Gilboa E (1990a). Retroviral vector-mediated γ-interferon gene transfer into tumor cells generates potent and long lasting antitumor immunity. Cancer Res **50**, 7820-7825.

Gansbacher B, Zier K, Daniels B, Cronin K, Bannerji R and Gilboa E (1990b). Interleukin 2 gene transfer into tumor cells abrogates tumorigenicity and induces protective immunity. J Exp Med **172**, 1217-1224.

Glas R, Sturmhöfel K, Hämmerling GJ, Kärre K and Ljundgren HG (1992). Restoration of a tumorigenic phenotype by β2-microglobulin transfection to EL-4 mutant cells. J Exp Med **175**, 843-846.

Greenberg PD, Cheever MA and Fefer A (1981). Eradication of disseminated murine leukemia by chemoimmunotherapy with cyclophosphamide and adoptively transferred immune syngeneic Lyt-1+2- lymphocytes. J Exp Med **154**, 952-963.

Harding FA, McArthur JG, Gross JA, Raulet DH and Allison JP (1992). CD28-mediated signalling co-stimulates murine T cells and prevents induction of anergy in T-cell clones. Nature **356**, 607-609.

Hock H, Dorsch M, Kunzendorf U, Qin Z, Diamantstein T and Blankenstein T (1993a). Mechanisms of rejection induced by tumor cell-targeted gene transfer of interleukin 2, interleukin 4, interleukin 7, tumor necrosis factor, or interferon gamma. Proc Natl Acad Sci USA **90**, 2774-2778.

Hock H, Dorsch M, Kunzendorf U, Uberla K, Qin Z, Diamantstein T and Blankenstein T (1993b). Vaccinations with tumor cells genetically engineered to produce different cytokines: Effectivity not superior to a classical adjuvant. Cancer Res **53**, 714-716.

Itaya T, Yamagiwa S, Okada F, Oikawa T, Kuzumaki N, Takeichi N, Hosokawa M and Kobayashi H (1987). Xenogenization of a mouse lung carcinoma (3LL) by transfection with an allogeneic class I major histocompatibility complex gene (H-2L^d). Cancer Res **47**, 3136-3140.

Klein G (1991). Immunovirology of transforming viruses. Current Opin Immunol **3**, 665-673.

Kourilsky P and Claverie JM (1986). The peptidic self model: A hypothesis on the molecular nature of the immunological self. Ann Inst Pasteur/Immunol **137D**, 3-21.

Kourilsky P, Claverie JM and Prochnicka-Chalufour A, Spetz-Hagberg AL and Larsson-Sciard EL (1989). How important is the direct recognition of polymorphic MHC residues by TCR in the generation of the T-cell repertoire? Cold Spring Harbor Symp Quant Biol **54**, 93-103.

Kourilsky P, Jaulin C and Ley V (1991). The structure and function of MHC molecules. Possible implications for the control of tumor growth by MHC-restricted T cells. Sem Cancer Biol **2**, 275-282.

Leclerc JC and Cantor H (1980). T-cell-mediated immunity to oncornavirus-induced tumors. II. Ability of different T cell sets to prevent tumor growth in vivo. J Immunol **124**, 851-854.

Ley V, Roth C, Langlade-Demoyen P, Larsson-Sciard EL and Kourilsky P (1990). A novel approach to the induction of specific cytolytic T cells in vivo. Res Immunol **141**, 855-863.

Ley V, Langlade-Demoyen P, Kourilsky P and Larsson-Sciard EL (1991). Interleukin 2-dependent activation of tumor-specific cytotoxic T lymphocytes in vivo. Eur J Immunol **21**, 851-854.

Linsley PS, Brady W, Grosmaire L, Aruffo A, Damle NK and Ledbetter JA (1991). Binding of the B cell activation antigen B7 to CD28 costimulates T cell proliferation and interleukin 2 mRNA accumulation. J Exp Med **173**, 721-730.

Lotze MT, Matory YL, Ettinghausen SE, Rayner AA, Sharrow SO, Seipp CAY, Custer MC and Rosenberg SA (1985). In vivo administration of purified human interleukin 2. II. Half life, immunologic effects, and expansion of peripheral lymphoid cells in vivo with recombinant IL-2. J Immunol **135**, 2865-2869.

Lurquin C, Van Pel A, Mariame B, De Plaen E, Szikora JP, Janssens C, Reddehase MJ, Lejeune J and Boon T (1989). Structure of the gene of tum$^-$ transplantattion antigen P91A: The mutated exon encodes a peptide regognized with L^d by cytolytic T cells. Cell **58**, 293-303.

Melief CJ, Vasmel WL, Offringa R, Sijts EJ, Matthews EA, Peters PJ, Meloen RH, Van der Eb AJ and Kast WM (1989). Immunosurveillance of virus-induced tumors. Cold Spring Harbor Symp Quant Biol **54**, 597-603.

Melief CJ (1992). Tumor eradication by adoptive transfer of cytotoxic T lymphocytes. Adv Cancer Res **58**, 143-175.

Milich DR, McLachlan A, Raney AK, Houghten R, Thornton GB, Maruyama T, Hughes JL and Jones JE (1991). Autoantibody production in hepatitis B e antigen transgenic mice elicited with a self T-cell peptide and inhibited with nonself peptides. Proc Natl Acad Sci USA **88**, 4348-4352.

Ostrand-Rosenberg S, Thakur A and Clements V (1990). Rejection of mouse sarcoma cells after transfection of MHC class II genes. J Immunol **144**, 4068-4071.

Pannetier C, Delassus S, Darche S, Saucier C and Kourilsky P (1993a) Quantitative titration of nucleic acids by enzymatic amplification reactions run to saturation. Nucl Acids Res **21**, 577-583.

Pannetier C, Cochet M, Darche S, Casrouge A, Zöller M and Kourilsky P (1993b). The sizes of the CDR3 hypervariable regions of the murine T-cell receptor β chains vary as a function of the recombined germ-line segments. Proc Natl Acad Sci USA **90**, 4319-4323.

Plata F, Langlade-Demoyen P, Abastado JP, Berbar T and Kourilsky P (1987). Retrovirus antigens recognized by cytolytic T lymphocytes activate tumor rejection in vivo. Cell **48**, 231-240.

Rosenberg SA, Packard BS, Aebersold PM, Solomon D, Topalian SI, Toy ST, Simon P, Lotze MT, Yang JC, Seipp CA, Simpson C, Carter C, Bock S, Schwartzentruber D, Wei JP and White DE (1988) Use of tumor-infiltrating lymphocytes and interleukin-2 in the immunotherapy of patients with metastatic melanoma. A preliminary report. N Engl J Med **319**, 1676-1680.

Roth C, Mir LM, Cressent M, Quintin-Colonna F, Ley V, Fradelizi D and Kourilsky P (1992). Inhibition of tumor growth by histoincompatible cells expressing interleukin-2. Int Immunol **4**, 1429-1436.

Rudensky AYU, Preston-Hurlbort P, Hong SC, Barlow A and Janeway CA (1991). Sequence analysis of peptides bound to MHC class II molecules. Nature **353**, 622-627.

Russell SJ, Eccles SA, Flemming CL, Johnson CA and Collin MKL (1991). Decreased tumorigenicity of a transplantable rat sarcoma following transfer and expression of an IL-2 cDNA. Int J Cancer **47**, 244-251.

Schwartz RH (1992). Costimulation of T lymphocytes: The role of CD28, CTLA-4, and B7/BB1 in interleukin-2 production and immunotherapy. Cell **71**, 1065-1068.

Sibille C, Chomez P, Wildmann C, Van Pel A, DePlaen E, Maryanski JL, De Bergeyck V and Boon T (1990). Structure of the gene of Tum- transplantation antigen P198: A point mutation generates a new antigenic peptide. J Exp Med **172**, 35-45.

Townsend AR, Rothbard J, Gotch FM, Bahadur G, Wraith D and McMichael AJ (1986). The epitopes of influenza nucleoprotein recognized by cytotoxic T lymphocytes can be defined with short synthetic peptides. Cell **44**, 959-968.

Van den Eynde B, Lethe B, Van Pel A, DePlaen E and Boon T (1991). The gene coding for a major tumor rejection antigen of tumor P815 is identical to the normal gene of syngeneic DBA/2 mice. J Exp Med **173**, 1373-1384.

14

IMMUNOGENICITY OF MURINE RENAL CANCER CELLS EXPRESSING VARIOUS CYTOKINE GENES

J. L. Franco, K. L. Komschlies, M. E. Gruys, T. C. Back, R. G. Fenton and R. H. Wiltrout
Laboratory of Experimental Immunology and Biological Response Modifiers Program, NCI, and Biological Carcinogenesis and Development Program, PRI/DynCorp, Inc., NCI-FCRDC, Frederick, MD, USA

Introduction

Approximately 25000 new cases of renal cell cancer are diagnosed each year in the United States (Silverberg and Lubera, 1989). Renal cell adenocarcinoma is devastating because many patients have metastatic disease at the time of presentation and treatment with chemotherapeutic drugs is largely ineffective. Some BRM, such as IFN-α and IL-2, have shown therapeutic efficacy as good as or better than chemotherapy (Clark and Longo, 1987) raising the possibility that this disease induces an immune response under some circumstances. However, attempts to establish a clear association between augmentation of selected BRM-induced immune parameters and therapeutic responses in renal cancer patients have been elusive (Rosenberg et al., 1993; Parkinson, 1990). Although much recent effort has been focused on the therapeutic effects of endogenously induced or adoptively transferred broadly reactive IL-2 induced LAK cells (Parkinson, 1990; Rosenberg et al., 1993), this approach has resulted in only limited success (Parkinson, 1990; Sosman et al., 1991). Even in mice, the antitumor effects of LAK activity are usually characterized by incomplete reductions in the number of experimental pulmonary or hepatic metastases (Mule et al., 1987).

The limited success using antigen nonspecific approaches has served to refocus the dialogue toward the use of antigen specific T lymphocytes for the immunotherapy of murine and human cancer (Rosenberg et al., 1988; Chang and Shu, 1992). A particular concern with this approach has been that autologous tumor cells may be inadequate in the way that they present antigen to T cells because of a lack of costimulatory signals (i.e. B7), resulting in the subsequent induction of anergy (Mueller et al., 1989). Also the failure of the tumor to express class I or class II MHC antigens, to

CYTOKINE-INDUCED TUMOR IMMUNOGENICITY
ISBN 0–12–261520–4

process antigen adequately, or to trigger the appropriate local cytokine milieu for the expansion and/or activation of $CD4^+$ or $CD8^+$ T cells could all contribute to an unproductive immune response. In fact, defective $CD8^+$ T-cell responses can sometimes be attributed to a lack of class II MHC induced $CD4^+$ helper cells that provide necessary cytokines required by $CD8^+$ cells in the local tumor site. One approach to circumvent defective Th responses has been recently proposed (Tepper et al., 1989; Fearon et al., 1990; Golumbek et al., 1991), whereby the genes for IL-2 or IL-4 have been introduced into poorly immunogenic tumor cells to test the hypothesis that the local paracrine production of costimulating cytokines could bypass or correct a defective Th cell response and lead to tumor regression. The introduction of Th cytokine genes into tumor cells induces a beneficial class I MHC restricted T-cell response against the transduced tumor cells and results in resistance to rechallenge by nontransduced parental tumor cells and, in some, regression of early established parental tumors. These data support the previous observations of Forni and colleagues (Forni et al., 1985; Bosco et al., 1990) that the local injection of very small amounts of IL-2 and IL-4 into tumors induced profound antitumor effects. Other studies have demonstrated that murine tumor cells transfected with other cytokine genes including IL-1 (Douvdevani et al., 1992), IL-6 (Mullen et al., 1992), IL-7 (Aoki et al., 1992), TNF-α (Asher et al., 1991), IFN-γ (Restifo et al., 1992) and G-CSF (Colombo et al., 1991) are generally rejected following implantation into immunocompetent mice, while cytokine transfected human renal tumors also do not grow in immunodeficient mice (Belldegrun et al., 1993). Overall, a variety of transduced cytokine genes have been shown to render progressively growing murine tumors immunologically responsive. However, there remains little information regarding the optimal cytokine(s) of choice for various tumors, a clear unifying hypothesis as to why various cytokines work in this manner, and whether this approach is truly feasible for the treatment of established tumors in hosts that have been conditioned or altered by the growth of their cancer.

Our studies are designed to determine whether cytokine transduced Renca cells or fibroblasts can be used for the therapy of preexisting Renca tumor and the mechanism(s) by which these effects occur.

Materials and methods

Mice. BALB/c mice were obtained from the animal production area of the NCI-FCRDC, and were maintained under specific pathogen free conditions. Mice were routinely used at 8-10 weeks of age.

Experimental model. The Renca tumor arose spontaneously in BALB/c mice and was isolated by Dr Sarah Stewart at the NCI, Bethesda, MD. The growth and progression of this tumor occurred in most mice following transfer of as few as 50 viable cells (Murphy and Hruskesky, 1980). Renca spontaneously metastasizes from an intrarenal implant to the regional lymph nodes, then preferentially to the lungs, and later the liver. The immunogenicity of Renca has been determined to be low to moderate although some protection to rechallenge with viable Renca cells after immunization with a membrane vaccine preparation has been reported (Huben et al., 1983). In our hands, the Renca tumor can be implanted in the orthotopic site (kidney), from whence it progressively invades locally and forms metastases in the lungs and liver (Salup et al., 1985). Tumor bearing mice routinely die in 30-40 days. In addition, the Renca tumor becomes highly vascularized as it develops and its progressive growth occurs in a similar manner to that generally described for human renal cell cancer (Bassil et al., 1985) with most mice exhibiting pulmonary metastases as the most frequent extrarenal site of tumor progression (Wiltrout, 1991).

Transfected tumor cells and fibroblasts. The transfectants used in these studies are described in Table 14.1. The $Renca^{IL\text{-}4}$ and $Renca^{IL\text{-}2}$ secreting lines were established as previously described (Golumbek et al., 1991) by $Renca^{par}$ cells with a murine IL-4 cDNA in a bovine papilloma virus vector containing the hygromycin resistance gene (pBCMG hygro). BALB/c 3T3 cells were obtained from the American Type Culture Collection and transfected in our laboratory with the mIL-4 gene in the same manner as that described above. Both IL-4 producing lines were maintained in media containing 400 μg/ml hygromycin. The $Renca^{IFN\text{-}\gamma}$ and $Renca^{TNF\text{-}\alpha}$ cell lines were obtained by transfection with the mIFN-γ or hTNF-α cDNAs, respectively, in the pSV2 vector containing the neomycin resistance gene.

Experimental design. All cell lines were monitored for growth in vitro, and for tumor formation and progressive growth in vivo. Those mice that survived the initial implant were subsequently challenged with $Renca^{par}$ cells in order to determine whether resistance had been induced. In some experiments various numbers of $Renca^{IL\text{-}4}$, $3T3^{IL\text{-}4}$ or control 3T3 cells were co-injected s.c. with 1×10^5 $Renca^{par}$ cells. Tumor growth and survival of mice was then monitored.

Results

In vivo growth of cytokine transfected Renca sublines

Various cytokine genes have been transduced into the $Renca^{par}$ tumor line. The in vivo growth characteristics of these cytokine secreting tumor cells were then studied and observed to be quite different. The data in Fig. 14.1A demonstrate that Renca transfected with the mouse IFN-γ gene ($Renca^{IFN\text{-}\gamma}$) rapidly formed tumors in most (> 80%) BALB/c mice

Table 14.1. *Summary of transfectants used for immunogenicity studies*

Cell line transfected	Gene	Vector	Cytokine production[a]
Renca	mIL-2	pBCMG-Hygro	≥ 3,000 U/ml
	mIL-4	pBCMG Hygro	≥ 15,000U/ml
	mIFN-γ	pBCMG-Neo	≥ 200 U/ml
	hTNF-α	pSV2-Neo	≥ 512 U/ml
	None	pBCMG-Hygro	None
	None	pSV2-Neo	None
BALB/c 3T3	mIL-4	pBCMG-Hygro	≥ 1,500 U/ml
	mIL-2	pBCMG-Neo	≥ 200 U/ml
	None	pBCMG-Hygro	None

[a]Cytokine production is based on the number of U of biological activity/ml produced by 1 x 10^6 cells in 24 h

after a s.c. injection. The Renca$^{IFN-\gamma}$ tumors grew progressively in most mice and ultimately killed 60% and 80% after the injection of 5 x 10^5 or 4 x 10^6 Renca$^{IFN-\gamma}$, respectively (Fig. 14.1B). During the course of this study, the Renca$^{IFN-\gamma}$ bearing mice exhibited a generally poor appearance suggesting some toxicity related to the production of mIFN-γ. It should be noted that in a minority of mice there was no tumor growth observed.

However, where the tumor grew initially there was no evidence of any subsequent tumor regression. Similar effects were obtained using a variant of Renca that had been transfected with the human TNF-α gene (Renca$^{TNF-\alpha}$) (data not shown). These results suggested to us that neither IFN-γ or TNF-α production was sufficient to elicit a strong antitumor response.

Additional studies have also been performed using Renca cells transfected with either the mIL-2 gene (Renca^{IL-2}) or the mIL-4 gene (Renca^{IL-4}). The growth and immunogenicity of these transfectants has been previously characterized by Golumbek et al. (1991). Our studies support their conclusions in that although both of these transfectants grow initially after a s.c. injection, they ultimately regress in virtually all mice. The results of an experiment using Renca^{IL-2} are shown in Fig. 14.2. Almost all BALB/c mice injected with 4 x 10^6 Renca^{IL-2} cells exhibit an initial tumor that rapidly regresses (Fig. 14.2A). All these mice survived (Fig. 14.2B) and were rechallenged with 5 x 10^6 Renca^{par} cells on day 50.

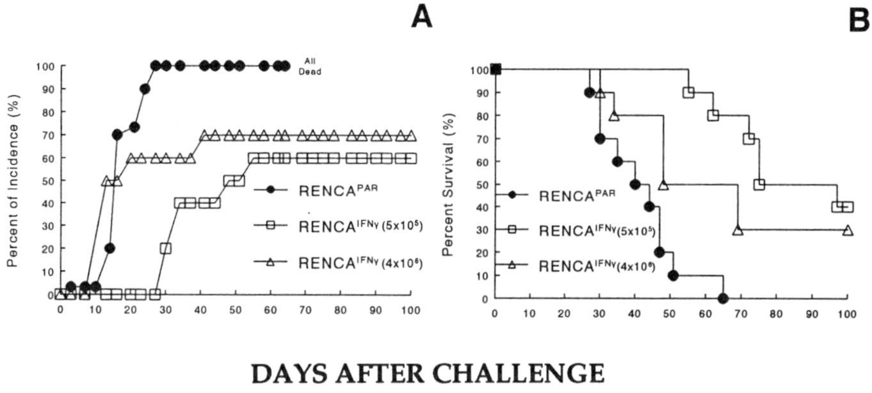

Fig. 14.1. *Growth of Renca*$^{IFN-\gamma}$ *in vivo.* BALB/c mice were injected s.c. with either 5 x 10^5 or 4 x 10^6 Renca$^{IFN-\gamma}$ cells and monitored for tumor incidence and tumor progression (Panel A) and survival (Panel B). These growth patterns are compared to the characteristic growth pattern for 5 x 10^5 Renca^{Par}.

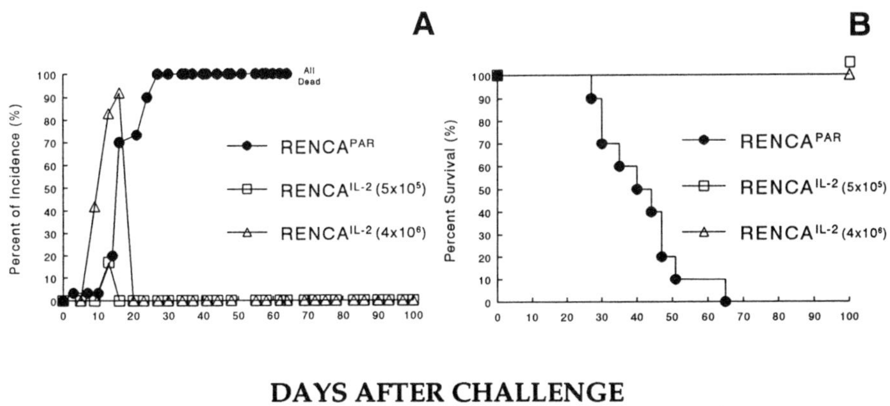

Fig. 14.2. *Growth of Renca*$^{IL-2}$ *in vivo.* BALB/c mice were injected s.c. with either 5 x 10^5 or 4 x 10^6 Renca^{IL-2} cells and monitored for tumor incidence and tumor progression (Panel A) and survival (Panel B). These growth patterns are compared to the characteristic growth of Renca^{Par}.

Most of the mice that were rechallenged subsequently died of progressive Renca^{par} tumor (data not shown). Very similar results were observed for Renca^{IL-4} whereby mice injected with 4 x 10^6 cells exhibited a transient small tumor (Fig. 14.3A) that rapidly regressed (Fig. 14.3B). Most of the mice were also not resistant to rechallenge with Renca^{par} cells (data not shown).

Inhibition of Renca^{par} tumor growth by co-administration of Renca^{IL-4} or 3T3^{IL-4}

Because the Renca^{IL-4} and Renca^{IL-2} tumor lines grew so transiently and did not induce protective immunity for most mice, we designed studies whereby the cytokine producing tumor cells were coinjected with Renca^{par} cells. For these studies we selected Renca^{IL-4} as the transfectant to be evaluated. Since antigen was to be supplied in the form of Renca^{par}, we also evaluated antitumor effects of 3T3^{IL-4} in order to determine whether cytokine production by an unrelated cell in close proximity to the tumor was sufficient to induce local antitumor effects and systemic immunity. The results of these studies shown in Fig. 14.4 demonstrate that most mice injected with either the Renca^{IL-4} or 3T3^{IL-4} transfectants alone survived, while most mice injected s.c. with viable Renca^{par} died. In contrast, when 1 x 10^5 or 1 x 10^6 Renca^{IL-4} was coinjected with 1 x 10^5 Renca^{par}, 60% of the mice survived and were tumor free (Fig. 14.4A). Similar protective results were obtained when the 3T3^{IL-4} was coinjected with Renca^{par}, although this effect was not observed when control 3T3 cells were coinjected with Renca^{par} (Fig. 14.4B). These results demonstrate that local production of IL-4 was sufficient to cause the regression of Renca^{par} irrespective of whether it was produced by Renca cells or fibroblasts.

Induction of systemic immunity following transfectant induced regression of Renca^{par}

Our previous studies (Fig. 14.3) had shown that most mice that were injected with Renca^{IL-4} developed only a very small tumor that rapidly regressed. Most of these mice were not resistant to rechallenge, suggesting that some nonspecific local response was able to induce rejection of these transfectants in the absence of the induction of specific T cell mediated immunity. However, during the performance of the experiment described above in Fig. 14.4, we noted that tumors persisted for longer times in mice that were coinjected with Renca^{par}, and either Renca^{IL-4} or 3T3^{IL-4}, although they were then later rejected. Therefore, we speculated that this subsequent rejection of a larger tumor burden could be consistent with a T-cell mediated response that should also have induced immunity to rechallenge. Thus, all of the surviving mice from the

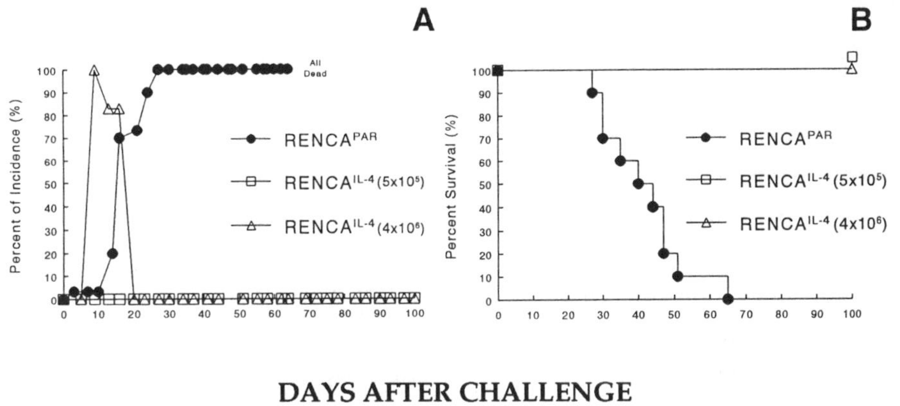

Fig. 14.3. *Growth of Renca^{IL-4} in vivo.* BALB/c mice were injected s.c. with either 5 x 10^5 or 4 x 10^6 Renca^{IL-4} cells and monitored for tumor incidence and tumor progression (Panel A) and survival (Panel B). These growth patterns are compared to the characteristic growth of Renca^{par}.

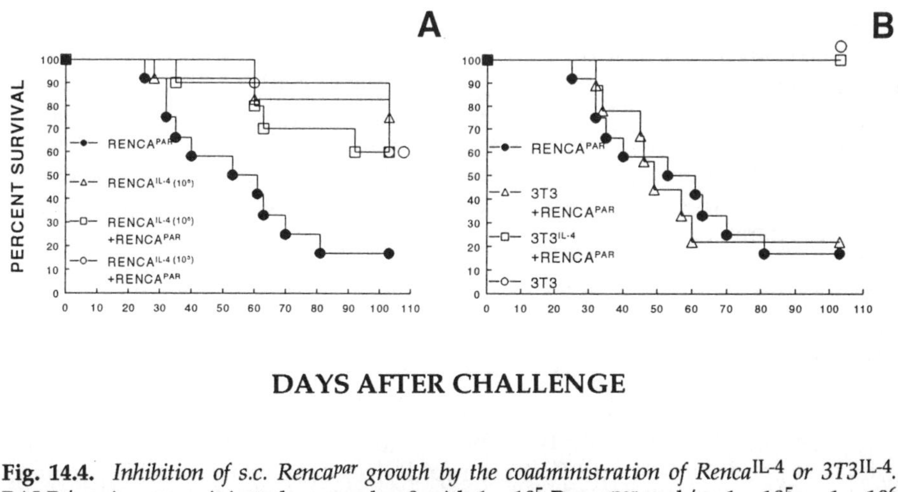

Fig. 14.4. *Inhibition of s.c. Renca^{par} growth by the coadministration of Renca^{IL-4} or 3T3^{IL-4}.* BALB/c mice wer e injected s.c. on day 0 with 1 x 10^5 Renca^{par} and/or 1 x 10^5 or 1 x 10^6 Renca^{IL-4}, 3T3, or 3T3^{IL-4} cells. The mice were monitored for tumor appearance, tumor size and tumor progression. Moribund mice were euthanized.

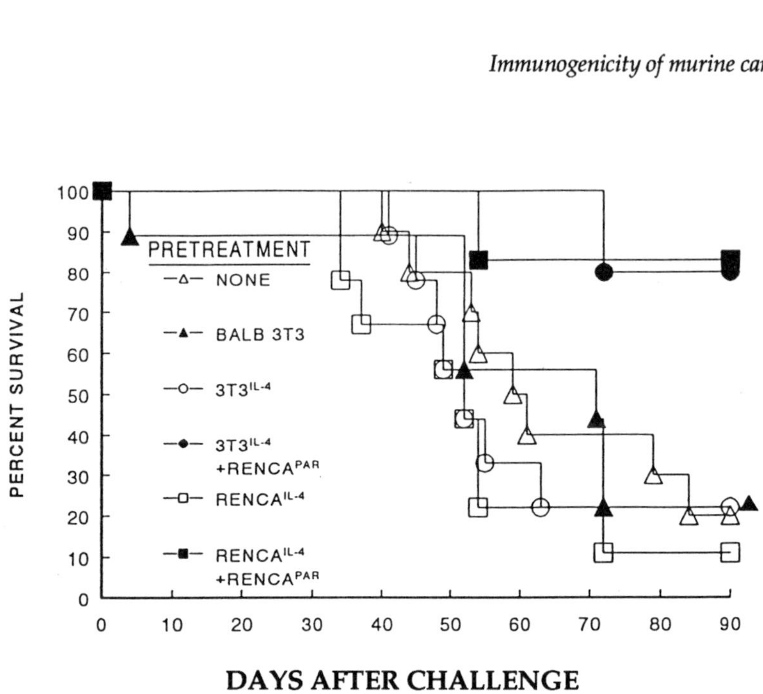

Fig. 14.5. *Induction of resistance to rechallenge by Renca^{par} by the previous coinjection of Renca^{IL-4} or 3T3^{IL-4} and Renca^{par}.* Those mice from Fig. 14.4 that survived the previous implantation of Renca^{IL-4}, 3T3, or 3T3^{IL-4} alone or in combination with Renca^{par} cells were rechallenged with 1 x 10^3 Renca^{par} cells and monitored for the appearance and progression of tumor. Moribund mice were euthanized.

experiment shown in Fig. 14.4 were rechallenged with 1 x 10^3 viable Renca^{par} cells and monitored for tumor growth and survival. As expected, none of the mice that had survived the implantation of control 3T3 cells, 3T3^{IL-4} cells or Renca^{IL-4} cells exhibited any protection to rechallenge when compared to normal mice challenged with 1 x 10^3 Renca^{par} cells (Fig. 14.5). In contrast, 80% of the mice that had previously rejected Renca^{par} in combination with either Renca^{IL-4} or 3T3^{IL-4} were resistant to rechallenge with Renca^{par}. This result suggests the successful induction of systemic resistance rechallenge, probably through a T-cell dependent mechanism.

Discussion

The results of this study demonstrate that transfected Renca tumor cells that produce IL-2 and IL-4 are rapidly rejected by immunocompetent syngeneic mice, while Renca transfectants that produce TNF-α or IFN-γ are not effectively rejected. Further, coinjection of Renca^{IL-4} with viable Renca^{par} results in the rejection of Renca^{par} with the induction of resistance to a subsequent rechallenge with Renca^{par}. A similar effect was achieved when 3T3^{IL-4} was coinjected with Renca^{IL-4}, demonstrating that the local production of cytokine need not be by the tumor cells themselves but can be supplied by unrelated cells in close proximity. In addition, the induction of systemic resistance to rechallenge may be dependent on the presence of some threshold tumor burden that is not effectively controlled by a local inflammatory response, thereby providing an appropriate stimulus for triggering an immune response. This conclusion is supported by previous studies by Tepper et al. (1989 & 1992) and Golumbek et al. (1991) who demonstrated an early infiltration of small IL-4 producing tumors by eosinophils. These infiltrating eosinophils may to be critical for the overall antitumor effects of locally secreted IL-4. The ability to induce a beneficial T-cell mediated response by local production of IL-4 remains somewhat controversial and may be dependent on the tumor model used.

Overall, the application of gene therapy to cancer treatment is in its infancy. It remains unclear how effective such approaches will be against advanced cancer in compromised hosts. It is also unclear whether the antitumor effects of tumor cells secreting different cytokines are mediated via a common pathway. Further, the requirement for isolating, expanding and successfully transfecting each patient's own tumor cells remains expensive and time consuming, and the reintroduction of autologous genetically engineered tumors remains a concern. Thus, one future goal of our studies is designed to determine the therapeutic efficacy of Renca^{IL-4} against advanced Renca^{par} primary tumors and visceral metastases.

To date there are no reports demonstrating that cytokine transfected cells can induce an immune response against a well established tumor. Preliminary studies from our laboratory have suggested that the repeated administration of Renca^{IL-4} can have some antitumor effects against established Renca^{par}, but these effects are only evident in the delay of tumor progression. One potential complication of such an approach could be tumor induced alterations in TCR mediated signal transduction recently described by Mizoguchi et al. (1992). Such preexisting changes may complicate the ability of cytokines to induce a productive antitumor immune response. This concern may be relevant to the Renca tumor model since T lymphocytes from Renca bearing mice do exhibit altered levels of several key components of TCR mediated signaling (Franco et al., 1993).

The use of nontumorigenic cytokine transfected fibroblasts for the therapy of Renca^{par} is also being studied in order to determine their possible utility in gene therapeutic approaches to cancer treatment. Such an approach may minimize the requirements for obtaining large numbers of cells, and transfecting each patient's own tumor. In our initial studies the 3T3^{IL-4} transfected fibroblasts work as well as Renca^{IL-4} for inducing the regression of coinjected Renca^{par} and for inducing resistance to rechallenge by Renca^{par}. Preliminary studies also suggest that the repeated administration of 3T3^{IL-4} may have some antitumor effects against established Renca (Mizoguchi et al., 1992). An overall analysis of the literature suggests that successful gene therapy with cytokine induced tumor cells may first be dependent on the recruitment of appropriate APC and T-cell subsets to the site of tumor antigen. Pardoll has recently reported that the local production of GM-CSF may be useful for mobilizing professional APC in the form of dendritic cells (Pardoll, 1993). In addition, Taub et al. (1993) have suggested that the chemokines MIP-1α and MIP-1β may preferentially attract $CD8^+$ and $CD4^+$ T cells, respectively. Thus, the use of these cytokines may also be necessary to attract appropriate T cells to the tumor site. Ultimately, the local induction of T-cell stimulating cytokines may then successfully be used to induce a productive antitumor response. This type of scenario suggests that such a successful vaccine type approach will be difficult both because of the complexities inherent in inducing an immune response against tumor antigens and the possible altered state of TCR mediated signaling in tumor conditioned hosts.

Acknowledgments

The authors thank Dr Drew Pardoll for his generous gift of the $Renca^{IL-2}$ and $Renca^{IL-4}$ cell lines. The authors also thank Drs Dan Longo and John Ortaldo for their review of this manuscript.

By acceptance of this article, the publisher or recipient acknowledges right of the U.S. Government to retain a nonexclusive, royalty free license in and to any copyright covering the article. The contents of this publication do not necessarily reflect the views or policies of the Department of Health and Human Services, nor does mention of trade names, commercial products, or organizations imply endorsement by the U.S. Government. Animal care was provided in accordance with the procedure outlined in "The Guide for Care and Use of Laboratory Animals" (NIH Publication No. 86-23, 1985).

References

Aoki T, Tashiro K, Miyatake S, Kinashi T, Nakano T, Oda Y, Kikuchi H and Honji T (1992). Expression of murine interleukin 7 in a murine glioma cell line results in reduced tumorigencity in vivo. Proc Natl Acad Sci USA **89**, 3850-3864.

Asher AL, Mule JJ, Kasid A, Restifo NP, Salo JC, Reichert CM, Jaffe G, Fendly B, Kriegler M and Rosenberg SA (1991). Murine tumor cells transduced with the gene for tumor necrosis factor-α. J Immunol **146**, 3227-3234.

Bassil B, Dosoretz DE and Prout Jr GR (1985). Validation of the tumor, nodes and metastasis classification of renal cell carcinoma. J Urol **134**, 450-454.

Belldegrun A, Tso CL, Sakata T, Brunda MJ, Barsky SH, Chai J, Kaboo R, Lavey RS and McBride DH (1993). Human renal carcinoma line transfected with interleukin-2 and/or interferon alpha gene(s): Implications for live cancer vaccines. J Natl Cancer Inst **85**, 207-216.

Bosco MC, Giovarelli M, Forni M, Modesti A, Scarpa S, Masuelli L and Forni G (1990). Low doses of interleukin-4 injected perilymphatically in tumor bearing mice inhibit the growth of poorly and non-immunogenetic tumors and induce a tumor-specific immune memory. J Immunol **145**, 3136-3143.

Chang AE and Shu S (1992). Immunotherapy with sensitized lymphocytes. Cancer Invest **10**, 357-369.

Clark JW and Longo DL (1987). Interferons in Cancer Therapy. In "Cancer Statistics: Principles and Practice of Oncology" (eds VT DeVita Jr, S Hellman and SA Rosenberg), pp 1-16. Lippincott, Philadelpia.

Colombo MP, Ferrari G, Stoppacciaro A, Parenza M, Rodolfo M, Mavilio F and Parmiani G (1991). Granulocyte colony stimulating factor gene transfer suppresses tumorigenicity of a murine adenocarcinoma in vivo. J Exp Med **173**, 889-897.

Douvdevani A, Huleihel M, Zoller M, Segal S and Apte RN (1992). Reduced tumorigenicity fibrosarcomas which constitutively generate IL-1 alpha either spontaneously or following IL-1 alpha gene transfer. Int J Cancer **51**, 822-830.

Fearon ER, Pardoll DM, Itaya T, Golumbek P, Levitsky H.I., Simons JW, Karasuyama H, Vogelstein B and Frost P (1990). Interleukin-2 production by tumor cells bypasses T helper function in the generation of an antitumor response. Cell **60**, 397-403.

Forni G, Giovarelli M and Santoni A (1985). Lymphokine-activated tumor inhibition in vivo. The local administration of interleukin-2 triggers nonreactive lymphocytes from tumor bearing mice to inhibit tumor growth. J Immunol **134**, 1305-1311.

Franco JL, Wiltrout RH and Mizoguchi (1993). Treatment of established murine renal carcinoma with flavone-8-acetic acid (FAA) and rhIL 2 correlates with reversal of tumor induced T cell signal transduction alterations. J Cell Biochem **170**, 131.

Golumbek PT, Lazenby AJ, Levitsky HI, Jaffee LM, Karasuyama H, Baker M and Pardoll DM (1991). Treatment of established renal cancer by tumor cells engineered to secrete interleukin-4. Science **254**, 713-716

Huben RP, Connelly R and Godrosen MH (1983). Immunotherapy of a murine renal cancer. J Urol **129**, 1075-1078.

Mizoguchi H, O'Shea JJ, Longo DL, Loeffler CM, Mc Vicar DW and Ochoa A (1992). Alterations in signal transduction molecules in T lymphocytes from tumor-bearing mice. Science **258**, 1795-1798.

Mueller DL, Jenkins MK and Schwartz RH (1989). Clonal expansion versus functional clonal inactivation: A costimulatory signaling pathway determines the outcome of T cell antigen receptor occupancy. Ann Rev Immunol **7**, 445-480.

Mule JJ, Yang JC, Lafreniere RL, Shu SY and Rosenberg SA (1987). Identification of cellular mechanisms operational in vivo during the regression of established pulmonary metastases by the systemic administration of high-dose recombinant interleukin 2. J Immunol **139**, 285-294.

Mullen CA, Coale MM, Levy AT, Stetler-Stevenson WG, Liotta LA, Brandt S and Blaese RM (1992). Fibrosarcoma cells transduced with the IL-6 gene exhibit reduced tumorigenicity, increased immunogenicity and decreased metastatic potential. Cancer Res **52**, 6020-6024.

Murphy GP and Hruskesky WJ (1980). A murine renal cell carcinoma. J Natl Cancer Inst **50**, 1013-1016.

Pardoll DM (1993). Dissection of the immune response induced by tumors engineered to secrete lymphokines. J Cell Biochem **170**, 96.

Parkinson DR (1990). Interleukin-2: Further progress through greater understanding J Natl Cancer Inst **82**, 1374-1386.

Restifo NP, Spiess PJ, Karp SE, Mule JJ and Rosenberg SA (1992). A nonimmunogenic sarcoma transduced with the cDNA for interferon-γ elicits $CD8^+$ T cells against the wild-type tumor: Correlation with antigen presentation capability. J Exp Med **175**, 1423-1431.

Rosenberg SA, Packard BS, Aebersold PM, Solomon P, Topalian SL, Toy ST, Simon P, Lotze MT, Yang JC, Seip CA, Simpson GC, Carter C, Bock S, Schwartzetruber S., Wei JP and White DE (1988). Use of tumor-infiltrating lymphocytes and interleukin 2 in the immunotherapy of patients with metastatic melanoma. N Engl J Med **319**, 1676-1680.

Rosenberg SA, Lotze MT, Yang JC, Topalian SL, Chang AE, Schwartzentruber DJ, Aebershold P, Leitman S, Linehan WM and Seipp CA (1993). Prospective randomized trial of high-dose interleukin-2 alone or in conjunction with lymphokine-activated killer cells for the treatment of patients with advanced cancer. J Natl Cancer Inst **85**, 622-632.

Salup RR, Herberman RB and Wiltrout RH (1985). Role of natural killer activity in development of spontaneous metastases in murine renal cancer. J Urol **134**, 1236-1241.

Silverberg BS and Lubera JA (1989). Cancer statistics. Cancer J Clin **39**, 3-20.

Sosman JA, Hank JA, Moore KH, Borchert A, Schell K, Kohler PC, Goldstein D, Bechhofer R, Storer B and Albertini MR (1991). Prolonged interleukin 2 (IL-2) treatment can augment immune activation without enhancing antitumor activity in renal cell carcinoma. Cancer Invest **9**, 35-48.

Taub DD, Conlon K, Lloyd AR, Oppenheim JJ and Kelvin DJ (1993). Preferential migration of activated $CD4^+$ T cells in response to MIP-1. Science **260**, 355-358.

Tepper RI, Pattengale PK and Leder P (1989). Murine interleukin-4 displays potent anti tumor activity in vivo. Cell **57**, 503-512.

Tepper RI, Coffman RL and Leder P (1992). An eosinophil dependent mechanism for the antitumor effect of interleukin-4. Science **257**, 548-551.

Wiltrout RH (1991). The immunotherapy of murine renal cancer. Neuroend Immunol J **4**, 154-160.

Wiltrout RH, Komschlies KL and Franco JL (1993). Cytokine transfected 3T3 fibroblasts inhibit the progression of murine renal cancer. J Cell Biochem **170**, 121.

15

ANTITUMOR REACTIVITY ELICITED BY IL-4 TRANSDUCED TUMOR CELLS

E. Goillot and R. I. Tepper
MGH Cancer Center, Massachusetts General Hospital and Harvard Medical School, Charlestown, MA, USA

Introduction

In vitro analysis has demonstrated a wide range of functions of IL-4 on B cells, T cells, macrophages, on hematopoietic precursor cells and also on cells from nonhematopoietic origin, such as stromal cells.

Over the last several years, IL-4 has been demonstrated to have a potent antitumor activity in vivo, although the means through which IL-4 limits tumor cell growth remains controversial. IL-4 has been shown to induce both lymphoid independent inflammatory cell mechanisms and CTL mediated specific immune responses, depending upon the experimental model.

The biological basis for the antitumor mechanisms of IL-4 will be described, and some approaches for research investigations and clinical applications of the action of IL-4 will be discussed in this report.

An IL-4 induced inflammatory response results in potent antitumor activity

IL-4 has been demonstrated to have a potent antitumor activity by experiments in which the gene for IL-4 has been introduced into murine tumor cell lines (Tepper et al., 1989). Tumor cells are then transplanted into recipient hosts, and the growth of the cytokine secreting tumor cell lines are compared to mock transfected tumor cells.

Using this type of assay, which has been termed the "tumor-cytokine transplantation assay" (Tepper, 1992), it has been shown that IL-4 inhibits the growth of tumors of diverse histologic types, including lymphoid malignancies, plasmacytomas, sarcomas, various adenocarcinomas, melanomas, and malignant gliomas. These experiments have demonstra-

CYTOKINE-INDUCED TUMOR IMMUNOGENICITY
ISBN 0-12-261520-4

ted that the action of IL-4 is dose dependent, with complete killing of tumor requiring a high localized concentration of cytokine.

For example, the degree of in vitro growth inhibition of IL-4 transfected clones of the murine mammary adenocarcinoma line K485 increases with the concentration of IL-4 produced. At 50 days, the mean volume of the control tumor is 3000 mm^3, whereas the tumor size of clones producing 1500 U of IL-4/10^6 cells in 24 h is reduced to approximately 200 mm^3. The growth of most tumor cell lines expressing more than 5000 U of IL-4 are completely inhibited, regardless of the histologic type (Tepper et al., 1989; Tepper et al., 1992). The correlation between IL-4 production and tumor growth inhibition is further strengthened by experiments showing that this inhibition can be partially or fully reversed by in vivo delivery of varying concentrations of the neutralizing anti-IL-4 11B11 mAb (Tepper et al., 1992).

Evidence that the antitumor effect of IL-4 is not mediated by the action of IL-4 on tumor cells itself, but by host effector cells, is provided by:

- the absence of inhibitory effect or toxicity of IL-4 on the growth of most tumor cell lines in vitro (Tepper et al., 1989)
- the observation that murine IL-4 can inhibit the growth of human cell lines in nu/nu mice (Yu et al., 1993), despite the inability of murine IL-4 to bind to human IL-4 receptor (Mosley et al., 1989).

Eosinophils as a key effector cell for IL-4 induced tumor cytotoxicity

The fact that the antitumor effect of IL-4 is observed in nu/nu mice, as well as bg/bg mice (NK cell deficient), bg/nu/xid mice (B, T and NK cell deficient), and scid mice (B and T cell deficient), provides evidence that various T-cell subsets, as well as other lymphoid populations, are not required for IL-4 mediated killing (Tepper et al., 1992).

Histologic analysis of the injection site performed at different time points after administration of IL-4 transfected tumor cells reveals a host inflammatory reaction consisting initially of infiltration with a large number of eosinophils (within 18 h), followed by macrophage infiltration (48-96 h) and then fibroblastic scarring (Tepper et al., 1989). The majority of tumor cells are destroyed within the first 48 h following inoculation, at a time when the inflammatory infiltrate is almost exclusively comprised of eosinophils. Lymphocytes remain sparse at any stage of the host inflammatory response, although T lymphocytes can be observed in the infiltrate by two weeks after tumor cell inoculation (Golumbek et al., 1991). The significance of these $CD4^+$ and $CD8^+$ T cells with respect to immune stimulation against the parental tumor cell lines remains unclear at the present time.

The principal role of eosinophils in tumor killing has been directly demonstrated by depleting animals of mature granulocytes (neutrophils and eosinophils) with the mAb RB6-8C5. This antibody is specifically cytotoxic for granulocytes. Pretreatment of mice with this antibody resulted in rapid restoration of tumor immunogenicity of IL-4 producing cells. The kinetics of tumor formation was similar to that observed with the parental tumor cell line. Histological analysis confirmed tumor viability with numerous mitoses, with a greater than 95% reduction in the number of infiltrating eosinophils. With high IL-4 producing tumor cell lines, synergy was observed using an antibody against IL-4 together with the RB6-8C5 anti-granulocyte antibody (Tepper et al., 1992). Moreover, the neutralization of IL-5, a cytokine known to participate in the differentiation of eosinophils from myeloid precursors in the bone marrow, was not effective in restoring the tumorigenicity of IL-4 producing cells, suggesting that, in the presence of localized IL-4 production, a potent stimulus for the migration of preformed eosinophils into tissues may still occur (Tepper et al., 1992).

Relatively little is known about the mechanism by which eosinophils accumulate and are activated at local sites of tissue inflammation. The activation of endothelial cells by TNF-α and IL-1 induces the adherence of all granulocyte types (Butcher, 1991) by increasing the expression of three adhesion molecules on the surface of endothelial cells: ELAM-1 (Pober et al., 1986; Wellicome et al., 1990), ICAM-1 (Dustin et al., 1986) and VCAM-1 (Osborn et al., 1989). Each of these adhesion molecules has been shown to be capable of binding purified eosinophils (Weller et al., 1991). IL-4 has been shown to stimulate alone or synergistically with IL-1 or TNF-α the expression of VCAM-1 on human endothelial cells, while ICAM-1 and ELAM-1 are unaffected by the action of IL-4 (Thornhill et al., 1991; Walsh et al., 1991; Schleimer et al., 1992).

Eosinophil adherence induced by IL-4 is largely inhibitable by treatment with antibodies against VCAM-1 and VLA-4 (the leukocyte cell surface receptor for VCAM-1); in vitro, these two antibodies have no effect on neutrophil adherence (Dobrina et al., 1989; Walsh et al., 1991; Schleimer et al., 1992).

The fact that eosinophils, but not neutrophils, express VLA-4 might explain the selectivity of the action of IL-4 on eosinophil transmigration into tissues (Walsh et al., 1991). Moreover, by inhibiting the expression of ELAM-1 and ICAM-1 induced by TNF-α and IL-1 in vitro, IL-4 may impose the selective expression of VCAM-1 (Thornhill and Haskard, 1990). These findings therefore suggest that the characteristic inflammatory infiltrate induced by IL-4 may be related to the specific induction of VCAM-1 on the endothelial cell surface.

The expression of VCAM-1 in the inflammatory lesions found at the site of injections of IL-4 producing tumors, by immunostaining or by detection of specific RNA, and the utilization of neutralizing mAb in vivo against this adhesion molecule or its integrin counterreceptor, can provide information as to the importance of VCAM-1 induction in IL-4 induced inflammation and tumor cytotoxicity.

Does IL-4 induce a T lymphocyte dependent memory response against tumors?

Using the tumor cytokine transplantation assay, at least five cytokines (IL-2, IFN-γ, IL-6, IL-7 and GM-CSF) have been found to reduce tumorigenicity, ostensibly by stimulating specific immune responses through their action on the expansion of Th and/or CTL populations (Watanabe et al., 1989; Fearon et al., 1990; Gansbacher et al., 1990; Hock et al., 1991; Jicha et al., 1991; Porgador et al., 1992; Dranoff et al., 1993). In the reports involving IL-2 and IFN-γ, however, lack of important controls assessing the immune enhancing potential of growth inactivated, parental (non-cytokine producing) cells makes it difficult to ascribe the observed immune potentiation to the expression of the cytokine. This is especially important given that the cell types used in these studies are known to possess some degree of inherent immunogenicity in syngeneic hosts.

In two reports, IL-4 has been described to play a role in the augmentation of tumor specific immunity. In the first report, IL-4 was shown to induce memory specific immunity by daily local s.c. administration of small amounts (0.00001 to 1000 pg/day) of IL-4 around tumor draining lymph nodes. In this model, IL-4 induced appreciable inhibition of the growth of the tumor cells (using the dose of 0.1 pg/day) and protected 1/3 of the mice upon contralateral tumor challenge 60 days after the initial immunization with tumor plus IL-4. Adoptive transfer experiments showed that Thy1$^+$ lymphocytes and in particular the CD4$^+$ cell subpopulation was required for the immune memory (Bosco et al., 1990).

The other report showed the generation of antigen specific antitumor immunity for a spontaneous arising murine renal cell tumor, RENCA, engineered to secrete high levels of IL-4 (1500 U/10^5 cells in 24 h). Tumor cells were rejected in a T-cell dependent manner upon subsequent challenge with the parental tumor (Golumbek et al., 1991).

However, these studies again provided no information regarding the inherent immunity of the tumor model used. Certain murine and human tumors have been shown to be inherently immunogenic (Van Pel et al., 1983; Van der Bruggen et al., 1991; Van den Eynde et al., 1991). Dranoff et al. (1993) showed that at the vaccine and challenge doses used in previous studies with RENCA and other cell lines (for example, those used in the IL-4, IL-2 and IFN-γ studies noted above), irradiated cells alone were as immunogenic as live cytokine transduced cells.

These results suggest that the antitumor activity of some cytokines, previously described using this type of assay, may be due predominantly to their capacity to promote the destruction of the tumor cells at the site of injection and not as a result of the augmentation of specific immune effector cells.

Thus, in experiments evaluating the immune capabilities of IL-4 or any other cytokines, more information on the immunogenicity of inactivated cells alone (inactivation achieved, for example, by irradiation or treatment by mitomycin C) are needed, in order to accurately assess the ability of a cytokine to augment tumor immunity. Although IL-4 has been shown to be an autocrine factor for some Th cells and can augment CTL proliferation in vitro, the role of this cytokine in the induction or augmentation of tumor immunity remains an important unanswered question.

Possible therapeutic implications of the antitumor activity of IL-4

Initial studies of IL-4 have focused on the failure of tumors to develop in the presence of the cytokine. Given the non cell-autonomous effect of IL-4 (Tepper et al., 1989) and the ability of IL-4 to affect the regression of established tumors (Tepper et al., 1992), a potential clinical utility of this cytokine is worthy of investigation.

We have demonstrated that the activity of IL-4 in large doses can inhibit the growth of established tumors, both by direct administration of purified recombinant cytokine into the tumor mass or by injection at the tumor site of IL-4 producing cells (Tepper et al., 1992). Moreover, we have recently shown that the associated antitumor activity of IL-4 mediated by eosinophils is not restricted to s.c. tissue sites, but it can be induced in a number of other tissues tested including the peritoneal cavity (Tepper, unpublished), the brain (inhibition of the growth of a human glioblastoma in situ in nude mice) (Yu et al., 1993) and experimental metastases established in the lungs and in the liver (Goillot and Tepper, unpublished). Since the localized antitumor effect is not achievable by systemic delivery of the cytokine, the major problem which remains is the need to direct IL-4 to the site of the tumor. Experimental approaches such as direct instillation of cytokine or cytokine producing cells (e.g. in the case of brain tumors) or the targeted delivery of cytokine using viral vectors are currently under investigation.

References

Bosco M, Giovarelli M, Forni M, Modesti A, Scarpa S, Masuelli L and Forni G (1990). Low doses of IL-4 injected perilymphatically in tumor-bearing mice inhibit the growth of poorly and apparently nonimmunogenic tumors and induce a tumor-specific immune memory. J Immunol **145**, 3136-3143.

Butcher EC (1991). Leukocyte-endothelial cell recognition: Three (or more) steps to specificity and diversity. Cell **67**, 1033-1039.

Dobrina A, Schwartz BR, Carlos TM, Ochs HD, Beatty PG and Harlan JM (1989). CD11/CD18-independent neutrophil adherence to inducible endothelial-leucocyte adhesion molecules (ELAM) in vitro. Immunology **67**, 502-508.

Dranoff G, Jaffee E, Lazenby A, Golumbek P, Levitsky H, Brose K, Jackson V, Hamada H, Pardoll D and Mulligan RC (1993). Vaccination with irradiated tumor cells engineered to secrete murine granulocyte-macrophage colony-stimulating factor stimulates potent, specific, and long-lasting antitumor immunity. Proc Natl Acad Sci USA **90**, 3539-3543.

Dustin ML, Rothlein R, Bhan AL, Dinarello CA and Springer TA (1986). Induction by IL1 and interferon-γ: Tissue distribution, biochemistry, and function of a natural adherence molecule, (ICAM-1). J Immunol **137**, 245-254.

Fearon ER, Pardoll DM, Itaya T, Golumbek P, Levitsky HI, Simons JW, Karasuyama H, Vogelstein B and Frost P (1990). Interleukin-2 production by tumor cells bypassess T helper function in the generation of an antitumor response. Cell **60**, 397-403.

Gansbacher B, Zier K, Daniels B, Cronin K, Bannerji R and Gilboa E (1990). Interleukin 2 gene transfer into tumor cells abrogates tumorigenicity and induces protective immunity. J Exp Med **172**, 1217-1224.

Golumbek P, Lazenby A, Levitsky HI, Jaffee LM, Karasuyama H, Baker M and Pardoll DM (1991). Treatment of established renal cancer by tumor cells engineered to secrete interleukin-4. Science **254**, 713-716.

Hock H, Dorsch M, Diamantstein T and Blankenstein T (1991). Interleukin 7 induces $CD4^+$ T cell-dependent tumor rejection. J Exp Med **174**, 1291-1298.

Jicha DL, Mule JJ and Rosenberg SA (1991). Interleukin 7 generates antitumor cytotoxic T lymphocytes against murine sarcomas with efficacy in cellular adoptive immunotherapy. J Exp Med **174**, 1511-1515.

Mosley B, Beckmann P, March CJ, Idzerda RL, Gimpel SD, Van den Bos T, Friend D, Alpert A, Anderson D, Jackson J, Wignall JM, Smith C, Gallis B, Sims JE, Urdal D, Widmer MB, Cosman D and Park LS (1989). The murine interleukin-4 receptor: Molecular cloning and characterization of secreted and membrane bound forms. Cell **59**, 335-348.

Osborn L, Hession C, Tizard R, Vassalo C, Luhowsky S, Chi-Rosso G and Lobb R (1989). Direct expression cloning of vascular cell adhesion molecule 1, a cytokine-induced endothelial protein that binds to lymphocytes. Cell **59**, 1203-1211.

Pober JS, Bevilacqua MP, Mendrick DL, Lapierre LA, Fiers W and Gimbrone MA (1986). Two distinct monokines, interleukin-1 and tumor necrosis factor, each independently induce biosynthesis and transient expression of the same antigen on the surface of cultured human vascular endothelial cells. J Immunol **136**, 1680-1684.

Porgador A, Tzehoval E, Katz A, Vadai E, Revel M, Feldman M and Eisenbach L (1992). Interleukin 6 gene transfection into Lewis lung carcinoma tumor cells suppresses the malignant phenotype and confers immunotherapeutic competence against parental metastatic cells. Cancer Res **52**, 3679-3686.

Schleimer RP, Sterbinsky SA, Kaiser J, Bickel CA, Klunk DA, Tomioka K, Newman W, Luscinskas FW, Gimbrone MA, McIntyre BW and Bochner BS (1992). IL-4 induces adherence of human eosinophils and basophils but not neutrophils to endothelium. J Immunol **148**, 1086-1092.

Tepper RI, Pattengale PK and Leder P (1989). Murine interleukin-4 displays potent antitumor activity in vivo. Cell **57**, 503-512.

Tepper RI (1992). The tumor-cytokine transplantation assay and the antitumor activity of interleukin-4. Bone Marrow Transpl **9** (suppl 1), 177-181.

Tepper RI, Coffman RL and Leder P (1992). An eosinophil-dependent mechanism for the antitumor effect of IL-4. Science **257**, 548-551.

Thornhill MH and Haskard DO (1990). IL-4 regulates endothelial cell activation by IL-1, tumor necrosis factor, or IFN-γ. J Immunol **145**, 865-872.

Thornhill MH, Wellicome SM, Mahiouz DL, Lanchbury JSS, Kyan-Aung U and Haskard DO (1991). Tumor necrosis factor combines with IL-4 or IFN-γ to selectively enhance endothelial cell adhesiveness for T cells. The contribution of vascular cell adhesion molecule-1-dependent and -independent binding mechanisms. J Immunol **146**, 592-598.

Van der Bruggen P, Traversari C, Chomez P, Lurquin C, DePlaen E, Van den Eynde B, Knuth A and Boon T (1991). A gene encoding an antigen recognized by cytolytic T lymphocytes on a human melanoma. Science **254**, 1643-1647.

Van den Eynde B, Lethe B, Van Pel A, De Plaen E and Boon T (1991). The gene coding for a major tumor rejection antigen of tumor P815 is identical to the normal gene of syngeneic DBA/2 mice. J Exp Med **173**, 1373-1384.

Van Pel A, Vessiere F and Boon T (1983). Protection against two spontaneous mouse leukemias conferred by immunogenic variants obtained by mutagenesis. J Exp Med **157**, 1992-2001.

Walsh GM, Mermod JJ, Hartnell A, Kay AB and Wardlaw AJ (1991). Human eosinophil, but not neutrophil, adherence to IL-1-stimulated human umbilical vascular endothelial cells is α4β1 (very late antigen-4) dependent. J Immunol **146**, 3419-3423.

Watanabe Y, Kuribayashi K, Miyatake S, Nishihara K, Nakayama E, Taniyama T and Sakata T (1989). Exogenous expression of mouse interferon γ cDNA in mouse neuroblastoma C1300 cells results in reduced tumorigenicity by augmented anti-tumor immunity. Proc Natl Acad Sci USA **86**, 9456-9460.

Weller PF, Rand TH, Goelz SE, Chi-Rosso G and Lobb RR (1991). Human eosinophil adherence to vascular endothelium mediated by binding to vascular cell adhesion molecule 1 and endothelial leukocyte adhesion molecule 1. Proc Natl Acad Sci USA **88**, 7430-7433.

Wellicome SM, Thornhill MH, Thomas DS, Pitzalis JS, Lanchbury S, Panayi GS and Haskard DO (1990). A monoclonal antibody that detects a novel antigen on endothelial cells that is induced by TNF, IL-1 or LPS. J Immunol **144**, 2558-2562.

Yu JS, Wei MX, Chiocca EA, Martuza RL and Tepper RI (1993). Treatment of glioma by engineered interleukin 4-secreting cells. Cancer Res **53**, 3125-3128.

16

ANTITUMOR RESPONSES ELICITED BY MOUSE ADENOCARCINOMA CELLS ENGINEERED TO RELEASE IL-2 AND IL-4

F. Cavallo, M. P. Colombo, A. Allione, S. Bertini, M. Consalvo, F. Novelli and M. Giovarelli
CNR Center for Immunogenetics and Experimental Oncology at the Institute of Microbiology, University of Turin, Turin, Italy and Istituto Nazionale Tumori, Division of Experimental Oncology D, Milan, Italy

Introduction

Many features of both nonspecific and specific immune responses depend upon cytokines. Their interplay guides granulocyte, macrophage and NK activities during inflammatory responses, the progression of T and B lymphocytes following antigen recognition, the deflection of specific responses towards Th1 or Th2 type reactions, and the antibody isotype switching (Smith, 1990). Consequently, when an antigen enters the immune system the local cytokine setting plays a pivotal role in regulating the features of the ensuing response.

These reasons form the rationale for the local exploitation of cytokines in modulating immune responses (Forni et al., 1988). Even if their multifunctional activities still makes it difficult to use them to pilot a separate immune function, they can be used as a new form of adjuvant able to make antigens more immunogenic (Colombo and Forni, 1994) and overcome tolerance (Malkowsky et al., 1985).

Conventional adjuvants activate several effector functions directly and elicit inflammatory responses by promoting the release of a large repertoire of cytokines. The functions activated and the cytokines induced are often too many to be studied, and their efficacy is evaluated empirically. The use of local cytokines results in a slightly tighter system, the features of which can be understood with greater precision at the gene and cell level (Audibert and Lise, 1993).

Much detailed information has been gathered with regard to the molecular aspects of the action of cytokines. By contrast, our knowledge of their major activities in vivo is certainly no more extensive than that of conventional adjuvants; it is thus not surprising that they are still being employed in a crude fashion and hence do not seem to offer any greater efficacy (Hock et al., 1993). Understanding of the mechanisms that are

CYTOKINE-INDUCED TUMOR IMMUNOGENICITY
ISBN 0-12-261520-4

activated in vivo by cytokines will help us to use them more appropriately.

Several data have shown that the relationship between a tumor and the host immune system can be altered by the local presence of cytokines. These do not affect tumor proliferation directly, but they activate a strong antitumor reaction (see Colombo et al., 1992 for a review). Interestingly, this reaction is the result of a continuous integration between inflammatory and specific immune responses leading to both local tumor inhibition and systemic immunity. The sequential passage from local, antigen nonspecific inflammatory reactions of great power in inhibiting established tumors to a systemic response able to hamper tumor recurrence and metastasis even against poorly immunogenic tumors, makes the use of cytokines of particular interest.

The features of both inflammatory and specific reactions can be shaped in a characteristic way by distinct cytokines (Hock et al., 1993; Musiani et al., 1994). Here we compare the features of the immune reactions elicited by the cells of a spontaneous murine adenocarcinoma engineered to release a typical Th1 or Th2 cytokine, namely IL-2 or IL-4.

Materials and methods

Mice. Seven-week-old female BALB/cnAnCr (BALB/c) (H-2^d, Mls^b) mice (Charles River Lab., Calco, Italy) were treated in accordance with the European Community guidelines. When sublethal whole body irradiation was required, mice received 5 Gy from a ^{137}Cs source (dose rate: 0.5 Gy/min). Starting 1 h after challenge, mice received 10 daily i.p. injections of 75 mg/kg of cyclosporin A (Sandimmun, Sandoz, Basel, Switzerland) diluted in 0.2 ml of Hank's basal salt solution (HBSS). Two days before, and 4 h, 3, 7, 10 and 13 days after challenge, other mice received 6 i.p. injections of 0.2 ml of HBSS containing a 1/20 dilution of anti-asialo GM1 rabbit antiserum (Wako Chemicals GmbH, Dusseldorf, Germany), or 100 μg anti-CD4 (GK1.5 hybridoma, L3T4, American Type Culture Collection, ATCC, Rockville, MD), anti-CD8 (TIB-105 hybridoma, Lyt 2, ATCC), anti-granulocyte mAb (RB6-8C5 hybridoma kindly provided by Dr R.L. Coffman, DNAX Inc., Palo Alto, CA), or normal rat Ig purified from serum or ascitic fluid by passage through an anionic exchange column (DE 52, Whatman Ltd., Maidstone, England) (Cavallo et al., 1992). Cytofluorimetric analysis of the residual blood and spleen cells (Spc) from mice receiving these antibodies showed that target leukocytes were selectively decreased to <1/5000 of PBL during treatment.

Tumor cells. Cell cultures were performed at 37° C in a humidified 5% CO_2 atmosphere, using RPMI 1640 medium supplemented with 10% FCS, 100 U/ml penicillin, 100 U/ml streptomycin, 50 μg/ml gentamycin and 2.5 x 10^{-5} M 2β-mercaptoethanol (Flow Laboratories, Opera, Italy). TS/A parental cells (TS/A-pc) were obtained from the 8th in vitro passage of a tumor cell line established from the first in vivo transplant of a moderately differentiated mammary adenocarcinoma which arose spontaneously in a BALB/cnAnCr female mouse. TS/A-pc express MHC class I, but not class II molecules and do not stimulate a syngeneic antitumor response in vitro nor in vivo (Forni et al., 1988). Four x 10^4 cells are the minimal 100% tumor inducing dose in BALB/c mice. When necessary, TS/A-pc were treated with mitomycin-C (Mit-C, Sigma, St. Louis, MO) (60 μg/ml) in a cell suspension of 1 x 10^7 cells/ml for 30 min at 37°C (Cavallo et al., 1992).

In vivo evaluation of tumor growth. Mice were challenged s.c. in the left or right flank with 0.2 ml of a suspension containing the specified number of tumor cells. For 120 days, mice were inspected twice weekly and neoplastic masses were measured with calipers in the two perpendicular diameters in a fashion blind to the group in which they had been treated. Latency and survival times were considered as the periods (in days) between challenge and the growth of a neoplastic mass >3 mm and >10 mm mean diameter, respectively. Mice were euthanized for humane reasons when the tumor exceeded this maximum. Mice free of tumor at day 120 were classed as survivors.

Transduction of cytokine genes. TS/A-pc were transfected by electroporation with the linearized BCMGneo (neomycin resistance gene) (NEO cells) or BCMGneo-IL-2 plasmids (Karasuyama et al., 1989), cloned and cultured in selective medium as previously described (Cavallo et al., 1992). For IL-4 transduction, the L4SN vector was constructed by introducing the murine IL-4 cDNA into the HPA I site of the LXSN vector (Miller and Rosman, 1989). The IL-4 cDNA was the 480 bp Bam HI-Sca I fragment contained in the plasmid pD2-E3 (ATCC, # 37561). Transfected cells were selected with 0.8 mg/ml of G418 (Gibco, Grand Island, NY) and then used to generate virus containing supernatants at a titer of 10^5 colony forming U/ml. TS/A-pc were infected for 2 h with 5 ml of this supernatant in the presence of 8 mg/ml of polybrene, grown for an additional 24 h, and then selected in the presence of 1.2 mg/ml of G418. Resistant cells were cloned as previously described (Cavallo et al., 1992).

Cytokine titration. Six representative IL-2 secreting clones (B clones) and four representative IL-4 secreting clones (D clones) were used. The last number of their denomination indicates the mean units of IL-2 or IL-4 released by 1 x 10^5 cells of each clone cultured for 48 h in 1 ml of complete medium, evaluated as the ability to support the growth of the mouse cell line HT-2, which is both IL-2 and IL-4 responsive (Lichtman et al., 1987). The titer that gave 50% of the maximal ^{3}H-thymidine uptake was determined by a probit analysis computer program and calibrated against the Biological Response Modifiers Program Reference reagents human IL-2 or murine IL-4.

Cellular cytotoxicity. CTL were generated by culturing 1 x 10^7 responder Spc and 5 x 10^5 Mit-C stimulator tumor cells for 6 days in mixed lymphocyte-tumor cell cultures (MLTC). Cytotoxicity was assayed in a 48 h ^{3}H-thymidine release assay (Giovarelli et al., 1982) by mixing various concentrations of effector cells with 5 x 10^3 labeled target cells at 50:1, 25:1, 12:1, and 6:1 effector to target (E:T) ratios in 96-well microtiter plates in triplicate. Lysis was determined as previously described in detail (Cavallo et al., 1992), and values were expressed as lytic units $(LU)_{20}/1 \times 10^7$ effector cells calculated according to the equation of Pross (Trinchieri, 1989). Proliferation was evaluated by culturing in triplicate 4 x 10^5 Spc/well, alone or with different amounts of Mit-C stimulator cells, for 5 days in flat-bottomed microtest plates using the basic medium supplemented with 4% FCS only. Each well was pulsed with 1 µCi of ^{3}H-thymidine for 8 h before harvesting. ^{3}H-thymidine uptake was assayed in a scintillation spectro-photometer, and the data were expressed as total cpm less spontaneous uptake of responder and stimulator cells (Δ cpm).

Statistical analysis. All in vitro experiments were performed in triplicate, and repeated independently at least three times. A representative experiment is shown. All in vivo experiments were performed three or more times with groups of 4-6 mice, and were cumulated as they gave homogeneous results. The significance of differences in tumor takes was determined by Fisher's exact method, whereas those in latency and survival time and in vitro data were determined by a two-tailed Wilcoxon's signed test.

Results

Characteristics of B clones

Six clones displaying different amounts of the IL-2 specific mRNA as shown by RNA blotting analysis were established from the bulk culture of TS/A-pc transfected with the IL-2 gene (Cavallo et al., 1992). Two of them secreted low (B1.30, B5.500), two intermediate (B2.2000, B6.3600), and two high (B4.6000, B3.6800) amounts of IL-2. A G418-resistant cell population transfected with the BCMGneo vector (without the IL-2 gene) neither displayed IL-2 mRNA nor released IL-2 and was used as a negative control (NEO).

Characteristics of D clones

Two clones secreting small amounts (D1.5, D2.5) and two larger amounts (D4.20, D5.40) of IL-4 were established from the bulk culture of TS/A-pc infected with the L4SN vector containing the murine IL-4 cDNA.

Table 16.1. *Oncogenicity mice of B and D clones in syngeneic Balb/c*

Challenge[a]	Tumor takes/ total mice	Latency time (days)	Survival time (days)
TS/A-pc	20/20 (100%)	11 ± 2	23 ± 2
NEO	20/20 (100%)	14 ± 3	25 ± 4
NEO-i	20/20 (100%)	12 ± 3	22 ± 2
B1.30	13/20[b] (65%)	25 ± 4[b]	44 ± 6[b]
B5.500	0/20[b] (0%)	-	-
B2.2000	0/20[b] (0%)	-	-
B6.3600	0/20[b] (0%)	-	-
B4.6000	0/20[b] (0%)	-	-
B3.6800	0/20[b] (0%)	-	-
D2.5	16/20 (80%)	26 ± 5[b]	40 ± 3[b]
D1.5	15/20 (75%)	24 ± 3[b]	33 ± 2[b]
D4.20	11/20[b] (55%)	25 ± 2[b]	45 ± 4[b]
D5.40	4/20[b] (20%)	28 ± 3[b]	47 ± 7[b]

[a]Mice were challenged with 1 x 10^5 cells
[b]Value significantly different ($p < 0.01$) from that of mice challenged with TS/A-pc

The G418-resistant cell population infected with LXSN (NEO-i) did not release IL-4 and was used as negative control. The in vitro doubling times of the IL-2 and IL-4 releasing clones, TS/A-pc, NEO and NEO-i cells were similar, suggesting that IL-2 and IL-4 secretion has no direct cytotoxic or cytostatic effect on TS/A cells (data not shown).

Oncogenicity of B and D clones

To evaluate whether constitutive secretion of IL-2 or IL-4 affects the ability to form tumor, mice were challenged with 1 x 10^5 cells from the variously gene engineered and control TS/A cells. All mice challenged with TS/A-pc, NEO, and NEO-i cells developed tumor with a similar progression pattern (Table 16.1). By contrast, TS/A clones releasing more than 30 U of IL-2 were promptly rejected by all the mice, whereas 60% of mice challenged with B1.30 displayed tumors. With D clones, 80%, 73%, 53% and 20% of mice challenged with D2.5, D1.5, D4.20 and D5.40 cells

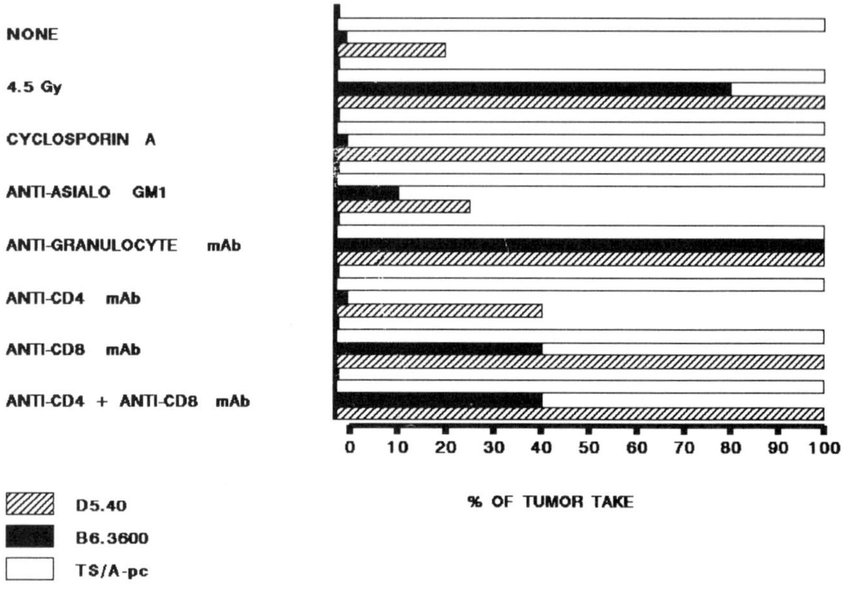

Fig. 16.1. *Effect of selective immunosuppression of recipient BALB/c mice on the ability of TS/A-pc, B6.3600 and D5.40 cells to form progressively growing tumors.* Mice were challenged in the left flank with 1 x 10^5 cells and treated as described in Materials and methods before and after the challenge.

respectively displayed tumors. Tumors formed by B1.30 and D clone cells were characterized by delayed growth patterns suggesting that the oncogenicity of transfected cells inversely correlates with the amount of IL-2 or IL-4 secreted.

Distinct host reactivity mechanisms restrain the oncogenicity of B and D clones

To evaluate the weight of host immune mechanisms in limiting the oncogenicity of B and D clones, recipient mice were variously immunosuppressed (Fig. 16.1). Sublethal total body irradiation allowed B6.3600 to grow in 80% of mice and D5.40 to grow in all mice, showing that both clones are potentially oncogenic, but their growth is hampered by radiation sensitive mechanisms activated by the very cytokines they produce. Treatment with cyclosporin A did not increase B6.3600 cell takes, whereas it allowed D5.40 cells to grow in all mice, suggesting that the active release of secondary cytokines is an important feature in the IL-4 elicited reaction only. The action of cyclosporin A is quickly reversible, however, and one-third of the D5.40 tumors progressively growing during cyclosporin A treatment regressed when this was interrupted.

A marginal increase in tumor takes was found with both B6.3600 and D5.40 cells after the removal of asialo $GM1^+$ cells. By contrast, B6.3600 and D5.40 cells grew and killed all mice treated with anti-granulocyte mAb, confirming the major role of local IL-2 and IL-4 in promoting tumor killing by granulocytes (Forni et al., 1988; Tepper et al, 1992; Modesti et al., 1993). While depletion of $CD4^+$ lymphocytes did not affect significantly B6.3600 nor D5.40 cell takes, depletion of $CD8^+$ lymphocytes partially increased B6.3600 takes but allowed D5.40 cells to grow in all mice.

Distinct ability of B6.3600 and D5.40 cells to stimulate T lymphocyte proliferation and cytotoxicity

TS/A-pc, NEO, NEO-i and D5.40 cells were unable to stimulate a proliferative response in either Spc from untreated mice or Spc from tumor bearing mice (TB-mice) displaying a 5 mm mean diameter tumor (Table 16.2). The Spc recovered from these MLTC were tested for their cytotoxicity to TS/A-pc. Poor cytotoxicity (8-10 LU) was evident in Spc from untreated mice stimulated by TS/A-pc, NEO or NEO-i cells. No cytotoxicity was found following stimulation with D5.40 cells, whereas a strong cytotoxicity ensued after B6.3600 stimulation (Table 16.2). Spc from TS/A TB-mice displayed a moderate (33-41 LU) cytotoxic response following stimulation with TS/A-pc, NEO or NEO-i cells. Again, no cytotoxicity was found following D5.40 stimulation, whereas high cytotoxicity ensued after B6.3600 stimulation. This inability of D5.40 cells to induce either a proliferative or a cytotoxic response is clearly not dependent on low cytokine production since both types of response are elicited also by B1.30, which produces even less. It can thus be deduced that D5.40 cells inhibit, whereas B clone cells stimulate cytotoxic responses.

Ability of B6.3600 and D5.40 cells to preimmunize

To determine whether the rejection of cytokine transduced cells leads to an immune memory against TS/A-pc, mice with no tumor following the challenge with IL-2 and IL-4 secreting cells performed 30 days earlier in the left flank were challenged with TS/A-pc in the right flank.

No protection was conferred by the previous injection of Mit-C TS/A-pc. Little protection was found in mice that had rejected the low IL-2 producer B1.30 clone. By contrast, rejection of B clones releasing higher amounts of IL-2 conferred protection in 55-75% of mice (Fig. 16.2).

Table 16.2. *Proliferative and cytotoxic responses of Spc from normal or TS/A-pc TB-mice stimulated by B6.3600 and D5.40 cells*

Responder cells from mice challenged with	Stimulator cells[a]	^{3}H-thymidine uptake (Δ cpm x 10^{-3})	Cytotoxicity to[b]	
			TS/A-pc	CE-2
Nothing	TS/A-pc	-0.8	10	2
	NEO	-0.7	9	1
	NEO-i	-0.8	8	0
	B1.30	35.1	92	1
	B6.3600	92.0	230[c]	1
	D5.40	-0.1	0[c]	1
TS/ A-pc	TS/A-pc	-7.0	41	2
	NEO	-6.0	33	1
	NEO-i	-7.2	35	2
	B1.30	35.1	92	1
	B6.3600	84.0	602[c]	1
	D5.40	-8.4	0[c]	1

[a]Responder Spc were cultured with Mit-C tumor cells at a 20:1 responder : stimulator ratio

[b]Cytotoxicity was evaluated in a 48h ^{3}H-thymidine release test and expressed as $LU_{20}/10^7$

[c]Values significantly different from those observed when responder lymphocytes were stimulated by all the other cells

A striking difference in the ability to preimmunize was found with D clones. Total protection was observed after rejection of all four D clones, irrespective of the amount of IL-4 secreted by each of them.

To evaluate a few requirements for the induction of this memory, groups of mice were immunosuppressed during the first challenge with transduced cells. As shown in Fig. 16.1, immunosuppression with anti-CD4 mAb or cyclosporin A did not impair their ability to reject a B6.3600 challenge, and all mice survived. By contrast, these immunosuppressive treatments blocked the induction of a tumor specific memory against TS/A-pc (Table 16.3), showing that CD4+ lymphocytes and the release of additional cytokines were not required for the rejection of B6.3600 cells, whereas they played a central role in memory induction. In the case of D5.40 cells, tumor free mice following the interruption of cyclosporin A treatment (Fig. 16.1) were fully resistant to the subsequent TS/A-pc challenge (Table 16.3). Immunosuppression with anti-CD4, which slightly

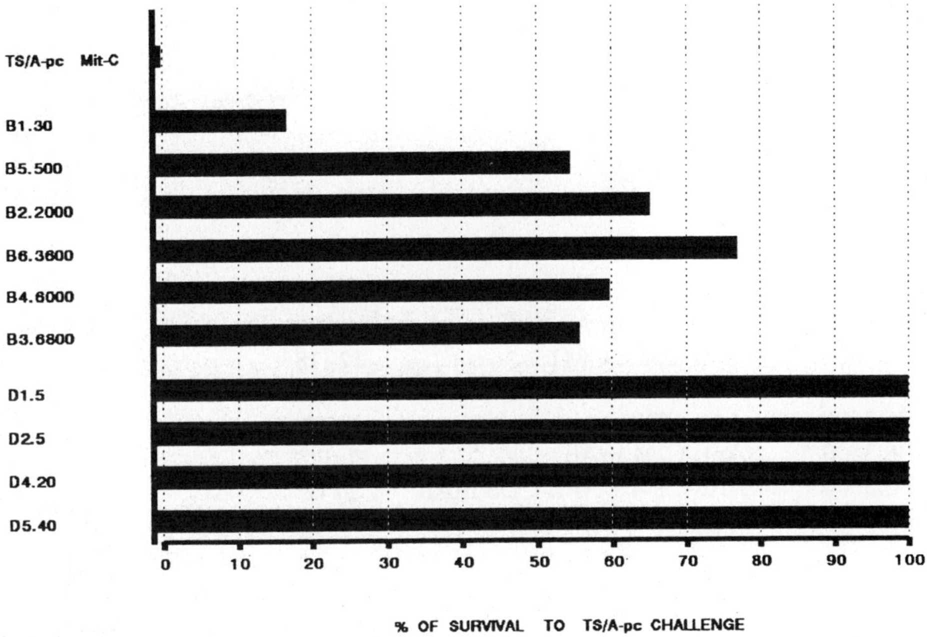

Fig. 16.2. *Ability of IL-2 and IL-4 transduced cells to induce protection against a contralateral challenge of TS/A-pc.* Mice that had received, at day 0, 1 x 10^5 cells of TS/A-pc Mit-C, IL-2, and IL-4 transduced clones injected in the left flank were challenged in the right flank with 4 x 10^4 TS/A-pc at day 30. Mice that were tumor free 120 days after challenge were classified as survivors. All transduced clones provided significant protection ($p<0.001$) as compared to TS/A-pc Mit-C.

impaired first D5.40 cell rejection (Fig. 16.1), significantly affected the ability to acquire an immune memory to TS/A-pc (Table 16.3).

Ability of B6.3600 and D5.40 cells to cure TB-mice

To evaluate whether B6.3600 or D5.40 cells can also affect the growth of 1, 7, or 14 day old tumors, groups of 20-24 mice challenged s.c. in the middle of the left flank or i.v. with TS/A-pc at day 0, subsequently received in the right flank single or multiple s.c. injections of 1 x 10^5 B6.3600 or D5.40 cells, starting at various times after challenge (Fig. 16.3).

Multiple B6.3600 injections cured a significant number of mice bearing contralateral TS/A-pc tumors, thought their efficacy decreased from 35-30% for 1-day to 15-20% and 8-18% for 7- and 14-day tumors, respectively. Significant protection against 1- and 7-day old lung metastases was also provided.

Table 16.3. *Requirement for the acquisition of an immune memory to TS/A-pc following the rejection of B6.3600 and D5.40 cells*

First challenge		Second challenge with TS/A-pc	
Tumor	Immunosuppressive treatment	Surviving/ total mice	% of survival
B6.3600	None	13/20	65
B6.3600	Cyclosporin A	0/20	0
B6.3600	Anti-asialo GM1	12/18	66
B6.3600	Normal rabbit serum	14/20	70
B6.3600	Anti-CD8 mAb	7/12	58
B6.3600	Anti-CD4 mAb	1/20	5
B6.3600	Anti-CD4 + Anti-CD8 mAb	2/12	12
B6.3600	Normal rat Ig	12/20	60
D5.40	None	16/16	100
D5.40	Cyclosporin A	12/12	100
D5.40	Anti-asialo GM1	15/15	100
D5.40	Normal rabbit serum	14/14	100
D5.40	Anti-CD4 mAb	8/12	60
D5.40	Normal rat Ig	16/16	100

Discussion

Establishment of a tumor may also be regarded as the outcome of the absence or ineffectiveness of an immune response on the part of the host. There are various reasons for failures of this kind, and a sort of blindness vis-à-vis neoplastic cells is by no means the least important. It has, however, been shown that in some cases sight can be restored by helping the immune system to recognize a tumor and mount an effective reaction against it. One way of doing this is offered by the use of cytokines constitutively released by genetically engineered tumor cells. Comparison of the literature data on this subject is hampered by the fact that different tumor models have been employed. This paper, therefore, provides a comparative evaluation of the antitumor response elicited by IL-2 and IL-4 releasing cells.

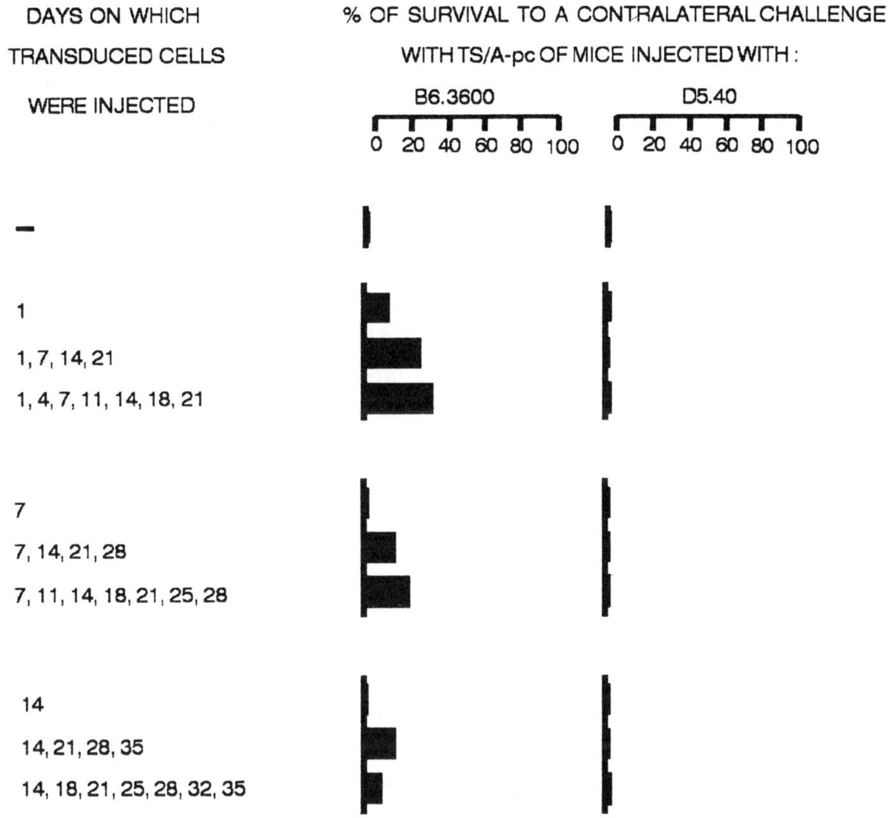

Fig. 16.3. *Ability of systemic immunity elicited by B6.3600 or D5.40 s.c. injections to protect against subsequent TS/A-pc challenge and to cure established TS/A-pc tumors.* Mice challenged s.c. in the middle of the left flank or i.v. with TS/A-pc at day 0 received in the right flank single or multiple injections of 1 x 10^5 B6.3600 or D5.40 cells starting at various times before or after challenge.

Two types of immune response mechanism were observed. The first is nonspecific. It is responsible for the rejection of engineered cells, and both its power and its effectiveness depend on the type and amount of cytokine produced. B (IL-2 transduced) clones were mainly rejected via a reaction mediated by neutrophils, whereas D (IL-4 transduced) clones mainly by eosinophils. The first mechanism is the more effective of the two in limitating the oncogenicity of the engineered cells.

As a consequence of the rejection of engineered cells, some mice were protected against a subsequent tumor challenge. Here, too, the features of the response depend on the type of cytokine. B clones offer less protection. The different efficacy of IL-2 and IL-4 is even more evident when curing TB-mice. Repeated B clone injections cure an appreciable percentage of animals, and D clones are totally ineffective. Clone B6.3600, indeed, cures not only 1-day tumors, but also a few well-structured 7-day and 14-day forms. It also arrests experimentally induced metastasis. Can a reason be given for this marked difference? The in vitro data show that B clones induce CTL capable of killing TS/A-pc cells, whereas D clones actually inhibit the production of CTL. In addition, there is preliminary evidence that rejection of D5.40 is followed by the production of IgG1 and IgE specific for TS/A-pc. It may thus be suggested that they are involved in the secondary rejection of this tumor through induction of ADCC, as has been observed in vitro (Pericle et al., submitted). This marked difference in the type of immune reaction and in vivo protection induced by B and D clones suggests that the mechanisms associated with Th1 interleukins, such as IL-2, are by no means the same as those associated with Th2 interleukins, e.g. IL-4. Exploitation of these differences could perhaps lead to the design of immunotherapeutic protocols better suited to the single phases of tumor progression.

Acknowledgments

We wish to thank Dr Federica Pericle who began the first experiments with IL-4 clones, Prof Guido Forni for advice and criticism, Dr John Iliffe for careful review of the manuscript and Paola Feraiorni for skillful technical help. This work was supported by grants from the Istituto Superiore di Sanita, Italy-USA special projects on Tumor Therapy and AIDS, the Italian Association for Cancer Research (AIRC), the National Research Council (CNR PF-ACRO) and the Italian Ministry of education (MPI 40% and 60%).

References

Audibert FM and Lise LD (1993). Adjuvants: Current status, clinical perspectives and future prospects. Immunol Today **6**, 281-284.

Cavallo F, Giovarelli M, Gulino A, Vacca A, Stoppacciaro A, Modesti A and Forni G (1992). Role of neutrophils and $CD4^+$ T lymphocytes in the primary and memory response to nonimmunogenic murine mammary adenocarcinoma made immunogenic by IL-2 gene. J Immunol **149**, 3627-3635.

Colombo MP, Modesti A, Parmiani G and Forni G (1992). Local cytokine availability elicits tumor rejection and systemic immunity through granulocyte-T-lymphocyte cross-talk. Cancer Res **52**, 4853-4857.

Colombo MP and Forni G (1994). Cytokine gene transfer in tumor inhibition and tumor therapy: Where are we now? Immunol Today, in press.

Forni G, Fujiwara H, Martino F, Hamamoka T, Jemma C, Caretto P and Giovarelli M (1988). Helper strategy in tumor immunology: Expansion of helper lymphocytes and utilization of helper lymphokines for experimental and clinical immunotherapy. Cancer Metastasis Rev **7**, 289-296.

Giovarelli M, Landolfo S, Whitmore AC and Forni G (1982). Rous sarcoma virus induced tumors in mice. I. Macrophage mediated natural cytotoxicity. Eur J Cancer Clin Oncol **18**, 307-315.

Hock H, Dorsch M, Kunzendorf U, Uberla K, Qin Z, Diamantstein T and Blankenstein T (1993). Mechanisms of rejection induced by tumor cell-targeted gene transfer of interleukin 2, interleukin 4, interleukin 7, tumor necrosis factor or interferon γ. Proc Natl Acad Sci USA **90**, 2774-2778.

Karasuyama H, Tohyama N and Tada T (1989). Autocrine growth and tumorigenicity of interleukin-2-dependent helper T cell-transfected with the IL-2 gene. J Exp Med **169**, 13-19.

Lichtman AK, Kurt-Jones EA and Abbas AK (1987). B-cell stimulatory factor-1 and not interleukin 2 is the autocrine growth factor for some helper T lymphocytes. Proc Natl Acad Sci USA **84**, 824-830.

Malkowsky M, Medawar PM, Thacher DR, Toy J, Hunt L, Rayfield S and Dore C (1985). Acquired immunological tolerance of foreign cell is impaired by recombinant interlelukin 2 or vitamin A. Proc Natl Acad Sci USA **82**, 536-541.

Miller DA and Rosman GJ (1989). Improved retroviral vectors for gene transfer and expression. Biotechniques **7**, 980-795.

Modesti A, Masuelli L, Modica A, D'Orazi G, Scarpa S, Bosco MC and Forni G (1993). Ultrastructural evidence of the mechanisms responsible for interleukin-4-activated rejection of a spontaneous murine adenocarcinoma. Int J Cancer **53**, 988-993.

Musiani P, Modesti A, Brunetti M, Modica A, Gulino A, Bosco MC, Colombo MP, Nanni P, Cavallo F, Pericle F, Giovarelli M, Cavallo R and Forni G (1994). The nature and potential of the reactive response to mouse mammary adenocarcinoma cells engineered with IL-2, IL-4 or IFN-γ gene. Nat Immun, **13**, 93-101.

Smith KA (1990). Cytokines in the nineties. Eur Cytokine Net **1**, 7-12.

Tepper RI, Coffman R and Leder P (1992). An eosinophil-dependent mechanism for the anti-tumor effect of interleukin-4. Science **257**, 548-551.

Trinchieri G (1989). Biology of natural killer cells. Adv Immunol **47**, 187-195.

17

REJECTION OF IL-7 GENE TRANSFECTED TUMORS

T. Blankenstein, H. Hock, M. Dorsch and T. Diamantstein
Institute of Immunology, Klinikum Steglitz, Freie Universitat Berlin, Berlin, Federal Republic of Germany

Introduction

Recently, a series of studies has been published assessing the role of various cytokines, including IL-7, and their interference with tumor growth when provided locally and continuously at the tumor site. The experimental approach consisted in the transfer and expression of cytokine genes in cultured tumor cells in an effort to explore whether secretion of the transfected cytokine by the tumor cells altered their tumorigenicity and which local and systemic immunological effects could be observed in mice.

Tumor suppressive effects of cytokines have been detected (e.g. IL-7), and these experiments demonstrated the high efficiency of some cytokines in inducing tumor rejection; in some cases this occurred even with coinjected nontransfected tumor cells. They also showed the failure of other cytokines to act suppressively on tumor growth.

The close proximity of cytokines at the tumor site is important since this seemed to permit a more specific attraction and activation of immune cells at the site of antigen. Moreover, cytokine toxicity could be avoided.

Systemic application of cytokines for immunotherapy requires high doses due to their short half life in vivo. The massive doses necessary to reach an effective concentration at the tumor site are often accompanied by various toxic symptoms. So far, this has limited the therapeutic value of this strategy. Because the rejection of tumors transfected with various cytokine genes was always immunologically mediated, this represents a suitable model to better understand the function of cytokines in vivo and to elucidate the quality of "physiologically" applied cytokines as therapeutic agents for cancer immunotherapy.

The results obtained with IL-7 gene transfected tumors are reviewed in this chapter.

CYTOKINE-INDUCED TUMOR IMMUNOGENICITY
ISBN 0-12-261520-4

IL-7

Only few natural sources of IL-7 have been described. Thus far, bone marrow derived stromal cell lines (Namen et al., 1988a; Namen et al., 1988b) and thymic epithelial cells (Gutierrez and Palacios, 1991) are known to produce IL-7. IL-7 was identified, and the cDNA cloned, by its ability to support the growth of pre-B cells (Namen et al., 1988b); in the meantime IL-7 has been shown to act on various cell lineages. It is a growth factor for thymocytes (Conlon et al., 1989; Okazaki et al., 1989; Watson et al., 1989) and mature $CD4^+$ and $CD8^+$ T cells (Chazen et al., 1989; Morrissey et al., 1989; Welch et al., 1989; Grabstein et al., 1990). The stimulatory effect of IL-7 on T-cell proliferation seems to occur both directly and via an IL-2 dependent pathway (Widmer et al., 1990). IL-7 also activates CTL and LAK cells (Alderson et al., 1990; Hickman et al., 1990; Lynch and Miller, 1990), and it induces tumoricidal activity and cytokine production by human monocytes (Alderson et al., 1991). Some indication on the function of IL-7 in vivo was obtained from either systemic IL-7 administration (Morrissey et al., 1991; Damia et al., 1992) or from the generation of IL-7 transgenic mouse lines (Samaridis et al., 1991; Rich et al., 1993). These experiments showed that IL-7 increased the number of pre-B cells, thymocytes and $CD4^+$ and $CD8^+$ T cells. The role of IL-7 during defined immune reactions is, however, poorly understood.

IL-7 gene transfected tumors are rejected

Upon gene transfer and sufficient expression of IL-7, tumor cells can completely lose their tumorigenicity. This is exemplified in Fig. 17.1 with plasmacytoma J558L cells transfected to produce IL-7. After an initial growth phase in syngeneic mice the cells were completely rejected, whereas parental or mock-transfected cells grew to a large tumor within 2 weeks (Hock et al., 1991). If, however, the IL-7 transfectants were injected into mice which, in parallel, received anti-IL-7 antibody, they grew as tumors like the parental cells for as long as the Ig was active in the mice; subsequently, the tumor stopped growing and finally disappeared completely (Fig. 17.1). Thus, the IL-7 producing cells had not lost the ability to proliferate in vivo consistent with their unaltered doubling time after IL-7 gene transfer in vitro, but the loss of tumorigenicity seemed to result from the interaction between the secreted IL-7 and the tumor bearing host. This experiment also showed that IL-7, once it could develop its activity in the mice, was able to induce regression of a considerable tumor mass.

The tumor suppressive effect of IL-7 was dose dependent since the efficiency of tumor rejection was proportional to the amount of IL-7 secreted by the tumor. The tumor suppressive effect of IL-7 has been

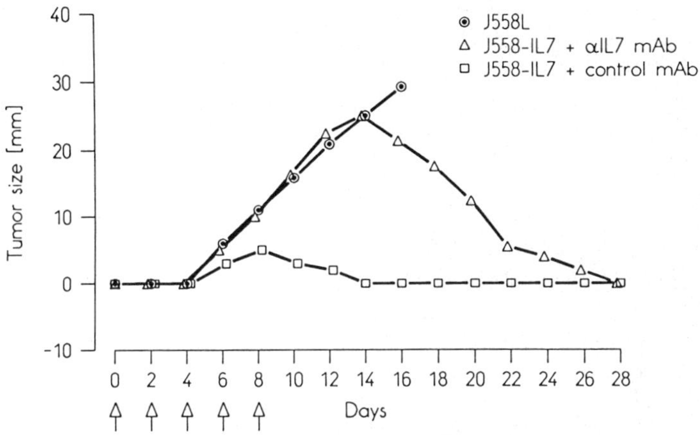

Fig. 17.1. *IL-7 gene transfected tumor cells (plasmacytoma J558L) have lost tumorigenicity in syngeneic BALB/c mice.* In this experiment, groups of mice were injected s.c. with 4 x 10^6 J558-IL-7 cells which produced 50 U/ml of IL-7 in a thymocyte proliferation assay, or with parental J558L cells. Mice injected with J558-IL-7 cells additionally received i.v. five consecutive times either 200 µg of anti-IL-7 or an isotype matched control mAb (indicated by the arrows). In the presence of the control mAb, J558-IL-7 cells were rapidly rejected. In anti-IL-7 mAb treated animals, tumors grew for 2 weeks with kinetics similar to that of parental J558L cells. Then, most likely due to the clearance of the mAb, tumors stopped growing and, subsequently, regressed in all 5 mice (with permission from Hock et al., 1991).

confirmed in several other tumor cell lines transfected to produce IL-7 (see below).

Mechanism of IL-7 mediated tumor rejection

Several lines of evidence have shown that rejection of IL-7 producing tumors resulted from the induction of an antitumor immune response. Firstly, the IL-7 secreted by the tumors induced a strong inflammatory reaction. This could be demonstrated because rejection of IL-7 producing tumors was preceded by an initial growth phase which allowed reisolation and characterization of the tumor tissue. By immunohistologic analysis substantial numbers of T cells, both $CD4^+$ and $CD8^+$, were detected in IL-7 producing tumors (Table 17.1).

Table 17.1. *Cellular infiltrate in IL-7 producing tumors (J558L cells) resulting from immunohistologic analysis in syngeneic mice and mice depleted of T-cell subpopulations*

	Tumor infiltrate			
mAb treatment	$CD4^+$ T cells	$CD8^+$ T cells	$CR3^+$ (Mac 1^+) cells	Eosinophils
--	+	+	+	+
anti-CD4	-	+	-	-
anti-CD8	+	-	+	+

T cells were present early after tumor cell injection; they increased in number throughout the rejection process, but they were not detected in the tumor tissue of parental cells.

Additionally, IL-7 producing tumors contained increased numbers of both macrophages and eosinophils as determined by specific antibodies and morphology (Hock et al., 1991; Hock et al., 1993a). Interestingly, both cell types were absent in the tumors from mice which were depleted of $CD4^+$ T cells. Therefore, the arrival of macrophages and eosinophils does not seem to be a direct effect of tumor derived IL-7, but rather an infiltration of the tumor in a $CD4^+$ T-cell dependent fashion (Table 17.1). However, $CD8^+$ T cells were still present in tumors from $CD4^+$ T-cell depleted mice and vice versa; $CD4^+$ T cells and accompanying macrophages and eosinophils were present in IL-7 tumors from mice depleted of $CD8^+$ cells. Thus, IL-7 producing tumors contain a complex pattern of infiltrating cells: $CD4^+$ and $CD8^+$ cells accumulate independently from each other, but the nonspecific inflammatory response is mediated by IL-7 induced $CD4^+$ T cells.

In order to correlate the occurrence of individual infiltrating cell types in IL-7 tumors with their involvement in tumor rejection, the ability of IL-7 producing cells to grow in nude mice was analyzed. Although rejected in almost all cases in syngeneic BALB/c mice, IL-7 producing tumor cells grew as a tumor in nude mice as did the parental cells (Fig. 17.2).

A very similar result was obtained in mice depleted of $Thy1^+$ cells (Hock et al., 1991). This demonstrated that the antitumor effect of IL-7 is entirely dependent on T cells, at variance with what was observed with other cytokines (see below).

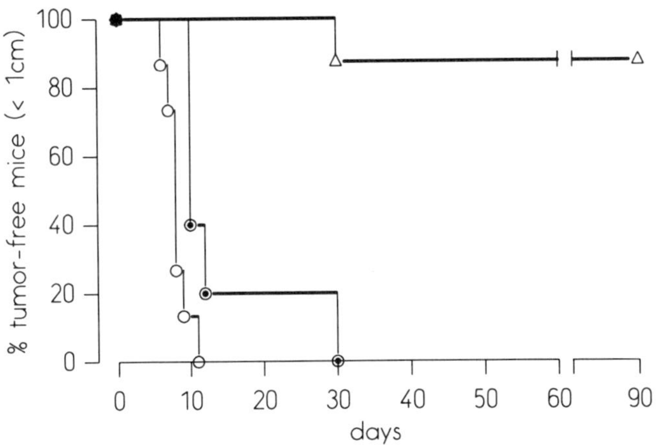

Fig. 17.2. *IL-7 gene transfected tumors are rejected in syngeneic, but not in T-cell deficient (nude) mice.* 4 x 10^6 J558-IL-7 cells were injected s.c. into BALB/c mice (Δ) or BALB/nu/nu mice (⊙) and tumor growth was compared to that of parental J558L cells injected in BALB/c mice (O) (modified with permission from Hock et al., 1993a).

Remarkably, both CD4+ and CD8+ T cells participated in the antitumor response; however, this occurred at different time points of tumor rejection. This became obvious from the observation of tumor growth of IL-7 producing cells in mice which had been long-term depleted of either CD4+ or CD8+ T cells with appropriate antibodies. In the absence of CD4+ T cells, IL-7 had almost completely lost its antitumor activity (Fig. 17.3) even if tumors grew with a slight delay compared to those which grew in Thy 1.2+-cell depleted animals (Hock et al., 1991).

Because the depletion of CD4+ T cells also resulted in the absence of macrophages and eosinophils in IL-7 producing tumors, it seems as if CD4+ T cells primarily served to induce an inflammatory reaction.

A similar involvement of CD4+ T cells in tumor eradication has been observed in other models and has been explained by a delayed type hypersensitivity reaction (Fujiwara et al., 1984; Greenberg et al., 1985). It cannot be excluded, however, that CD4+ T cells also had a direct tumoricidal activity. The CD8+ T cells which were present in IL-7 producing tumors from CD4+ T-cell depleted mice were obviously not capable of eliminating the tumor.

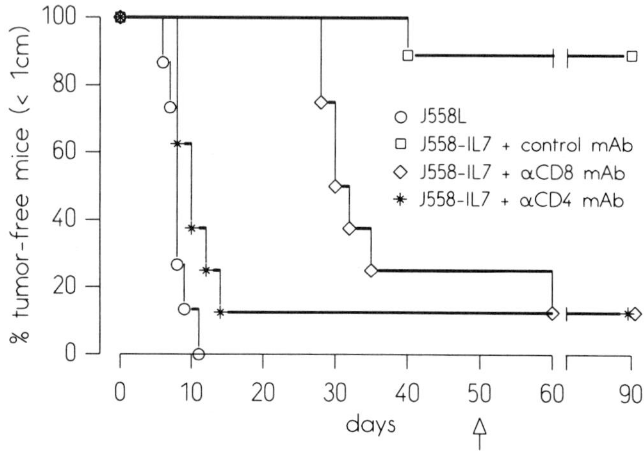

Fig. 17.3. *CD4+ and CD8+ T cells participate in IL-7 mediated tumor rejection.* BALB/c mice were s.c. injected with 4 x 10^6 J558-IL-7 cells and were treated weekly for 50 days with the indicated mAbs to deplete the respective T-cell populations as indicated by the arrow (with permission from Hock et al., 1993a).

We first observed that the depletion of CD8+ T cells in mice bearing IL-7 producing tumors did not interfere with the antitumor response in the early phase of tumor rejection (Hock et al., 1991). In subsequent experiments with long-term CD8+ T-cell depleted mice (>7 weeks), we confirmed that IL-7 producing tumors were strongly suppressed at the beginning but, ultimately, most mice developed a tumor (Fig. 17.3). Therefore, the synergistic action of CD4+ and CD8+ T cells was neccessary for complete tumor eradication. CD4+ T cells (and accompanying cells) effectively prevented rapid expansion of the tumor but, in most cases, they were unable to completely eliminate the tumor. In contrast, CD8+ T cells were ineffective in an early control of tumor growth but, in the long run, they were indispensable for complete tumor eradication. It is currently not known whether CD8+ T cells in IL-7 producing tumors require paracrine stimulation by CD4+ T cells. It is also not yet clear how much they need the tumor cell derived IL-7, particularly since their role was to eliminate those tumor cells which, in the late rejection phase, escaped detection by the CD4+ T-cell dependent mechanism.

The role of macrophages and eosinophils in IL-7 producing tumors, the presence of which was dependent on CD4+ T cells, has not been completely resolved. However, application of an anti-CR3 (Mac1) antibody which interferes with migration of CR3+ cells to mice bearing IL-7 producing tumors partially restored tumor growth (Hock et al., 1991). In tumors from these mice macrophages were not absent but clearly reduced in number suggesting their involvement in tumor regression. To what extent eosinophils, which also bear the CR3 antigen, were responsible for this effect has still to be determined.

The mode of rejection of IL-7 producing tumors seems to be influenced by the tumor cells

In addition to the plasmacytoma J558L, three other tumor lines have been shown to have a strongly reduced tumorigenicity following IL-7 gene transfer (Hock et al., 1991; Aoki et al., 1992; McBride et al., 1992). They are of different cellular origin (mammary adenocarcinoma, fibrosarcoma, glioblastoma) and derived from different mouse strains (Table 17.2.). Although limited data are available on the suppressive mechanism and although the cell types infiltrating IL-7 producing tumors are essentially the same, based on the comparison of the different tumor models it seems that the participation of individual cell populations in the process of tumor rejection may be influenced by the particular tumor cells. Thus, a fibrosarcoma tumor transfected to produce IL-7 contained a cellular infiltrate composed of high numbers of CD4+ and CD8+ T cells and eosinophils (McBride et al., 1992). Additionally, an increased number of B cells was observed. Macrophages were not found to be increased in this model, possibly because the parental tumor was spontaneously infiltrated by a substantial number of macrophages. T cells are the effector cells responsible for tumor suppression, as shown by the absence of an antitumor effect in nude mice and by cytolytic activity of TIL in vitro.

In another model with an IL-7 gene transfected glioblastoma cell line only CD8+, but not CD4+, T cells were determined to be the effector cells responsible for tumor suppression (Aoki et al., 1992). The reason for this apparent contradiction when compared to the results described above with an IL-7 gene transfected plasmacytoma (which required CD4+ and CD8+ T cells for rejection) is not known. Differences in class II MHC expression should not account for this discrepancy, since the plasmacytoma cell line did not express detectable class II MHC molecules (Hock et al., 1993a). A further characterization of the tumor cells should permit an identification of those phenotypical features responsible for this difference in the rejection mechanisms.

Table 17.2. *Immune response against IL-7 transfected tumor cells*

Cell line	Tumor type	Mouse strain	Effector cells	Tumor infiltrating cells	References
J558L	Plasmacytoma	BALB/c	$CD4^+$ T cells $CD8^+$ T cells Macrophages Eosinophils	$CD4^+$ T cells $CD8^+$ T cells Macrophages	Hock et al., 1991 Hock et al., 1993a Hock et al., 1993b
TS/A	Mammary adenocarinoma	BALB/c	n.a.[a]	n.a.	Hock et al., 1991
FSA	Fibrosarcoma	C3H	T cells	$CD4^+$ T cells $CD8^+$ T cells Eosinophils B cells	McBride et al., 1992
203-glioma	Glioblastoma	C57BL/6	$CD8^+$ T cells	n.a.	Aoki et al., 1992

[a]n.a., not assessed

Comparison between the rejection mechanisms induced by IL-7 and other cytokines

In addition to IL-7, several other cytokines have been shown to elicit an effective antitumor response when produced locally after gene transfer. Among these are IL-2, IL-4, TNF and IFN-γ. Their rejection mechanisms have partially been resolved (for reviews see Blankenstein et al., 1991b; Colombo et al., 1992 and various chapters in this volume). As shown above, a comparison of the rejection mechanisms induced by different cytokines can lead to false conclusions, due to the influence of the tumor model.

For these reasons we have recently compared the cellular requirements for tumor suppression and the immunohistologic infiltrate specific for different cytokines (IL-2, IL-4, IL-7, TNF and IFN-γ) when expressed by the same tumor cells (Hock et al., 1993a). The results are only briefly discussed here with respect to some unique characteristics of IL-7 and some similarities between IL-7 and other cytokines. All five cytokines have in common tha fact that, upon gene transfer, tumors were rejected in immunocompetent mice. IL-7 was found to be unique in that it was the only cytokine not acting in a tumor suppressive manner in immunodeficient mice, most likely due to the fact that only IL-7, and not IL-2, IL-4, TNF or IFN-γ, required $CD4^+$ T cells for tumor suppression. Remarkably, the rejection of all five cytokine producing cell lines was in a very similar way dependent on the presence of $CD8^+$ T cells.

Thus, in $CD8^+$ T-cell depleted mice all five cytokine producing cell lines showed an initial tumor growth suppression, but ultimately in most cases mice developed a tumor. This apparent redundancy in the rejection mechanisms induced by the different cytokines can be explained in two ways. Either all transfected cytokines were able to paracrinely stimulate cytolytic activity of $CD8^+$ T cells as originally proposed for rejection of IL-2 transduced tumors (Fearon et al., 1990), and/or the $CD8^+$ T-cell independent tumor suppression in the early rejection phase allowed the $CD8^+$ T cells to develop the cytolytic activity required for complete tumor eradication without beeing directly stimulated by the transfected cytokine.

A further similarity between IL-7 and the other cytokines was the infiltration of all cytokine producing tumors by macrophages. It is not always clear whether they are involved in tumor suppression and whether their presence is a direct or indirect effect of the transfected cytokine. At least for TNF it has been shown that an anti-CR3 antibody interfered with tumor suppression (Blankenstein et al., 1991a) similar to what was observed with IL-7. Finally, the $CD4^+$ T-cell dependent macrophages and eosinophils in IL-7 producing tumors resemble the infiltrate in IL-4 producing tumors. Thus, IL-4 attracts macrophages and eosinophils (Tepper et al., 1989) and there is indication that both cell populations contribute to tumor suppression (Platzer et al., 1992; Tepper et al., 1992). The eosinophils in IL-4 producing tumors may be indirectly mediated by IL-5 (Platzer et al., 1992) and, similarly, the supportive effect of IL-7 on eosinophils in vitro can be blocked by anti-IL-5 antibodies (Vellenga et al., 1992). This raises the question of how much the nonspecific infiltrate in IL-7 producing tumors is caused by CD4+ T cells released cytokines such as IL-4 and IL-5.

Intratumoral injection of IL-7

Direct repeated injection into or close to the tumor may be an alternative to gene transfection approaches. McBride et al. (1992) have observed tumor suppressive effects by twice daily injections of 20 μg of IL-7 into the tumor for a period of 10 days. However, unlike what had been observed with IL-7 gene transfected tumor cells, only in a proportion of the mice tumors completely regressed. At least from these limited data it seems as if direct injection of IL-7 can not substitute for comparatively low continuous IL-7 production by the tumor.

Effect of IL-7 producing tumors on systemic immunity

It has been shown in a number of cases in different tumor models and with various cytokines that mice which had rejected cytokine producing tumors resisted a subsequent challenge with parental tumor cells (reviewed in Blankenstein et al., 1991b; Colombo et al., 1992; Pardoll, 1992). This has been explained by the acquisition of tumor immunity during rejection of the primary tumor and has stimulated new efforts to use cytokine/tumor cell vaccines for immunotherapy. However, systemic immunity could not clearly be attributed to the secreted cytokine rather than to exposure of live tumor cells to the immune system and the strength of systemic immunity had not been compared with established immunotherapeutic protocols.

We have recently compared the efficiency of various cytokine (IL-2, IL-4, IL-7, TNF, IFN-γ) producing J558L cells to induce systemic immunity in the course of primary tumor rejection (Hock et al., 1993b). "Challenge" experiments showed that mice which had rejected any of the cytokine producing cell lines were capable of rejecting parental tumor cells at a higher percentage than mice which had been immunized with irradiated tumor cells. However, in comparison to tumor cells admixed with *Corynebacterium parvum* (an adjuvant extensively tested clinically) as vaccine, none of the cytokine producing cell lines nor any combination of two producer lines were more effective in protecting mice from a challenge tumor. In contrast, Connor et al. (1993) very recently showed that IL-2 gene modified bladder tumor cells were more potent in the

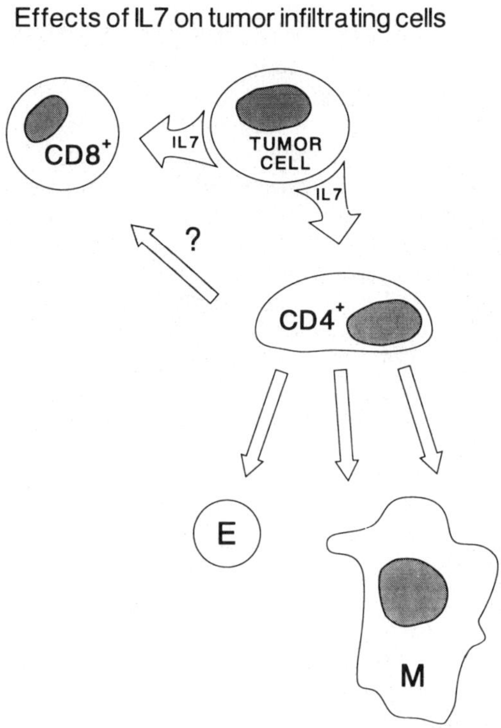

Fig. 17.4. *Effect of IL-7 on tumor infiltrating cells.* $CD4^+$ and $CD8^+$ T cells most likely accumulate in IL-7 producing tumors in direct response to IL-7. Macrophages (M) and eosinophils (E) infiltration is dependent on $CD4^+$ T cells. Involvement in tumor suppression has been shown for $CD4^+$, $CD8^+$ T cells and macrophages, while the participation of eosinophils is not yet clear.

treatment of an orthotopic growing tumor than conventional (cisplatin) therapy.

Thus, much more empirical work with different tumor models and cytokines is needed to be able to judge under which conditions local cytokines are most effective for immunotherapy.

Conclusions

IL-7 very effectively suppresses tumor growth. This has been demonstrated by gene transfer and expression of IL-7 in several tumor cell lines, which resulted in loss or, depending on the amount of IL-7 secreted, in reduced tumorigenicity. This effect is the result of a complex immunological reaction which is summarized in Fig. 17.4. $CD4^+$ and $CD8^+$ T cells accumulate independently in IL-7 producing tumors. Additionally, $CD4^+$ T cells cause infiltration by macrophages and eosinophils possible by secondary cytokines such as IL-4 and IL-5 which still have to be identified. Tumor suppression by IL-7 is entirely dependent on T cells. $CD4^+$ T cells and recruited macrophages and possibly eosinophils (whose involvement in tumor suppression has still to be proven) are required to prevent rapid tumor burden, but in most cases are unable to completely eradicate the tumor. In contrast, $CD8^+$ T cells are not involved in the early phase of tumor suppression, but are indispensable for long-term tumor eradication. Although IL-7 reliably acted suppressively in several tumor models, the precise mechanism of rejection induced by IL-7 can be influenced by the particular tumor. The immunotherapeutic potential of IL-7 gene transfected tumors as vaccine, in particular in comparison with other cytokines or clinically established protocols, has to be the subject of further investigations.

Acknowledgments

Supported by the Deutsche Krebshilfe, Mildred-Scheel Stiftung.

References

Alderson MR, Sassenfeld HM and Widmer MB (1990). Interleukin 7 enhances cytolytic T lymphocyte generation and induces lymphokine activated killer cells from human peripheral blood. J Exp Med **172**, 577-587.

Alderson MR, Tough TW, Ziegler SF and Grabstein KH (1991). Interleukin-7 induces cytokine secretion and tumoricidal activity by human peripheral blood monocytes. J Exp Med **173**, 923-930.

Aoki T, Tashiro K, Miyatake SI, Kinashi T, Nakano T, Oda Y, Kikuchi H and Honji T (1992). Expression of murine interleukin 7 in a murine glioma cell line results in reduced tumorigenicity *in vivo*. Proc Natl Acad Sci USA **89**, 3850-3854.

Blankenstein T, Qin Z, Uberla K, Muller W, Rosen H, Volk HD and Diamantstein T (1991a). Tumor suppression after tumor cell targeted tumor necrosis factor alpha gene transfer. J Exp Med **173**, 1047-1052.

Blankenstein T, Rowley DA and Schreiber H (1991b). Cytokines and cancer: Experimental systems. Current Opin Immunol **3**, 694-698.

Chazen GD, Pereira GMB, LeGros G, Gillis S and Shevach EM (1989). Interleukin 7 is a T-cell growth factor. Proc Natl Acad Sci USA **86**, 5923-5927.

Colombo MP, Modesti A, Parmiani G and Forni G (1992). Local cytokine availability elicits tumor rejection and systemic immunity through granulocyte-T-lymphocyte cross-talk. Cancer Res **52**, 4853-4857.

Conlon PJ, Morrissey PJ, Nordan RP, Grabstein KH, Prickett KS, Reed SG, Goodwin R, Cosman D and Namen AE (1989). Murine thymocytes proliferate in direct response to interleukin-7. Blood **74**, 1368-1373.

Connor J, Bannerji R, Saito S, Heston W, Fair W and Gilboa E (1993). Regression of bladder tumors in mice treated with interleukin 2 gene-modified tumor cells. J Exp Med **177**, 1127-1134.

Damia G, Komschlies KL, Faltynek CR, Ruscetti FW and Wiltrout RH (1992). Administration of recombinant human interleukin-7 alters the frequency and number of myeloid progenitor cells in the bone marrow and spleen of mice. Blood **79**, 1121-1129.

Fearon ER, Pardoll DM, Itaya T, Golumbek P, Levitsky HI, Simons JW, Karasuyama H, Vogelstein B and Frost P (1990). Interleukin 2 production by tumor cells by passes T helper function in the generation of an antitumor response. Cell **60**, 397-403.

Fujiwara H, Fukuzawa M, Yoshioka T, Nakajima H and Hamaoka T (1984). The role of tumor-specific LYT-1^+2^- T cells in eradicating tumor cells in vivo. J Immunol **133**, 1671-1676.

Grabstein KH, Namen AE, Shanebeck K, Voice RF, Reed SG and Widmer MB (1990). Regulation of T cell proliferation by IL-7. J Immunol **144**, 3015-3020.

Greenberg PD, Kern DE and Cheever MA (1985). Therapy of disseminated murine leukemia with cyclophosphamide and immune LYT-1^+2^- T cells. J Exp Med **161**,1122-1134.

Gutierrez JC and Palacios R (1991). Heterogeneity of thymic epithelial cells in promoting T-lymphocyte differentiation in vivo. Proc Natl Acad Sci USA **88**, 642-646.

Hickman CJ, Crim JA, Mostowski HS and Siegel JP (1990). Regulation of human cytotoxic T lymphocyte development by IL-7. J Immunol **145**, 2415-2420.

Hock H, Dorsch M, Diamantstein T and Blankenstein T (1991). Interleukin 7 induces CD4+ T cell-dependent tumor rejection. J Exp Med **174**, 1291-1298.

Hock H, Dorsch M, Kunzendorf U, Qin Z, Diamantstein T and Blankenstein T (1993a). Mechanisms of rejection induced by tumor cell-targeted gene transfer of interleukin 2, interleukin 4, interleukin 7, tumor necrosis factor, or interferon γ. Proc Natl Acad Sci USA **90**, 2774-2778.

Hock H, Dorsch M, Kunzendorf U, Uberla K, Qin Z, Diamantstein T and Blankenstein T (1993b). Vaccinations with tumor cells genetically engineered to produce different cytokines: Effectivity not superior to a classical adjuvant. Cancer Res **53**, 714-716.

Lynch DH and Miller RE (1990). Induction of murine lymphokine-activated killer cells by recombinant IL-7. J Immunol **145**, 1983-1990.

McBride WH, Thacker JD, Comora S, Economu JS, Kelley D, Hogge D, Dubinnet SM and Dougherty GJ (1992). Genetic modification of a murine fibrosarcoma to produce interleukin 7 stimulates host cell infiltration and tumor immunity. Cancer Res **52**, 3931-3937.

Morrissey PJ, Goodwin RG, Nordan RP Anderson D, Grabstein KH, Cosman D, Sims J, Lupton S, Acres B, Reed SG, Mochizuki D, Eisenman J, Conlon PJ and Namen AE (1989). Recombinant interleukin 7, pre-B cell growth factor, has costimulatory activity on purified mature T cells. J Exp Med **169**, 707-716.

Morrissey PJ, Conlon P, Charrier K, Braddy S, Alpert A, Williams D, Namen AE and Mochizuki D (1991). Administration of IL-7 to normal mice stimulates B-lymphopoiesis and peripheral lymphadenopathy. J Immunol **147**, 561-568.

Namen AE, Lupton S, Hjerrild K, Wignall J, Mochizuchi DY, Schmierer A, Mosley B, March CJ, Urdal D, Gillis S, Cosman D and Goodwin RG (1988a). Stimulation of B-cell progenitors by cloned murine interleukin-7. Nature **333**, 571-573.

Namen AE, Schmierer AE, March CJ, Overell RW, Park LS, Urdal DI and Mochizuki DY (1988b). B cell precursos growth-promoting activity. J Exp Med **167**, 988-1002.

Okazaki H, Ito M, Sudo T, Hattori M, Kano S, Katsura Y and Minato N (1989). IL-7 promotes thymocyte proliferation and maintains immunocompetent thymocytes bearing α β or γ δ T cell receptors in vivo: Synergism with IL-2. J Immunol **143**, 2917-2922.

Pardoll D (1992). New strategies for active immunotherapy with genetically engineered tumor cells. Current Opin Immunol **4**, 619-623.

Platzer C, Richter G, Uberla K, Hock H, Diamantstein T and Blankenstein T (1992). Interleukin 4 mediated tumor suppression in nude mice involves interferon-gamma. Eur J Immunol **22**, 1729-1733.
po
Rich BE, Campos-Torres J, Tepper RI, Moreadith RW and Leder P (1993). Cutaneous lymphoproliferation and lymphomas in interleukin 7 transgenic mice. J Exp Med **177**, 305-316.

Samaridis J, Casorati G, Traunecker A, Iglesias A, Gutierrez JC, Muller U and Palacios R (1991). Development of lymphocytes in interleukin 7 transgenic mice. Eur J Immunol **21**, 453-460.

Tepper RI, Pattengale PG and Leder P (1989). Murine interleukin 4 displays potent anti-tumor activity in vivo. Cell **57**, 503-512.

Tepper RI, Coffman RL and Leder P (1992). An eosinophil-dependent mechanism for the anti-tumor effect of interleukin 4. Science **257**, 548-551.

Vellenga E, Esselink MT, Straaten J, Stupl BK, de Wolf JTM, Brons R, Giannotti J, Smit JW and Halie MR (1992). The supportive effects of IL-7 on eosinophil progenitors from human bone marrow cells can be blocked by anti-IL-5. J Immunol **149**, 2992-2995.

Watson JD, Morrissey PJ, Namen AE, Conlon PJ and Widmer MB (1989). Effect of IL-7 on the growth of fetal thymocytes in culture. J Immunol **143**, 1215-1222.

Welch PA, Namen AE, Goodwin RG, Armitage R and Cooper MD (1989). Human IL-7: A novel T cell growth factor. J Immunol **143**, 3562-3567.

Widmer MB, Morrissey PJ, Namen AE, Voice RF and Watson JD (1990). Interleukin 7 stimulates growth of fetal thymic precursors of cytolytic cells: Induction of effector function by interleukin 2 and inhibition by interleukin 4. Int Immunol **2**, 1055-1061.

18

TUMORIGENICITY AND IMMUNOGENICITY OF IFN-γ PRODUCING TUMOR CELLS

Y. Watanabe
Department of Molecular Microbiology, Faculty of Pharmaceutical Sciences, Kyoto University, Sakyo-ku, Kyoto, Japan

Introduction

IFN-γ, produced in nature by activated T lymphocytes and NK cells, is not primarily an antiviral but rather a multifunctional cytokine (for a review see De Maeyer and De Maeyer-Guignard, 1988). It plays an important role in host defence mechanisms, including immune responses against neoplasia in which it attacks tumor cells both indirectly and directly. IFN-γ is, therefore, an important cytokine in the therapeutic approach to cancer (Kurzrock, 1992). The antitumor activity of IFN-γ is not fully elucidated. This is due to several reasons, but mainly to its short half life in vivo.

In order to examine comprehensively the physiological roles as well as the antitumor potential of IFN-γ, ideally the cytokine should be closely and effectively supplied to tumor cells, which in animals are usually localized (Giovarelli et al., 1986). In experimental tumors, neoplastic cells genetically engineered to produce constitutively IFN-γ by gene transfer represent ideal models because, following injection into animals, the engineered cells should continue to supply the cytokine around the tumor deposit for as long as they live. Subsequently, the tumor cell derived IFN-γ should allow stimulation of a local antitumor response without inducing systemic toxic side effects. This notion led us to device an efficient IFN-γ gene (cDNA) transfer system using a retroviral vector to investigate the effect of IFN-γ gene transfer on a wide variety of tumor cell types (Watanabe and Sakata, 1988).

In 1989, we first reported that the exogenous expression of IFN-γ cDNA in C1300, a malignant mouse neuroblastoma cell line, was capable of enhancing class I MHC antigen expression and of augmenting specific antitumor immunity in the recipient (Watanabe et al., 1989).

This striking effect of IFN-γ gene transfer on tumorigenicity, soon confirmed by several groups including ours using other tumor cell types (see other chapters in this book), indicated that this approach could shed light on the role of IFN-γ in antitumor immunity and could also provide

CYTOKINE-INDUCED TUMOR IMMUNOGENICITY
ISBN 0–12–261520–4

experimental systems for developing tumor cell targeted IFN-γ gene therapy of cancer (Watanabe, 1992). In many tumor models substantial tumor suppression was observed that was mediated by specific or nonspecific antitumor mechanisms, or by other mechanisms; the precise antitumor mechanisms involved are different from one another in the various tumor systems and still remain to be fully analyzed. In this chapter, further recent developments will be concisely reviewed, focusing in particular on the cellular alterations, tumorigenicity and immunogenicity of the IFN-γ gene transduced tumor cells and also on the therapeutic potential of the cells.

Table 18.1 will serve as an overview of the subject; fourteen different tumor cell lines derived from various mouse strains and transduced with the IFN-γ gene are tabulated with some experimental results which include IFN-γ production, class I MHC antigen expression, and tumorigenicity in the syngeneic mouse.

Cellular and subcellular alterations of IFN-γ gene transduced cells

IFN-γ production levels

IFN-γ production levels by transduced tumor cells were generally moderate or rather low: at the most 300 IU/ml, except for 1200 IU/ml in J558L, as reported in Table 18.1. Secretion of IFN-γ from tumor cells exerted various effects on the cells in vitro in an autocrine manner through functional IFN-γ receptors that are ubiquitously expressed on almost all cells including tumor cells (Ucer et al., 1986; Valente et al., 1992).

Several effects will be hereby surveyed from the point of view of tumor immunology.

(Table 18.1, notes)

[a]S.c. tumor in the syngeneic mice

[b]Maximal level in the culture fluid

[c]Unpublished observation

[d]The survival period of the tumor bearing mice was prolonged

[e]Even parental cells displayed a high level of expression

[f]Not clearly documented

[g]Calculated as the specific activity of 10^7 IU/mg protein

[h]Not examined

[i]Manuscript submitted

Table 18.1. *IFN-γ production, class I MHC expression and tumor suppression of various mouse tumor cell lines transduced with the IFN-γ gene*

Cell line	Tumor type[a]	Mouse strain	IFN-γ yield (IU/ml)[b]	Class I MHC expression	Tumor suppression	References
C1300	Neuroblastoma	A/J	50	upregulation	specific	Watanabe et al., 1989
CMS-5	Fibrosarcoma	BALB/c	25	upregulation	specific	Gansbacher et al., 1990
Pam-T	Squamous	BALB/c	200	upregulation[c]	none[d]	Maruguchi et al., 1991
SP1	Mammary	CBA	256	upregulation	nonspecific	Esumi et al., 1991
CT-26	Colon tumor	BALB/c	8	no alteration[e]	specific[f]	Esumi et al., 1991
MCA101	Fibrosarcoma	C57BL/6	> 5[f]	upregulation	none[f]	Restifo et al., 1992
3LL-D122	Lung carcinoma	C57BL/6	200	upregulation	specific	Porgador et al., 1993
J558L	Plasmacytoma	BALB/c	1200[g]	n.e.[h]	specific	Hock et al., 1993
B16	Melanoma	C57BL/6	200[g]	n.e.	none	Dranoff et al., 1993
MOPC104E	Plasmacytoma	BALB/c	300	no alteration[e]	specific	Teramura et al., 1993
SC115	Mammary	DS	50	no alteration[e]	specific	Teramura et al., 1993
LP92-T	Lung carcinoma	B6C3F1	20	upregulation	specific	Yanagihara et al., 1994
OVHM/3	Ovarian tumor	B6C3F1	130	upregulation	none	Yanagihara et al., 1994
MBT-2	Bladder tumor	C3H	10	upregulation	specific	Hiura et al., 1994[i]

In vitro cell growth

An inhibitory effect on cell proliferation is a representative biological activity of IFN-γ. The in vitro growth of tumor cells transduced to produce IFN-γ can be so markedly affected that it might be difficult to dissect the in vivo tumor suppressive effect of IFN-γ. This is true for some tumor cell lines, like the B16 melanoma (Watanabe and Sakata, 1988), the OVHM/3 ovarian tumor (Yanagihara et al., 1994) and the CT-26 colon tumor (unpublished observations). Many of the tumor cell lines so far examined were, however, rather refractory to the antiproliferative effects of IFN-γ, allowing the evaluation of the genuine in vivo antitumor potential of IFN-γ using IFN-γ producing tumor cells.

Expression of MHC antigens

The most prominent cellular alteration occurring in the IFN-γ producing cells was the enhanced expression of class I MHC antigens (Table 18.1) that play an essential role in the specific T-cell response against tumor cells (Tanaka et al., 1985). Interestingly, the enhancement was maximal in spite of the low titer of IFN-γ secreted and did not depend on the levels of IFN-γ produced by the cells (C1300, CMS-5, SP1). Autocrine action of IFN-γ was probably fully efficient in increasing class I expression (Watanabe and Sakata, 1988). Cells that originally exhibited a high level of class I expression were, however, not affected by IFN-γ production (CT-26, MOPC104E, SC115).

Although IFN-γ is also a potent inducer of MHC class II antigens on APC, a marked augmentation of MHC class II expression by the IFN-γ gene transduced tumor cells has not been demonstrated. In the IFN-γ secreting neuroblastoma C1300, only a marginal induction was detected by fluorescence-activated cell sorter analysis (Burrows et al., 1992). The expression of MHC class II seems, in general, more strictly regulated.

Other possible modifications

Inasmuch as IFN-γ is capable of activating various cellular genes, other possible modifications in the IFN-γ producing tumor cells that could be relevant to host defense mechanisms against cancer can be hypothesized. For example, cell adhesion molecules such as ICAM-1 (Dustin et al., 1986), other cytokine(s) and/or receptor(s) (Ruggiero et al., 1986; Doukas and Pober, 1990), proteasomes (Yang et al., 1992), and tumor-associated antigens (Carrel et al., 1985; Giacomini et al., 1985). In the squamous tumor Pam-T, IFN-γ producing cells increased cell attachment to culture substrates (Maruguchi et al., 1991).

Tumorigenicity and immunogenicity of IFN-γ producing tumor cells

Tumorigenicity of IFN-γ producing tumor cells

The in vivo antitumor effect of tumor cell targeted IFN-γ gene transfer has been examined with several types of tumor cell lines obtained from various mouse strains (Table 18.1). In many of the tumor models, s.c. injected IFN-γ producing tumor cells C1300 (Watanabe et al., 1989), CMS-5 (Gansbacher et al., 1990), SP1 and CT-26 (Esumi et al., 1991), J558L (Hock et al., 1993), MOPC104E and SC115 (Teramura et al., 1993), LT92P (Yanagihara et al., 1994) and MBT-2 (Hiura et al., submitted), could be rejected by syngeneic hosts, whereas the parental cells could not. Tumor growth of IFN-γ producing 3LL-D122 injected into the footpad was also profoundly reduced in syngeneic mice (Porgador et al., 1993). In some models (C1300, SP1, 3LL-D122), tumor suppression was not observed in low or non-IFN-γ producers, clearly pointing to an IFN-γ dependency. Interestingly, in CMS-5, at variance with C1300 or 3LL-D122, no symptom of initial tumor growth was observed with the use of the low IFN-γ producing tumor, while the high producers were rejected after early transient formation of a minute tumor: this characteristic feature of tumor regression was seen with MOPC104E, SC115 and LT92P cells. On the other hand, in the Pam-T, MCA101, B16 (Dranoff et al., 1993) and OVHM/3 tumor models, apparently no tumor suppression was observed with the IFN-γ transduced cells.

The suppression of tumor growth of IFN-γ producing tumor cells did not imply the loss of tumorigenicity per se but was most likely mediated by a specific T-cell immunity induced by the action of the tumor-derived IFN-γ because it was abrogated by the simultaneous i.p. administration of anti-IFN-γ antibodies (C1300, MOPC104E, SC115), or anti-CD8 antibodies (C1300, J558L, MOPC104E, SC115), or by irradiation of the tumor cell recipient animals (LT92P). Interestingly, the suppression of the IFN-γ producing SC115 tumor was impeded also by anti-CD4 antibodies (Teramura et al., 1993). Following rejection of the IFN-γ producing cells, the mouse could not be rechallenged by the parental tumor, while being rechallengeable by unrelated tumor cells even derived from the same strain (C1300, CMS-5, 3LL-D122, J558L, MOPC104E, SC115). Again, this points to the induction of a specific antitumor immunity. Besides tumor-specific mechanisms, nonspecific mechanisms could also contribute to tumor suppression: e.g. macrophages or NK cells, since their tumoricidal activity is boosted by IFN-γ. In the tumor systems mentioned above, however, NK cells were not decisive for tumor regression.

In contrast with those models, suppression of the SP1 tumor was exceptionally mediated by some nontumor specific mechanisms: mice which rejected the IFN-γ producing tumor could not block the challenge by the parental cells, and their splenic cytotoxic activity was not diminished after depletion of CD8+ as well as of CD4+ cells (Esumi et al., 1991).

Several of the IFN-γ producing tumor lines, such as C1300, 3LL-D122, J558L, LT92P and OVHM/3, displayed reduced tumor growth in athymic nude mice. Other lines, such as SP1, MOPC104E and SC115, did not. The tumor suppression in nude mice was strongly associated with the action of NK effector cells probably triggered by tumor-secreted IFN-γ since the i.p. administration of anti-asialo ganglioside GM1 antibody (Kasai et al., 1981) is capable of preventing tumor suppression. The IFN-γ producing OVHM/3 cells, while being nonimmunogenic, were apparently susceptible to the NK activity of nude mice (Yanagihara et al., 1994). By contrast, both IFN-γ producing cells from MOPC104E and SC115 seemed to be resistant to such activity (Teramura et al., 1993).

Pam-T (Maruguchi et al., 1991) is a highly tumorigenic variant spontaneously derived from the squamous cell carcinoma Pam212. When s.c. injected, IFN-γ producing cells exhibited similar in vivo tumor growth compared to the parental cells, but tumor invasion of surrounding tissues was more retarded in the former. The mean survival of the mice with the IFN-γ producing tumor was significantly longer compared with control mice. Interestingly, the adhesion of IFN-γ producing cells to culture substratum was increased compared to parental cells. IFN-γ might affect the process of tumor cell invasion in vivo by altering the adhesion or motility of tumor cells, although the precise mechanisms remain to be investigated.

Metastatic ability of IFN-γ producing tumor cells

Since tumor progression often causes lethal metastasis, it is important to ascertain whether IFN-γ gene transfer can affect the metastatic properties of tumor cells. In the metastatic 3LL-D122 tumor, metastasis of IFN-γ secreting cells injected into the footpad was more strongly suppressed compared to IFN-γ nonsecreting, MHC class I gene transduced or to the parental cells, without changing the sensitivity to in vitro NK lysis (Porgador et al., 1993). Thus, the suppression of metastasis depended on IFN-γ production rather than on MHC class I expression.

Inhibition of metastasis was seen also in other lines, LT92P and OVHM/3 (more markedly in the former), following IFN-γ gene transfer (Yanagihara et al., 1993). The suppression of metastasis to lung tissue via vascular routes was reduced in irradiated mice, so it seemed likely that the suppression did not represent an intrinsic loss of the metastatic properties of the tumor cells, but rather that it was due to some host mechanisms as yet unclear.

The apparent discrepancy between these results and those previously reported, demonstrating that IFN-γ treatment of tumor cells enhanced their metastatic ability in relation to the elevation of MHC class I expression (Kelly et al., 1991), may be explained in several ways:

- different tumor types
- the potential of secreted IFN-γ to activate tissue-specific macrophages, NK cells and tumor specific CTL.

Immunogenicity of IFN-γ producing tumor cells

It has been demonstrated that tumor cell specific CTL can be elicited in spleen cells of mice immunized with IFN-γ producing cells (CMS-5, 3LL-D122, MOPC104E, SC115), but not with the parental tumor cells (SP1, 3LL-D122).

In the CMS-5 tumor model, the parental cells elicited a CTL activity similar to that of IFN-γ producing cells (Gansbacher et al., 1990). However, mice injected with the latter manifested a longer lasting protective immune response against the parental tumor. Furthermore, the IFN-γ producing cells were more effectively lysed in vitro by splenic CTL induced by the parental tumor cells. The enhanced susceptibility of the IFN-γ producing tumor cells to specific CTL was also observed in the 3LL-D122 tumor model (Dranoff et al., 1993). Since IFN-γ plays a role in the functional maturation of CTL (Maraskovsky et al., 1989; Novelli et al., 1991), it is not surprising that IFN-γ secretion in the proximal neighborhood of the tumor deposit may affect the outcome of T-cell mediated immune responses against the tumor. In addition, IFN-γ producing tumor cells may present more effectively tumor antigen(s) in association with MHC class I molecules to immune competent cells and may also be more susceptible to lysis by CTL.

In order to obtain a regression of IFN-γ producing C1300 tumor, elevation of the levels of MHC class I expression alone was not sufficient, and IFN-γ secretion over certain levels was crucial (Watanabe et al., 1989). It was, however, not clear whether IFN-γ secretion is indeed essential for CTL induction or whether the elevation of MHC class I is sufficient for the generation of a T-cell immunity.

This issue was investigated in the 3LL-D122 tumor model (Porgador et al., 1993). IFN-γ nonsecretors that produced undetectable amounts of IFN-γ expressed a high level of MHC class I antigens but were still highly malignant; MHC class I gene-transfectants could induce CTL activity similar to that of the IFN-γ secretors. The potential to generate protective immunity against the parental D122 tumor in syngeneic mice by preimmunization was similar in the IFN-γ gene transduced cells regardless of their different IFN-γ production levels, and it was higher than that of the MHC class I gene transfectants or of the parental cells. Thus, MHC class I expression was the primary cause of CTL generation, while IFN-γ secretion and probably the expression of other IFN-γ inducible genes were necessary for their long-lasting protective effects.

In both MOPC104E and SC115 tumor cells the surface expression of MHC class I antigen was originally at a high level and was not affected by the IFN-γ gene transfer (Teramura at al., 1993). Similarly to C1300, the s.c. tumor growth of the IFN-γ producing cells was suppressed after transient formation of a small tumor at the site of inoculum. In vitro lytic activities of splenic CTL elicited by IFN-γ producing tumor cells were directed not only to the gene-modified cells used for immunization but also to the unmodified parental cells. It was, however, unclear whether CTL recognized the same molecule(s) on both cells. In the case of MOPC104E both cytolyses were blocked by the unlabelled (cold) parent or the cold IFN-γ producer, so that CTL recognized common antigens shared by the parental and IFN-γ producer cells.

While the cytolysis of the parent SC115 was inhibited by either the parent or the IFN-γ producer SC115, the lysis of IFN-γ producer is blocked by the target cells but not by the parent. This suggested that the CTL elicited by the SC115 consisted of at least two populations, one which recognized antigens shared by both the parent and the IFN-γ producer, and the other directed against different antigens present on SC115 but not on parental SC115. This implies that IFN-γ gene transfer might modify the expression of a surface antigen that is recognized by CTL. In this context, it is of interest that proteasomes may play a role in antigen processing (Rivett, 1993) because the expressions of several proteasomal subunits, including the ones encoded within the MHC region, are specifically regulated by IFN-γ (Yang et al., 1992).

Our ongoing studies show that the IFN-γ producing C1300 cells display an enhanced gene expression of a proteasome subcomponent which is accompanied by catalytic activity. These genes are different from those induced by exogenous IFN-γ. Although the physiological meaning of this phenomenon is at present unknown, continuous stimulation of particular genes might transiently produce distinct multimeric proteasomes. MCA101 was extremely aggressive and IFN-γ gene transduced cells still generated tumors. Nevertheless, gene transduction resulted in an increased expression of MHC class I antigens (Restifo et al., 1992). MCA101 was also a poor presenter of endogenously generated viral antigens in the context of MHC class I molecules to $CD8^+$ CTL. Following infection with an endogenous vaccinia virus containing the influenza nucleoprotein N-gene, MCA101 was unable to present the endogenous N-antigen. This defect was reversed by incubation with exogenous IFN-γ or by IFN-γ gene transduction, and the cell subsequently became susceptible to killing by CTL. Moreover, in harmony with this, difficulties in the generation of TIL from wild type MCA101 tumors could be overcome by utilizing a high MHC class I expressing clone of IFN-γ gene transduced tumor cells. These TIL were tumor-specifically cytolytic and inhibitory against pulmonary metastasis.

Antitumor immunotherapy using IFN-γ producing tumor cells

IFN-γ gene transfer into tumor cells and its expression

Gene transfer methods to generate IFN-γ producing tumor cells are not discussed here since this issue is being covered elsewhere in this book. It is, however, worth recalling that for several reasons retroviral vectors at present appear the most advantageous tool for this purpose: all tumor cell lines reported in Table 18.1 were transduced with the IFN-γ gene using retroviral vectors except for SP1 and CT-26, for which calcium-phosphate coprecipitation or lipofection were employed (Esumi et al., 1991).

The levels of IFN-γ released by the IFN-γ gene modified tumor cells is another issue which deserves attention. Very low levels of production might not induce an immune response in the proximity of the IFN-γ producing tumor cells. Inversely, exceedingly high levels of IFN-γ may cause a heavy inflammatory reaction far from the tumor area or a systemic induction of inflammation which would be harmful. IFN-γ producing CHO cells that secreted 20,000 to 50,000 U/ml of culture fluids caused a severe complex syndrome and cachexia in the tumor cell recipient nude mice (Matthys et al., 1991).

Another approach to IFN-γ gene transduction contemplates the possibility of transferring the gene directly into tumor cell masses of patients without removing the cells in an attempt to stimulate immunity against the tumor cells. If such direct IFN-γ gene transfer proves effective, it could offer a much simpler and less costly alternative to ex vivo gene transfer methods.

Knowing the way in which IFN-γ affects tumor growth does not necessarily help us to predict the ways in which other cytokine gene transfers might work. Tumor growth of the B16 melanoma was not very sensitive to IFN-γ gene transfer but was completely suppressed by IL-2 gene transfer. Of the different cytokine genes analyzed, GM-CSF appeared to induce the strongest antitumor activity (Dranoff et al., 1993). However, GM-CSF expression can enhance the metastatic potential of some tumor cell lines (Takeda et al., 1991). On the other hand, certain tumor cells sensitive to IFN-γ gene transfer might be also affected by the transduction of other cytokine genes; plasmacytoma J558L cells transfected to produce IL-2, IL-4, IL-7 or TNF-α were, similarly to IFN-γ producers, rejected in the syngeneic mice by a $CD8^+$ cell mediated immune response (Hock et al., 1993) but not by IL-6 gene transfectants (Blankenstein et al., 1991).

Tumor cell targeted IFN-γ gene therapy against cancer

Tumor vaccine. In a number of mouse tumor models so far examined, IFN-γ gene transduced neoplastic cells acquired a high potency to induce antitumor immunity that impeded subsequent challenges by parental tumors. This implies that the IFN-γ producing tumor cells could work as an efficient tumor vaccine. This was shown in the mouse tumor cell line 3LL-D122 that metastasized into the lung from the footpad (Porgador et al., 1993). IFN-γ secreting cells were the most protective against pulmonary metastasis. Tumor derived IFN-γ seemed to augment the antitumor potential of various effector cells; furthermore, IFN-γ producing tumor cells might be more susceptible to the attack of such effector cells, although the precise mechanisms remain to be elucidated. Moreover, as shown in the MCA101 tumor (Restifo et al., 1992), IFN-γ gene transduced cells may be effective in generating antitumor $CD8^+$ effector cells.

IFN-γ delivery system. The titers of IFN-γ produced by IFN-γ gene transduced tumor cells were relatively low but still sufficient to induce s.c. tumor suppression. The IFN-γ concentration in the proximal area around the tumor cells tends to be high, and the constitutive supplementation of low levels of IFN-γ around the tumor deposit are, therefore, important. The supply of IFN-γ to tumor cells could be carried out by mixing the cells with adequate IFN-γ producing cells. When injected as a mixture with apparently nonmalignant IFN-γ producing LT92P cells, some types of syngeneic tumor cells reduced their tumorigenic potential (Yanagihara et al., 1994). Such benign IFN-γ producers or irradiated producer cells would be effective as an IFN-γ delivery vehicle for cancer treatment.

IFN-γ producing human tumor cells. Further studies are still required to establish the most effective and optimal protocols of IFN-γ gene therapy for the treatment of human cancer. This experimental approach has recently been extended to human melanoma (Gansbacher et al., 1992) and renal cell cancer lines (Gastl et al., 1992) using human IFN-γ expressing retroviral vectors. Similarly to mouse tumor cells, IFN-γ gene transduced human tumor cells produced moderate amounts of IFN-γ and increased the surface expression of MHC class I antigens. After γ irradiation, the engineered cells could continue to secrete the cytokine for several weeks. The results encourage the possible use of irradiated IFN-γ producing autologous tumor cells in vaccination protocols for cancer patients.

The potential of the host antitumor activity seems to be rather high and most efficient when the effector cells, including tumor-specific CTL, NK cells and macrophages, are properly stimulated by IFN-γ. IFN-γ may, therefore, play a more important role in the process of immunosurveillance against neoplasia than hitherto thought. The in vivo administration of IFN-γ producing tumor cells appears to enable a continuous and locoregional supply of adequate amounts of the cytokine in vivo near the tumor. Engineered cells administered in vivo should continue to supply IFN-γ in the vicinity of the cell as long as they live and would cease upon their depletion.

Conclusion

The transduction of an exogenous IFN-γ gene into tumor cells is a simple method for testing the effect of the cytokine on the tumorigenicity without systemic side effects. This method is also proving useful in elucidating the role of IFN-γ in tumor immunity. It could help in defining the precise role of IFN-γ in inducing an immune response against weakly immunogenic tumors.

Tumor cell targeted IFN-γ gene transfer induces a brief perturbation of the physiological conditions of the cytokine: from transient, T-cell restricted expression to long term, tumor cell derived expression. This seems to cause numerous effects on host antitumor immunity as well as on the tumor cell itself, as examined in the mouse tumor models. Most of the results obtained to date clearly indicate that IFN-γ is capable of playing a role in tumor immunity greater than that recognized so far.

Furthermore, tumor suppression by IFN-γ producing tumor cells leads to a promising clinical application of the engineered cells for cancer treatment in the hope that the localized, tumor cell targeted IFN-γ can enhance the antitumor potential of the host defence mechanisms, including specific immune responses, in an attempt to prevent future disease recurrence. The evidence that the injection of IFN-γ producing cells can generate a long lasting immunity against the parental unmodified tumor cells and that the cytokine-secreting tumor cells (so far examined only in the 3LL-D122 tumor) can effectively stimulate the destruction of previously injected parental tumor cells, further stress the possible role of tumor cell targeted IFN-γ gene therapy for cancer.

It is important to note that in many experimental models the level of IFN-γ production was too low to be harmful to the recipient mice, while being still sufficient to stimulate host antitumor responses. This is one obvious point in favor of IFN-γ gene therapy against cancer.

Tumor cell targeted IFN-γ gene therapy may have a rationale and a likelihood of success only in some cancers, particularly when applied to patients at an early stage of their disease when they still possess potential high immune response to IFN-γ. These studies, together with work on genes such as IL-2, IL-4, TNF-α and others (Russel, 1990; Miller, 1992; Cournoyer and Caskey, 1993; Pardoll, 1993), ensure that cytokine gene therapy will be an important treatment modality for cancer patients in the coming years.

References

Blankenstein T, Qin Z, Uberla K, Muller W, Rosen H, Volk HD and Diamantstein T (1991). Tumor suppression after tumor cell-targeted tumor necrosis factor α gene transfer. J Exp Med **173**, 1047-1052.

Burrows FJ, Watanabe Y and Thorpe PE (1992). A murine model for antibody-directed targeting of vascular endothelial cells in solid tumors. Cancer Res **52**, 5954-5962.

Carrel S, Schmidt-Kessen A and Giuffre L (1985). Recombinant interferon-γ can induce the expression of HLA-DR and -DC on -DR negative melanoma cells and enhance the expression of HLA-ABC and tumor-associated antigens. Eur J Immunol **15**, 118-123.

Cournoyer D and Caskey CT (1993). Gene therapy of the immune system. Ann Rev Immunol **11**, 297-329.

De Maeyer E and De Maeyer-Guignard J (1988). "Interferons and Other Regulatory Cytokines". John Wiley, New York.

Doukas J and Pober J (1990). IFN-γ enhances endothelial activation induced by tumor necrosis factor but not IL-1. J Immunol **145**, 1727-1733.

Dranoff G, Jaffee E, Lazenby A, Golumbek P, Levitsky H, Brose K, Jackson V, Hamada H, Pardoll D and Mulligan R (1993). Vaccination with irradiated tumor cells engineered to secrete murine granulocyte-macrophage colony-stimulating factor stimulates potent, specific, and long-lasting anti-tumor immunity. Proc Natl Acad Sci USA **90**, 3539-3543.

Dustin ML, Rothlein R, Bhan AK, Dinarello CA and Springer TA (1986). Induction by IL 1 and interferon-γ: Tissue distribution, biochemistry and function of a natural adherence molecule (ICAM-1). J Immunol **137**, 245-254.

Esumi N, Hunt B, Itaya and Frost P (1991). Reduced tumorigenicity of murine tumor cells secreting γ-interferon is due to nonspecific host responses and is unrelated to class I major histocompatibility complex expression. Cancer Res **51**, 1185-1189.

Gansbacher B, Bannerji R, Daniels B, Zier K, Cronin A and Gilboa E (1990). Retroviral vector-mediated γ-interferon gene transfer into tumor cells generates potent and long lasting antitumor immunity. Cancer Res **50**, 7820-7825.

Gansbacher B, Zier K, Cronin K, Hantzpoulos PA, Bouchard B, Houghton A, Gilboa E and Golde P (1992). Retroviral gene transfer induced constitutive expression of interleukin-2 or interferon-γ in irradiated human melanoma cells. Blood **180**, 2817-2825.

Gastl G, Finstad CL, Guarini A, Bosl G, Gilboa E, Bander NH and Gansbacher B (1992). Retroviral vector-mediated lymphokine gene transfer into human renal cancer cells. Cancer Res **52**, 6229-6236.

Giacomini P, Imberti L, Aguzzi A, Fisher PB, Trinchieri G and Ferrone S (1985). Immunochemical analysis of the modulation of human melanoma-associated antigens by DNA recombinant immune interferon. J Immunol **135**, 2887-2894.

Giovarelli M, Cofano F, Vecchi A, Forni M, Landolfo S and Forni G (1986). Interferon-activated tumor inhibition in vivo. Small amounts of interferon-γ inhibit tumor growth by eliciting host systemic immunoreactivity. Int J Cancer **37**, 141-148.

Hock H, Dorsch M, Kunzendorf U, Qin Z, Diamantstein T and Blankenstein T (1993). Mechanisms of rejection induced by tumor cell-targeted gene transfer of interleukin 2, interleukin 4, interleukin 7, tumor necrosis factor, or interferon-γ. Proc Natl Acad Sci USA **90**, 2774-2778.

Kasai M, Yoneda T, Habu S, Maruyama Y, Okumura K and Tokunaga T (1981). In vivo effect of anti-asialo GM1 antibody on natural killer activity. Nature **291**, 334-335.

Kelly SA, Gschmeissner S, East N and Balkwill FR (1991). Enhancement of metastatic potential by γ-interferon. Cancer Res **51**, 4020-4027.

Kurzrock R (1992). Interferon-gamma. In "Cytokine Therapy" (eds DW Galvani and JC Cawley), pp 131-151. Cambridge University, Cambridge.

Maraskovsky E, Chen WF and Shortman K (1989). IL-2 and IFN-γ are two necessary lymphokines in the development of cytolytic T cells. J Immunol **143**, 1210-1214.

Maruguchi Y, Toda KI, Fujii K, Imamura S and Watanabe Y (1991). Survival period of tumor-bearing mice is prolonged after the interferon-γ-producing gene transfer. Cancer Lett **60**, 41-49.

Matthys P, Dijkmans R, Proost P, Van Damme J, Hermans H, Sobis H and Billiau A (1991). Severe cachexia in mice inoculated with interferon-γ-producing tumor cells. Int J Cancer **49**, 77-82.

Miller AD (1992). Human gene therapy comes of age. Nature **357**, 455-460.

Novelli F, Giovarelli M, Reber-Liske R, Virgallita G, Garotta and Forni G (1991). Blockade of physiologically secreted IFN-γ inhibits human lymphocyte and natural killer cell activation. J Immunol **147**, 1445-1452.

Pardoll DM (1993). Cancer vaccines. Immunol Today **14**, 310-316.

Porgador A, Bannerji R, Watanabe Y, Feldman M, Gilboa E and Eisenbach L (1993). Antimetastatic vaccination of tumor-bearing mice with two types of IFN-γ gene-inserted tumor cells. J Immunol **150**, 1458-1470.

Restifo NP, Spiess PJ, Karp SE, Mule J and Rosenberg SA (1992). A nonimmunogenic sarcoma transduced with the cDNA for interferon-γ elicits CD8+ cells against the wild-type tumor: Correlation with antigen presentation capability. J Exp Med **175**, 1423-1431.

Rivett AJ (1993). Proteasomes: Multicatalytic proteinase complexes. Biochem J **291**, 1-10.

Ruggiero V, Tavernier J, Fiers W and Baglioni C (1986). Induction of the synthesis of tumor necrosis factor receptors by interferon-γ. J Immunol **136**, 2445-2450.

Russell SJ (1990). Lymphokine gene therapy for cancer. Immunol Today **11**, 196-200.

Takeda K, Hatakeyama K, Tsuchiya Y, Rikiishi H and Kumagai K (1991). A correlation between GM-CSF gene expression and metastases in murine tumors. Int J Cancer **47**, 413-420.

Tanaka K, Isselbacher K, Khoury G and Jay G (1985). Reversal of oncogenesis by expression of a major histocompatibility complex class I gene. Science **228**, 26-30.

Teramura Y, Watanabe Y, Kan N, Masuda T and Kuribayashi K (1993). Interferon-γ-producing tumor induces host tumor-specific T cell responses. Jpn J Cancer Res **84**, 689-696.

Ucer U, Bartsch M, Scheurich P, Berkovic D, Ertel C and Pfizenmaier K (1986). Quantitation and characterization of γ-interferon receptors on human tumor cells. Cancer Res **46**, 5339-5343.

Valente G, Ozmen L, Novelli F, Guena M, Palestro G, Forni G and Garotta G (1992). Distribution of interferon-γ receptor in human tissues. Eur J Immunol **22**, 2403-2412.

Watanabe Y and Sakata T (1988). Highly efficient action of autocrine mouse interferon-γ expressed via a retroviral vector. Eur J Immunol **18**, 1627-1630.

Watanabe Y, Kuribayashi K, Miyatake S, Nishihara K, Nakayama E, Taniyama T and Sakata (1989). Exogenous expression of mouse interferon-γ cDNA in mouse neuroblastoma C1300 cells results in reduced tumorigenicity by augmented anti-tumor immunity. Proc Natl Acad Sci USA **86**, 9456-9460.

Watanabe Y (1992). Transfection of interferon-γ gene in animal tumors; a model for local cytokine production and tumor immunity. Seminars in Cancer Biol **3**, 43-46.

Yanagihara K, Seyama T and Watanabe Y (1994). Antitumor potential of interferon-γ: Retroviral expression of mouse interferon-γ cDNA in two kinds of highly metastatic mouse tumor line reduces their tumorigenicity. Nat Imm **13**, 102-112.

Yang Y, Waters JB, Fruh K and Peterson PA (1992). Proteasomes are regulated by interferon-γ : Implication for antigen processing. Proc Natl Acad Sci USA **89**, 4928-4932.

19

ANTITUMOR RESPONSE ELICITED BY IL-2 AND IFN-γ TRANSDUCED TUMOR CELLS

J. J. Lewis, K. Cronin, F. Rosenthal, A. Guarini, K. Zier and B. Gansbacher
Department of Surgery and Division of Hematologic Oncology, Memorial Sloan-Kettering Cancer Center, New York, NY, USA, Department of Medicine, Mount Sinai School of Medicine, New York, NY, USA and Department of Biomedical Sciences and Human Oncology, University of Turin, Turin, Italy

Introduction

In the past decade it has become increasingly apparent that effector cells of the immune system play an important role in the recognition and elimination of neoplastic cells. A wide array of cells, including CTL, Th cells, NK cells, LAK cells, macrophages and granulocytes, have both in vitro and in vivo cytostatic or cytolytic activities for tumor cells. It is incompletely understood how the immune system, with its powerful armamentarium, often fails to reject tumor cells. Multiple theories exist, including possible lack of immunogenic tumor associated antigens, heterogeneity or lack of MHC expression, or the absence of CTL precursors. In addition, the presence of suppressor cells or factors, induction of anergy subsequent to TCR occupancy by tumor antigen ligands without a costimulatory signal, and the presence of T cells with abnormal TCR in a cancer bearing host have been postulated. Because of their integral role in regulating several of these processes, cytokines have been extensively studied as a potential therapy to augment antitumor immune responses.

Cytokines are soluble factors secreted by diverse cell types that are responsible for communication between cells of the immune system. Many have been identified and cloned in recent years. The production and purification of the recombinant forms of several of these cytokines have made both in vitro and in vivo studies possible. Initial trials focused on the systemic administration of IL-2 because of its ability to stimulate CTL, NK and LAK cells, all of which can kill tumor cells (Hwu et al., 1993; Long et al., 1993). The first evidence that systemic administration of a cytokine could induce an antitumor response was reported in 1985 following the administration of IL-2 together with LAK cells (Rosenberg

CYTOKINE-INDUCED TUMOR IMMUNOGENICITY
ISBN 0–12–261520–4

and Mule, 1985; Rosenberg, 1988a; Rosenberg, 1990). In these studies, almost 20% of patients with advanced melanoma or renal cell carcinoma experienced complete remission of their disease. However, because of its short half-life in vivo, high doses of IL-2 had to be used leading to severe side effects (Lotze et al., 1985; Rosenberg, 1988a; Rosenberg et al., 1989). Furthermore, the clinical responses were often only transient.

The second era in cytokine therapy began after developments in recombinant DNA technology enabled a more refined approach to biological therapy. This culminated with the ability to use highly efficient retroviral vectors to transfer genes into cells, thus opening up a novel approach to cytokine based immunotherapy. Transduction of immunocompetent cells with cytokine genes using retroviral gene transfer enabled the continuous production of cytokines in a paracrine fashion and avoided the toxicity of high systemic doses. T cells, LAK cells and NK cells have been targeted in an attempt to amplify cell mediated immunity and improve antitumor responses (Rosenberg et al., 1986; Rosenberg et al., 1990; Rosenberg, 1991). These pioneering studies were important because they showed that it was feasible to introduce foreign genetic material into lymphocytes without any major toxicity or alteration in their function (Kasid et al., 1990). However, both technical problems and concerns about constitutive production of cytokines in vivo subsequent to tumor rejection, have limited the clinical application of this approach. It therefore occurred to several investigators, including one of us (B.G.), that the tools of molecular biology could be utilized to reverse the natural order and cause tumor cells to secrete immunoregulatory cytokines (Tepper et al., 1989; Watanabe et al., 1989; Fearon et al., 1990; Gansbacher et al., 1990a; Gansbacher et al., 1990b; Asher et al., 1991; Blankenstein et al., 1991; Golumbek et al., 1991; Gansbacher et al., 1992; Hock et al., 1993). This approach was attractive because it would facilitate local paracrine secretion of cytokine that, following destruction of the tumor cell, would terminate.

Retroviral mediated gene transfer

Retroviral mediated gene transfer is at present the most suitable approach for transferring genes into cells for clinical use (Gilboa, 1986). It is a promising technique for two reasons. First, because of its potential high efficiency, it is the only current system available to introduce the gene of interest into a large proportion of the target cells. This is in sharp contrast to other gene transfer systems such as DNA transfection, protoplast fusion and electroporation. Second, retroviral vectors have a broad host range, enabling the introduction of genes not only into adherent cells but also into suspension cells.

Retroviruses are RNA viruses in which the genes are encoded in a single stranded RNA molecule. Reverse transcriptase converts the viral RNA into double stranded DNA that integrates into the infected host cell genome. The integrated DNA or provirus is the template for expression of the viral genes and synthesis of progeny virion RNA. The viral gene products and progeny RNA assemble into a virion that exits the cell by budding through the cell membrane. The presence of the viral genome in the cell, expression of its genes, and the formation of progeny virus have no apparent effects on the viability of the infected cell. The cell is chronically infected, remaining otherwise healthy, while continuing to secrete virus. The suitability of retroviruses for gene transfer is predicated on their mode of replication. Thus, by replacing the structural viral genes with the gene of interest and utilizing the efficient viral infection process, the gene is transferred into the target cell as if it were a viral gene.

The efficiency of retroviral gene transfer is determined by several factors, including the titer of the recombinant virus and the cycling status of the target cells. The titer of vector is dependent on vector size (the smaller the vector, the higher the titer), the length of the *gag* region, whether the inserted gene contains sequences that destabilize the vector genome, and the quality of the packaging cell line (Miller, 1993). Packaging cell lines are cells that contain encapsidation defective viral genomes that produce all the viral structural proteins, but cannot be packaged into viral particles because they lack the necessary regulatory sequences. In our studies, we have used N2 based retroviral vectors because of the high titers of virus they yield, possibly due to the presence of some of the *gag* coding sequence 5' to the neomycin gene (Armentano et al., 1987). We have used these vectors successfully to transfer genes into human cells including the myeloid cell line K562, the T-cell line HUT-78, melanoma cell lines, renal carcinoma cell lines, and normal primary T lymphocytes as well as several murine tumor cell lines, including the fibrosarcoma CMS-5, the fibroblast line NIH 3T3, and the mouse lymphoma 38C13 (Gansbacher et al., 1990a; Gansbacher et al., 1990b; Foa et al., 1992; Gansbacher et al., 1992; Gastl et al., 1992). In the remainder of this chapter we will review our experience with IL-2 gene transfer into CMS-5 cells and IL-2 and IFN-γ gene transfer into melanoma cells.

IL-2 gene transfer into CMS-5 cells

Chemically-induced tumors in animals possess tumor specific antigens (TSA), as demonstrated by the ability of animals transplanted with a syngeneic tumor to reject a second challenge of the same tumor (Foley, 1953; Prehn and Main, 1957; Old, 1981). The relevance of these observations to human neoplasia has been questioned, since spontaneously arising tumors in mice are essentially nonimmunogenic suggesting that the existence of highly immunogenic tumors in nature is quite rare (Hewitt et al., 1976). While it now appears that at least some spontaneous tumors can be recognized by the immune system, even these elicit a weak immune response of little or no protective value (Key et al.,

1984). In several instances, however, weak antitumor responses generated in animals bearing "nonimmunogenic" tumors could be potentiated, protecting the animal from a subsequent tumor challenge (Forni and Santoni, 1984; Key et al., 1984). The growth of some tumors in experimental animals, as well as in human patients, is often associated with the establishment of an active state of immune suppression (North, 1982). Using chemically induced tumors in mice similar to the tumor system used in our studies, North showed that tumor progression in the animal induced a population of Lyt $1^{+}2^{-}$ T cells which suppressed the emerging antitumor response mediated by Lyt $1^{-}2^{+}$ T cells.

IL-2 is a growth factor which stimulates the proliferation of CTL, Th T cells, NK cells and LAK cells, all of which can participate in the antitumor response. In the mouse, systemic administration of IL-2 induces the expansion and long-term persistence of adoptively transferred tumor specific T cells. To test whether "locally" secreted cytokines were able to influence the generation of antitumor immune responses we used retroviral vectors to introduce the gene for human IL-2 into CMS-5 cells. CMS-5, a methylcholantrene induced tumor in BALB/c mice, is a weakly immunogenic tumor devoid of viral antigens (DeLeo et al., 1979; DeLeo et al., 1980; Srivastava et al., 1986). CMS-5 is weakly immunogenic when compared to CMS-13 or Meth A, two non crossreacting methylcholanthrene induced tumors of the same genetic background (Srivastava et al., 1986). CMS-5, CMS-13 and Meth A cells were adapted to culture and grown in DMEM medium. BALB/c mice were obtained from a colony bred at Sloan-Kettering Institute (New York) from which the CMS-5, CMS-13 and Meth A tumors originated. Tumor cell injections were carried out using freshly prepared tumor cells.

Generation of tumor cell lines expressing IL-2

We cloned the cDNA for human IL-2 into different sites of the N2 retroviral vector. Transcription was driven by either the cytomegalovirus or thymidine kinase promoter. Vector DNA was transfected into a packaging cell line, and neomycin resistant colonies were selected with G418. Drug resistant colonies were isolated, expanded to cell lines and virus containing cell free supernatant was used to infect CMS-5 cells. DNA blot analysis of G418 resistant CMS-5 cells was used to show that an intact provirus was present in the infected cells. Expression of the IL-2 gene in the CMS-5 cells transduced with the retroviral vectors was determined by measuring the secretion of the cytokine into the cell supernatant first using a bioassay and then confirming these results with an enzyme linked immunosorbent assay.

Clonal isolates of CMS-5 cells were found to vary in their expression and secretion of IL-2 depending on the particular retroviral vector used. In each case, human IL-2 activity could be neutralized by an appropriate mAb. Secretion of human IL-2 had no discernible effect on cell morphology or on the growth rate in culture compared to parental CMS-5 cells or CMS-5 cells transduced with a control N2 vector.

These results underscore the importance of retroviral vector design and of the choice of expression system in these kinds of studies. Furthermore, they provided an opportunity to test the effects of varying levels of cytokine production on tumorigenesis.

The effect of IL-2 secretion on tumor growth in vivo

We then tested whether the local secretion of IL-2 from the genetically modified cells would affect their tumorigenicity in vivo in two ways. First, equal numbers of unmodified or vector modified CMS-5 cells were injected intradermally into the back of syngeneic BALB/c mice and the appearance and size of the resulting tumor was monitored over time.

The results of our experiment indicated that mice injected with either unmodified CMS-5 cells or with cells transfected with a control vector, which expresses the human ADA gene product, developed tumors at the site of injection and had to be sacrificed by day 40. In contrast, in mice injected with various IL-2 producing CMS-5 cells, tumor growth correlated inversely with the amount of IL-2 secreted by the tumor cells. No tumors were induced in mice injected with high IL-2 producing cells. In mice injected with CMS-5 cells that secreted intermediate levels of IL-2, tumor growth was significantly retarded and mice survived for over 75 days.

The second approach we used was to measure the effect of cytokine production on tumorigenicity by evaluating the number of IL-2 secreting tumor cells which could be injected without causing tumors. While injection of only 2×10^5 CMS-5 cells into BALB/c mice caused the appearance of large tumors in 25 out of 25 animals, injection of as many as 10^6 IL-2 producing CMS-5 cells failed to result in tumors. Thus, localized IL-2 secretion altered the ability of CMS-5 cells to cause tumors in vivo.

In vivo immune response generated by cytokine producing CMS-5 cells

The association between cytokine secretion and growth of modified tumor cells could be a result of altered growth properties of the tumor cells following genetic manipulation, or of an immune response triggered by the cytokines. We excluded the first possibility, since our results showed no change in cell morphology, viability, or growth rate in culture of vector modified tumor cells when compared to parental CMS-5 cells.

To test the more intriguing possibility that increased amounts of cytokine secretion played a role in the reduced outgrowth of tumor cells, we performed an experiment in which 1 x 10^6 of the high IL-2 producing cells were mixed with increasing numbers of unmodified tumor cells and then injected into mice. We found that coinjection of 1 x 10^6 IL-2 producing cells was able to induce the rejection of up to 8 x 10^5 unmodified tumor cells. Indeed, injection of 2 x 10^5 unmodified tumor cells alone generated a larger tumor then injection of 1.6 x 10^6 tumor cells mixed with the IL-2 producing cells. These observations demonstrated that local secretion of IL-2 influenced the growth of unmodified tumor cells and were consistent with the hypothesis that an immune response was induced which was responsible for the loss of tumorigenicity of CMS-5 cells.

Further support for this hypothesis was obtained when mice which had rejected an IL-2 producing CMS-5 tumor were challenged after 6 weeks with 2 x 10^5 CMS-5 cells, a number which would usually cause tumor formation. All mice failed to show tumors at the site of the challenge. Furthermore, Meth A and CMS-13 were not rejected. Both of these cell lines had been grown in vitro in a fashion identical to CMS-5 cells, thereby also eliminating the possibility that rejection of CMS-5 cells was due to sensitization to serum proteins. Since all control mice developed tumors, protection was specific. Thus, the abrogation of tumorigenicity was characterized by a process which was CMS-5 specific, had memory, and was best explained by the existence of a protective immune response against the parental tumor.

In vitro lytic activity of spleen cells derived from animals injected with parental and cytokine producing CMS-5 cells

Next we tested whether a CTL response might be at least partially responsible for the failure of IL-2 producing tumor cells to grow. Since we reasoned that an ongoing antitumor immune response would be best detectable in animals exhibiting reduced tumor growth, spleens were taken from mice which had been inoculated with IL-2 producing tumor cells which grew at a significantly reduced rate, or from mice injected with unmodified CMS-5 cells. The nonadherent T-cell enriched populations were then cocultured in vitro with irradiated, parental CMS-5 cells for 6 days and tested in a standard ^{51}Cr release assay for their ability to lyse various targets, including parental CMS-5 cells, an IL-2 producing CMS-5 line, K562, an NK sensitive cell line, and CMS-13.

Mice injected with unmodified CMS-5 cells were able to lyse parental CMS-5 cells and IL-2 producing CMS-5 cells, if effector spleen cells were taken at day 11 after injection of tumor cells. Neither K562 nor CMS13 cells were lysed. In other experiments, YAC-1 cells, another NK sensitive cell line, were also shown to be resistant to lysis by the activated spleen cells. In contrast, if spleen cells were taken from mice inoculated with the unmodified tumor 28 days after the injection of CMS-5 cells, there was no longer detectable lysis of unmodified CMS-5 cells. This observation is consistent with the transient nature of the CTL response against CMS-5 and several other mouse tumors, which is followed by the emergence of a state of immune suppression mediated by a population of tumor specific T suppressor cells. In contrast, spleen cells derived as late as 75 days after injection of IL-2 producing CMS-5 cells still demonstrated a strong CTL response, similar to that observed with spleen cells taken 11 days after injection of unmodified CMS-5 cells. Thus, the normal transient CTL response was not seen in the mice injected with IL 2 secreting tumors, but rather was replaced by one which was long-lived.

We attempted to investigate in vitro whether secretion of IL-2 by the tumor cells could influence T-cell activity by testing the ability of parental CMS-5 cells and their IL-2 producing counterparts to serve as targets for CTL. Effector cells were produced by inoculating mice with parental CMS-5 cells. After 18 days spleen cells were isolated, stimulated in vitro for 6 days with irradiated unmodified CMS-5 cells and used as effectors in a CML assay against a panel of targets. As expected, given the immune suppression in mice inoculated with tumor cells close to 3 weeks prior to the harvest of spleen cells (North, 1982), only low levels of lysis were observed when unmodified CMS-5 cells were used as targets. In contrast, efficient lysis was observed when IL-2 producing CMS-5 cells were used as targets. Moreover, the efficiency of lysis of various targets was directly proportional to the level of IL-2 secreted by the target cells. Even parental CMS-5 cells could be made susceptible to lysis by the addition of exogenous IL-2 during the 18 h of incubation with the activated CTL. We estimated that 1-3 U/ml of IL-2 accumulate during the lysis reaction when 10000 N2/IL-2 vector containing CMS-5 cells are used as targets. In other experiments, 5 and 2 U/ml of exogenous IL-2 were effective at inducing lysis of CMS-5 cells, indicating that IL-2 secretion from the tumor cells could indeed be responsible for the increase in lysis seen. Additional support was obtained in experiments in which the addition of an anti-IL-2, but not a control mAb to the overnight CML reaction inhibited the lysis of N2/IL-2/CMS-5 target cells by approximately 50%.

Expression of IL-2 and IFN-γ in human melanoma cells

Despite many attempts to use chemotherapy for metastatic melanoma, results remain suboptimal suggesting that new treatment strategies will be necessary to improve survival rates. Therapeutic approaches that aim to amplify the patient's own immune response have received considerable attention given evidence in animal models of protective immune responses to a tumor cell challenge following appropriate immunization (Shu and Rosenberg, 1985). To date, vaccination protocols in melanoma patients have not shown any clear overall impact on survival. Evidence in several trials, however, suggests that patients who develop immune responses have a significant improvement in disease free survival and overall survival compared to non responders (Livingston et al., 1983; Livingston et al., 1985a; Livingston et al., 1985b; Livingston et al., 1989; Livingston, 1991). While antibody or cellular immune responses directed against FCS, alloantigens or known surface glycoproteins have been described, they have had no impact on the clinical response (Livingston, 1991). A major aim of vaccination trials has been to induce a CTL response, since these are the effector cells that have specificity and immunological memory. Blood from melanoma patients contains circulating cytotoxic precursors with restricted reactivity for autologous melanoma. TIL both from melanomas (Belldegrun et al., 1989; Crowley et al., 1990; Kawakami et al., 1992) and renal cell carcinomas (Belldegrun et al., 1988a; Belldegrun et al., 1988b; Belldegrun and Rosenberg, 1989) consist of similar cells. Why various vaccination protocols have not induced or amplified these CTL is unclear. Several hypotheses exist, including the fact that tumor cells are poor antigen presenters, have low or absent MHC molecule expression on the cell surface, secrete suppressor factors, or that TCR engagement by tumor associated antigens without a costimulatory signal induces anergy. IL-2 can reverse two of these mechanisms. First, TGF-β secretion by tumor cells induces inhibition of a CTL response and IL-2 can reverse this suppression (Inge et al., 1992). Second, engagement of the TCR without a proliferative signal can induce anergy, but ligand interaction with CD28 or the presence of IL-2 can prevent this anergy from occurring (Harding et al., 1992; Liu and Janeway, 1992).

CTL clones specific for autologous melanoma cells and renal carcinoma cells can be generated in patients by stimulating PBL cells with IL-2 and tumor cells. Recently, three human melanoma specific genes have been cloned from a cosmid library made from DNA of tumor cells that expressed CTL defined determinants using a CTL stimulation assay (Van der Bruggen et al., 1991). This study and others provide evidence to support the hypothesis that shared melanoma specific epitopes exist which are recognized by T cells, implying that in some cases allogeneic melanoma cells might be suitable for use in vaccination protocols (Knuth et al., 1989; Wolfel et al., 1989; Kawakami et al., 1992). Based upon evidence that there are shared tumor antigens in melanoma (Knuth et al., 1989; Wolfel et al., 1989; Crowley et al., 1990; Kawakami et al., 1992), that cytokine secretion by tumor cells decreases tumorigenicity (Tepper et al., 1989; Watanabe et al., 1989; Fearon et al., 1990; Gansbacher et al., 1990a; Gansbacher et al., 1990b), that the frequency of CTL precursors in melanoma patients can be as high as 1:900 (Coulie et al., 1992), and that in vivo secretion of IL-2 by tumor cells increases the frequency of circulating cytotoxic precursors and of CTL specific for the immunizing tumor in a murine model (Ley et al., 1991), we decided to use retroviral mediated transduction to induce melanoma cells to secrete IL-2 or IFN-γ.

Cell lines and retroviral vector tumor cell infection

Five human melanoma cell lines were selected for this study. The cell lines SK-MEL-19,-29,-131,-245,-256 were established from primary tumor specimens or metastatic lesions and were obtained from our cell bank (Houghton et al., 1982; Houghton et al., 1984; Houghton et al., 1987). The retroviral vectors used in this study were identical to the ones used in our murine CMS-5 tumor model. In addition, we constructed viruses containing the cDNA for IFN-γ.

Secretion of IL-2 or IFN-γ by melanoma cells transduced with the retroviral vectors was determined using a bioassay and ELISA for IL-2 and a bioassay and RIA for IFN-γ. Despite the fact that all the selected clones demonstrated G418 resistance, the IL-2 produced by single clonal populations varied widely ranging from 691 U/ml per 10^6 cells to 0 U/ml per 10^6 cells (Table 19.1).

We studied several clonal populations for each construct and found this to be a reproducible finding. As expected from previous results, the cell lines varied in their expression and secretion of cytokines depending on the particular retroviral vector used.

In all cases human IL-2 or IFN-γ activity could be neutralized by an appropriate mAb. Melanoma cells transduced with the control vector, DCA, did not secrete detectable amounts of IL-2 or IFN-γ. Secretion of human IL-2 or IFN-γ had no discernible effect on cell morphology or on the growth rate in culture compared to parental cells or cells transduced with DCA. Some of these transduced cell lines have been in culture longer than 6 months and continue to secrete the expected amounts of IL-2 or IFN-γ.

Irradiated melanoma cells continue to secrete IL-2

In the context of immunotherapy, transduced cells injected intradermally could serve both as a source of continuous cytokine production and of tumor associated-antigen.

Although cytokine secreting tumor cells exhibit decreased tumorigenesis and would most likely not grow in vivo, we decided to test whether irradiating them as a further precaution would affect cytokine secretion. We found that following irradiation with up to 10000 rad, cytokine production persisted for 3-4 weeks. Between day 5 and 15 post irradiation IL-2 secretion increased, most likely due to increased leakage from radiation damaged cells. All five transduced tumor cell lines continued to secrete IL-2 for several weeks after radiation. The duration of

Table 19.1. *IL-2 production by transduced cloned melanoma cells*

Cell line	Clone	IL-2 U/ml
SK-MEL-19/DC/TK/IL-2	# 4	61
	# 3	36
	# 6	0
SK-MEL-29/DC/AD/R/IL-2	# 6	620
	# 8	168
	# 12	0
SK-MEL-131/DC/TK/IL-2	# 1	691
	# 6	314
	# 14	0

IL-2 secretion was influenced by the amount of radiation delivered to the cells and in this way defined by the radiosensitivity of each cell line. The irradiated cell lines were incapable of growth in culture, as demonstrated by the decreasing cell number over time. In addition, the irradiated IL-2 secreting tumor cells were not able to form tumors in nude mice. This has been a consistent finding in different tumor types (B.G. unpublished results). IFN-γ secretion also persists for several weeks after irradiation (Gastl et al., 1992). These results illustrate that it is possible to lethally irradiate the cells prior to injection into patients and continue to have cytokine secretion for several weeks.

Up-regulation of MHC class I by IFN-γ

IFN-γ is able to up-regulate surface expression of MHC antigens on a variety of cell types (Wallach et al., 1982; Gastl et al., 1992). Since antigen is recognized by T cells only when presented in the context of class I or II MHC, it is believed that an important mechanism by which IFN-γ exerts its antitumor effect may be to augment tumor cell MHC expression. To test whether there was a biological effect of IFN-γ on the melanoma cells we examined class I MHC antigen expression after IFN-γ transduction. We observed a variable up-regulation of class I MHC on the IFN-γ secreting melanoma lines. Similar to our observation with IL-2 we noted a wide variation in IFN-γ expression. Other investigators, using one of the IFN-γ vectors that we have used, have observed a similar finding in their experimental system (Rosa et al., 1983). Class I MHC was expressed on parental SK-MEL-131 tumor cells before infection and levels were up-regulated following IFN-γ gene transduction. SK-MEL-245 and SK-MEL-256 express HLA class I molecules, and IFN-γ transduction did not induce up-regulation. SK-MEL-19 does not express class I, and IFN-γ transduction had no effect. Thus, up-regulation of class I molecules after IFN-γ transduction is variable and must be tested on an individual basis.

Growth in nude mice

The mechanism by which IL-2 leads to decreased tumorigenesis is mediated by activation of T cells and NK cells at the tumor site. In order to evaluate the range in which IL-2 secretion is effective in vivo we chose to inject SK-MEL-29 and SK-MEL-245 cells, which secrete 40 and 2 U IL-2/ml per 10^6 cells respectively, into nude mice. Four to 6 week old BALB/c nu/nu mice were injected s.c. with 2.5×10^5, 5×10^5 and 1×10^6 cells of parental or the corresponding IL-2 secreting melanoma cell populations. We observed that all mice injected with parental cells developed tumors within 3 weeks, while all mice injected with IL-2 secreting tumor cells remained tumor free during an observation period of 3 months. In addition, irradiated IL-2 or IFN-γ secreting tumor cells also failed to form tumors. Nonirradiated IFN-γ secreting cells formed tumors as expected because IFN-γ action is species specific (Langer and Pestka, 1988). Identical findings have been observed with IL-2 secreting human renal carcinoma cells (Gastl et al., 1992). This suggests that the secreted IL-2 is biologically active and able to activate effector cells in vivo.

Conclusion

Our results, in CMS-5 cells, demonstrate that the modification of weakly immunogenic tumor cells to secrete IL 2, effectively abrogated their ability to form tumors in mice. By comparison, only partial regression of tumor growth has been achieved via the systemic administration of very high doses of IL-2 (Rosenberg, 1988b). We have also demonstrated that the immune reaction induced prevents highly tumorigenic doses of coinjected unmodified tumor cells from establishing a tumor, and that IL-2 secretion by the tumor leads to a long lasting state of T-cell immunity as measured by persistence of a specific cytotoxic activity, not seen after injection of unmodified tumor cells. In vitro data suggest that IL-2 secretion into the immediate surrounding of the tumor cells leads to a much more efficient lysis of the cytokine secreting CMS-5 cells.

We have transduced human melanoma cells with the IL-2 or IFN-γ gene using retroviral vectors and induced them to secrete these cytokines constitutively. Expression of a variety of cytokine genes including IL-2, IL-4, IFN-γ, GM-CSF , G-CSF and IL-7 has been shown by us and others to exert potent antitumor effects in animal tumor models (Tepper et al., 1989; Watanabe et al., 1989; Fearon et al., 1990; Gansbacher et al., 1990b; Dranoff et al., 1993; Hock et al., 1993). Several of these studies have demonstrated a role for CTL in tumor rejection. We decided to work in the melanoma system, with the aim of using these cells in a human vaccine protocol, for several reasons. Melanoma cells were the first human cancer cells shown to express tumor associated antigens (Van der Bruggen et al., 1991) and these antigens have epitopes recognized by CTL (Coulie et al., 1992). Indeed, Boon and his group, have used CTL to screen cosmid libraries to identify and clone MAGE-1, a melanoma tumor specific antigen (Van der Bruggen et al., 1991).

Recently, it was reported that a minimum of six determinants can be recognized by CTL in an MHC restricted fashion, either together with HLA-A1 or with HLA-A2 (Knuth et al., 1989; Wolfel et al., 1989; Crowley et al., 1990; Van der Bruggen et al., 1991; Kawakami et al., 1992). The genes that code for these determinants are tumor specific in the sense that their expression has not been detected in normal tissue. Most importantly from a therapeutic perspective, one of the genes cloned from a human melanoma cell was expressed on a variety of tumor cells, including non-melanoma tumor cells, but not on normal cells with the exception of the testes (Van der Bruggen et al., 1991). CTL have been shown to exert a tumoricidal effect on melanoma cells (Mukherji and MacAlister, 1983; De Vries and Spits, 1984). Coulie et al. (1992) reported that CTL precursors could be detected in the peripheral blood of melanoma patients, with a frequency ranging from 1/900 to 1/33000, which is in the range of that observed after successful immune responses against viruses such as cytomegalovirus, herpes simplex and varicella zoster (Hickling et al., 1987; Borysiewicz et al., 1988). These findings are important because they suggest vaccination protocols with the goal of increasing the frequency of CTL, rather than inducing them. Evidence that this can occur was reported in the P815 tumor model where in vivo secretion of IL-2 by the tumor cells increased the frequency of circulating tumor specific CTL 2-4 fold (Ley et al., 1991).

We chose IL-2 or IFN-γ constructs because of their previously demonstrated efficacy in decreasing tumorigenesis of transfected cells. In addition, they act on both effector and tumor cells, possibly in a synergistic fashion. IL-2 stimulates T and NK cell proliferation and differentiation. IFN-γ acts to induce up-regulation of MHC and adhesion molecules on the surface of tumor cells. Predictably, cells transfected with our recombinant vectors varied in the amounts of cytokines secreted. Within bulk infected cells we found clones secreting widely varying amounts of IL-2. If such cells are used in vaccination protocols two options are available. First, if transduced tumor cells express a known tumor specific antigen such as MAGE-1 one could select the highest IL-2 and MAGE-1 expressor clone in an attempt to maximize clonal expansion of the MAGE-1 specific CTL population in vivo. Second, if the tumor associated antigen is not known it is reasonable to use bulk selected cytokine secreting tumor cells as a vaccine, in order to maintain antigenic heterogeneity.

We have also demonstrated that following irradiation with 10,000 rad, transduced cells continue to produce significant amounts of IL-2 and IFN-γ for 3-4 weeks. Not only did irradiation not abolish cytokine secretion for the first 3-5 weeks, but it actually caused increased cytokine release for a short period. This may have been due to increased leakage of cytokines from dying cells and may be significant in clinical applications where one would like to avoid injecting live tumor cells into patients.

The results of our experiments have demonstrated a biological effect in vivo, the IL-2 secreting cells were rejected in BALB/c nude mice while the normal counterparts formed tumors. Since nude mice presumably lack T cells but do have NK cell activity we assume that NK cell activation induced tumor rejection. Clearly this supports only a statement that the secreted IL-2 is biologically active and does not predict the effect on a CTL population. We used human IFN-γ secreting tumor cells as a control to exclude other variables that may be induced by retroviral transfection which could be responsible for tumor rejection. Since IFN-γ action is species specific (Langer and Pestka, 1988), melanoma cells secreting human IFN-γ formed tumors identically to the parental melanoma cells. We have observed similar findings with CMS-5/IL-2 transduced cells that fail to grow in scid/scid mice unless they are NK cell depleted (Alosco et al., 1993).

While autologous cells would be preferable to allogeneic cells for vaccination protocols, in only 20-30% of melanoma patients can tumor cell lines be established. Since shared tumor associated antigens are class I MHC restricted (Crowley et al., 1990; Van der Bruggen et al., 1991), it is possible to use already established allogeneic cell lines transduced with cytokine genes for vaccination. In melanoma, the two class I MHC molecules that have been described to present tumor associated antigens are HLA-A1 and HLA-A2. Since 40% of the Caucasian population express the HLA-A2 molecule, the use of allogeneic melanoma cells that express HLA-A2 is an attractive strategy. Thus, HLA-A2 matched allogeneic irradiated tumor cells which secrete IL-2 have the potential to activate and amplify cytotoxic effector cells that recognize on their autologous tumor cells the same peptide presented to them by the allogeneic tumor cells. Recently published data demonstrates an 80 fold more potent expansion of the antigen specific clone if the antigen presenting cells express not only the antigen but a costimulatory signal as well (Liu and Janeway, 1992).

It is of interest that one of the cell lines, SK-MEL-29, has already been used as an allogeneic vaccine in 17 melanoma patients (Livingston et al., 1985b). SK-MEL-29 expresses four melanocyte specific differentiation antigens. None of the vaccinated patients developed an antibody response against these differentiation antigens. This documents that irradiated tumor cells alone are inadequate as stimulators of a host antitumor response. Induction of a high affinity IgG response requires CD4 T-cell help. Thus, high affinity IgG production against the melanocyte differentiation antigens would already be a first indication of an effective way to activate the host T-cell arm.

Thus, we have demonstrated that retroviral vectors can be used to introduce cytokine genes into both murine and human tumor cells. The transduced tumor cell lines can be maintained in culture for at least several months during which time they continue to secrete cytokines in a stable manner. Rejection of IL-2 secreting tumors by nude mice and up-regulation of class I MHC genes on some of the tumor cells demonstrates that the secreted cytokines are biologically active. These results provide scientific and practical support for active vaccination protocols using retroviral cytokine gene transfer in an attempt to increase the frequency of preexisting cytotoxic precursors in the circulation of tumor bearing patients.

Clinical applications and future directions

Progress in the development of retroviral vectors to replace defective genes, mark hematopoietic precursors, and introduce cytokine genes into effector or tumor cells in murine models has led to the development of several experimental gene therapy protocols that can be applied to the treatment of cancer in humans. The advances made in the murine tumor models using cytokine gene transfer, in particular the observation that the IL-2 gene introduced into tumor cells results in decreased tumorigenicity, have contributed to this renewed enthusiasm. Our results, as well as those of others, will certainly promulgate further attempts to develop cancer vaccines.

Protocols have been developed at a number of centers, including Memorial Sloan-Kettering, to express cytokine genes in cancer cells for use in human patients. We have chosen melanoma as our first model, and have begun immunizing HLA-A2 positive patients with the SK-MEL-29 (HLA-A2 positive) melanoma cell line transduced with the IL-2 gene. These cells are being irradiated and injected s.c. into patients with stage IV melanoma in a phase I study. In addition, we have a protocol for a similar approach in renal cell carcinoma, utilizing the HLA-A2 positive SK-RC-39 cell line, transduced with IL-2. We have currently immunized a total of 14 patients. In each protocol, 6 patients have completed the low dose (1×10^7) injection of irradiated tumor cells with no major complications. In addition, two patients in the melanoma protocol have completed the higher dose (5×10^7) immunization. While it is premature to evaluate the immune effects of vaccination, our hypothesis that the allogeneic response would induce a significant inflammatory reaction at the site of vaccination is seemingly correct. Indeed, several patients have had severe local inflammatory reactions. We are hopeful that this impressive cellular infiltrate, at the site of vaccination, will conceivably increase the frequency of T cells with tumor specific TCR and promote a clonal expansion of these effector cells.

Following these initial phase I studies to test the safety of this novel approach, the use of these modalities will be tested in patients with minimal residual disease who are the least immunocompromised and in whom the tumor load is minimal. Particular cytokine genes will be selected based on results from in vitro or animal studies. Further refinements, including the introduction of tumor antigens with retroviral vectors, possibly together with cytokine genes, into professional antigen presenting cells, could help to optimize the stimulation of the immune response.

References

Alosco T, Croy BA, Gansbacher B, Wang HQ, Rao U and Bankert R (1993). Antitumor response independent of functional B or T lymphocytes induced by the local and sustained release of interleukin-2 by the tumor cells. Cancer Immunol Immunother **36**, 363-372.

Armentano D, Yu SF and Kantoff PW (1987). Effect of internal viral sequences on the utility of retroviral vectors. J Virol **61**, 1647-1653.

Asher AL, Mule JJ, Kasid A, Restifo NP, Salo JC, Reichert CM, Jaffe G, Fendly B, Kriegler M and Rosenberg SA (1991). Murine tumor cells transduced with the gene for tumor necrosis factor-alpha. Evidence for paracrine immune effects of tumor necrosis factor against tumors. J Immunol **146**, 3227-3234.

Belldegrun A, Muul LM and Rosenberg SA (1988a). Interleukin 2 expanded tumor-infiltrating lymphocytes in human renal cell cancer: Isolation, characterization, and antitumor activity. Cancer Res **48**, 206-214.

Belldegrun A, Uppenkamp I and Rosenberg SA (1988b). Anti-tumor reactivity of human lymphokine activated killer (LAK) cells against fresh and cultured preparations of renal cell cancer. J Urol **139**, 150-155.

Belldegrun A, Kasid A, Uppenkamp M, Topalian SL and Rosenberg SA (1989). Human tumor infiltrating lymphocytes. Analysis of lymphokine mRNA expression and relevance to cancer immunotherapy. J Immunol **142**, 4520-4526.

Belldegrun A and Rosenberg SA (1989). Adoptive immunotherapy of urologic tumors. Cancer Treat Rep **46**, 213-233.

Blankenstein T, Qin ZH, Uberla K, Muller W, Rosen H, Volk HD and Diamantstein T (1991). Tumor suppression after tumor cell-targeted tumor necrosis factor alpha gene transfer. J Exp Med **173**, 1047-1052.

Borysiewicz LK, Graham S, Hickling JK, Mason PD and Sissons JGP (1988). Human cytomegalovirus-specific cytotoxic T cells: Their precursor frequency and stage specificity. Eur J Immunol **18**, 269-273.

Coulie PG, Somville M, Lehmann F, Hainaut P, Brasseur F, Devos R and Boon T (1992). Precursor frequency analysis of human cytolytic T lymphocytes directed against autologous melanoma cells. Int J Cancer **50**, 289-297.

Crowley NJ, Slinghuff CL and Darrow TL (1990). Generation of human autologous tumor specific cytotoxic T cells using HLA-A2 matched allogenic melanoma. Cancer Res **50**, 492-497.

Dranoff G, Jaffee E, Lazenby A, Golumbek P, Levitsky H, Brose K, Jackson V, Hamada H, Pardoll D and Mulligan RC (1993). Vaccination with irradiated tumor cells engineered to secrete murine granulocyte macrophage colony stimulating factor stimulates potent, specific, and long lasting antitumor immunity. Proc Natl Acad Sci USA **90**, 3539-3543.

DeLeo AB, Jay G, Appella E, Dubois GC, Law LW and Old LJ (1979). Detection of a transformation-related antigen in chemically induced sarcomas and other transformed cells of the mouse. Proc Natl Acad Sci USA **76**, 2420-2424.

DeLeo AB, Jay G, Appella E, Dubois GC, Law LW and Old LJ (1980). Cell surface antigens of chemically induced sarcomas of inbred mice. Transplant Proc **12**, 65-69.

De Vries JE and Spits H (1984). Cloned human cytotoxic T lymphocyte (CTL) lines reactive with autologous melanoma cells. J Immunol **132**, 510-514.

Fearon ER, Pardoll DM, Itaya T, Golumbek P, Levitsky HI, Simons JW, Karasuyama H, Vogelstein B and Frost P (1990). Interleukin-2 production by tumor cells bypasses T helper function in the generation of an antitumor response. Cell **60**, 397-403.

Foa R, Guarini A and Gansbacher B (1992). IL2 treatment for cancer: From biology to gene therapy. Br J Cancer **66**, 992-998.

Foley EJ (1953). Antigenic properties of methylcholanthrene-induced tumors in mice of the strain of origin. Cancer Res **13**, 835-837.

Forni G and Santoni I (1984). Immunogenicity of "nonimmunogenic" tumors. J Biol Res Mod **3**, 128-133.

Gansbacher B, Bannerji R, Daniels B, Zier K, Cronin K and Gilboa E (1990a). Retroviral vector-mediated gamma-interferon gene transfer into tumor cells generates potent and long lasting antitumor immunity. Cancer Res **50**, 7820-7825.

Gansbacher B, Zier K, Daniels B, Cronin K, Bannerji R and Gilboa E (1990b). Interleukin 2 gene transfer into tumor cells abrogates tumorigenicity and induces protective immunity. J Exp Med **172**, 1217-1224.

Gansbacher B, Zier K, Cronin K, Hantzopoulos PA, Bouchard B, Houghton A, Gilboa E and Golde D (1992). Retroviral gene transfer induced constitutive expression of interleukin-2 or interferon-gamma in irradiated human melanoma cells. Blood **80**,2817-2825.

Gastl G, Finstad CL, Guarini A, Bosl G, Gilboa E, Bander NH and Gansbacher B (1992). Retroviral vector-mediated lymphokine gene transfer into human renal cancer cells. Cancer Res **52**, 6229-6236.

Gilboa E (1986). Retrovirus vectors and their use in molecular biology. Bioassay **5**, 252-257.

Golumbek PT, Lazenby AJ, Levitsky HI, Jaffee LM, Karasuyama H, Baker M and Pardoll DM (1991). Treatment of established renal cancer by tumor cells engineered to secrete interleukin-4. Science **254**, 713-716.

Harding FA, McArthur JG, Gross JA, Raulet DH and Allison JP (1992). CD28-mediated signalling co-stimulates murine T cells and prevents induction of anergy in T-cell clones. Nature **356**, 607-609.

Hewitt HB, Blake ER and Walder AS (1976). A critique of the evidence for active host defence against cancer, based on personal studies of 27 murine tumours of spontaneous origin. Br J Cancer **33**, 241-259.

Hickling JK, Borysiewicz LK and Sissons JGF (1987). Varicella-zoster virus specific cytotoxic T Lymphocytes (Tc): Detection and frequency analysis of HLA class I restricted in human peripheral blood. J Virol **61**, 3463-3465.

Hock H, Dorsch M, Kunzendorf U and Qin Z (1993). Mechanisms of rejection induced by tumor cell-targeted gene transfer of interleukin 2, interleukin 4, interleukin 7, tumor necrosis factor, or interferon gamma. Proc Natl Acad Sci USA **90**, 2774-2778.

Houghton AN, Eisinger M and Albino A (1982). Surface antigens of melanocytes and melanomas: Markers of melanocyte differentiation and melanoma subsets. J Exp Med **156**, 1755-1759.

Houghton AN, Thomson TM, Gross D, Oettgen HF and Old LJ (1984). Surface antigens of melanoma and melanocytes. Specificity of induction of 1a antigens by human alpha-interferon. J Exp Med **160**, 255-260.

Houghton AN, Real FX, Davis LJ, Cordon-Cardo C and Old LJ (1987). Phenotypic heterogeneity of melanoma. Relation to the differentiation program of melanoma cells. J Exp Med **164**, 812-816.

Hwu P, Shafer GE, Treisman J, Schindler DG, Gross G, Cowherd R, Rosenberg SA and Eshhar Z (1993). Lysis of ovarian cancer cells by human lymphocytes redirected with a chimeric gene composed of an antibody variable region and the Fc receptor gamma chain. J Exp Med **178**, 361-366.

Inge TH, Hoover SK, Susskind BM, Barrett SK and Bear HD (1992). Inhibition of tumor-specific cytotoxic T-lymphocyte responses by transforming growth factor-beta. Cancer Res **52**, 1386-1390.

Kasid A, Morecki S, Aebersold P, Cornetta K, Culver K, Freeman S, Director E, Lotze MT, Blaese RM, Anderson WF and Rosenberg SA (1990). Human gene transfer: Characterization of human tumor-infiltrating lymphocytes as vehicles for retroviral-mediated gene transfer in man. Proc Natl Acad Sci USA **87**, 473-477.

Kawakami Y, Zakut R, Topalian SL, Slotter H and Rosenberg SA (1992). Shared human melanoma antigens: Recognition by tumor infiltrating lymphocytes in HLA-A2.1-transfected melanomas. J Immunol **148**, 638-643.

Key ME, Brandhorst JS and Hanna MG (1984). More on the relevance of animal tumor models: Immunogenicity of transplantable leukemias of recent origin in syngeneic strain 2 guinea pigs. J Biol Response Mod **3**, 359-365.

Knuth A, Wolfel T, Klehmann E, Boon T and Meyer zum Buschenfelde KH (1989). Cytolytic T-cell clones against an autologous human melanoma: Specificity study and definition of three antigens by immunoselection. Proc Natl Acad Sci USA **86**, 2804-2808.

Langer AJ and Pestka S (1988). Interferon receptors. Immunol Today **9**, 393-398.

Ley V, Langlade-Demoyen P, Kourilsky P and Larsson-Sciard EL (1991). Interleukin 2-dependent activation of tumor-specific cytotoxic T lymphocytes in vivo. Eur J Immunol **21**, 851-856.

Liu Y and Janeway CA (1992). Cells that present both specific ligand and costimulatory activity are the most efficient inducers of clonal expansion of normal CD4 T cells. Proc Natl Acad Sci USA **89**, 3845-3850.

Livingston PO, Takeyama H, Pollack MS, Houghton AN, Albino A, Pinsky CM, Oettgen HF and Old LJ (1983). Serological responses of melanoma patients to vaccines derived from allogeneic cultured melanoma cells. Int J Cancer **31**, 567-575.

Livingston PO, Albino AP, Chung TJ, Real FX, Houghton AN, Oettgen HF and Old LJ (1985a). Serological response of melanoma patients to vaccines prepared from VSV lysates of autologous and allogeneic cultured melanoma cells. Cancer **55**, 713-720.

Livingston PO, Kaelin K, Pinsky CM, Oettgen HF and Old LJ (1985b). The serologic response of patients with stage II melanoma to allogeneic melanoma cell vaccines. Cancer **56**, 2194-2200.

Livingston PO, Ritter G and Calves MJ (1989). Antibody response after immunization with the gangliosides GM1, GM2, GM3, GD2 and GD3 in the mouse. Cancer Immunol **29**, 179-184.

Livingston P (1991). Active specific immunotherapy in the treatment of patients with cancer. Immunol Allergy Clin North Am **11**, 401-423.

Long JP, Walther MM, Alexander RB, Linehan WM and Rosenberg SA (1993). The management of isolated renal recurrence of renal cell carcinoma following complete response to interleukin-2 based immunotherapy. J Urol **150**, 176-178.

Lotze MT, Frana LW, Sharrow SO, Robb RJ and Rosenberg SA (1985). In vivo administration of purified human interleukin 2. I. Half-life and immunologic effects of the Jurkat cell line-derived interleukin 2. J Immunol **134**, 157-166.

Miller AD (1993). Retrovirus packaging cells. Hum Gene Ther **1**, 5-10.

Mukherji B and MacAlister TJ (1983). Clonal analysis of cytotoxic T-cell response against human melanoma. J Exp Med **158**, 240-243.

North RJ (1982). Cyclophosphamide-facilitated adoptive immunotherapy of an established tumor depends on elimination of tumor-induced suppressor T cells. J Exp Med **155**, 1063-1074.

Old LJ (1981). Cancer immunology: The search for specificity - G. H. A. Clowes Memorial lecture. Cancer Res **41**, 361-375.

Prehn RT and Main JM (1957). Immunity to methylcholanthrene-induced sarcomas. J Natl Cancer Inst **18**, 769-778.

Rosa F, Hata D, Abadie A, Wallach D, Revel M and Fellous M (1983). Differential regulation of HLA-DR mRNAs and cell surface antigens by interferon. EMBO J **2**,1585-1589.

Rosenberg SA and Mule JJ (1985). Immunotherapy of cancer with lymphokine-activated killer cells and recombinant interleukin-2. Surgery **98**, 437-444.

Rosenberg SA, Spiess P and Lafreniere R (1986). A new approach to the adoptive immunotherapy of cancer with tumor-infiltrating lymphocytes. Science **233**, 1318-1321.

Rosenberg SA (1988a). The development of new immunotherapies for the treatment of cancer using interleukin-2. A review. Ann Surg **208**, 121-135.

Rosenberg SA (1988b). Immunotherapy of cancer using interleukin 2: Current status and future prospects. Immunol Today **9**, 58-62.

Rosenberg SA, Lotze MT, Yang JC, Linehan WM, Seipp C, Calabro S, Karp SE, Sherry RM, Steinberg S and White DE (1989). Combination therapy with interleukin-2 and alpha-interferon for the treatment of patients with advanced cancer. J Clin Oncol **7**, 1863-1874.

Rosenberg SA (1990). Adoptive immunotherapy for cancer. Sci Am **262**, 62-69.

Rosenberg SA, Aebersold P, Cornetta K, Kasid A, Morgan RA, Moen R, Karson EM, Lotze MT, Yang JC and Topalian SL (1990). Gene transfer into humans-immunotherapy of patients with advanced melanoma, using tumor-infiltrating lymphocytes modified by retroviral gene transduction. N Engl J Med **323**, 570-578.

Rosenberg SA (1991). Immunotherapy and gene therapy of cancer. Cancer Res **51**, 5074s-5079s.

Shu S and Rosenberg SA (1985). Adoptive immunotherapy of a newly induced sarcoma: Immunologic characteristics of effector cells. J Immunol **135**, 2895-2903.

Srivastava PK, DeLeo AB and Old LJ (1986). Tumor rejection antigens of chemically induced sarcomas of inbred mice. Proc Natl Acad Sci USA **83**, 3407-3411.

Tepper RI, Pattengale PK and Leder P (1989). Murine interleukin-4 displays potent anti-tumor activity in vivo. Cell **57**, 503-512.

Van der Bruggen P, Traversari C, Chomez P, Lurquin C, De Plaen E, Van den Eynde B, Knuth A and Boon T (1991). A gene encoding an antigen recognized by cytolytic T lymphocytes on a human melanoma. Science **254**, 1643-1647.

Wallach D, Fellous M and Revel M (1982). Preferential effect of interferon-gamma on the synthesis of HLA antigens and their mRNAs in human cells. Nature **299**, 833-835.

Watanabe Y, Kuribayashi K, Miyatake S, Nishihara K, Nakayama E, Taniyama T and Sakata T (1989). Exogenous expression of mouse interferon gamma cDNA in mouse neuroblastoma C1300 cells results in reduced tumorigenicity by augmented anti-tumor immunity. Proc Natl Acad Sci USA **86**, 9456-9460.

Wolfel T, Klehman E and Muller C (1989). Lysis of human melanoma cells by autologous cytolytic T cell clones. Identification of human histocompatibility leukocyte antigen A2 as a restriction element for three different antigens. J Exp Med **170**, 797-802.

20

REDUCED ONCOGENICITY AND ENHANCED METASTATIC SPREAD OF IFN-γ TRANSFECTED TUMOR CELLS: THERAPEUTIC IMPLICATIONS

P. L. Lollini, M. C. Bosco, C. De Giovanni, L. Landuzzi, G. Nicoletti, G. Palmieri, C. Riccardi and P. Nanni
Cancer Research Institute, University of Bologna; IST-Genoa, Bologna Biotechnology Section, Bologna, Italy; Department of Experimental Medicine, University La Sapienza, Rome, Italy; Institute of Medical Pharmacology, University of Perugia, Perugia, Italy and Laboratory of Molecular Immunoregulation, BRMP, NCI-FCRDC, Frederick, MD, USA

Introduction

Tumor cells transduced with cytokine genes are currently being investigated as new therapeutic tools in the fight against cancer. From a tumor biologist's point of view, cytokines can be broadly divided into two groups:

- cytokines for which (solid) tumor cells generally do not have appropriate receptors, e.g. IL-2
- cytokines directly active on many tumor types, e.g. IFNs.

It is conceivable that the scenario will be more complex following cytokine gene transduction, since this may induce new autocrine or paracrine circuits among tumor cells. The same broad categories can be applied to the production of cytokines: some are not commonly synthesized by solid tumors (IL-2 is again a good example), while others are released by many neoplastic cells, e.g. colony stimulating factors (Nicoletti et al., 1989). In principle, it is not clear whether cytokines which are spontaneously produced by tumors might be usefully exploited for therapeutic purposes. In practice, however, only a few molecules are produced by all tumors: for example, type I IFNs are produced by many murine fibrosarcomas, less frequently by carcinomas (Nanni et al., 1992).

For each tumor, cytokines can, thus, be divided into four classes (not taking into account inducible genes):

- **class I:** cytokines which are not produced by tumor cells and whose receptors are not expressed by tumor cells

CYTOKINE-INDUCED TUMOR IMMUNOGENICITY
ISBN 0-12-261520-4

- **class II:** cytokines which are not produced by tumor cells that do, however, express a functional receptor
- **class III:** cytokines which are released by tumor cells that do not express the receptor
- **class IV:** autocrine cytokines.

Class I cytokines offer the most straightforward model system, because cytokine production is the only difference between transduced tumor cells and parent cells. Class II cytokines are more intriguing: the phenotype of transduced tumor cells might be substantially altered with respect to parent cells, and their biology must be thoroughly investigated. In return, one might get effects unattainable with class I cytokines.

We have transfected the murine mammary adenocarcinoma TS/A with a class II cytokine gene, murine IFN-γ. Here, we will show that the release of IFN-γ modified several phenotypic features of TS/A cells, and we will suggest how this may be utilized in a therapeutic protocol.

Materials and methods

Cells. Cells were cultured in Dulbecco's Modified Eagle Medium (DMEM, GIBCO, Paisley, Scotland) supplemented with 10% heat-inactivated FCS (GIBCO); cultures were maintained at 37°C in a humidified atmosphere of 5% CO_2 in air. TS/A is a cell line from the first in vivo transplant of a moderately differentiated mammary adenocarcinoma that arose spontaneously in a 20-month-old multiparous BALB/c mouse (Nanni et al., 1983). The parental TS/A (TS/A-pc) cell line used in this work was reestablished in vitro from a tumor induced in a BALB/c mouse with the 8th passage TS/A cells. The cells employed were always >90% viable, as judged by trypan blue dye exclusion. For the determination of growth curves, 5 x 10^5 cells were seeded in 25 cm^2 flasks (Falcon, Oxnard, USA) and harvested daily.

Cell transfection. Transfection of TS/A-pc was performed by lipofection as described by Lollini et al. (1993), using a murine IFN-γ cDNA expression vector (BCMGNeo-IFN-γ) constructed by inserting a cDNA encoding IFN-γ into the bovine papilloma virus-based plasmid BCMGNeo (Karasuyama et al., 1989), and kindly provided by Dr D. Pardoll, Johns Hopkins University, Baltimore, MD, with the permission of Dr H. Karasuyama, University of Tokyo, Tokyo, Japan and Dr W. Fiers, State University of Ghent, Ghent, Belgium. Control cells were obtained by transfection with the plasmid BCMGNeo alone. The presence of the IFN-γ gene in the genome of transfected cells was demonstrated by Southern blot analysis with a murine IFN-γ probe. The presence of mRNA transcripts was demonstrated by reverse transcriptase PCR reactions (Lollini et al., 1993).

IFN-γ titration. The IFN-γ released after 72 h culture by 5 x 10^5 cells seeded in 25 cm^2 flasks with 5 ml DMEM-FCS was evaluated by inhibition of the cytopathic effect of vesicular stomatitis virus on mouse L929 cells (Giovarelli et al., 1986). Titers are the reciprocal of the highest dilution giving a 50% inhibition. An internal laboratory standard calibrated against the NIH G-002-904511 standard was included in each assay, in order to express the titers in IU/ml.

Expression of surface antigens. The expression of surface antigens was determined by flow cytometry (FACScan, Becton Dickinson, Mountain View, CA) after indirect immunofluorescence, as described (De Giovanni et al., 1991). The following mAb were used: H-$2K^d$ and H-$2D^d$: mAb 31-3-4S and 34-5-8S, respectively (Cedarlane, Hornby, Canada); Ly-6A/E: mAb D7 (a kind gift from Dr E. Shevach, NIAID, Bethesda, MD); NCAM: mAb H28-123-16 (Boehringer Mannheim, Milan, Italy); A-CAM: mAb GC-4 (Sigma, Milan, Italy); ICAM-1: mAb YN1 (a kind gift from Dr F. Takei, Terry Fox Laboratory, Vancouver, Canada); CD44: mAb IM7 (Pharmingen, San Diego, CA); integrin subunit α4: MAb R1-2 (Pharmingen); subunit β4: mAb 346-11A (a kind gift from Dr A. Sacchi, Istituto Regina Elena, Rome, Italy). FITC conjugated secondary antibodies were obtained from Technogenetics, Milan, Italy (anti-mouse Ig) and from DAKO, Milan, Italy (anti-rat Ig). The results shown are representative of at least three independent experiments.

Mice. Seven-week-old female BALB/cAnNCrlBR (BALB/c) mice and 5-week-old female nu/nu mice on Swiss CD-1 background were obtained from Charles River Laboratories (Calco, Italy); SCID mice (C.B-17/Icr scid/scid) were from Bomholtgaard Breeding and Research Centre Ltd, Bommice (Ry, Denmark). Mice were allowed to rest for 1 week before any treatment. Immunodepressed mice were fed and maintained under specific pathogen free conditions, and received sterilized food pellets and tap water ad libitum. Immunodepressive treatments were performed, as described by Lollini et al. (1993).

In vivo evaluation of tumor growth. Mice were challenged s.c. with 0.2 ml of a single cell suspension containing the specified number of tumor cells. The incidence and growth of tumors were evaluated twice weekly. Neoplastic masses were measured with calipers. Tumor volume was calculated as: $1/6\ \pi\ (ab)^{3/2}$ where a and b are the two perpendicular diameters. Latency was considered as the period (in days) between challenge and the growth of a neoplastic mass >15 mm^3 (corresponding to a mean tumor diameter >3 mm). In control mice, TS/A-pc induced tumors that overcame the latency threshold grew progressively, and no spontaneous regressions were observed. For morphologic analysis, groups of 3 mice were killed 7, 14, 21, 28 and 35 days after challenge.

Lung metastases. Spontaneous metastases in syngeneic BALB/c mice bearing s.c tumors were evaluated either 1 month after challenge or when the tumor volume exceeded 5 cm^3. Experimental metastases were evaluated 19-21 days after the i.v. injection of 2.5 x 10^4 to 1 x 10^5 cells suspended in phosphate-buffered saline in a lateral tail vein. In some experiments NEO control cells were cultured in the presence of 50 IU/ml of recombinant murine IFN-γ (kindly provided by Dr G.R. Adolf, Ernst-Boehringer-Institut, Wien, Austria) for 4 days before i.v. injection (Lollini et al., 1990). Lung nodules were contrasted with black india ink; all metastasis counts were performed on dissected lung lobes under a stereoscopic microscope (De Giovanni et al., 1991). The experiments shown are representative of 2-3 independent experiments.

Self-targeting antimetastatic therapy. All mice received 5 x 10^4 live TS/A-pc cells i.v.; starting 6 h after challenge, mice received twice weekly a total of seven i.v. injections of 10^6 mitomycin treated (40 μg/ml) cells. Mice were sacrificed 28 days after the challenge.

In vitro cytotoxicity tests. NK and LAK activity were assayed against YAC-1 and P815 mouse target cells, respectively in a 4 h ^{51}Cr sodium chromate release assay; cytotoxic activity of a peritoneal macrophage-enriched population was determined in a 48 h ^3H-thymidine release assay (Lollini et al., 1993). The release values are expressed as lytic units (LU) calculated according to the equation of Pross et al. (1981). One LU is here defined as the number of effector cells needed to kill 20% of the target cells (LU_{20}).

Retention of radiolabeled tumor cells. Cells were incubated with DMEM-FCS containing 5-[^{125}I]iodo-2'-deoxyuridine overnight. Thereafter, 10^6 cells in 0.4 ml were injected i.v. in BALB/c mice. Aliquots (0.4 ml) were retained to determine the radioactivity of the cells injected. Groups of 6 mice were killed 10 min, 1, 6, and 24 h after injection; lungs and kidneys were removed and counted in a γ-counter.

Statistics. Statistical evaluation was performed by Student's t test and by a nonparametric Wilcoxon rank sum test.

Results

Characterization of IFN-γ clones

Transfection of the murine IFN-γ gene into TS/A-pc cells yielded clones releasing between 2 and 6000 IU of IFN-γ per ml of culture medium; the name of each clone contains, after the point, the number of IFN-IU/ml determined in a standard bioassay (Table 20.1). For example, clone I5.1000 released 1000 IU/ml of IFN-γ. IFN-γ production was quite stable over several months.

Table 20.1. *IFN-γ production and in vivo proliferation*

Clone	IFN-γ (IU/ml)	Doubling time (h)
NEO	0	13
I1.2	2	14
I2.4	4	12
I3.500	500	13
I4.500	500	13
I5.1000	1000	15
I6.6000	6000	12

To analyze the consequences of IFN-γ gene transfection, we studied the in vitro cell growth, and the expression of H-2 class I membrane glycoproteins and of the GPI anchored molecule Ly-6, which is more resistant to IFN-γ induction than H-2 (Lollini et al., 1992), and of several cell adhesion molecules.

We have previously observed that TS/A derivatives are insensitive to the antiproliferative effect of IFN-γ, but they readily increase H-2 class I and Ly-6 expression after exposure to IFN-γ. All IFN-γ clones showed an increase of comparable intensity in the expression of H-2, while the increase in Ly-6 appeared more correlated to the amount of IFN-γ produced by each clone (Fig. 20.1). No significant variation was observed in the in vitro growth of IFN-γ clones (Table 20.1), and the expression of all adhesion molecules tested (Fig. 20.2) was similar to that of control TS/A-pc and of NEO cells.

Tumorigenicity of IFN-γ clones

The growth of tumors induced by the s.c. injection of IFN-γ clones was severely hampered in syngeneic immunocompetent BALB/c mice (Fig. 20.3). The inhibition of tumor growth in vivo was directly proportional to the amount of IFN-γ released in vitro. It should be noted that no IFN-γ clone was nontumorigenic, and no tumor regression was ever observed. Sporadic deaths occurred among mice bearing small I6.6000 tumors, and other mice became severely cachectic and had to be killed.

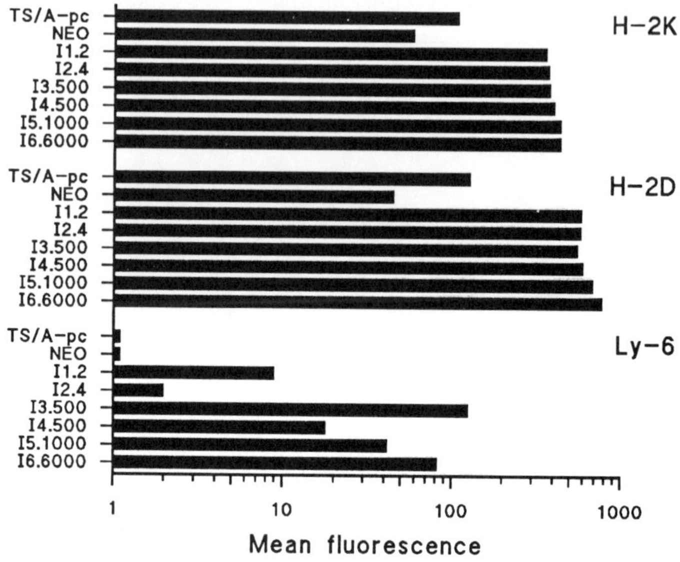

Fig. 20.1. *Expression of H-2 class I and Ly-6 membrane glycoproteins.* Mean fluorescence intensity in arbitrary units is shown.

High levels of serum IFN-γ (>256 IU/ml serum) were measured in mice with I6.6000 tumors; since tumor load did not appear life-threatening, we attributed the illness of mice to the systemic effects of the IFN-γ released by the tumor cells.

Analysis of tumor inhibition

The morphology and cellular infiltrate of tumors produced by IFN-γ clones were studied from 7 to 35 days after cell injection. The tumors displayed necrotic areas directly proportional to the amount of IFN-γ released. A marked infiltrate formed of macrophages, several lymphocytes and some degranulating eosinophils was already evident around and between IFN-γ producing clones, whereas very few reactive cells were present in TS/A-pc and NEO tumors.

To investigate the mechanisms leading to the in vivo inhibition of cells releasing IFN-γ we studied their growth in immunodepressed mice. We used nude and SCID mice and BALB/c mice sublethally irradiated or repeatedly treated with anti-CD4 and/or anti-CD8 mAb, or with anti-asialo GM1 serum, as well as BALB/c mice subjected to splenectomy, irradiation and anti-asialo combined treatment (SIA mice). In all cases, IFN-γ clones gave rise to a slower tumor growth compared to control TS/A-pc and NEO tumors (data not shown). These results seem to rule out a contributory role of T and B lymphocytes or of NK cells.

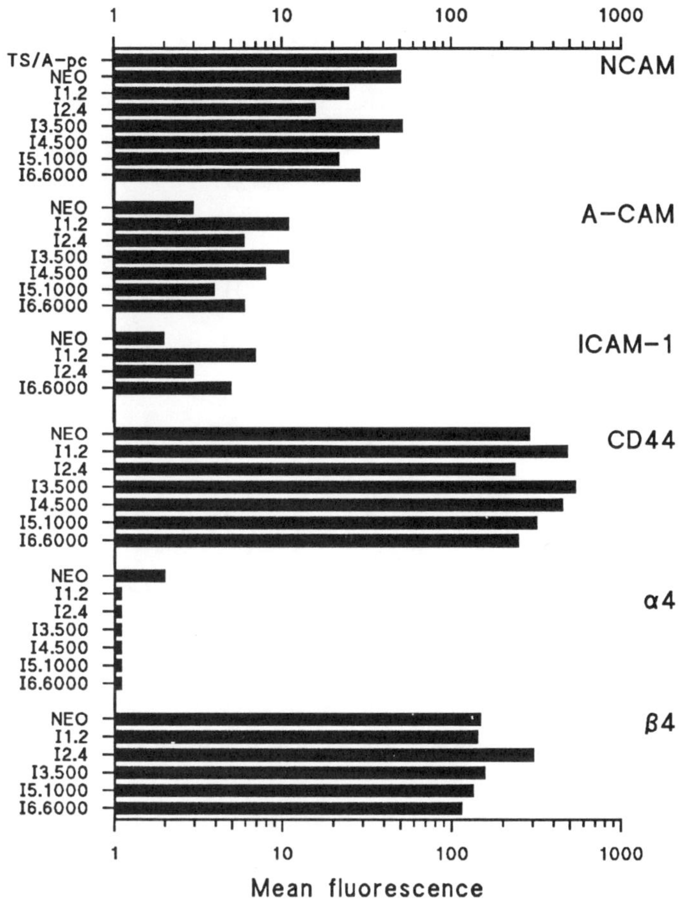

Fig. 20.2. *Expression of cell adhesion molecules.*

The analysis of in vitro macrophage activity showed a very high activity against the IFN-γ clones (Table 20.2). Altogether, these results suggest that macrophages were the cell population mainly responsible for the reduced tumorigenicity of IFN-γ clones.

Metastatic spread

When mice bearing s.c. tumors were killed one month after cell injection, the analysis of lung metastasization showed that low IFN-γ producers were more metastatic than control cells, while high IFN-γ producers were almost nonmetastatic (data not shown). Since tumor volumes were not comparable (see Fig. 20.3), we evaluated the metastatic spread in mice sacrificed when the tumor reached a predefined

Table 20.2. *Cytotoxic activity (LU_{20}) against IFN-γ clones*

Target	Macrophages	NK	NK boosted	LAK
TS/A-pc	12	12	27	10
NEO	15	4	15	6
I2.4	894	4	31	6
I3.500	1548	4	20	11
I4.500	415	2	20	4
I5.1000	1587	2	4	4
I6.6000	618	4	44	8
YAC[a]	n.d.	102	180	49

[a]NK sensitive Moloney virus induced leukemia

Table 20.3. *Spontaneous metastasis in BALB/c mice sacrificed when tumor volume reached 5 cm^3,*[a]

Clone	Days to sacrifice (median)	Lung metastases		
		Incidence	Median	Range
NEO[b]	37	5/5	43	8-56
I1.2	43	5/5	>200	>200
I2.4	47	5/5	>200	37->200
I3.500	55	5/5	47	37-66
I4.500	60	5/5	83	35-108
I5.1000	75	3/3	54	17-155
I6.6000	-[c]	2/3	1	0-4

[a]From Lollini et al., 1993

[b]Mice received 10^5 cells s.c.

[c]Mice were sacrificed when cachectic (47-77 days after cell injection)

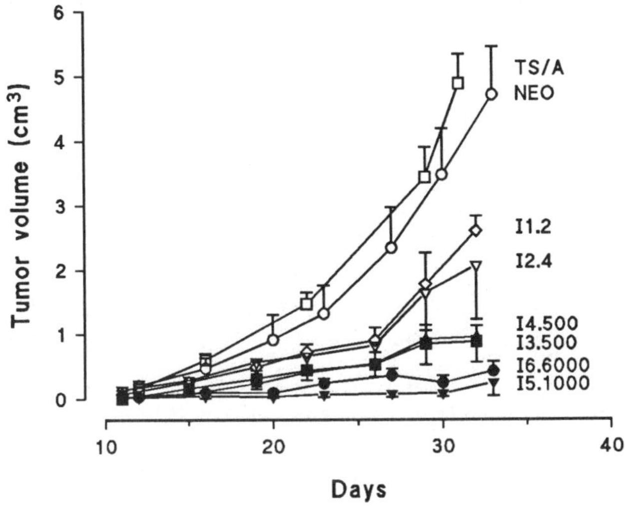

Fig. 20.3. *Tumor growth induced by s.c. injection of 10^5 cells.*

threshold (5 cm^3). Table 20.3 shows that under these conditions all IFN-γ clones were metastatic. The low IFN-γ producers I1.2 and I2.4 were significantly more metastatic than NEO cells, the high producer clones I3.500, I4.500, and I5.1000 were as metastatic as control cells. I6.6000 tumors induced a life-threatening cachexia and did not attain the predefined volume, thus it was not determined whether or not they maintained the metastatic potential of parent cells.

When cells were injected i.v. all IFN-γ clones, including I6.6000, produced more lung colonies than TS/A-pc or NEO cells (Table 20.4). It is known that a pretreatment in vitro with IFN-γ can enhance experimental metastasis (Lollini et al., 1987; Kelly et al., 1991). To confirm that the high metastatic capacity of IFN-γ clones was indeed mediated by a direct effect of the cytokine on tumor cells, we treated control NEO cells with exogenous IFN-γ before i.v. injection: the metastatic ability was enhanced and became similar to that of IFN-γ clones (data not shown).

In order to further characterize the early events of i.v. dissemination, we studied the fate of ^{125}IUdR-labeled cells during the first 24 h after i.v. injection. Fig. 20.4 shows that 6 h after injection radioactivity in the lungs was lower for NEO cells than for IFN-γ transfectants. Analogous results were obtained after 24 h, except that the I6.6000 clone was similar to NEO cells. At any time negligible radioactivity was found in a control organ (kidney) (data not shown).

NK sensitivity and NK deficient mice

The main reason for the higher metastatic ability of tumor cells treated in vitro with IFN-γ could be the enhancement of resistance to

Table 20.4. *Experimental metastases after i.v. injection of 10^5 cells in syngeneic BALB/c mice*

	Lung colonies		
Clone	Incidence	Median	Range
TS/A-pc	4/5	22	0-72
NEO	4/5	5	0-28
I1.2	5/5	>200	>200
I2.4	5/5	114	75-190
I3.500	5/5	>200	180->200
I4.500	5/5	>200	>200
I5.1000	5/5	118	96-175
I6.6000	5/5	56	31-85

Table 20.5. *Self targeting antimetastatic therapy*

	Lung colonies[a]			
Cells used for i.v. therapy[b]	Incidence	Median	Range	Significance[c]
None	5/5	15	5-19	
TS/A-pc	5/5	70	47-84	$p < 0.05$
I2.4	5/5	34	1-61	n.s.
I3.500	5/5	3	2-8	$p < 0.05$
I6.6000	5/5	3	2-4	$p < 0.05$

[a]Mice received 5 x 10^4 TS/A-pc cells i.v.

[b]Seven injections of 10^6 mitomycin C-treated cells

[c]Significance of the difference versus untreated mice (n.s., not significant)

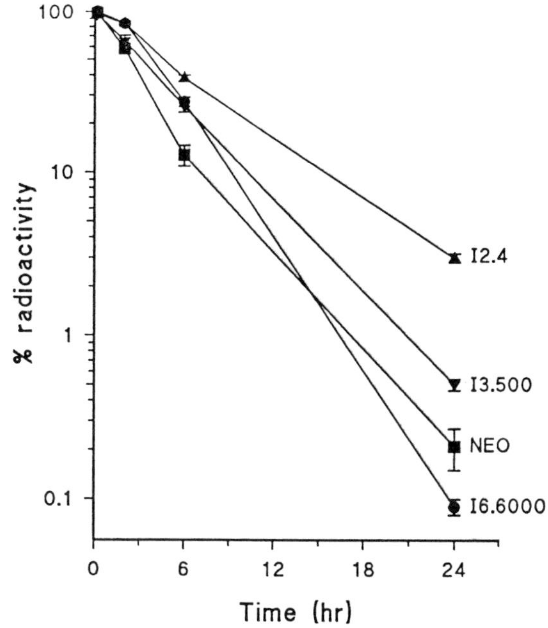

Fig. 20.4. *Kinetics of retention by lungs of radiolabeled cells injected i.v.*

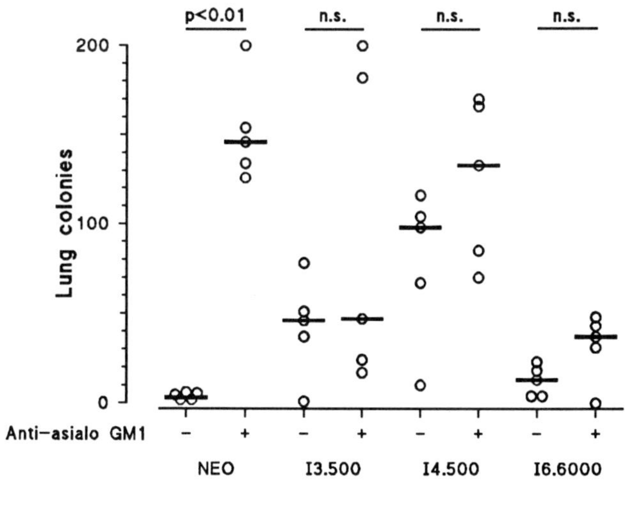

Fig. 20.5. *Experimental metastases induced by i.v. injections of 2 x 10^4 cells in normal and anti-asialo GM1 treated mice.*

NK effectors (Trinchieri and Santoli, 1978). Thus, we investigated the alterations in NK sensitivity induced by IFN-γ transfection. In vitro, the parent cells resulted quite resistant to NK, thus no increase in resistance could be measured (Table 20.2).

To obtain an estimate of the importance of NK resistance for the metastatic spread of IFN-γ clones, we evaluated the difference in metastatic capacity of cells injected in normal hosts or in mice rendered NK deficient by means of anti-asialo GM1 serum. As shown in Fig. 20.5, NEO cells were more metastatic in NK deficient hosts, while no significant variation was observed for IFN-γ clones. The independence of IFN-γ clones to NK status of the host suggests that the release of IFN-γ stimulated NK resistance in these cells.

Self targeting antimetastatic therapy

IFN-γ clones showed a greatly reduced ability to form tumors, but maintained the metastatic capacity of parent TS/A cells. The organ specificity was also maintained, since both TS/A and IFN-γ clones selectively metastasize to the lungs. We devised a therapeutic protocol that exploits the metastatic capacity of IFN-γ clones to stimulate host defenses right at the site of the metastatic lesions. Mice received an i.v. injection of TS/A-pc cells; therapy began 6 h later and consisted of seven i.v. injections of mitomycin treated cells administered twice a week. The administration of mitomycin treated TS/A-pc cells caused an increase in the number of lung metastases (Table 20.5); this might be due to a "feeder-layer effect".

The low IFN-γ producer clone I2.4 did not significantly alter the situation. A significant inhibition in the development of metastases was observed when clones I3.500 and I6.6000 were used for therapy.

Discussion

Transfection of a metastatic murine tumor with the murine IFN-γ gene strongly impaired tumorigenicity, but enhanced metastatic spread.

The effects of IFN-γ on tumor growth and on metastasis appeared to follow different dose response curves: tumorigenicity showed a steady decline inversely proportional to the quantities of IFN-γ released, while spontaneous metastasis had a peak at low IFN-γ levels and returned to the level of controls in clones secreting high amount of IFN-γ.

Since IFN-γ transfectants were highly metastatic after i.v. injection, the dose response curve of spontaneous metastasis probably derives from a combination between an increase in local tumor control and a reduction in the control of the events that follow intravasation of metastatic cells.

The residual tumorigenicity of cells releasing even more than 1000 IU/ml of IFN-γ indicates that some type of proliferation control (irradiation, mitomycin, suicide genes) is required in order to use these cells for therapeutic purposes. Different cell types transfected with the IFN-γ gene show a strong reduction in tumorigenicity (Watanabe et al., 1989; Gansbacher et al., 1990; Esumi et al., 1991; Maruguchi et al., 1991; Porgador et al., 1993), but a residual ability to grow in vivo can be retained by some tumors: Lewis lung carcinoma (Porgador et al., 1993), CT-26 carcinoma (Esumi et al., 1991), PAM-T squamous cell tumor (Maruguchi et al., 1991). It is, therefore, important to keep in mind that even very high doses of IFN-γ do not abolish the growth of truly aggressive tumors.

Since TS/A cells transfected with a different cytokine, IL-2, lose their ability to form tumors (Cavallo et al., 1992), the type and the entity of the response elicited by the cytokine are major determinants in the control of primary tumor growth elicited by these tumors. TS/A clones releasing various amounts of IL-2 following IL-2 gene transfection are quickly rejected through the combined action of neutrophils and $CD8^+$ T lymphocytes, as shown by previous morphologic and functional data (Cavallo et al., 1992). By contrast, the s.c. growth of IFN-γ clones is not enhanced in T deficient SCID mice nor in T and NK deficient splenectomized, sublethally irradiated nude mice treated with anti-asialo GM1 antibodies, suggesting that neither T lymphocytes nor NK cells have a major role in this inhibition. Morphologic data show that the inhibition of IFN-γ clones is associated with a capable of massive infiltrate, deeply penetrating into the tumor mass and mainly formed of macrophages, several lymphocytes and a few eosinophils. While macrophages and eosinophils apparently bind and kill tumor cells, the rôle of T lymphocytes is intriguing, since they are a significant component of the reactive infiltrate and display a few characteristics of activated lymphocytes. There is, however, no evidence that they display lytic activity.

We observed the death of some mice bearing small tumors induced by clone I6.6000. Since neither tumors nor metastases appeared to have killed these mice, we are currently investigating the systemic effects of IFN-γ they released. We suggest that in vivo the area under the time/concentration curve (AUC) reached with transduced cells might be several orders of magnitude higher than the AUC obtained with the infusion of recombinant cytokines. Thus we expect that novel types of toxicity will be detected, as observed when recombinant cytokines first entered clinical trials (Creekmore et al., 1991). Animal model systems of the type described here can provide both preclinical data on the usefulness of a new therapeutic approach and on its toxicity.

The metastatic ability of our IFN-γ clones was very high. Experimental metastases were clearly enhanced over parental cells; spontaneous metastases of two IFN-γ clones were increased, while all other clones remained as metastatic as control cells. It is interesting to note that the two clones with a higher metastatic ability were the lowest producers of IFN-γ. Also the survival of labeled cells suggested that the low producer I2.4 was more aggressive than either parental cells or the high IFN-γ producers. Lewis lung carcinoma cells transduced with the IFN-γ gene yielded some clones with an IFN-γ production which was below the detection threshold (Porgador et al., 1993); these clones proved more metastatic than parental cells and than high IFN-γ producers. It may be hypothesized that the enhancement of metastases mediated by IFN-γ follows a bell-shaped curve with a maximum corresponding to very low levels of IFN-γ release. We studied the expression of several molecules which are known to be involved in the metastatic processes (Albelda, 1993). The adhesion molecules studied did not show an appreciable variation in comparison to control cells. The increase in resistance to NK cells in vivo and the higher survival during the early phases after i.v. injection suggest that the enhancement of metastatic ability of IFN-γ clones was mainly mediated by the increase in resistance to NK cells.

The alterations induced in the phenotype of TS/A cells by the IFN-γ gene were very similar to those obtained after in vitro treatment with exogenous IFN-γ of TS/A derivatives (Lollini et al., 1988) or of other tumor types (Lollini et al., 1987; Taniguchi et al., 1987; Zoeller et al., 1988; Kelly et al., 1991). This consideration leads to the apparently trivial conclusion that IFN-γ is IFN-γ, and that cell responses do not change whether its source is exogenous or endogenous. In practice, this indicates that if exogenous IFN-γ enhances the metastatic capacity of a tumor, then the ability of IFN-γ transfectants to selectively reach metastatic sites will be equal to, or better than, that of parent cells.

To exploit this property of IFN-γ clones we devised a therapeutic protocol in which metastatic deposits to the lungs produced by the parental tumor were significantly reduced by the i.v. administration of IFN-γ producing clones. We hypothesize that similar protocols might be applied to other cytokines which are known to enhance the metastatic ability of tumor cells, such as TNF-α (Lollini et al., 1990; Malik et al., 1990; Palmieri et al., 1992; Orosz et al., 1993), IL-1 (Giavazzi et al., 1990), and CSFs (Ishikawa et al., 1986). As outlined in the Introduction, the biological effects of class II cytokines are necessarily complicated, but once these effects are better understood they may be exploited in specific therapeutic approaches.

Acknowledgments

We thank Dr D. Pardoll, Dr H. Karasuyama and Dr W. Fiers for the supply of the expression vector BCMGNeomIFN-γ. This work was supported by grants from Associazione Italiana per la Ricerca sul Cancro, from the Italian National Research Council, Special Project ACRO, and from the Italian Ministry for University and Research. This work was also supported by United States Federal funds from The Department of Health and Human Services.

References

Albelda SM (1993). Role of integrins and other cell adhesion molecules in tumor progression and metastasis. Lab Invest **68**, 4-17.

Cavallo F, Giovarelli M, Gulino A, Vacca A, Stoppacciaro A, Modesti A and Forni G (1992). Role of neutrophils and CD4+ T lymphocytes in the primary and memory response to nonimmunogenic murine mammary adenocarcinoma made immunogenic by IL-2 gene. J Immunol **149**, 3627-3635.

Creekmore SP, Longo DL and Urba WJ (1991). Principles of the clinical evaluation of biologic agents. In "Biologic Therapy of Cancer" (eds VT De Vita Jr, S Hellman and SA Rosenberg), pp 67-86. Lippincott, Philadelphia.

De Giovanni C, Palmieri G, Nicoletti G, Landuzzi L, Scotlandi K, Bontadini A, Tazzari PL, Sensi M, Santoni A, Nanni P and Lollini PL (1991). Immunological and nonimmunological influence of H-2Kb gene transfection on the metastatic ability of B16 melanoma cells. Int J Cancer **48**, 270-276.

Esumi N, Hunt B, Itaya T and Frost P (1991). Reduced tumorigenicity of tumor cells secreting gamma-IFN is due to nonspecific host responses and is unrelated to class I major histocompatibility complex expression. Cancer Res **51**, 1185-1189.

Gansbacher B, Bannerji R, Daniels B, Zier K, Cronin K and Gilboa E (1990). Retroviral vector-mediated γ-IFN gene transfer into tumor cells generates potent and long lasting antitumor immunity. Cancer Res **50**, 7820-7825.

Giavazzi R, Garofalo A, Bani MR, Abbate M, Ghezzi P, Boraschi D, Mantovani A and Dejana E (1990). Interleukin 1-induced augmentation of experimental metastases from a human melanoma in nude mice. Cancer Res **50**, 4771-4775.

Giovarelli M, Cofano F, Vecchi A, Forni M, Landolfo S and Forni G (1986). IFN-activated tumor inhibition in vivo. Small amounts of IFN-gamma inhibit tumor growth by eliciting host systemic immunoreactivity. Int J Cancer **37**, 141-148.

Ishikawa M, Koga Y, Hosokawa M and Kobayashi H (1986). Augmentation of B16 melanoma lung colony formation in C57BL/6 mice having marked granulocytosis. Int J Cancer **37**, 919-924.

Karasuyama H, Tohyama N and Tada T (1989). Autocrine growth and tumorigenicity of interleukin-2-dependent helper T cells transfected with IL-2 gene. J Exp Med **169**, 13-25.

Kelly SA, Gschmeissner S, East N and Balkwill FR (1991). Enhancement of metastatic potential by gamma-IFN. Cancer Res **51**, 4020-4027.

Lollini PL, De Giovanni C, Del Re B, Nicoletti G, Prodi G and Nanni P (1987). IFN-mediated enhancement of metastasis. Are MHC antigens involved? Clin Exp Metastasis **5**, 277-287.

Lollini PL, De Giovanni C, Del Re B, Nicoletti G, Prodi G, Scotlandi K and Nanni P (1988). IFN-mediated modulation of metastasis and MHC antigens. In "Cancer Metastasis - Biological and Biochemical Mechanisms and Clinical Aspects" (eds G Prodi, LA Liotta, PL Lollini, S Garbisa, S Gorini and K Hellmann), pp 129-139. Plenum Press, New York.

Lollini PL, De Giovanni C, Nicoletti G, Bontadini A, Tazzari PL, Landuzzi L, Scotlandi K and Nanni P (1990). Enhancement of experimental metastatic ability by tumor necrosis factor-alfa alone or in combination with IFN-gamma. Clin Exp Metastasis **8**, 215-224.

Lollini PL, Landuzzi L, Nicoletti G, De Giovanni C, Giovarelli M, Lalli E, Facchini A and Nanni P (1992). Ly-6A/E gene is widely expressed among transformed nonhematopoietic cells. Autocrine modulation by IFN. Anticancer Res **12**, 2245-2252.

Lollini PL, Bosco MC, Cavallo F, De Giovanni C, Giovarelli M, Landuzzi L, Musiani P, Modesti A, Nicoletti G, Palmieri G, Santoni A, Young HA, Forni G and Nanni P (1993). Inhibition of tumor growth and enhancement of metastasis after transfection of the gamma-IFN gene. Int J Cancer **55**, 320-329, 1993.

Malik STA, Naylor MS, East N, Oliff A and Balkwill FR (1990). Cells secreting tumor necrosis factor show enhanced metastasis in nude mice. Eur J Cancer **26**, 1031-1035.

Maruguchi Y, Toda K-I, Fujii K, Imamura S and Watanabe Y (1991). Survival period of tumor-bearing mice is prolonged after the IFN-gamma-producing gene transfer. Cancer Lett **60**, 41-49.

Nanni P, De Giovanni C, Lollini PL, Nicoletti G and Prodi G (1983). TS/A: A new metastasizing cell line from a BALB/c spontaneous mammary adenocarcinoma. Clin Exp Metastasis **1**, 373-380.

Nanni P, Landuzzi L, Nicoletti G, De Giovanni C, Giovarelli M, Lalli E, Facchini A and Lollini PL (1992). Control of H-2 expression in transformed nonhaemopoietic cells by autocrine IFN. Br J Cancer **66**, 479-482.

Nicoletti G, De Giovanni C, Lollini PL, Bagnara GP, Scotlandi K, Landuzzi L, Del Re B, Zauli G, Prodi G and Nanni P (1989). In vivo and in vitro production of haemopoietic colony-stimulating activity by murine cell lines of different origin: A frequent finding. Eur J Cancer Clin Oncol **25**, 1281-1286.

Orosz P, Echtenacher B, Falk W, Rueschoff J, Weber D and Mannel DN (1993). Enhancement of experimental metastasis by tumor necrosis factor. J Exp Med **167**, 1391-1398.

Palmieri G, Morrone S, Lollini PL, De Giovanni C, Nicoletti G, Nanni P, Frati L and Santoni A (1992). TNF impairs in vivo and in vitro natural killer (NK) susceptibility of B16 melanoma cells. Scand J Immunol **35**, 279-287.

Porgador A, Bannerji R, Watanabe Y, Feldman M, Gilboa E and Eisenbach L (1993). Antimetastatic vaccination of tumor-bearing mice with two types of IFN-gamma gene-inserted tumor cells. J Immunol **150**, 1458-1470.

Pross HF, Baines NG, Rubin P, Shragge P and Patterson MS (1981). Spontaneous human lymphocyte-mediated cytotoxicity against tumor target cells. IX. The quantitation of natural killer cell activity. J Clin Immunol **1**, 51-56.

Taniguchi K, Petersson M, Hoglund P, Kiessling R, Klein G and Karre K (1987). IFN gamma induces lung colonization by intravenously inoculated B16 melanoma cells in parallel with enhanced expression of class I major histocompatibility complex antigens. Proc Natl Acad Sci USA **84**, 3405-3409.

Trinchieri G and Santoli D (1978). Anti-viral activity induced by culturing lymphocytes with tumor-derived or virus-transformed cells. Enhancement of human natural killer cell activity by IFN and antagonistic inhibition of susceptibility of target cell to lysis. J Exp Med **147**, 1314-1333.

Watanabe Y, Kuribayashi K, Miyatake S, Nishihara K, Nakayama E, Taniyama T and Skata T (1989). Exogenous expression of mouse IFN gamma cDNA in mouse neuroblastoma C1300 results in reduced tumorigenicity by augmented anti-tumor immunity. Proc Natl Acad Sci USA **86**, 9456-9460.

Zoeller M, Strubel A, Haemmerling G, Andrighetto G, Raz A and Ben-Ze'ev A (1988). IFN-gamma treatment of B16 melanoma cells: Opposing effects for non-adaptive and adaptive immune defense and its reflection by metastatic spread. Int J Cancer **41**, 256-266.

21

ANTITUMOR RESPONSE IN MICE TRANSPLANTED WITH IFN-α PRODUCING TUMOR CELLS

M. Ferrantini, E. Proietti, L. Santodonato, L. Gabriele, M. Peretti, T. Kaido, I. Gresser and F. Belardelli
Laboratory of Virology, Istituto Superiore di Sanita, Rome, Italy and Laboratory of Viral Oncology, Institut de Recherches Scientifiques sur le Cancer, Villejuif, France

Introduction

Cytokine gene transfer and expression into tumor cells represents a novel approach for investigating antitumor activities of cytokines in experimental models. This approach has been successfully used to investigate the antitumor effects of IL-2 (Bubenik et al., 1990; Fearon et al., 1990; Gansbacher et al., 1990a; Russell et al., 1991; Cavallo et al., 1993), IL-4 (Tepper et al., 1989; Li et al., 1990; Golumbek et al., 1991; Tepper et al., 1992), IFN-γ (Watanabe et al., 1989; Gansbacher et al., 1990b; Esumi et al., 1991), TNF-α (Asher et al., 1991; Blankenstein et al., 1991; Teng et al., 1991), IL-7 (Hock et al., 1991; Aoki et al., 1992; McBride et al., 1992), G-CSF (Colombo et al., 1991) and IL-1 (Douvdevani et al., 1992).

In several of these studies it has been shown that injection of cytokine secreting tumor cells can elicit a lasting immunity responsible for the rejection of the parental tumor. It is becoming increasingly evident that the antitumor responses induced by different cytokines are mediated by characteristically different effector mechanisms and that various tumors exhibit different responses to a given cytokine therapy. As the expression of different cytokines in the same tumor may even exert opposite effects with regard to tumor suppression (Blankenstein et al., 1991), it is useful to analyze the effects exerted by the constitutive expression of different cytokines in the same tumor model (Hock et al., 1993a). Moreover, well characterized tumor models (extensively used for therapy experiments with exogenous cytokines) may be especially useful for evaluating some specific features of tumor cell targeted cytokine gene therapy as compared to systemic therapy with cytokines. In fact, tumor cell targeted gene transfer approach has recently been considered as a possible new strategy for cancer treatment (Anderson, 1992; Miller, 1992). The major advantage of tumor cell targeted cytokine gene transfer is the

CYTOKINE-INDUCED TUMOR IMMUNOGENICITY
ISBN 0-12-261520-4

possibility of delivering effectively high doses of the cytokine itself to tumor sites, thus circumventing practical drawbacks of systemic cytokine administration such as the short half life, the difficulty of obtaining the optimal doses at the tumor site and toxicity. Moreover, the constant secretion of a given cytokine at or near the site of tumor growth may represent a "physiologic" stimulus for the efficient activation of a host immune response to the tumor. Although the possible clinical applications of tumor cell targeted cytokine therapy are still to be defined, the antitumor effects obtained by means of this strategy stimulate further efforts to perform systematic studies in experimental models to evaluate the potential use of cytokine gene transfer in cancer therapy.

Recently we have reported the first example of tumor cell targeted gene therapy with IFN-α in an experimental tumor model (Ferrantini et al., 1993). In this study, we have used a metastatic tumor cell line which had been widely used in our previous studies on the antitumor effects of IFN-α/β. In this chapter, we review our data on the antitumor response in mice transplanted with IFN-α1 producing tumor cells, and we discuss the potential interest in the use of genetically modified IFN producing cells in cancer therapy.

The tumor model: metastatic Friend erythroleukemia cells (FLC)

Metastatic FLC have been derived from an IFN-α/β-resistant FLC clone (3Cl-8) (Affabris et al., 1982) after serial i.p. passages into syngeneic DBA/2 mice. After in vivo passages, FLC acquired the capacity of growing as ascites in the peritoneum (Belardelli et al., 1982a) and formed tumor metastases in the liver and spleen when injected s.c. or i.v. (Belardelli et al., 1984). This highly malignant behavior was a stable characteristic of these cells (Belardelli et al., 1984).

In previous studies, we have used mice transplanted with highly metastatic FLC to investigate the antitumor effects of different cytokines (Belardelli et al., 1982a; Belardelli et al., 1982b; Belardelli et al., 1983; Gresser et al., 1986; Gresser et al., 1988; Proietti et al., 1988; Gresser et al., 1990; Belardelli et al., 1991; Ciolli et al, 1991; Gresser et al., 1991a). In particular, repeated injections of high doses of IFN-α/β were highly effective in inhibiting tumor growth in mice injected i.p., s.c. or i.v. with FLC (Belardelli et al., 1982a; Belardelli et al., 1983; Gresser et al., 1988). Since these effects were exerted against IFN-α/β resistant FLC, IFN was not acting directly on the tumor cells but through IFN induced host mechanisms (Belardelli et al., 1982b; Gresser et al., 1990; Gresser, 1991b).

Because of this marked response to IFN-α/β therapy, the FLC system appeared to be a useful model for investigating the role of local IFN-α production on the FLC tumorigenicity and the therapeutic efficacy of

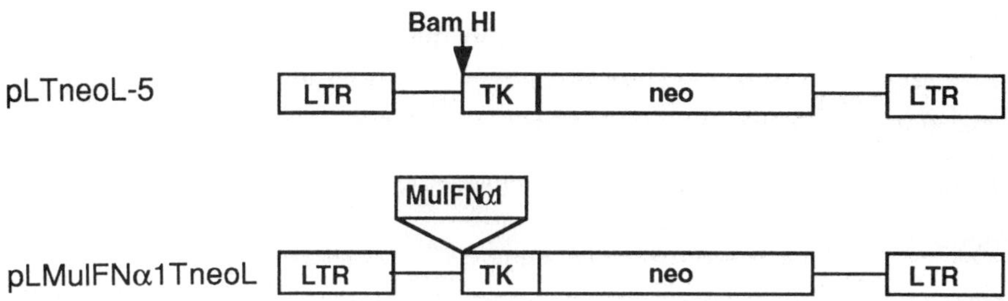

Fig. 21.1. *Structure of retroviral vectors transfected into 3Cl-8 FLC.*

treating established FLC tumors with genetically modified IFN producing tumor cells.

Isolation of IFN producing 3Cl-8 clones and analysis of their tumorigenic properties

The retroviral vectors utilized for the genetic transfer into IFN-α/β resistant 3Cl-8 cells are schematically depicted in Fig. 21.1 (Ferrantini et al., 1993). Plasmid pLTneoL-5 containing only the neomycin resistance selectable marker was used as transfection control plasmid. Insertion in this construct of the coding sequence for murine IFN-α1 (Shaw et al., 1983) produced plasmid pLMuIFN-α1TneoL, which was utilized for expression of the IFN-α1 gene. Both plasmids were transferred into parental 3Cl-8 FLC by electroporation, and individual G418 resistant clones were isolated and cultured.

Table 21.1 shows individual clones chosen for in vitro characterization and for in vivo experiments. One clone (3Cl-8-TC2), selected after transfection of pLTneoL-5, was used as a transfection control cell line. Individual clones obtained after transfection of the pLMuIFN-α1TneoL plasmid were tested for their ability to secrete constitutively biologically active IFN into the culture supernatant (Belardelli et al., 1987). As indicated in Table 21.1, there was a considerable variability in the amount of IFN produced among the different IFN-α1 transfected clones. The control transfected cells did not produce any IFN activity. The extent of IFN production was a stably acquired phenotype of each clone which did not vary significantly after continuous in vitro culture. Cell

Table 21.1 *IFN production by transfected clones*

Cell clone	Transfected plasmid	IFN production (IU/ml)
3 Cl-8 (parental cells)	none	< 4
TC-2 (transfection control)	pLTneoL-5	< 4
IFN-α1-Cl-11	pLMuIFN-α1TneoL	256-512
IFN-α1-Cl-3	pLMuIFN-α1TneoL	64-128
IFN-α1-Cl-4	pLMuIFN-α1TneoL	8- 16

morphology and in vitro growth properties (assessed by measuring the cloning efficiency in methylcellulose) were not affected by the constitutive secretion of IFN in the selected clones with respect to the parental and transfection control IFN resistant FLC. IFN-α1 producing 3Cl-8-FLC exhibited a markedly decreased tumorigenicity when injected into syngeneic DBA/2 mice.

Table 21.2 summarizes the results of experiments in which DBA/2 mice were injected s.c. or i.v. with different numbers of control transfected cells (TC-2) or IFN producing cells (IFN-α1-Cl-11). TC-2 cells showed tumorigenic properties typical of the parental 3Cl-8 cells in that TC-2 cells were capable of inducing s.c. tumors, and all the injected mice died with extensive liver and spleen metastases.

Both TC-2 and parental FLC had the same LD50 (i.e. approximately 10^2 cells). In contrast, injection of 2 x 10^5 IFN-α1-Cl-11 cells did not result in any tumor growth. In mice inoculated with 2 x 10^6 IFN producing FLC, tumor development was observed in approximately 40% of the injected animals. Interestingly, some of the tumors from these mice underwent complete regression 3-4 weeks after tumor cell injection.

When IFN producing FLC were injected i.v., a dramatic decrease in their tumorigenic potential compared to control transfected cells was also observed, further indicating the ability of the locally secreted IFN to inhibit metastases formation. As shown in Fig. 21.2, high titers of IFN were found in the tumor extracts as well as in the sera from mice injected s.c. with high numbers of IFN-α1-Cl-11 FLC, whereas no detectable IFN activity was found in the corresponding tissues from mice injected with control transfected cells. The IFN activity found in either tumor extracts or sera from mice injected with IFN-α1-Cl-11 FLC was completely neutralized by mAb to IFN-α1 (data not shown).

Table 21.2 *Tumorigenicity of IFN-α1-Cl -11 and TC-2 FLC when injected s.c. or i.v. into DBA/2 mice*

Route of injection[a]	No. of cells injected	TC-2 FLC Dead mice/ total injected	TC-2 FLC Day of death[b] (Mean ± S.E.)	IFN-α1 Cl-11 FLC Dead mice/ total injected	IFN-α1 Cl-11 FLC Day of death[b] (Mean ± S.E.)
s.c.	2×10^6	23/23	33.9 ± 1.5	13/30	60.4 ± 7.2[c]
s.c.	2×10^5	6/6	35.5 ± 2.7	0/6	>120[d]
i.v.	5×10^6	3/3	11.0 ± 0	3/3	35.3 ± 1.7
i.v.	5×10^5	3/3	12.0 ± 0	0/3	>133[d]

[a]Six-week-old male DBA/2 mice were injected with either transfection control or IFN producing FLC, as indicated
[b]Calculated only for dead mice
[c]Values significantly different ($p < 0.001$), as compared to mice injected with the same number of TC-2 FLC
[d]On the indicated days, surviving mice were sacrificed and found to be tumor-free

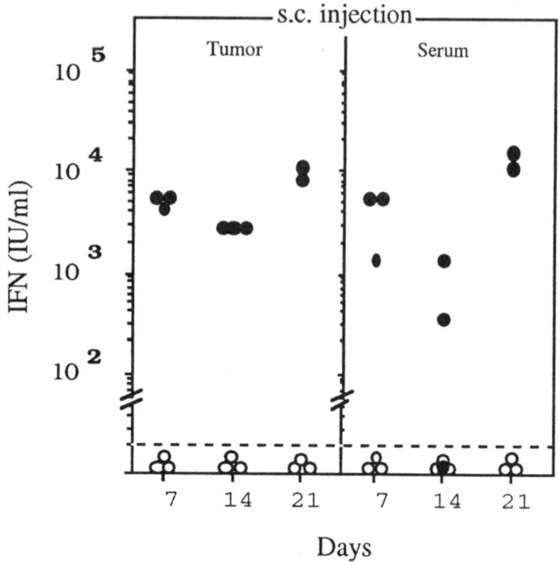

Fig. 21.2. *In vivo production of IFN in mice transplanted with INF-α1-Cl-11 FLC.* Six-week-old DBA/2 mice were given s.c. injections of TC-2 or IFN-α1-Cl-11 FLC (2 x 10^6). At different times after tumor injection, three mice from each group were sacrificed. Serum and tumor extracts were tested for IFN activity. (○) Mice given injections of TC-2 cells; (●) mice given injections of IFN-α1-Cl-11 cells.

Antitumor responses elicited in the host by injection of IFN producing FLC

To rule out the possibility that the rejection of IFN producing 3Cl-8 cells was due to a loss of the in vivo growth capAcity of these cells, we tested the tumorigenicity of IFN producing 3Cl-8 FLC when transplanted in totally immunosuppressed animals, i.e.: splenectomized, irradiated, anti-asialo GM1 treated (SIA) nude mice (Ferrantini et al., 1993).

As shown in Table 21.3, the injection of IFN producing FLC into immunodeficient SIA nude mice led to rapid tumor growth and subsequent death of all mice with a time course comparable to that observed in mice injected with transfection control cells. These results indicated that rejection of IFN producing FLC was dependent upon an IFN induced host mediated response in the immunocompetent mouse.

We then investigated whether the injection of IFN producing 3Cl-8 cells could elicit an immune response which was tumor specific and long-lasting. DBA/2 mice were injected s.c. with IFN producing FLC (2 x 10^6 cells) and 60 days later tumor free mice were challenged with 2 x 10^6 parental 3Cl-8 or unrelated ESb lymphoma cells (Schirrmacher et al., 1979). All control untreated mice developed rapidly growing tumors and

Table 21.3. *Tumorigenicity of IFN producing FLC in SIA nude mice*

Cell clone	No. of dead mice[a]/ total mice	Day of death (Mean ± S.E.)
TC-2	6/6	18.8 ± 0.5
IFN-α1-Cl-11 FLC	6/6	20.2 ± 0.5

[a]SIA nude mice were given s.c. injections of 2 x 10^6 TC-2 or IFN α1-C11-11. There were ten mice in each group

died with liver and spleen metastases 25-30 days after tumor challenge. In contrast, mice preinjected with IFN producing cells were completely resistant to the growth of parental 3Cl-8 cells but not to the growth of unrelated ESb tumor cells (Fig. 21.3).

Sera recovered from mice 30-60 days after injection of 3Cl-8IFN cells exhibited high titers of antibodies to FLC (ranging between 1:640 and 1: 1280), as assessed by radioimmunoassay (data not shown). These results indicated that IFN producing 3Cl-8 cells were capable of inducing a tumor specific host immune response for at least 2 months after the original injection.

In order to ascertain the existence of host immune cells responsible for the antitumor activity, we performed "Winn assays" by mixing parental metastatic cells with splenocytes from control mice or mice injected with IFN producing cells before s.c. inoculation into recipient animals. No antitumor activity was observed when spleen cells from mice injected with control transfection cells were used. In contrast, spleen cells from mice injected with IFN producing 3Cl-8 cells were capable of protecting 60% of the recipient mice against the growth of parental 3Cl-8 cells (Fig. 21.4). Studies are in progress to characterize the active immune splenocytes by selective in vivo depletion with specific antibodies.

Antitumor effects of IFN producing 3Cl-8 cells: inhibition of parental FLC growth and metastasis formation in mixed tumor transplantation assays and in mice with established parental tumors

The capability of IFN producing FLC of exerting antitumor activities against the parental tumor was first tested in mixed tumor transplantation assays. Fig. 21.5 summarizes the results of a typical experiment in which

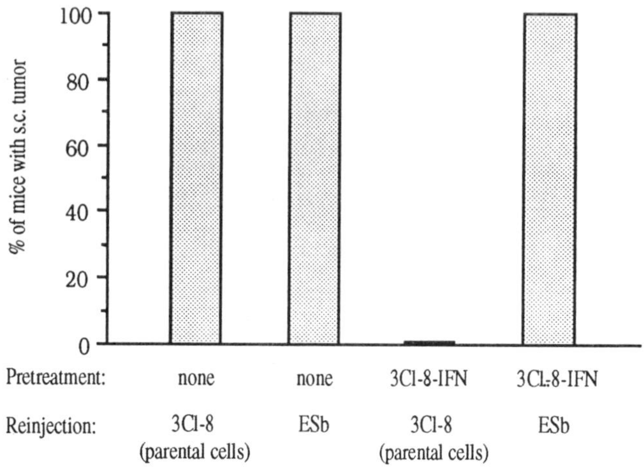

Fig. 21.3. *Tumorigenicity of homologous or heterologous tumors in mice previously given injections of IFN producing 3Cl-8.* Male DBA/2 mice, 6-7 weeks old, were given s.c. injections of 2 x 10^6 IFN-α1-Cl-11 cells or kept uninjected. After 60 days, some mice from each group were injected s.c. with either 2 x 10^6 parental 3Cl-8 cells or 2 x 10^6 ESb cells. On day 30 surviving mice exhibited no s.c. tumor. Mice were then sacrificed and found to be tumor free by gross examination of visceral organs.

DBA/2 mice were given an s.c. injection of a suspension containing 2 x 10^5 3Cl-8 parental cells (about 1000 LD50) and 2 x 10^6 IFN producing or transfection control FLC. TC-2 cells did not inhibit the growth of the parental tumor cells. In contrast, IFN producing FLC efficiently protected the mice from the growth of the parental 3Cl-8 cells resulting in a 56% survival rate. These results indicated that IFN producing 3Cl-8 cells were capable of inhibiting the growth of metastatic parental cells.

A challenging issue in adoptive immunotherapy is the attempt to cure established tumors. We first determined whether single injections of IFN producing FLC at different times after inoculum of parental metastatic cells could exert an antitumor effect. A single contralateral injection of 2 x 10^6 IFN producing cells 1 or 3 days after the injection of 2 x 10^5 parental cells proved to be very effective in delaying the growth of the parental tumor, resulting in an increased survival compared to controls (Table 21.4, Exp. 1). Single injections of IFN producing FLC on day 5 or 8 were less effective, but they still resulted in a significant increase in the mean survival time ($p < 0.001$, versus control FLC injected mice).

Furthermore, repeated contralateral injections of IFN producing FLC were highly effective in inducing an antitumor response in mice with 3-day metastatic FLC tumors, and 10% of the mice were tumor free 120 days after tumor cell injection.

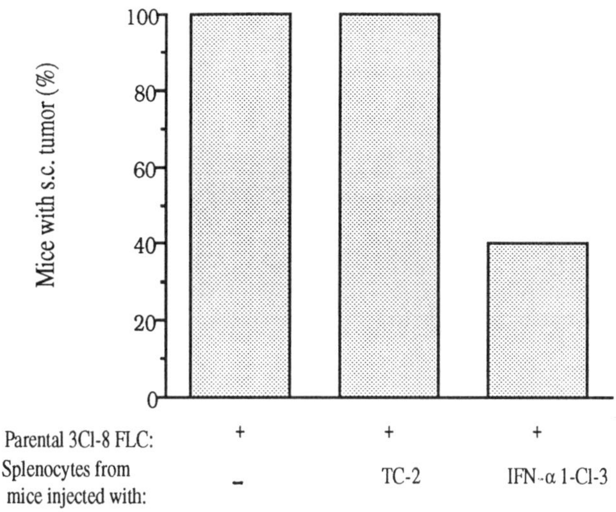

Fig. 21.4. *Six-7 week-old male DBA/2 mice were given s.c. injections of 2 x 10^6 TC-2 or IFN-α1-Cl-3 cells.* On day 14, three mice from each group were sacrificed, and the spleen cells from each mouse were mixed with 5 x 10^4 parental 3Cl-8 cells (effector/target cell ratio = 20:1) and injected s.c. into five recipient DBA/2 mice. The number of recipient mice developing a tumor in each group was determined 30 days after the mixed cell injection.

Conclusions and perspectives

In this article, we have reviewed our recent data on the antitumor response in mice injected with IFN-α producing FLC. Although IFN-α was the first cytokine to be used in clinical trials with cancer patients (Strander, 1986; Gresser, 1989), our data represent the first study on a cell targeted gene therapy with this cytokine in an experimental model. Our results demonstrate that highly metastatic FLC genetically modified to secrete IFN-α1 exhibit a marked loss of their tumorigenic potential when injected s.c. or i.v. into syngeneic mice. Likewise, these genetically modified IFN producing FLC inhibited the growth of metastatic parental cells in a mixed transplantation assay. It is of interest that high titers of IFN were found in the sera and tumor extracts of mice injected with IFN producing FLC, indicating that a remarkably high level of IFN production was maintained in vivo in these animals. The relevance of this in vivo production of IFN in the inhibition of the in vivo growth of IFN producing FLC is supported by the finding that injection of anti-IFN-α/β antibodies resulted in rapid tumor growth in these mice (Ferrantini et al., 1993).

The finding that IFN producing FLC were highly tumorigenic when injected into immunosuppressed nude mice indicates that these gene-

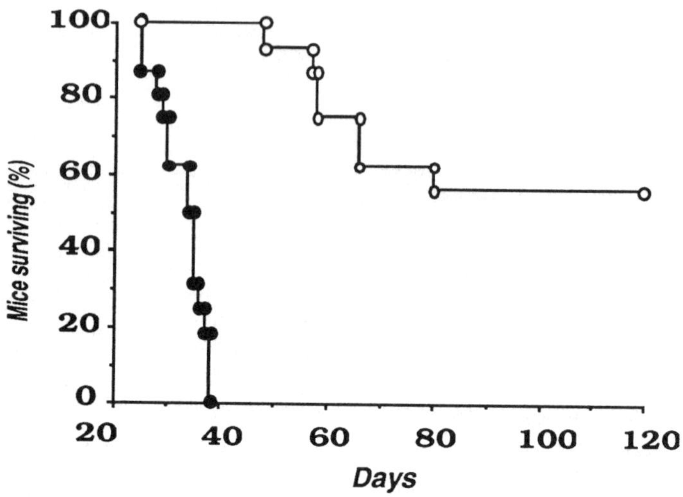

Fig. 21.5. *Inhibition of the parental tumor growth by IFN producing FLC ("mixed tumor transplantation assay").* Parental 3Cl-8 cells (2 x 10^5) were mixed with 2 x 10^6 TC-2 or IFN-α1-Cl-3 and injected s.c. into 6-7 week-old male DBA/2 mice. Some control mice were given s.c. injections of only IFN-α1-Cl-3 (2 x 10^6). On day 120 all the surviving mice were sacrificed and found to be tumor free. None of the mice given injections only of IFN-α1-Cl-3 cells developed a tumor. For statistical analyses, an arbitrary survival time of 120 days was considered for all surviving mice. (●) mice given injections of parental FLC and TC-2 cells (16 mice); (○) mice given injections of parental FLC and IFN-α1-Cl-3 cells (16 mice).

tically modified tumor cells had not lost their intrinsic ability to proliferate in vivo, but their injection into immunocompetent mice induced a host immune response to the tumor. Moreover, mice injected with IFN producing FLC developed a tumor specific immune resistance, which was associated with the development of a specific antibody response to FLC and the generation of spleen cells with potent anti-FLC activity.

In recent studies, we have extensively investigated the host immune mechanisms involved in the IFN-α/β mediated suppression of metastatic tumor growth in mice injected with FLC. In particular, IFN-α/β was less effective in inhibiting the development of FLC metastases in immunodeficient nude mice than in immunocompetent mice (Gresser et al., 1990). Moreover, spleen cells and anti-FLC antibodies from IFN treated tumor bearing mice exhibited a marked antitumor effect in adoptive transfer immunotherapy experiments (Gresser et al., 1991a). We may postulate, therefore, that the host immune mechanisms activated in mice injected with IFN producing cells are similar to those previously observed in tumor bearing mice treated with exogenous IFN-α/β. Further

Table 21.4. *Antitumor activity by single or repeated treatment with IFN producing FLC in mice transplanted with metastatic parental cells*

Exp. #	Day of IFN-α1-Cl-11 injections	No. of mice with tumor/ total mice	Day of death (Mean ± S.E.)
1[a]	--	10/10	29.3 ± 0.6
	1	10/10	47.4 ± 2.4
	3	10/10	46.7 ± 2.5
	5	10/10	38.1 ± 2.3
	8	10/10	37.5 ± 1.4
2[b]	--	10/10	27.2 ± 1.7
	3,6,11,14,18,21	9/10	56.5 ± 5.2

[a]Fifty 6-7-week-old male DBA/2 mice were given s.c. injections of 2 x 10^6 parental 3Cl-8 cells and divided into 5 groups of 10 mice. Mice were subjected to a single contralateral s.c. injection of 2 x 10^6 IFN-α1-Cl-11 cells, on days 1 (Group B), 3 (Group C), 5 (Group D), or 8 (Group E), respectively, after injection of parental cells. One group (Group A) was given injections of parental 3Cl-8 FLC and not treated with IFN producing FLC. For groups B, C, D, and E, the p values versus Group A were < 0.001

[b]Six-week-old male DBA/2 mice were given s.c. injections of 2 x 10^6 parental 3Cl-8 cells. On day 3 mice were divided into 2 groups of 10 mice each. One group was subjected to a contralateral s.c. injection of 2 x 10^6 IFN-α1-C1-11FLC, on days 3, 6, 11, 14, 18 and 21 as indicated. The other group was left untreated. On day 120 the surviving mice in the treated group were sacrificed and found to be tumor free (for statistical analysis, an arbitrary survival time of 120 days was considered for this surviving mice). The p value of the untreated group versus the treated group was <0.0001

studies are necessary to verify this assumption.

Tumor cell targeted cytokine gene transfer has proven to be a useful and practical approach to investigate the antitumor effects of a variety of cytokines in experimental models. However, the possible application of this strategy to cancer therapy is still a matter of debate (Anderson, 1992; Miller, 1992). The possibility of obtaining suitable cytokine production at the tumor cell site by means of tumor cell targeted cytokine gene transfer has focused the attention of investigators and clinicians on the evaluation of this strategy as an alternative to systemic administration of cytokines in cancer therapy. But to establish the possible clinical interest of cytokine gene therapy some important requirements need to be fulfilled.

Generation of a long lasting immunity

To evaluate the possible usefulness of cytokine producing tumor cells as tumor vaccines in adjuvant therapy it is important to define the capability of cytokine secreting tumor cells to generate a lasting immunity against the parental tumor cells. Several reports have indicated that immunization with cytokine secreting tumor cells can effectively protect recipient animals against the challenge with unmodified parental tumor cells (Fearon et al., 1990; Gansbacher et al., 1990a; Golumbek et al., 1991; Cavallo et al., 1993; Ohe et al., 1993; Staib et al., 1993). In this regard, our data with IFN producing FLC strongly suggest that IFN-α may be very effective in obtaining a long lasting immunity (Ferrantini et al., 1993). However, further studies are necessary to compare the cytokine transfected tumor cells with the unmodified tumor cells as immunogens (Hock et al., 1993b).

Effects on established tumors

The efficacy of cytokine secreting tumor cells in vaccination protocols might be useful for defining immunotherapeutic strategies against established tumors (Porgador et al., 1993a; Porgador et al., 1993b). Although several studies are now available on the in vivo behavior of genetically modified cytokine producing tumor cells, so far only one study has described the inhibitory effects of viable cytokine producing tumor cells against established tumors (Golumbek et al., 1991), a property that would be particularly important in treatment of human cancers. Our results in the FLC system indicate that repeated s.c. injections of IFN producing cells in mice with established metastatic tumors result in a definite antitumor effect. The finding that the IFN-α1 gene therapy was effective in a highly metastatic tumor system (such as the FLC model) further supports the potential interest of this approach for the treatment of malignant tumors. However, comparative studies with various metastatic tumors exhibiting a different antitumor response to IFN will be helpful in evaluating the possible advantage of this approach, as compared with exogenous IFN therapy, in the treatment of established tumors.

Therapeutic perspectives

The major drawback in the use of viable IFN producing tumor cells for the therapy of established tumors is the fact that injection of high numbers of these cells may result in tumor growth. The definition and the optimization of antitumor cytokine gene therapy strategies making use of irradiated cytokine secreting tumor cells would therefore be very valuable with respect to the issue of safety. However, our preliminary experiments indicate that the treatment of established s.c. FLC tumors with irradiated IFN-α1-Cl-11 cells does not result in any significant therapeutic effect (data not shown). To overcome the possible risk of using tumor cells still endowed with an intrinsic growth capacity in an immunodepressed host, it might be possible to design appropriate retroviral vectors containing, in addition to the cytokine gene, a second gene conferring sensitivity to a cytotoxic drug. Infection of tumor cells or primary tumor tissue with such a vector might allow a controlled in vivo growth of tumor cells, which can be selectively eliminated upon administration of the drug.

IFN-α antitumor activities have been documented for 25 years. The effects exerted by this cytokine against different experimental tumors have been extensively demonstrated (Gresser, 1991b), and IFN-α was the first cytokine to be used in clinical trials and to prove effective in the therapy of certain types of human cancers (Strander, 1986). In spite of this intense investigation, the understanding of the mechanisms underlying IFN-α antitumor action has been elusive. This difficulty is probably due to the complex and multiple biologic effects that IFN exert on cells and on different cellular compartments in the organism. Nevertheless, conclusions can be drawn from the ensemble of experimental studies that indicate that antitumor therapy with IFN is particularly effective when:

- IFN is injected repeatedly, which means that its level is to be maintained quite constant over a certain period of time
- IFN is in close contact with the tumor cells.

Clinical studies also indicated an improved response of some human cancers to IFN-α in local versus systemic therapy (Strander, 1989; Berek, 1992). Such evidences, together with the adverse side effects observed in patients treated with high doses of IFN (Quesada et al., 1986), led us to consider IFN-α as a potential good candidate for an effective tumor cell targeted cytokine therapy. Although further experimental studies are required to define the necessary conditions for performing any cytokine gene therapy in cancer patients, our data provide the experimental basis for the use of IFN-α transfected tumor cells as a potent selective agent for the treatment of metastatic tumors.

Acknowledgments

We are indebted to Anna Ferrigno and Maria Cristina Ferrari for their valuable secretarial assistance and to C. Maury for her excellent technical assistance.

This work was supported in part by the Consiglio Nazionale delle Ricerche, special grant on Applicazioni Cliniche della Ricerca Oncologica, the Associazione Italiana Ricerca sul Cancro and the Association Recherche Cancer.

References

Affabris E, Jemma C and Rossi GB (1982). Isolation of interferon-resistant variants of Friend erythroleukemia cells: Effects of interferon and ouabain. Virology **120**, 441-452.

Anderson WF (1992). Human gene therapy. Science **256**, 808-813.

Aoki T, Tashiro K, Miyatake S, Kinashi T, Nakano T, Oda Y, Kikuch H and Honji T (1992). Expression of murine interleukin 7 in a murine glioma cell line results in reduced tumorigenicity in vivo. Proc Natl Acad Sci USA **89**, 3850-3854.

Asher AL, Mule JJ, Kasid A, Restifo NP, Salo JC, Reichert CM, Jaffe G, Fendly B, Kriegler M and Rosenberg SA (1991). Murine tumor cells transduced with the gene for tumor necrosis factor-α. J Immunol **146**, 3227-3234.

Belardelli F, Gresser I, Maury C and Maunoury MT (1982a). Antitumor effects of interferon in mice injected with interferon-sensitive and interferon-resistant Friend leukemia cells. I. Int J Cancer **30**, 813-820.

Belardelli F, Gresser I, Maury C and Maunoury MT (1982b). Antitumor effects of interferon in mice injected with interferon-sensitive and interferon-resistant Friend leukemia cells. II. Int J Cancer **30**, 821-825.

Belardelli F, Gresser I, Maury C, Duvillard P, Prade M and Maunoury MT (1983). Antitumor effects of interferon in mice injected with interferon-sensitive and interferon-resistant Friend leukemia cells. III. Inhibition of growth and necrosis of tumors implanted subcutaneously. Int J Cancer **31**, 649-653.

Belardelli F, Ferrantini M, Maury C, Santurbano L and Gresser I (1984). Biologic and biochemical differences between in vitro and in vivo passaged Friend erythroleukemia cells. Int J Cancer **24**, 389-395.

Belardelli F, Gessani S, Proietti E, Locardi C, Borghi P, Watanabe Y, Kawade Y and Gresser I (1987). Studies on the expression of spontaneous and induced interferons in mouse peritoneal macrophages by means of monoclonal antibodies to mouse interferons. J Gen Virol **68**, 2203-2212.

Belardelli F, Gabriele L, Proietti E, Sestili P, Peretti M, Rozera C and Gresser I (1991). Synergistic anti-tumor effects of combined IL-1/IFN-α/β therapy in mice injected with metastatic Friend erythroleukemia cells. Int J Cancer **49**, 274-278.

Berek JS (1992). Intraperitoneal immunotherapy for ovarian cancer with alpha interferon. Eur J Cancer **28A**, 719-721.

Blankenstein T, Qin Z, Uberla K, Muller W, Rosen H, Volk HD and Diamantstein T (1991). Tumor suppression after tumor cell-targeted tumor necrosis factor α gene transfer. J Exp Med **173**, 1047-1052.

Bubenik J, Simova J and Jandlova T (1990). Immunotherapy of cancer using local administration of lymphoid cells transformed by IL-2 cDNA and constitutively producing IL-2. Immunol Lett **23**, 287-292.

Cavallo F, Giovarelli M, Gulino A, Vacca A, Stoppacciaro A, Modesti A and Forni G (1993). Role of neutrophils and CD4+ T lymphocytes in the primary and memory response to nonimmunogenic murine mammary adenocarcinoma made immunogenic by IL-2 gene transfection. J Immunol **149**, 3627-3635.

Ciolli V, Gabriele L, Sestili P, Varano F, Proietti E, Gresser I, Testa U, Montesoro E, Bulgarini D, Mariani G, Peschle C and Belardelli F (1991). Combined IL-1/IL-2 therapy of mice injected with highly metastatic Friend leukemia cells: Host antitumor mechanisms and marked effects on established metastases. J Exp Med **173**, 313-322.

Colombo MP, Ferrari G, Stoppacciaro A, Parenza M, Rodolfo M, Mavilio F and Parmiani G (1991). Granulocyte colony-stimulating factor gene transfer suppresses tumorigenicity of a murine adenocarcinoma in vivo. J Exp Med **173**, 889-897.

Douvdevani A, Huleihel M, Zoller M, Segal S and Apte RN (1992). Reduced tumorigenicity of fibrosarcomas which constitutively generate IL-1 alpha either spontaneously or following IL-1 alpha gene transfer. Int J Cancer **51**, 822-830.

Esumi N, Hunt B, Itaya T and Frost P (1991). Reduced tumorigenicity of murine tumor cells secreting γ-interferon is due to non specific host responses and is unrelated to class I major histocompatibility complex expression. Cancer Res **51**, 1185-1189.

Fearon ER, Pardoll DM, Itaya T, Golumbek P, Levitsky HI, Simons JW, Karasuyama H, Vogelstein B and Frost P (1990). Interleukin-2 production by tumor cells bypasses T helper function in the generation of an antitumor response. Cell **60**, 397-403.

Ferrantini M, Proietti E, Santodonato L, Gabriele L, Peretti M, Plavec I, Meyer F, Kaido T, Gresser I and Belardelli F (1993). α1-Interferon gene transfer into metastatic Friend leukemia cells abrogated tumorigenicity in immunocompetent mice: Antitumor therapy by means of interferon-producing cells. Cancer Res **53**, 1107-1112.

Gansbacher B, Zier K, Daniels B, Cronin K, Bannerji R and Gilboa E (1990a). Interleukin 2 gene transfer into tumor cells abrogates tumorigenicity and induces protective immunity. J Exp Med **172**, 1217-1224.

Gansbacher B, Bannerji R, Daniels B, Zier K, Cronin K and Gilboa E (1990b). Retroviral vector-mediated γ-interferon gene transfer into tumor cells generates potent and long lasting antitumor immunity. Cancer Res **50**, 7820-7825.

Golumbek PT, Lazenby AJ, Levitsky HI, Jaffee LM, Karasuyama H, Baker M and Pardoll DM (1991). Treatment of established renal cancer by tumor cells engineered to secrete interleukin-4. Science **254**, 713-716.

Gresser I, Belardelli F, Maury C, Tovey MG and Maunoury MT (1986). Antitumor effects of interferon in mice injected with interferon-sensitive and interferon-resistant Friend leukemia cells. IV. Definition of optimal treatment regimens. Int J Cancer **38**, 251-257.

Gresser I, Maury C, Woodrow D, Moss J, Grutter MG, Vignaux F, Belardelli F and Maunoury MT (1988). Interferon treatment markedly inhibits the development of tumor metastases in the liver and spleen and increases survival time of mice after intravenous inoculation of Friend erythroleukemia cells. Int J Cancer **41**, 135-142.

Gresser I (1989). Antitumor effects of interferon. Acta Oncol **28**, 347-353.

Gresser I, Maury C, Carnaud C, De Maeyer E, Maunoury MT and Belardelli F (1990). Antitumor effects of interferon in mice injected with interferon-sensitive and interferon-resistant Friend erythroleukemia cells. VIII. Role of the immune system in the inhibition of visceral metastases. Int J Cancer **46**, 468-474.

Gresser I, Carnaud C, Maury C, Sala A, Eid P, Woodrow D, Maunoury MT and Belardelli F (1991a). Host humoral and cellular immune mechanisms in the continued suppression of Friend erythroleukemia metastases after interferon α/β treatment in mice. J Exp Med **173**, 1193-1203.

Gresser I (1991b). Antitumor effects of interferons: Past, present and future. Br J Haematol **79** (Supp 1), 1-5.

Hock H, Dorsch M, Diamantstein T and Blankenstein T (1991). Interleukin 7 induces CD4+ T cell-dependent tumor rejection. J Exp Med **174**, 1291-1298.

Hock H, Dorsch M, Kunzendorf U, Qin Z, Diamantstein T and Blankenstein T (1993a). Mechanisms of rejection induced by tumor cell-targeted gene transfer of interleukin 2, interleukin 4, interleukin 7, tumor necrosis factor, or interferon-γ. Proc Natl Acad Sci USA **90**, 2774-2778.

Hock H, Dorsch M, Kunzendorf U, Uberla K, Qin ZH, Diamantstein T and Blankenstein T (1993b). Vaccinations with tumor cells genetically engineered to produce different cytokines: Effectivity not superior to a classical adjuvant. Cancer Res **53**, 714-716.

Li W, Diamantstein T and Blankenstein T (1990). Lack of tumorigenicity of interleukin 4 autocrine growing cells seems related to the anti-tumor function of interleukin 4. Mol Immunol **27**, 1331-1337.

McBride WH, Thacker JD, Comora S, Economou JS, Kelley D, Hogge D, Dubinett SM and Dougherty GJ (1992). Genetic modification of a murine fibrosarcoma to produce interleukin 7 stimulates host cell infiltration and tumor immunity. Cancer Res **52**, 3931-3937.

Miller AD (1992). Human gene therapy comes of age. Nature **357**, 455-460.

Ohe Y, Podack ER, Olsen KJ, Miyahara Y, Ohira T, Miura K, Nishio K and Saijo N (1993). Combination effect of vaccination with IL-2 and IL-4 cDNA transfected cells on the induction of a therapeutic immune response against Lewis lung carcinoma cells. Int J Cancer **53**, 432-437.

Porgador A, Bannerji R, Watanabe Y, Feldman M, Gilboa E and Eisenbach L (1993a). Antimetastatic vaccination of tumor-bearing mice with two types of IFN-gamma gene-inserted tumor cells. J Immunol **150**, 1458-1470.

Porgador A, Gansbacher B, Bannerji R, Tzehoval E, Gilboa E, Feldman M and Eisenbach L (1993b). Anti-metastatic vaccination of tumor-bearing mice with IL-2-gene-inserted tumor cells. Int J Cancer **53**, 471-477.

Proietti E, Belardelli F, Carpinelli G, Di Vito M, Woodrow D, Moss J, Sestili P, Fiers W, Gresser I and Podo F (1988). Tumor necrosis factor α induces early morphologic and metabolic alterations in Friend leukemia cell tumors and fibrosarcomas in mice. Int J Cancer **42**, 582-591.

Quesada JR, Talpaz M, Rios A, Kurzrock R and Gutterman JU (1986). Clinical toxicity of interferons in cancer patients: A review. J Clin Oncol **4**, 234-244.

Russell SJ, Eccles SA, Flemming CL, Johnson CA and Collins MKL (1991). Decreased tumorigenicity of a transplantable rat sarcoma following transfer and expression of an IL-2 cDNA. Int J Cancer **47**, 244-251.

Schirrmacher V, Shantz G, Clauer K, Komitowski D, Zimmermann HP and Lohmann-Matthes ML (1979). Tumor metastases and cell-mediated immunity in a model system in DBA/2 mice. Tumor invasiveness in vitro and metastasis formation in vivo. Int J Cancer **23**, 233-244.

Shaw GD, Boll W, Taira H, Mantei N, Lengyel P and Weissman C (1983). Structure and expression of cloned murine IFN-α genes. Nucl Acid Res **11**, 555-573.

Staib L, Harel W and Mitchell MS (1993). Protection against experimental cerebral metastases of murine melanoma-B16 by active immunization. Cancer Res **53**, 1113-1121.

Strander HA (1986). Interferon treatment of human neoplasia. Adv Cancer Res **46**, 1-265.

Strander HA (1989). Clinical effects of interferon therapy with special emphasis on antitumor efficacy. Acta Oncol **28**, 355-362.

Teng MN, Park BH, Koeppen HKW, Tracey KJ, Fendly BM and Schreiber H (1991). Long-term inhibition of tumor growth by tumor necrosis factor in the absence of cachexia or T-cell immunity. Proc Natl Acad Sci USA **88**, 3535-3539.

Tepper RI, Pattengale PK and Leder P (1989). Murine interleukin-4 displays potent anti-tumor activity in vivo. Cell **57**, 503-512.

Tepper RI, Coffman RL and Leder P (1992). An eosinophil-dependent mechanism for the antitumor effect of interleukin-4. Science **257**, 548-551.

Watanabe Y, Kuribayashi K, Miyatake S, Nishihara K, Nakayama E, Taniyama T and Sakata T (1989). Exogenous expression of mouse interferon-gamma cDNA in mouse neuroblastoma C1300 cells results in reduced tumorigenicity by augmented anti-tumor immunity. Proc Natl Acad Sci USA **86**, 9456-9460.

22

ANTIGEN PRESENTATION BY IFN-γ RELEASING TUMOR CELLS

N. P. Restifo
Surgery Branch, Division of Cancer Treatment, National Cancer Institute, National Institutes of Health, Bethesda, MD, USA

Introduction

Immune recognition by T lymphocytes occurs via specialized molecules called TCR. Unlike antibodies, to which they bear a striking structural similarity, TCR generally do not recognize intact antigens. Instead, with few exceptions, they can only recognize fragments of antigen presented by molecules of the MHC (Townsend and Bodmer, 1989). Recognition of tumor cells by T cells is thought to be no exception to this paradigm. T lymphocytes have been shown to be capable of recognizing tumor associated antigens (TAA), but only after they have been "processed", i.e. broken down into peptide fragments of the size and sequence capable of binding stably to MHC molecules. These antigen/MHC complexes are then presented at the cell surface for recognition by T lymphocytes. How can the immune recognition of TAA be enhanced?

T cells and the antitumor response

TCR engage MHC molecules which are presenting a diverse set of antigenic peptides derived primarily from intracellular antigens. There are a few exceptions to the generalization that the TCR engages only MHC/peptide complexes, and these include superantigens (Marrack et al., 1993) and mucins (Finn, 1992). There are two major types of MHC molecules, designated class I and class II, which present antigens to two broad categories of mature T lymphocytes, designated here as T_{CD8+} and T_{CD4+}. These lymphocytes derive their names from molecules that they express on their cell surfaces called CD8 and CD4, which are coreceptors that bind to highly conserved regions of MHC molecules (Potter et al., 1989; Salter et al., 1990). The specificity of the interaction of the CD8

CYTOKINE-INDUCED TUMOR IMMUNOGENICITY
ISBN 0-12-261520-4

molecule with class I and the CD4 molecule with class II has important functional consequences. For example, initial studies of T_{CD8+} function were dominated by observations of their ability to lyse a target cell presenting an antigenic peptide presented by class I. Subsequent studies have borne out that most T_{CD8+} are CTL. Conversely, since the first observations made of T_{CD4+} cells concerned their ability to greatly increase the production of antibodies by B cells, T_{CD4+} are frequently termed "helper" T cells.

Fundamental structural differences in the two classes of MHC result in important differences in intracellular trafficking and biochemical behavior. Consequently, distinctly different antigen pools are presented to different T-cell subsets. Antigens presented by class I molecules are generally derived from endogenous proteins, while class II molecules present peptide antigens generally derived from exogenous sources. More precisely, however, antigens presented by class I molecules are derived from peptides intersecting with the endoplasmic reticulum, while those presented by class II molecules are derived from peptides intersecting the endocytic compartment. The cell biology of class I molecules will be discussed in more detail below.

The crystal structures of both class I and class II molecules are now known (Bjorkman et al., 1987a; Bjorkman et al., 1987b; Brown et al., 1993). At first glance, the molecules are virtually superimposable. However, a closer examination reveals important structural differences. Class I molecules have a binding region which is closed at both ends and has very deep pockets for binding amino acid side chains. This imposes very strict size limitations on bound peptides. Class II molecules, on the other hand, have a structure which is open on both ends and is virtually devoid of pockets to accommodate side chains. This appears to be the explanation for the apparently reduced stringency observed for class II molecule binding of peptide.

Another important difference to consider is the histologic distribution of these MHC molecules. Class I molecules are very widely distributed and can be found on virtually every nucleated cell with the exception of some neural and germ line cells. Although usually expressed at low levels, class I molecules are highly inducible, and cell surface levels of class I are greatly increased during many attacks by viruses and other intracellular pathogens. Class II molecules, more narrowly distributed, are found in high concentrations only on B cells, Langerhans cells, dendritic cells, and on bone marrow derived phagocytic cells (e.g. macrophages). Class II molecules are found in lesser quantities on other cells including subpopulations of thymocytes, peripheral T cells, and on certain epithelial cells. In mouse and in man most tumor types, like the tissue histologies from which they are derived, express class I molecules. Selected histologies, like some melanomas and T- and B-cell lymphomas, express class II molecules.

The successful immune response against a tumor is likely to involve an orchestrated attack by the immune system involving more than a single effector mechanism. In fact, B cells and the Ig they produce, macrophages, and NK cells all have been shown to play important roles in the immune rejection of tumor cells in one model or another. The conclusion that can be derived from large set of sometimes contradictory data concerning the roles of T-cell subsets are that T cells bearing either the CD4 or the CD8 markers can mediate complete tumor rejection.

This can occur even in the absence of the other T-cell subsets. Nevertheless, certain principles governing the mechanisms of tumor rejection by these T-cell subsets have emerged. There are now numerous examples, in both mouse and man, in which highly specific, class I restricted T_{CD8+} can lyse tumor cells in vitro (Itoh et al., 1988; Topalian et al., 1989). Pure populations of T_{CD8+} adoptively transferred in man and mouse can be therapeutically active in vivo (Rosenberg et al., 1986; Rosenberg et al., 1988; Schreiber et al., 1988; Greenberg, 1991). Similar findings have been obtained in mouse with the extension that cloned T_{CD8+} have been shown to be therapeutic as well. It is generally assumed that T_{CD8+} are effective because they directly recognize TAA presented by class I molecules on the surface of the tumor cell. It has recently been demonstrated, however, that macrophages can present exogenously provided antigens via the class I molecules on their cell surfaces. This opens the possiblity for indirect recognition of tumor cells by T_{CD8+}. Also unappreciated until recently is that T_{CD8+} destroy tumor not only by direct lysis via delivery of cytotoxic granules, but also by release of cytokines upon contact with tumor cells. Among the cytokines shown to be released by antitumor T_{CD8+} are IFN-γ, TNF, and GM-CSF (Barth et al., 1991; Schwartzentruber et al., 1991).

The existence of a subset of T_{CD4+} that are directly cytotoxic suggests that such cells could theoretically be used to directly lyse tumors that express class II molecules (Kaplan et al., 1984). However, most experimental mouse tumors and many human tumor cells do not express class II (Elliot et al., 1989). Lack of class II expression by a tumor, however, does not preclude specific antitumor immunotherapy since antigenic fragments of TAA are likely to be presented by class II expressing cells infiltrating the tumor. The T_{CD4+} cells present would then release cytokines which could trigger an inflammatory reaction resulting in the recruitment of immune cells capable of destroying tumor cells. Thus, evidence suggests that TAA presentation both by class I and by class II molecules to T_{CD8+} and T_{CD4+} respectively, has relevance to tumor immunology (Greenberg, 1991). The remainder of this chapter, however, will focus on the presentation of endogenous TAA by class I molecules.

Tumor antigens: what is currently known

The debate over the existence of TAA has persisted for several decades. One argument against the existence of TAA on spontaneous human tumors might be that every tumor that results in the death of its host has escaped immune eradication. Thus, some would argue that even though tumors may express aberrant gene products, either the immune response is simply not elicited as a result of insufficient antigenic stimuli, or the immune system is actively suppressed as a result of altered signal transduction in T lymphocytes (Mizoguchi et al., 1992). Others would argue that the most immunogenic of TAA are selected against by an immune response. Nevertheless, based on a growing body of evidence, there seems to be an emerging consensus amongst tumor immunologists that TAA do exist in some tumors. Numerous workers have shown, in a variety of systems, that tumors can be recognized, and in some cases destroyed, by elements of the immune system. Cellular immunologists have documented the existence of many antigens recognized by T cells, and in several cases the sequences of these antigens have been identified by molecular biologists and protein biochemists. For example, Boon and coworkers have recently cloned a series of TAA recognized by T_{CD8+}. In the mouse, TAA from the Tum^{-} mutants of P815 (De Plaen et al., 1988; Sibille et al., 1990; Szikora et al., 1990), as well as the wild type P815, were identified (Van den Eynde et al., 1991). A gene coding for a human melanoma TAA has also been identified and designated MAGE-1 (Van der Bruggen et al., 1991); more recently, a tumor antigen has been identified as the normal tyrosinase gene (Brichard et al., 1993). Each of these antigens has been identified at the level of a minimal antigenic determinant, i.e. an 8-10 amino acid residue fragment that binds to class I. These are examples of TAA recognized by T_{CD8+}.

What are the possible candidates for other TAA? Cellular transformation is thought to be a complex set of events, in most cases, with the requirement of multiple molecular steps. Given the sometimes bizarre genotypes observed in cancer cells, it could be argued that tumor cells are likely to contain many new antigens potentially recognizable by the immune system.

The possible reasons why tumor cells might express TAA are beginning to be understood. For example, TAA might be the result of the processes of carcinogenesis which are generally thought to involve damage to a large number of genes, some of them involved with the molecular machinery regulating cell growth and division (Cho and Vogelstein, 1992). This results in the uncontrolled cellular proliferation that defines the transformed cell. Other possible candidate molecules that might be TAA include oncogene products activated by mutation, deletion, truncation, duplication or rearrangement. A partial list might include p21*ras* (Skipper and Stauss, 1993) and *bcr abl* (Witte, 1993). The mutated tumor suppressor gene product p53 may be a good candidate for antitumor immunotherapy (Yanuck et al., 1993) as may other mutated cellular proteins not playing a functional role in the malignant phenotype (passenger mutations). Foreign proteins involved in tumorigenesis may be ideal tumor antigens in some cases. There are examples of viral gene products including, but not limited to, products of oncogenic viruses like the SV 40 T antigen (Dalianis, 1990; Tevethia, 1990; Knowles et al., 1991), EBNA of EBV, and E6 and E7 of the human papilloma virus (Chen et al.,

1992; Hu et al., 1992; Shepherd et al., 1992; Stauss et al., 1992; Steele et al., 1993).

Idiotypic epitopes such as Ig idiotypes or TCR idiotypes are potential antigens as well. Nonmutated cellular proteins may also be antigenic if they are expressed aberrantly (e.g. in an inappropriate subcellular compartment) or in higher than normal quantities (Schild et al., 1990). Finally, totally normal proteins could be recognized in a tumor by autoreactive T cells such as a viral *env* protein in a mouse transgenic for *env* (Hu et al., 1993) or the tyrosinase enzyme in patients with cancer (Dalianis, 1990; Anichini et al., 1993; Brichard et al., 1993).

How tumor cells process and present antigen to T cells

Based on significant evidence that potential tumor antigens exist, how are these TAA processed and presented for recognition by T_{CD8+}? The cell biology and the biochemistry of class I have recently been extensively reviewed (Yewdell and Bennink, 1992). Most notably for the present discussion, it has been shown that T_{CD8+} recognize MHC class I molecules bearing peptides of 8-10 residues (Rötzschke et al., 1990) derived from proteins located in the cytosol. Antigenic peptides are believed to be generated from a cytosolic pool of proteins. Association of these peptides with class I α chains and β_2-microglobulin (β_2m) occurs in the endoplasmic reticulum (ER) or in an early secretory compartment.

Generation of the peptide fragments of the ultimate length of 8-10 residues is likely to involve the unfolding of proteins containing antigenic epitopes, but this requirement for unfolding is entirely speculative. There is evidence that antigenic fragments presented by MHC class I molecules are generated by proteolysis, but very little is known about the molecular structures involved (Yewdell and Bennink, 1992). Ubiquitin targeted proteolysis may play a role here (Goldberg and Rock, 1992; Michalek et al., 1993).

There is circumstantial evidence that a large nuclear and cytosolic proteolytic structure (termed the proteasome), physically associated with two MHC gene products, called LMP-7 and LMP-2 (formerly known as RING 10 and RING 12, respectively), is involved in the production of antigenic peptides or their delivery to class I molecules (Kelly et al., 1991; Martinez and Monaco, 1991; Trowsdale et al., 1991). Two recent reports, however, indicate that the LMP-2 and LMP-7 gene products are not necessary for the presentation of some antigens (Arnold et al., 1992; Momburg et al., 1992). The function of these gene products in antigen processing, therefore, remains to be established.

Peptide production occurs mainly in the cytoplasm (Townsend et al., 1988), although as with many biological processes, there appear to be exceptions (the HIV-1 envelope protein has been reported to be processed in the ER (Hammond et al., 1993). In any case, the vast majority of peptide fragments are generated in the cytoplasm and associate with class I molecules in the ER. This raises the question of how these peptides are translocated across the ER membrane. Two gene products that are clearly involved are encoded in the class II region of the MHC, and they are now designated TAP-1 and TAP-2 (for *T*ransporter associated with *A*ntigen *P*rocessing) (Monaco, 1992). Based on sequence homology, these genes are members of the ABC transporter family. TAP-1 (previously known as RING 4, Y3, and PSF-1) and TAP-2 (previously known as RING 11, Y1, and PSF-2) are clearly needed for cells to efficiently process antigen (Deverson et al., 1990; Monaco et al., 1990; Spies et al., 1990; Trowsdale et al., 1990). Based on size and predicted structure, it seems likely that these proteins directly transport peptide from the cytosol. Consistent with this notion, it has recently been shown that the MHC encoded transporters are located in the endoplasmic reticulum and *cis*-Golgi (Kleijmeer et al., 1992).

MHC class I molecules clearly play some role in the generation of properly sized antigens (Falk et al., 1990; Rötzschke et al., 1990). Rammensee and his group have shown that the peptide fragments recovered from an antigen presenting cell are dependent on the class I molecules expressed by that cell. There are two competing explanations for this phenomenon: the first is that cytoplasmic proteins are broken down to generate extremely short-lived intermediates, and class I molecules "capture" the right fragments, protecting them from further degradation. Alternatively, cytoplasmic proteins could be broken down into longer peptide fragments that are held by class I molecules while cellular proteases trim them to the appropriate lengths. This could be called the "cookie cutter" hypothesis.

One member of the heat shock protein family of proteins (HSP), called gp96, has been proposed by Srivastava to be capable of binding peptides that could be subsequently presented by class I molecules (Srivastava and Heike, 1991). These HSP may help to stabilize unfolded class I heavy chain molecules in the ER, then transfer peptides to class I molecules, leading to the formation of the stable trimolecular complex of class I α-chain, β_2m and peptide antigen.

The genes coding for class I α-chains, the putative peptide transporters, and the proteasome component molecules appear to be very closely associated and are in the MHC region on chromosome 17 in the mouse, and on chromosome 6 in the human (Trowsdale et al., 1991). Importantly, each of these genes, along with β_2m and gp96 is inducible by IFN-γ (Trowsdale et al., 1990). Thus, IFN-γ could increase the processing and presentation of antigenic peptides by enhancing proteolysis via the proteasome, by increasing the transport of these antigenic peptide fragments into the ER, by enhancing the binding of these molecules to an intermediate molecule like gp96, and by increasing the expression of MHC class I and β_2m.

Many tumors process intracellular antigens poorly

Because of the requirement for antigen processing, some tumor cells may escape recognition by T cells because of a failure to process and present TAA. Evidence for this mechanism will be presented in detail below. Thus, a tumor might evade recognition by T cells by failing to present tumor antigens which were not absent, but rather "hidden" intracellularly and not presented on the cell surface. A tumor cell does not have to have an absolute inability to process and present antigen for it to escape from immune recognition; it simply has to fail to activate an immune response. The antigen density required for the stimulation or afferent limb of the immune response is clearly much higher than is required for the recognition of a target cell (efferent immune response).

Several molecular mechanisms by which tumor cells fail to process and present endogenous antigens to T_{CD8+} could be hypothesized (Table 22.1). Using isolates of tumor cells from patients at the National Institutes of Health to establish cell lines for in vitro studies and fresh tumor sections for immunohistochemical studies and in vivo correlates, we have identified that at least three different mechanisms are operative. The first mechanism, found in five different melanoma cell lines (N.P. Restifo, unpublished observations) involves the loss of functional β_2m expression, similar to that seen by Ferrone's group (D'Urso et al., 1991). This antigen presentation defect is entirely correctable in these cell lines by gene insertion of β_2m.

A second mechanism, also observed in melanoma, involves the loss of expression of particular MHC class I alleles (F. Marincola, unpublished observations), and is again correctable by gene insertion of the missing α-chain genes. The third type of defect observed, originally in the small cell lung cancer histology, involved the down-regulation of certain molecules that putatively process antigens, including TAP and MHC encoded proteasome components (Restifo et al., 1993). This mechanism is correctable with treatment with exogenous IFN-γ or with gene insertion of the IFN-γ gene. It is this final mechanism that will be discussed in detail below.

One very simple way to measure the antigen processing and presenting ability of a tumor cell is to infect the tumor cell with a virus producing high levels of endogenous antigens (e.g. influenza virus or vaccinia virus) and then to measure how efficiently the tumor cells present this antigen. In our attempts to examine the capacity of human tumor cell lines to process cytosolic antigens for T_{CD8+} recognition, we were faced with the problem of the massive polymorphism of the MHC. Since class I molecules are among the most polymorphic genes in the human genome, we were unable to consistently use one class I molecule to present a single antigenic epitope for recognition by a single population of effector cells to study the antigen processing machinery. Instead, early studies involved the cumbersome task of tailoring reagents for the particular MHC alleles involved.

Table 22.1. *Mechanisms of escape of human tumor cells from recognition by T_{CD8+} relating to antigen processing and presentation*

Gene lost, mutated or underexpressed	Result	Cell surface class I
TAP 1, 2 (LMP, proteases?)	Poor processing of peptide antigens	Decreased to normal
β_2m	α-chain instablity	Absent
α-chain alleles	No presentation	Absent to normal

We thus devised a method for screening a large number of tumor cell isolates that was independent of both the HLA type of the tumor and the presence, or absence, of specific cellular proteins. This method exploits the capacity of recombinant vaccinia virus (rVV) to express large quantities of a desired protein and to infect a wide variety of cells (Moss, 1991). Utilizing a rVV encoding the mouse H-2 K^d class I molecule (K^d-VV), we tested large panels of human tumor cell lines for presentation of viral antigens to mouse K^d restricted, VV specific T_{CD8+} populations. We could thus study antigen processing capabilities of human tumor cells per se, without dependence on the expression of any given class I by the tumor cell. We found that human tumor cell lines varied widely in their abilities to process and present viral antigens to T_{CD8+}. Notably, all six of the small cell lung carcinomas (SCLC) studied were consistently recognized at levels similar to, or even lower than, the antigen processing mutant T2 cells.

Pulse-chase methodology was utilized to study the biosynthesis and intracellular transport of class I molecules and revealed that class I molecules remained completely sensitive to digestion with endo H throughout an 80-minute chase period. This indicated that class I molecules were not transported through the Golgi complex to the cell surface. However, IFN-γ treatment of SCLC cells greatly altered the intracellular trafficking of class I molecules, inducing the transport of a substantial portion of K^d molecules from the early portion of the secretory pathway through the Golgi complex. This was shown to be the case since approximately half of the class I molecules immunoprecipitated became resistant to endo H digestion within 80 minutes of their synthesis. The acceleration of the intracellular transport of class I molecules was associated with an increased steady state class I expression on the cell surface.

Retention of class I molecules in an endo H sensitive form has been observed in cells that fail to express TAP or MHC proteasome subunit gene products. When the RNA from the SCLC lines was probed for the expression of message from the TAP and MHC proteasome subunit molecules, no mRNA could be detected. Our inability to detect these relevant mRNAs by Northern blot analysis did not indicate a deletion of the genes in these cells, since treatment of cells with IFN-γ induced the expression of large quantities of mRNA encoding the four genes. Most importantly, IFN-γ greatly enhanced the antigen presenting capacity of the SCLC, but not T2 cells, in functional assays of antigen presentation (Restifo et al., 1992).

Although this was only a model system, the functional effects that IFN-γ had on antigen processing in tumor cells suggested that this cytokine might be very useful in enhancing the ability of a tumor to present its own endogenous TAA.

Immunogenicity correlates with antigen presentation capability

Based on our findings in the human tumor cell setting, antigen processing and presentation by tumor cells was very poor in the worst cases and only marginally better in many other cases. In order to test a therapeutic intervention in animals, we studied cases of extremely poor antigen processing in order to learn about the biologic mechanisms potentially involved in the escape of tumor cells from the immune response. Of eight murine tumors that have been extensively characterized in our laboratory (Table 22.2), MCA 38, 105, 203, 205 and 207 were described as immunogenic since, after a simple immunization of tumor in *Corynebacterium parvum*, they protected against subsequent challenge with the same tumor. MCA 102 had an intermediate phenotype; it induced a CTL response, but it did not protect against subsequent challenge. At opposite ends of the distribution of class I expression were MCA 38 and MCA 101 or MCA 106, which exhibited a 100 fold difference in steady state levels of cell surface class I.

Similar results were obtained using either fresh tumor explants or cells that had been maintained in tissue culture, with the exception of MCA 106, which expressed higher levels of class I in vivo. MHC class II molecules were not detected on the surfaces of any of these tumors by FACS analysis. Thus, MCA 101 expressed very low levels of class I and neither induced CTL nor protected against tumor challenge.

In studies using viral model systems, in which tumor cell lines were infected with influenza A/PR/8/34 (PR8) virus, MCA 101 was found to be nearly incapable of presenting endogenous antigens to anti-viral CTL, despite the fact that MCA 101 was extremely well infected. The antigen processing ability of MCA 101 was found to be as poor as the antigen

Table 22.2. *Summary of in vitro and in vivo characteristics of MCA induced sarcomas*

Tumor	Antitumor CTL in vitro	Immunogen in vivo	Relative MHC class I	Processing of flu virus[a]
MCA 38	Yes	Yes	423 ± 12	37
MCA 207	Yes	Yes	249 ± 4	54
MCA 205	Yes	Yes	110 ± 3	48
MCA 203	Yes	Yes	64 ± 2	32
MCA 105	Yes	Yes	32 ± 9	21
MCA 102	Yes	No	28 ± 3	38
MCA 101	No	No	7 ± 1	4
MCA 106	Yes	Yes	4 ± 1	10

[a]Percentage of killing at a 30:1 effector to target ratio

processing ability of the mutant RMA-S, which was included as a negative control. Several of the other tumors tested presented endogenously generated viral antigens poorly. Since T_{CD8+} have a role in the immune response against MCA induced tumors, it was not surprising that many of the tumors tested had a reduced ability to present endogenous viral antigens. In fact, one would expect that tumor cells with antigen processing or presentation defects would enjoy a selective advantage. Importantly, as with the SCLC cell lines, the presentation of endogenous antigens by MCA 101 was made possible by up-regulating class I expression with IFN-γ (Restifo et al., 1991; Restifo et al., 1992). Thus, the murine tumor MCA 101 presented endogenous antigens poorly, grew rapidly and lethally in nonimmunosuppressed mice, and could neither act as an immunogen in vivo nor generate CTL in vitro (Mule et al., 1987; Restifo et al., 1991). The positive effects of exogenously provided IFN-γ, however, proved to be transient: class I levels returned to low levels after only 48 h. We thus needed a way of increasing the antigen processing capability of a tumor which lasted more than a couple of days.

Immunogenetherapy designed to correct deficient antigen processing

In the next series of experiments (Restifo et al., 1992), we hypothesized that IFN-γ could be used to enhance a cell line ability to process its own tumor antigens. This hypothesis was tested by transducing the cells with the cDNA for murine IFN-γ. Our aim with the gene insertion studies was to use IFN-γ, not for its direct effects on cells of immune lineage, but instead for its effects on the antigen presentation capabilities of the tumor cells before and after gene modification. We then hoped to correlate the in vitro antigen processing capabilities with the in vivo behavior of these cells as immunogens. Using this system, it was thus possible to address the question of whether MCA 101 was nonimmunogenic because of its failure to present some postulated, but as yet unidentified, TAA.

IFN-γ transduced MCA 101 expressed higher levels of surface class I molecules as measured by FACS than the wild-type MCA 101 in both fresh and cultured tumor lines. Importantly, IFN-γ gene modification of MCA 101 converted it from a poor presenter of a variety of influenza A antigens to a cell line capable of presenting these antigens as efficiently as the most immunogenic tumors studied. We then tested its ability to present its own tumor antigens in vivo. In order to obtain pure cultures of uniformly high class I expressing transductants, the bulk transfected MCA 101 (101.NAT) was cloned. We found that the insertion of the cDNA for IFN-γ caused a 100 fold increase of cell surface class I in some of the tumor clones tested. When clones were implanted s.c. into syngeneic C57BL/6 mice, no significant differences between the growth rates of high class I expressing and low class I expressing tumors in vivo were found.

Antitumor T cells could only be grown from high class I expressing transductants. This was done by harvesting s.c. tumors after 10-20 days, then immunobeading single cell suspensions of tumor digests using immunomagnetic beads coated with anti-Thy 1.2^{+} Ig. Like other murine T-cell cultures that originate from the tumor, sometimes called TIL, the TIL cultures used in vivo were found to be exclusively T_{CD8+} by FACS. These antitumor CTL were then expanded in vitro in 20 U/ml of recombinant IL-2. Most importantly, TIL generated from the murine IFN-γ gene modified MCA 101 tumor clones expressing high levels of class I were effective against pulmonary metastases, not only against the gene modified MCA 101, but also against metastases from the wild-type MCA 101 tumor as well.

TIL generated from IFN-γ gene modified tumor were tested for cytokine release after coincubation with appropriate tumor cells. Wild-type MCA 101 stimulated release of IFN-γ and TNF-α almost as well as IFN-γ gene modified MCA 101. Thus, TIL generated from murine IFN-γ gene modified tumor clones could be triggered to secrete, in a relatively tumor specific fashion, IFN-γ and TNF-α against not only high class I expressing tumor clones but also against the low class I expressing wild-type tumor. Interestingly, the MCA 101 tumor was not consistently lysed by therapeutic TIL. One possibility for this effect was that the measurement of cytokine release was more sensitive than assays of cytotoxicity. This finding was consistent with previous data which showed that $CD8^{+}$ TIL that were therapeutic in vivo were not in every case cytotoxic (Barth et al., 1991).

Mechanisms of action of IFN-γ on tumor cells

The enhanced antigen processing by IFN-γ transduced MCA 101 is likely to result in an increased antigen density on IFN-γ gene modified tumor cells. This may explain why a therapeutically useful T_{CD8+} response can be obtained from high class I expressing clones and not from low class I expressing MCA 101 clones. This increased antigen density may, in fact, be an immunologically "strong" afferent tumor stimulus that can elicit a response against an immunologically much "weaker" efferent stimulus. This effect has been previously demonstrated in another tumor system (LeGrue et al., 1989).

In exploring the possible mechanisms of the effect of IFN-γ in our model system, one of the difficult aspects is that it has pleiotropic effects. Our studies have focused on IFN-γ enhancement of antigen processing. IFN-γ has also been shown to increase surface levels of important adhesion molecules like ICAM-1 and to have important effects on the differentiation of both T_{CD8+} and T_{CD4+}.

In binding experiments with ^{125}I-labeled IL-2, allocultures set up in the presence of anti-IFN-γ mAb display a decreased number of low affinity and almost no high affinity IL-2 receptor as compared with control cultures (Giovarelli et al., 1988). Furthermore, blockade of physiologically secreted IFN-γ with mAb or with IFN-γ receptor has been shown to inhibit human T lymphocyte and NK cell activation.

IFN-γ genes have been introduced into a low H-2K^b class I expressing tumor cell (Porgador et al., 1993). Some high IFN-γ secretors and some low IFN-γ secreting clones were isolated. While both groups expressed high levels of class I, only the high secretors showed a decrease in tumorigenicity and metastatic growth. Both groups induced similar high levels of CTL and protection against challenge with the parental tumor. These authors observed reduced tumorigenicity in CD1 nude mice, thus providing evidence for the involvement of non-T cell effectors in the response to IFN-γ secreting tumors.

Another possibility is that IFN-γ could act primarily through up-regulation of class II expression on macrophages and dendritic cells, thereby enhancing antigen presentation to T helper cells. IFN-γ may alter the responsiveness of other immune cells. For example, the observed effects of tumor transduction with the gene for IFN-γ may be due partially or entirely to the enhancement of Th1 responses with a concomitant diminution of Th2 responses. This conversion has been shown to occur under the influence of IFN-γ in other systems (Coffman et al., 1991). The relative contributions of these mechanisms are currently being addressed in our laboratory.

Recent reports about IFN-γ knock-out mice (Dalton et al., 1993) and IFN-γ receptor knock-out mice (Huang et al., 1993) have been very helpful to investigators studying the role of IFN-γ in the generation of CTL. Importantly, in both of the new studies CTL were shown to be normal. Thus, IFN-γ is clearly not necessary for the differentiation of CTL. In fact, there is evidence that it can be harmful. For example, while the addition of IFN-γ to peritoneal exudate cells enhances their ability to activate T cells, the presence of IFN-γ during the culture period inhibits the proliferation of these same T-cell cultures (Duong et al., 1992). Liu and Janeway have shown that IFN-γ can induce T-cell death upon ligation of the TCR. This was shown to occur in the absence of costimulatory cells and can be prevented by the addition of adherent spleen cells (Liu and Janeway, 1990).

Current work and future directions

Other investigators have recently shown the antimetastatic effects of vaccination of tumor bearing mice by immunization with IFN-γ gene modified tumor cells. Furthermore, the use of tumor cells gene modified with the cDNA for IFN-γ has been extended, and partially confirmed, in our own laboratory (Shiloni et al., 1994). In these studies, one high (4JK) and one low (24JK) class I expressing clones of the nonimmunogenic MCA-102 fibrosarcoma were retrovirally transduced with the cDNA encoding for murine IFN-γ. Retroviral transduction of tumor cells with the cDNA encoding for IFN-γ resulted in a substantial up-regulation of class I surface expression in 24JK but little change in 4JK. In an attempt to generate antitumor lymphocytes, these gene modified cells were inoculated into mouse footpads, and draining lymph nodes (DLN) were removed, dispersed, and cultured in vitro for 10 days with irradiated tumor cells and IL-2. These DLN cells were used to treat recipients harboring 3-day pulmonary metastases induced by the parental, unmodified tumor.

Treatment with DLN cells obtained following the injection of 24JK tumor cells modified with the gene for IFN-γ significantly reduced the number of pulmonary metastases in four separate experiments, compared to groups treated by DLN cells generated from inoculation of either the unmodified, parental 24JK clone, or the same clone transduced with the Neo^R gene only. In contrast, DLN cells induced either by IFN-γ transduced 4JK (high MHC class I expressing) or an unmodified 4JK tumor (moderate MHC class I expressing) had significant but equal therapeutic efficacy.

Although the in vitro growth rate of tumor cell lines was unaffected by the insertion of the mouse IFN-γ cDNA, their in vivo (s.c.) growth rates were significantly slower than those of the nontransduced tumors. Thus, after retroviral transduction of the murine IFN-γ cDNA into a nonimmunogenic tumor with a very low level of surface expression of MHC class I, modified tumor cells could elicit therapeutic T cells from DLN capable of successfully treating established pulmonary metastases upon adoptive transfer. This strategy confirms previous observations on the potential therapeutic effects of gene modification of tumor cells with

IFN-γ. Significantly, these studies further define the circumstances in which IFN-γ immunogenetherapy might be useful.

How can the positive effects of IFN-γ be separated from the antiproliferative effects that IFN-γ has on T cells? Genetic or pharmaceutical therapies directed at the enhancement of antigen processing may exploit the beneficial effects of IFN-γ. It seems likely that the genes involved in antigen processing and presentation share common regulatory elements. When these regulatory elements are elucidated, new therapies could be developed for specific up-regulation of antigen processing in cancer and infectious disease. Conversely, down-regulation of these gene products may prove useful in tissue transplantation or in autoimmune disease.

DNA binding proteins may be one solution to this problem. There are several candidate molecules which may enhance the ability of tumor cells to present endogenous TAA to T_{CD8+}. For example, cotransfection of H-2RIIBP (RXRβ) with T3Rα or RARα synergistically enhanced MHC class I transcription in murine cell lines (Marks et al., 1992). Eisenbach's group has shown that stable transfection of a p65 expression vector into a low class I expressing clone of the Lewis lung carcinoma activated H-2 transcription (Plaaksin et al., 1993). Most importantly, this resulted in increased levels of H-2 cell surface expression. These studies suggested that KBF1 had the potential to down-regulate class I gene expression, whereas dimers containing the p65 subunit were activators of class I gene expression.

These studies are potentially interesting and important to the immunotherapy of cancer since they may have implications for the up-regulation of MHC class I molecules and possibly other molecules important in antigen processing, like the TAP gene products, without the use of IFN-γ. DNA binding proteins, not secreted into the extracellular matrix, could be used to have direct effects on tumor cells without having deleterious effects on the immune system. Perhaps such regulatory proteins could be targeted to one particular tissue. For example, tissue specific gene expression was targeted to melanoma cells in vivo and in vitro using the 5' flanking regions of the tyrosinase or the tyrosinase related protein 1 genes (Vile and Hart, 1993). This strategy may point the way toward in situ gene therapy.

In conclusion, new immunogenetherapeutic strategies may be based on a more detailed understanding of the molecular mechanisms used by tumors to escape from immune recognition. A more complete knowledge of the cell biology of MHC class I in health and disease can lead to new strategies aimed at enhancing processing and presentation of tumor antigens. These approaches may lead to the generation of immune responses against tumor histologies not previously thought to be susceptible to T-cell based immunotherapies of cancer.

Acknowledgements

I would like to thank Steven A. Rosenberg, Jack R. Bennink, Jonathan W. Yewdell, Eli Gilboa, Yutaka Kawakami, Akash Taggarse and Fernando Esquivel for creating and supplying reagents without which these studies could not have been performed. Most importantly, they have made a tremendous intellectual contribution to the work described above. I would also like to thank Steve Kishter and John Wunderlich for reading this manuscript.

References

Anichini A, Maccalli C, Mortarini R, Salvi S, Mazzocchi A, Squarcina P, Herlyn M and Parmiani G (1993). Melanoma cells and normal melanocytes share antigens recognized by HLA-A2-restricted cytotoxic T cell clones from melanoma patients. J Exp Med **177**, 989-998.

Arnold D, Driscoll J, Ardrolewicz M, Hughes E, Cresswell P and Spies T (1992). Proteasome subunits encoded in the MHC are not generally required for the processing of peptides bound by MHC class 1 molecules. Nature **360**, 171-174.

Barth RJ, Mule JJ, Spiess PJ and Rosenberg SA (1991). Interferon gamma and tumor necrosis factor have a role in tumor regressions mediated by murine CD8+ tumor-infiltrating lymphocytes. J Exp Med **173**, 647-658.

Bjorkman PJ, Saper MA, Samraoui B, Bennett WS, Strominger JL. and Wiley DC (1987a). Structure of the human class I histocompatibility antigen HLA-A2. Nature **329**, 506-512.

Bjorkman PJ, Saper MA, Samraoui B, Bennett WS, Strominger JL and Wiley DC (1987b). The foreign antigen binding site and T cell recognition regions of class I histocompatibility antigens. Nature **329**, 512-518.

Brichard V, Van Pel A, Wölfel T, Wölfel C, De Plaen E, Lethe B, Coulie P and Boon T (1993). The tyrosinase gene codes for an antigen recognized by autologous cytolytic T lymphocytes on HLA-A2 melanomas. J Exp Med **178**, 489-495.

Brown JH, Jardetzky TS, Gorga JC, Stern LJ, Urban RG, Strominger JL and Wiley DC (1993). Three-dimensional structure of the human class II histocompatibility antigen HLA-DR1. Nature **364**, 33-39.

Chen L, Mizuno MT, Singhal MC, Hu SL, Galloway DA, Hellstrom I and Hellstrom KE (1992). Induction of cytotoxic T lymphocytes specific for a syngeneic tumor expressing the E6 oncoprotein of human papillomavirus type 16. J Immunol **148**, 2617-2621.

Cho KR and Vogelstein B (1992). Genetic alterations in the adenoma-carcinoma sequence. Cancer **70**, 1727-1731.

Coffman RL, Varkila K, Scott P and Chatelain R (1991). Role of cytokines in the differentiation of CD4+ T-cell subsets in vivo. Immunol Rev **123**, 189-201.

D'Urso CM, Wang Z, Cao Y, Tatake R, Zeff RA and Ferrone S (1991). Lack of HLA class 1 antigen expression by cultured melanoma cells F0-1 due to a defect in β_2m gene expression. J Clin Invest **87**, 284-292.

Dalianis T (1990). Studies on the polyoma virus tumor-specific transplantation antigen (TSTA). Adv Cancer Res **55**, 57-85.

Dalton DK, Pitts-Meek S, Keshav S, Figari IS, Bradley A and Stewart TA (1993). Multiple defects of immune cell function in mice with disrupted interferon-gamma gene. Science **259**, 1739-1742.

De Plaen E, Lurquin C, Van Pel A, Mariamé B, Szikora J-P, Wölfel T, Sibille C. Chomez P and Boon T (1988). Immunogenic (tum-) variants of mouse tumor P815: Cloning of the gene of tum- antigen P91A and identification of the tum- mutation. Proc Natl Acad Sci USA **85**, 2274-2278.

Deverson E, Gow IR, Coadwell WJ, Monaco JJ, Butcher GW and Howard JC (1990). MHC class II region encoding proteins related to the multidrug resistance family of transmembrane transporters. Nature **348**, 738-741.

Duong TT, Finkelman FD and Strejan GH (1992). Effect of interferon-gamma on myelin basic protein-specific T cell line proliferation in response to antigen-pulsed accessory cells. Cell Immunol **145**, 311-323.

Elliot BE, Carlow DA, Rodricks A and Wade A (1989). Perspectives on the role of MHC antigens in normal and malignant cell development. Adv Cancer Res **53**, 181-245.

Falk K Rötzschke O and Rammensee HG (1990). Cellular peptide composition governed by major histocompatibility complex class I molecules. Nature **348**, 248-251.

Finn OJ (1992). Antigen-specific MHC-unrestricted T cells. Biotherapy **4**, 239-249.

Giovarelli M, Santoni A, Jemma C, Musso T, Giuffrida AM, Cavallo G, Landolfo S and Forni G (1988). Obligatory role of IFN-gamma in induction of lymphokine-activated and T lymphocyte killer activity but not in boosting of natural cytotoxicity. J Immunol **141**, 2831-2836.

Goldberg AL and Rock KL (1992). Proteolysis proteasomes and antigen presentation. Nature **357**, 375-379.

Greenberg PD (1991). Adoptive T cell therapy of tumors: Mechanisms operative in the recognition and elimination of tumor cells. Adv Immunol **49**, 281-355.

Hammond SA, Bollinger RC, Tobery TW and Siliciano RF (1993). Transporter-independent processing of HIV-1 envelope protein for recognition by CD8+ T cells. Nature **364**, 158-161.

Hu SL, Hellstrom I and Hellstrom KE (1992). Recent advances in antitumor vaccines. Biotechnology **20**, 327-343.

Hu J, Kindsvogel W, Busby S, Bailey MC, Shi Y-Y and Greenberg PD (1993). An evaluation of the potential to use tumor-associated antigens as targets for antitumor T cell therapy using transgenic mice expressing a retroviral tumor antigen in normal lymphoid tissues. J Exp Med **177**, 1681-1690.

Huang S, Hendriks W, Althage A, Hemmi S, Bluethmann H, Kamijo R, Vilcek J, Zinkernagel RM and Aguet M (1993). Immune response in mice that lack the interferon-gamma receptor. Science **259**, 1742-1745.

Itoh K, Platsoucas DC and Balch CM (1988). Autologous tumor-specific cytotoxic T lymphocytes in the infiltrate of human metastatic melanomas: Activation by interleukin 2 and autologous tumor cells and involvement of the T cell receptor. J Exp Med **168**, 1419-1441.

Kaplan DR, Griffith R, Braciale VL and Braciale TJ (1984). Influenza virus-specific human cytotoxic T cell clones: Heterogeneity in antigenic specificity and restriction by class II MHC products. Cell Immunol **88**, 193-206.

Kelly A, Powis SH, Glynne R, Radley E, Beck S and Trowsdale J (1991). Second proteasome-related gene in the human MHC class II region. Nature **353**, 667-668.

Kleijmeer MJ, Kelly A, Geuze HJ, Slot JW, Townsend A and Trowsdale J (1992). Location of MHC-encoded transporters in the endoplasmic reticulum and cis-Golgi. Nature **357**, 342-344.

Knowles BB, Faas S, Juretic A, Fox N, Crooke R, Hanahan D, Solter D and Jewett L (1991). SV40 T antigen transgenic mice: Cytotoxic T lymphocytes as a selective force in tumor progression. Basic Life Sci **57**, 111-124.

LeGrue SJ, Ananthaswamy HN and Simick WJ (1989). Afferent and efferent specificity in the induction and elicitation of parental cross-protective immunity by an immunogenic murine tumor variant: Associative recognition of a unique tumor-specific antigen on somatic cell hybrids. Cancer Res **49**, 4747-4758.

Liu Y and Janeway C (1990). Interferon gamma plays a critical role in induced cell death of effector T cells: A possible third mechanism of self-tolerance. J Exp Med **172**, 1735-1739.

Marks MS, Hallenbeck PL, Nagata T, Segars JH, Appella E, Nikodem VM and Ozato K (1992). H-2RIIBP (RXR-beta) heterodimerization provides a mechanism for combinatorial diversity in the regulation of retinoic acid and thyroid hormone responsive genes. EMBO J **11**, 1419-1435.

Marrack P, Winslow GM, Choi Y, Scherer M, Pullen A, White J and Kappler JW (1993). The bacterial and mouse mammary tumor virus superantigens; Two different families of proteins with the same functions. Immunol Rev **131** 79-92.

Martinez CK and Monaco JJ (1991). Homology of proteasome subunits to a major histocompatibility complex-linked LMP gene. Nature **353**, 664-667.

Michalek MT, Grant EP, Gramm C, Goldberg AL and Rock KL (1993). A role for the ubiquitin-dependent proteolytic pathway in MHC class I-restricted antigen presentation. Nature **363**, 552-554.

Mizoguchi H, O'Shea JJ, Longo DL, Loeffler CM, McVicar DW and Ochoa AC (1992). Alterations in signal transduction molecules in T lymphocytes from tumor-bearing mice. Science **258**, 1795-1798.

Momburg F, Ortiz-Navarrete V, Neefjes J, Goulmy E, Van de Wal Y, Spits H, Powis SJ, Butcher GW, Howard JC, Walden P and Hämmerling GJ (1992). Proteasome subunits encoded by the major histocompatibility complex are not essential for antigen presentation. Nature **360**, 174-177.

Monaco JJ, Cho S and Attaya M (1990). Transport protein genes in the murine MHC: Possible implications for antigen processing. Science **250**, 1723-1726.

Monaco JJ (1992). A molecular model of MHC class-I-restricted antigen processing. Immunol Today **13**, 173-179.

Moss B (1991). Vaccinia virus: A tool for research and vaccine development. Science **252**, 1662-1667.

Mule JJ, Yang JC, Lafreniere R, Shu S and Rosenberg SA (1987). Identification of cellular mechanisms operational in vivo during the regression of established pulmonary metastases by the systemic administration of high-dose recombinant interleukin 2. J Immunol **139**, 285-291.

Plaaksin D, Baeuerle PA and Eisenbach L (1993). KBF1 (p50 NF-kappaB homodimer) acts as a repressor of H-2K^b gene expression in metastatic tumor cells. J Exp Med **177**, 1651-1662.

Porgador A, Bannerji R, Wantanabe Y, Feldman M, Gilboa E and Eisenbach L (1993). Antimetastatic vaccination of tumor-bearing mice with two types of IFN-gamma gene-inserted tumor cells. J Immunol **150**, 1458-1470.

Potter TA, Rajan TV, Dick RF II and Bluestone JA (1989) Substitution at residue 227 of H-2 class I molecules abrogates recognition by CD8-dependent but not CD8-independent cytotoxic T lymphocytes. Nature **337**, 73-75.

Restifo NP, Esquivel F, Asher AL, Stötter H, Barth RJ, Bennink JR, Mulé JJ, Yewdell JW and Rosenberg SA (1991). Defective presentation of endogenous antigens by a murine sarcoma: Implications for the failure of an anti-tumor immune response. J Immunol **147**, 1453-1459.

Restifo NP, Spiess PJ, Karp SE, Mulé JJ and Rosenberg SA (1992). A nonimmunogenic sarcoma transduced with the cDNA for interferon gamma elicits CD8+ T cells against the wild-type tumor: Correlation with antigen presentation capability. J Exp Med **175**, 1423-1431.

Restifo NP, Esquivel F, Kawakami Y, Yewdell JW, Mulé JJ, Rosenberg SA and Bennink JR (1993). Identification of human cancers deficient in antigen processing. J Exp Med **177**, 265-272.

Rosenberg SA, Spiess P and Lafreniere R (1986). A new approach to the adoptive immunotherapy of cancer with tumor-infiltrating lymphocytes. Science **233**, 1318-1321.

Rosenberg SA, Packard BS, Aebersold PM, Solomon D, Topalian SL, Toy ST, Simon P, Lotze MT, Yang JC, Seipp CA, Simpson C, Carter C, Bock S, Schwartzentruber D, Wei JP and White DE (1988). Use of tumor infiltrating lymphocytes and interleukin-2 in the immunotherapy of patients with metastatic melanoma. Preliminary report. N Engl J Med **319**, 1676-1680.

Rötzschke O, Falk K, Deres K, Schild H, Norda M, Metzger J, Jung G and Rammensee H-G (1990). Isolation and analysis of naturally processed viral peptides as recognized by cytotoxic T cells. Nature **348**, 252-254.

Salter RD, Benjamin RJ, Wesley PK, Buxton SE, Garrett TPJ, Clayberger C, Krensky AM, Norment AM, Littman DR and Parham P (1990). A binding site for the T-cell co-receptor CD8 on the α_3 domain of HLA-A2. Nature **345**, 41-46.

Schild HO, Rötzschke O, Kalbacher H and Rammensee H-G (1990). Limit of T cell tolerance to self proteins by peptide presentation. Science **247**, 1587-1589.

Schreiber H, Ward PL, Rowley DA and Stauss HJ (1988). Unique tumor-specific antigens. Ann Rev Immunol **6**, 465-483.

Schwartzentruber DJ, Topalian SL, Mancini MJ and Rosenberg SA (1991). Specific release of granulocyte-macrophage colony-stimulating factor tumor necrosis factor-α and IFN-γ by human tumor-infiltrating lymphocytes after autologous tumor stimulation. J Immunol **146**, 3674-3681.

Shepherd PS, Tran TT, Rowe AJ, Cridland JC, Comerford SA, Chapman MG and Rayfield LS (1992). T cell responses to the human papillomavirus type 16 E7 protein in mice of different haplotypes. J Gen Virol **73**, 1269-1274.

Shiloni E, Karp SE, Custer MC, Shilyansky J, Restifo NP, Rosenberg SA and Mulé JJ (1994). Retroviral transduction of interferon-gamma cDNA into a nonimmunogenic murine fibrosarcoma: Generation of T cells in draining lymph nodes capable of treating established parental metastatic tumor. Cancer Immunol Immunother, in press.

Sibille C, Chomez P, Wildmann C, Van Pel A, De Plaen E, Maryanski JL, de Bergeyck V and Boon T (1990). Structure of the gene of tum-transplantation antigen P198: A point mutation generates a new antigenic peptide. J Exp Med **172**, 35-45.

Skipper J and Stauss HJ (1993). Identification of two cytotoxic T lymphocyte-recognized epitopes in the ras protein. J Exp Med **177**, 1493-1498.

Spies T, Bresnhan M, Bahram S, Arnold D, Blanck G, Mellins E, Pious D and DeMars R (1990). A gene in the human major histocompatibility complex class II region controlling the class I antigen presentation pathway. Nature **348** 744-747.

Srivastava PK and Heike M (1991). Tumor-specific immunogenicity of stress-induced proteins: Convergence of two evolutionary pathways of antigen presentation? Semin Immunol **3**, 57-64.

Stauss HJ, Davies H, Sadovnikova E, Chain B, Horowitz N and Sinclair C (1992). Induction of cytotoxic T lymphocytes with peptides in vitro: Identification of candidate T-cell epitopes in human papilloma virus. Proc Natl Acad Sci USA **89**, 7871-7875.

Steele JC, Stankovic T and Gallimore PH (1993). Production and characterization of human proliferative T-cell clones specific for human papillomavirus type 1 E4 protein. J Virol **67**, 2799-2806.

Szikora J-P, Van Pel A, Brichard V, Andre M, Van Baren N, Henry P, De Plaen E and Boon T (1990). Structure of the gene of tum- transplantation antigen P35B: Presence of a point mutation in the antigenic allele. EMBO J **9**, 1041-1050.

Tevethia SS (1990) Recognition of simian virus 40 T antigen by cytotoxic T lymphocytes. Mol Biol Med **7**, 83-96.

Topalian SL, Solomon D and Rosenberg SA (1989). Tumor-specific cytolysis by lymphocytes infiltrating human melanomas. J Immunol **142**, 3714-3725.

Townsend A, Bastin J, Gould K, Brownlee G, Andrew M, Coupar B, Boyle D, Chan S and Smith G (1988). Defective presentation to class I-restricted cytotoxic T lymphocytes in vaccinia-infected cells is overcome by enhanced degradation of antigen. J Exp Med **168**, 1211-1224.

Townsend A and Bodmer H (1989). Antigen recognition of class I restricted T lymphocytes. Ann Rev Immunol **7**, 601-624.

Trowsdale J, Hanson I, Mockridge I, Beck S, Townsend A and Kelly A (1990). Sequences encoded in the class II region of the MHC related to the "ABC" superfamily of transporters. Nature **348**, 741-744.

Trowsdale J, Ragoussis J and Campbell RD (1991). Map of the human MHC. Immunol Today **12**, 443-446.

Van den Eynde B, Lethe B, Van Pel A, De Plaen E and Boon T (1991). The gene coding for a major tumor rejection antigen of tumor P815 is identical to the normal gene of syngeneic DBA/2 mice. J Exp Med **173**, 1373-1384.

Van der Bruggen P, Traversari C, Chomez P, Lurquin C, De Plaen E, Van den Eynde B, Knuth A and Boon T (1991). A gene encoding an antigen recognized by cytolytic T lymphocytes on a human melanoma Science **254**, 1643-1647.

Vile RG and Hart IR (1993). In vitro and in vivo targeting of gene expression to melanoma cells Cancer Res **53**, 962-967.

Witte ON (1993). Role of the bcr-abl oncogene in human leukemia. Cancer Res **53**, 485-489.

Yanuck M, Carbone DP, Pendleton CD, Tsukui T, Winter SF, Minna JD and Berzofsky JA (1993). A mutant p53 tumor suppressor protein is a target for peptide-induced $CD8^+$ cytotoxic T-cells. Cancer Res **53**, 3257-3261.

Yewdell JW and Bennink JR (1992). Cell biology of antigen processing and presentation to MHC class I molecule-restricted T lymphocytes. Ann Rev Immunol **52**, 1-123.

23

GRANULOCYTE T-CELL COOPERATION IN THE REJECTION OF A COLON CARCINOMA TRANSDUCED WITH THE G-CSF GENE

M. P. Colombo, A. Stoppacciaro and L. Lombardi
Divisions of Experimental Oncology D and A, Istituto Nazionale Tumori, Milan, Italy and Department of Experimental Medicine, University La Sapienza, Rome, Italy

Introduction

G-CSF was first identified through its ability to induce the differentiation of the murine myelomonocytic leukemia cell line WEHI-3B, and then characterized as a potent differentiation inducing CSF, essential in granulopoiesis (Nicola et al., 1983). G-CSF shows a restricted action in the stimulation of hematopoietic progenitor cells since no eosinophil, erythroid or megakaryocyte colonies are induced in clonal culture assays (Metcalf and Nicola, 1983). Human and mouse G-CSF bind equally well to receptors of either species and show biological cross reactivity (Nicola et al., 1985). This made it possible to study the action of human G-CSF in mice, where it was shown capable of increasing splenic granulopoiesis, but also of stimulating erythropoiesis and megakaryocytopoiesis and of increasing the number of multipotential stem cells (CFU-S) (Moore et al., 1987; Tamura et al., 1987). Chronic exposure to the factor, studied in lethally irradiated mice reconstituted with syngeneic bone marrow infected with a retroviral vector expressing G-CSF cDNA, has been shown to induce nonneoplastic granulocytic and progenitor cell hyperplasia (Chang et al., 1989). Recombinant human G-CSF is widely used in clinical practice to sustain the granulocytic progenitor cell compartment in cancer patients undergoing cytoreductive treatment and for boosting granulocytic response after high dose chemotherapy (Gabrilove et al., 1988; Bronchud and Dexter, 1989).

G-CSF and granulocytes in cancer immunity

The immunological role of G-CSF in cancer immunity remained elusive until gene transfer experiments showed that its transduction into a

CYTOKINE-INDUCED TUMOR IMMUNOGENICITY
ISBN 0-12-261520-4

murine colon adenocarcinoma (C-26) was capable of inducing an immune response that inhibited tumor growth (Colombo et al., 1991). Furthermore, human recombinant G-CSF was found to inhibit the metastasis of hematogeneous and nonhematogeneous tumors in mice (Matsumoto et al., 1991). The antitumoral effect, in the case of C-26 cells transduced with the human G-CSF cDNA (C-26/G-CSF), is mediated by recruitment and targeting of neutrophilic granulocytes to G-CSF releasing cells and does not occur in irradiated mice. Thus, granulocytes, which are usually considered terminally differentiated cells active only during inflammation, have been found to exert antitumoral activity in vivo (Katano and Torisu, 1982; MacPherson and North, 1986; Midorikawa et al., 1990; Colombo et al., 1991). The ability to synthesize and release cytokines suggests their role as coregulators of the immune response (Lloyd and Oppenheim, 1992) through antigen presentation ability (Okuda et al., 1980; Del Pozo et al., 1992). PMN that express IL-1β, IL-1α (Tiku et al., 1986; Marucha et al., 1990; Lord et al., 1991) and TNF-α mRNA (Djeu et al., 1990; Dubravec et al., 1990) also synthesize and secrete the proteins when stimulated (Tiku et al., 1986; Djeu et al., 1990; Marucha et al., 1990; Lord et al., 1991). CSFs, IL-3, IL-8, TGF-α and TGF-β have been found to be expressed by either neutrophilic or eosinophilic PMN (reviewed by Lloyd and Oppenheim, 1992). More recently, IL-6, which is known to stimulate cell proliferation and cytotoxic function in T lymphocytes, has also been shown to be expressed by neutrophils (Melani et al., 1993).

G-CSF transduction into a colon carcinoma drives neutrophils to tumor cell contact

In order to transduce the G-CSF gene into the C-26 tumor, the human G-CSF cDNA (1.2 kb Hinc II/Bal I cDNA fragment) containing the complete coding region of human G-CSF (Tweardy et al., 1987) has been inserted into the Hind III cloning site of the NSV vector, a derivative from the N2 vector (Armentano et al., 1987). In this vector the G-CSF cDNA is under the control of the SV40 early promoter whereas Neo^R, used as a selectable marker, is driven by the Moloney murine leukemia virus long terminal repeat (LTR). Helper-free, virus-containing supernatants (titers ranging between 10^4 and 5×10^5 CFU/ml) were generated by transinfection using ψ2 echotropic and amphotropic PA317 packaging lines. One million C-26 target cells were infected for 4-12 h with undiluted supernatants containing 8 μg/ml polybrene grown for 48 h and then selected in 0.5 mg/ml G418. Individual clones were isolated, then expanded into cell lines, and injected into mice or subjected to further analysis.

Two G418 resistant clones were studied in detail. They were found to express the G-CSF gene weakly in Northern blot and to release G-CSF in the culture supernatant at quantities as low as 20 and 90 pg/ml per 10^6

cells in 48h, respectively (Colombo et al., 1991). Despite the low amount of G-CSF produced, these tumor cells do not form tumors when injected into syngeneic BALB/c mice, even at a dose 300-fold higher than the minimal lethal one, as determined with nontransduced cells. Loss of tumorigenicity was not due to impaired tumor proliferation; tumor nodules are, in fact, formed when injection is performed in sublethally irradiated mice. Tumor take inhibition occurs because of the recruitment and activation of PMN at the site of tumor injection (Fig. 23.1a). G-CSF is, thus, a chemotactic factor for PMN (Wang et al., 1988).

PMN at the tumor site are capable of expressing cytokine genes like IL-1α and IL-1β, and TNF-α, whereas IFN-γ and IL-4 mRNA were not detected, confirming the absence of T cells within the infiltrating leukocytes (Stoppacciaro et al., 1993).

When a mixture of virgin and G-CSF transduced C-26 were injected together the progressing tumors were formed exclusively by virgin C-26 cells. Genetic labelling of virgin C-26 cells with the β gal gene made it possible to demonstrate that PMN establish a cell contact only with G-CSF producing C-26 cell (Table 23.1) and that PMN remain within the tumor as long as G-CSF producing cells are present (7-10 days) (Colombo et al., 1992a). Furthermore, these cell mixture experiments suggest that PMN eliminate G-CSF producing C-26 cells in less than 10 days, a time that appears insufficient to recruit T lymphocytes and to mount a possible IgG response against human G-CSF.

This latter finding excludes ADCC as the mechanism of tumor destruction, while the former may explain why mice injected with C-26 producing G-CSF do not develop memory T cells.

G-CSF, via neutrophils, triggers a cascade of cytokines resulting in the rejection of an established tumor

An interesting feature of the C-26/G-CSF model is that these cells grow and form large tumors which progress until self reconstitution normalizes the leukocyte number in sublethally irradiated mice. At that time point (around 30 days after irradiation and tumor injection), tumor regression begins and proceeds to completion in about 70-80% of the treated mice (Colombo et al., 1991). We studied the mechanisms operating during tumor regression, thus comparing the inhibition of tumor take following injection of C-26/G-CSF cell/suspension (described above) and the rejection of an already established tumor (Stoppacciaro et al., 1993). During the growing phase the first cells which infiltrate the tumor are PMN (between day 15 and day 20), while macrophages and T cells become detectable after 20 and 30 days, respectively. Immunocytochemical analysis of the tumors at the onset of rejection shows that all leukocyte populations are present within the tumor (Fig. 23.2).

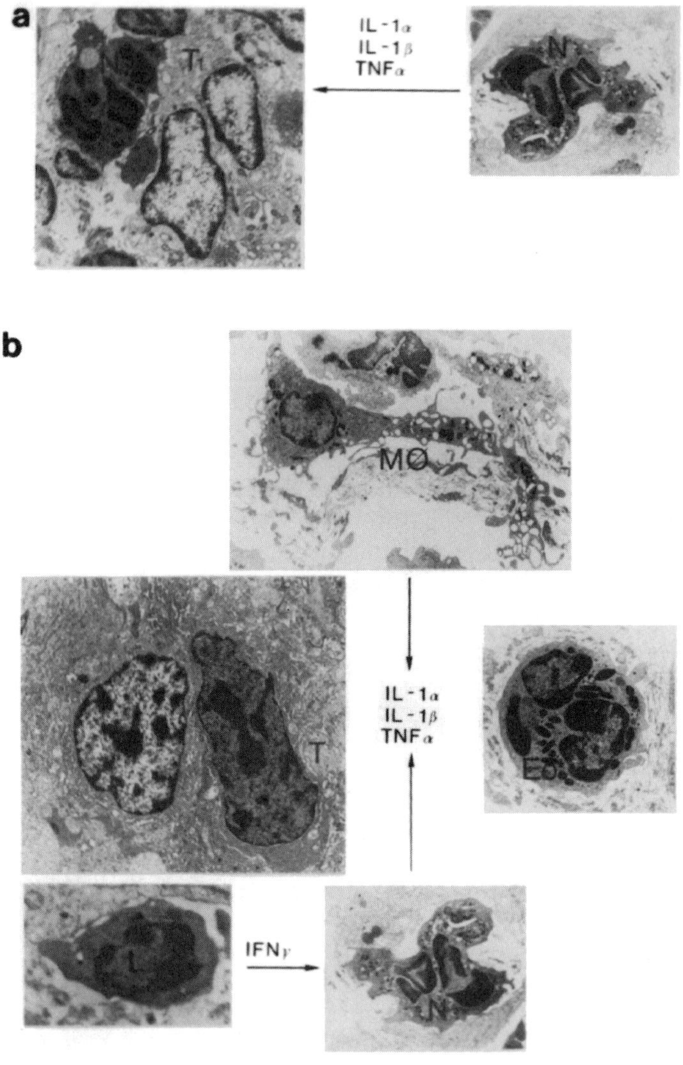

Figure 23.1. *Schematic representation of leukocyte infiltration and of the cytokines that are produced within the tumor environment during the rejection of C-26/G-CSF cells injected as a cell suspension (**a**), and during the regression of C-26/G-CSF cells initially grown in sublethally irradiated mice (**b**).* **T**, tumor; **N**, neutrophils; **Eo**, eosinophils; **L**, lymphocytes; **MO**, macrophages.

The presence of T lymphocytes has a functional role since T-cell depletion abrogated tumor rejection. In situ hybridization, performed to analyze cytokine gene expression in regressing tumors between day 38

Table 23.1. *Preferential contact between PMN and G-CSF producing (β gal⁻) tumor cells as detected by electron microscopy (EM)*

Tumor examined	No. of sections with informative areas examined[a]	No. of PMN in contact with β gal⁻ cells	No. of PMN in contact with β gal⁺ cells	No. of PMN without contact
1	8	6	0	4
2	7	8	0	2
3	10	9	0	4

[a]Informative areas are those where PMN, β gal⁻ and β gal⁺ cells are simultaneously visible

Methods: To prove that in our system PMN make intimate contact only with G-CSF producing cells leading to their growth inhibition, we marked the nonproducing C-26 cells by infection with a retroviral vector carrying the *Escherichia coli* LacZ gene (β gal) under the LTR promoter. In the presence of x-gal this gene gives the cells a blue color, and the electron dense indigo precipitate that localizes into the perinuclear cisterna and endoplasmic reticulum allows their detection by transmission EM. A mixture of G-CSF producing β gal⁻ and G-CSF nonproducing β gal⁺ cells at a 20:1 ratio was injected s.c. in syngeneic mice; EM was performed on samples collected at day 5, fixed and stained for β gal detection. Twenty-five sections, from three independent tumors, selected among those containing β gal⁺, β gal⁻ cells and PMN were analyzed. On a total number of 32 PMN counted in these sections, 23 were in intimate contact with β gal⁻, 9 were without contact and none was in contact with β gal⁺ cells. For details and figures, see Colombo et al. (1992a)

and day 43, identified neutrophils and macrophages expressing IL-1α, IL-1β and TNF-α mRNA, and cells with lymphoid morphology expressing mRNA for IFN-γ (Fig. 23.1b). IL-4 mRNA was not detected at the tumor site.

In order to identify the distinct role of CD8 and CD4 subpopulations, selective depletion with mAb was performed. Depletion of CD8⁺ cells abrogated tumor regression, and in situ hybridization showed the virtual absence of mRNA for IL-1α, IL-1β, TNF-α and IFN-γ in these tumors, suggesting that CD8⁺ T lymphocytes contribute some crucial function to the milieu of cytokine production by the TIL. Conversely, depletion of CD4⁺ cells did not appear to affect tumor regression. Immunocytochemical analysis of the CD4 antibody treated mice showed an infiltrate at the tumor site similar to that of non-CD4 depleted animals except for the absence of CD4⁺ cells. T-cell subset depletion suggested that CD8⁺ more than CD4⁺ T lymphocytes were expressing IFN-γ.

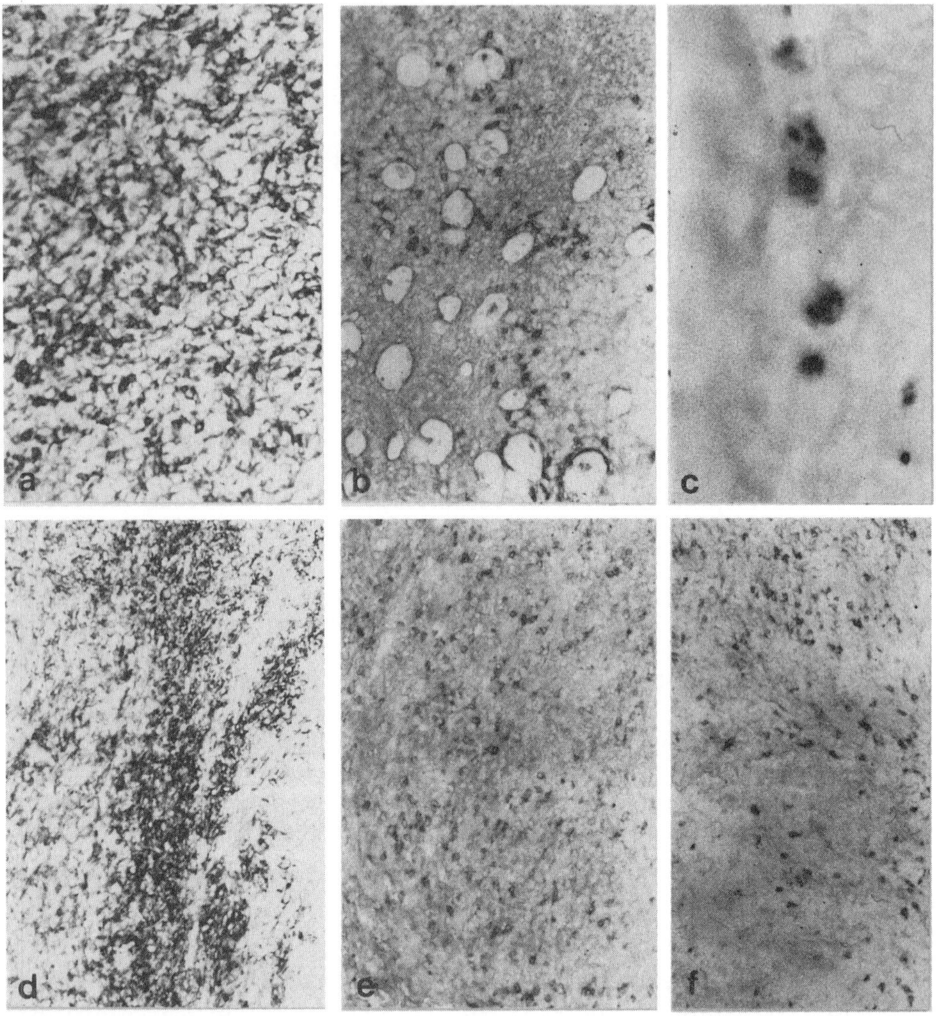

Figure 23.2. *Immunocytochemical characterization of the reactive infiltrate present in regressing C-26/G-CSF tumors during the regression phase between days 38 and 43 after tumor injection.* (**a**) CD45; (**b**) RB6-8C5; (**c**) eosinophils stained for endogenous peroxidase; (**d**) Mac3; (**e**) CD8; (**f**) CD4. (**a**) x300; (**b, d, f**) x250; (**c**) x400; not counter stained.

The difference which emerges between inhibition of tumor take and regression of an established tumor consists of the necessary participation, through the release of IFN-γ, of $CD8^+$ T lymphocytes. Based on the observed correlation between the presence of IFN-γ transcript and the effective regression of C-26/G-CSF tumors, we injected recombinant IFN-γ into irradiated mice bearing these tumors; immunohistological analysis performed 15 days after treatment showed large areas of necrosis, and a two-fold increase in the number of infiltrating cells. In situ hybridization revealed an increased number of cells expressing mRNA for IL-1α, IL-1β and TNF-α. Furthermore, injection of mAb to IFN-γ abrogated tumor regression (Fig. 23.3).

These data clearly show the crucial role of CD8 producing IFN-γ, although this feature remains dependent on the presence of PMN within the tumor; in fact, PMN depletion with the anti-granulocyte mAb RB6-8C5 reduced the number of CD8 to less than 30%, impaired cytokine expression by the remaining leukocytes and abrogated tumor regression.

PMN-T cell cross-talk without induction of memory T cells

Our analysis of the antitumor response initially triggered by G-CSF provides evidence of PMN-T lymphocyte cross-talk (Colombo et al., 1992b). Expression of IFN-γ mRNA by $CD8^+$ cells infiltrating regressing tumors suggests some role for IFN-γ in sustaining PMN survival and function (Perussia et al., 1987), while IL-1 and TNF initially produced by PMN may activate T cells as occurs following infection with microbial pathogens (Blanchard et al., 1986). Other examples of functional dependence of T cells from PMN are available in the literature. Depletion of rat PMN by mAb treatment resulted in abrogation of specific transplantation resistance to challenge with syngeneic, chemically induced tumors used for immunization (Midorikawa et al., 1990). In addition, tumors engineered to produce IL-4 or IL-2 are first infiltrated with PMN and subsequently with T lymphocytes (Golumbek et al., 1991; Cavallo et al., 1992). In a recent paper, Hock et al. (1993) compared the mechanism of rejection of the same tumor transduced with the IL-2, IL-4, IL-7, TNF-α or IFN-γ genes, and reached the conclusion that all these cytokines required $CD8^+$ T cells for long term tumor destruction, but also that an initial transient tumor suppression may occur in the absence of T cells. In the lag time between tumor injection and arrival of T cells other leukocytes can cause tumor inhibition through nonspecific mechanisms. In the IL-2 and IFN-γ systems, NK cells were suggested to be the players (Hock et al., 1993), but PMN may well be a more general leukocyte type involved in tumor rejection throughout an inflammatory like response that might create the favorable local environment for the recruitment and activation of T lymphocytes. In fact, neutrophils were found infiltrating tumors transduced with the IL-2 (Cavallo et al., 1992) and G-CSF (Colombo et al., 1991) genes, while eosinophils were found within tumors transduced with the IL-4 (Tepper et al., 1989; Golumbek et al., 1991) and IL-7 (Hock et al.,1993) genes. What is relevant is that PMN depletion abrogated the rejection of tumors transduced with the IL-4 (Tepper et al, 1992), G-CSF (Stoppacciaro et al., 1993) and IL-2 (G. Forni, personal communication) genes.

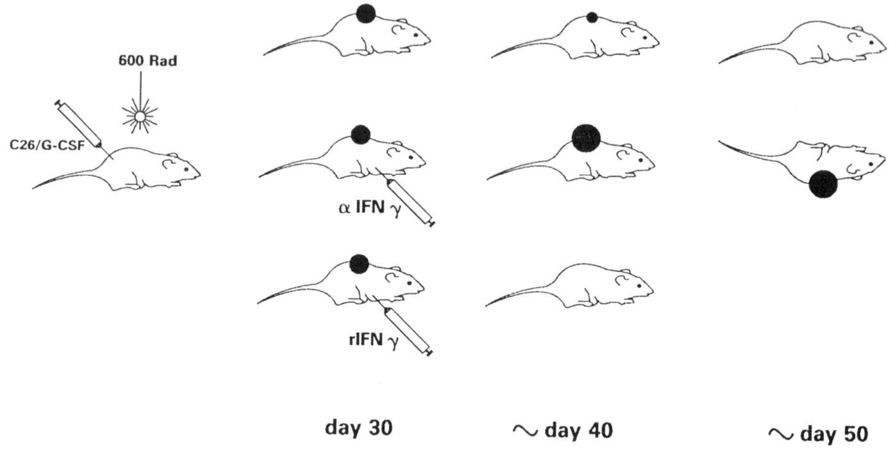

Figure 23.3. *Role of IFN-γ in the regression of C-26/G-CSF cells.* The drawing shows that C-26/G-CSF cells grow and regress in sublethally irradiated mice (top), that injection of a monoclonal antibody to IFN-γ abrogates tumor regression (center), and that the injection of recombinant IFN-γ accelerates the time of regression (bottom).

Our study of G-CSF transduced colon carcinoma suggests that IFN-γ, 2along with IL-1 and TNF-α, provides a favorable setting for analyzing tumor destruction. Although G-CSF production by neoplastic cells is not per se sufficient to induce the regression of an established tumor, it can inhibit tumor take and may well trigger the cascade of cytokine and leukocyte interactions potentially responsible for tumor destruction (Fig. 23.1).

However, mice which reject an established C-26/G-CSF tumor, as well as those which eliminate C-26/G-CSF cells injected as a cell suspension, remain susceptible to a challenge with a minimal lethal dose of C-26 cells. Lack of memory T cells in the latter case was explained by the absence of T cells within the infiltrating leukocytes; in the former, T cells were present but only $CD8^+$ T cells were found to be activated as inferred by their expression of IFN-γ. $CD4^+$ cells were without influence on tumor regression and IL-4 mRNA was not detectable within the regressing tumor. Thus, the lack of Th function may explain this phenomenon.

Conclusions

In conclusion, our work indicates that G-CSF behaves as a pro-inflammatory cytokine capable of chemotactic activity on neutrophils that became fully activated, in situ, by additional stimuli which may well include adhesion to appropriate extracellular matrix proteins or, even better, to tumor cells releasing G-CSF, as in our case. Activated neutrophils release IL-1α, IL-1β and TNF-α, among possible other cytokines, that tune up the inflammatory response and favor the recruitment of additional leukocytes. While this kind of inflammation when occurring at the tumor site may cause tumor rejection, it does not stimulate memory T cells which are expected to prevent tumor recurrence and to act at distant sites. For these reasons, G-CSF is not included in the group of cytokine genes proposed for gene transfer into tumor cells to provide tumor cell vaccines.

Acknowledgments

We thank Mr. Mario Azzini for the art work. This research has been supported by grants from the Italian Association for Cancer Research (AIRC), Milan and CNR, special project on Clinical Application of Cancer Research, Rome, Italy.

References

Armentano D, Yu S, Kantoff PH, Von Ruden T and Anderson WF (1987). Effect of internal viral sequences on utility of retroviral vector. J Virol **61**, 1647-1650.

Barth RJ, Mule JJ, Spiess PJ and Rosenberg SA (1991). Interferon gamma and tumor necrosis factor have a role in tumor regressions mediated by murine CD8+ tumor-infiltrating lymphocytes. J Exp Med **173**, 647-658.

Blanchard DK, Djeu JY, Klein TW, Friedman H and Stewart II WE (1986). Interferon-induction by lipopolysaccharide: Dependence on interleukin 2 and macrophages. J Immunol **136**, 963-970.

Bronchud MH and Dexter TM (1989). Clinical use of growth factors. Br Med Bull **45**, 590-599.

Cavallo F, Giovarelli M, Gulino A, Vacca A, Stoppacciaro A, Modesti A and Forni G (1992). Role of neutrophils and CD4+ T lymphocytes in the primary and memory response to nonimmunogenic murine mammary adenocarcinoma made immunogenic by IL-2 gene. J Immunol **149**, 3627-3635.

Chang JM, Metcalf D, Gonda TJ and Johnson GR (1989). Long-term exposure to retrovirally expressed granulocyte-colony-stimulating factor induces a non neoplastic granulocytic and progenitor cell hyperplasia without tissue damage in mice. J Clin Invest **84**, 1488-1496.

Colombo MP, Ferrari G, Stoppacciaro A, Parenza M, Rodolfo M, Mavilio F and Parmiani G (1991). Granulocyte colony-stimulating factor gene transfer suppresses tumorigenicity of a murine adenocarcinoma in vivo. J Exp Med **173**, 889-897.

Colombo MP, Lombardi L, Stoppacciaro A, Melani C, Parenza M, Bottazzi B and Parmiani G (1992a). Granulocyte colony-stimulating factor (G-CSF) gene transduction in murine adenocarcinoma drives neutrophil-mediated tumor inhibition in vivo. J Immunol **149**, 113-119.

Colombo MP, Modesti A, Parmiani G and Forni G (1992b). Local cytokine availability elicits tumor rejection and systemic immunity through granulocyte-T-lymphocyte cross-talk. Cancer Res **52**, 4853-4857.

Del Pozo V, De Andres B, Martin E, Cardaba B, Fernandez JC, Gallardo S, Tramon P, Leyva-Cobian F, Palomino P and Lahoz C (1992). Eosinophil as antigen-presenting cell: Activation of T cell clones and T cell hybridoma by eosinophils after antigen processing. Eur J Immunol **22**, 1919-1925.

Djeu YJ, Serbousek D and Blanchard DK (1990). Release of tumor necrosis factor by human polymorphonuclear leukocytes. Blood **76**, 1405-1409.

Dubravec DB, Spriggs DR, Mannick JA and Rodrick ML (1990). Circulating human peripheral blood granulocytes synthesize and secrete tumor necrosis factor α. Proc Natl Acad Sci USA **87**, 6758-6761.

Gabrilove JL, Jakubowski A, Scher H, Sternberg C, Wong G, Grous J, Yagoda A, Fain K, Moore MAS, Clarkson B, Oettgen H, Alton K, Welte K and Souza L (1988). Effect of granulocyte colony-stimulating factor on neutropenia and associated morbidity due to chemotherapy for transitional-cell carcinoma of the urothelium. N Engl J Med **318**, 1414-1422.

Golumbek PT, Lazenby AJ, Levitsky HI, Jaffe LM, Karsuyama H, Baker H and Pardoll DW (1991). Treatment of established renal cancer by tumor cells engineered to secrete interleukin-4. Science **254**, 713-716.

Hock H, Dorsch M, Kunzendorf U, Qin Z, Diamantstein T and Blankenstein T (1993). Mechanisms of rejection induced by tumor cell-targeted gene transfer of interleukin 2, interleukin 4, interleukin 7, tumor necrosis factor, or interferon γ. Proc Natl Acad Sci USA **90**, 2774-2778.

Katano M and Torisu M (1982). Neutrophil-mediated tumor cell destruction in cancer ascites. Cancer **50**, 62-68.

Lloyd AR and Oppenheim JJ (1992). Poly's lament: The neglected role of the polymorphonuclear neutrophil in the afferent limb of the immune response. Immunol Today **13**, 169-172.

Lord PCW, Wilmoth LMG, Mizel SB and McCall CE (1991). Expression of interleukin 1α and β genes by human blood polymorphonuclear leukocytes. J Clin Invest **87**, 1312-1321.

MacPherson GG and North RJ (1986). Endotoxin-mediated necrosis and regression of established tumors in the mouse. A correlative study of quantitative changes in blood flow and ultrastructural morphology. Cancer Immunol Immunother **21**, 209-216.

Marucha PT, Zeff RA and Kreutzer DL (1990). Cytokine regulation of IL-1β gene expression in the human polymorphonuclear leukocytes. J Immunol **145**, 2932-2937.

Matsumoto Y, Saiki I, Okuyama H, Tamura M and Azuma I (1991). Recombinant human granulocyte colony stimulating factor inhibits the metastasis of hematogenous and non-hematogenous tumors in mice. Int J Cancer **49**, 444-449.

Melani C, Mattia GF, Silvani A, Care A, Rivoltini L, Parmiani G and Colombo MP (1993). IL-6 expression in human neutrophil and eosinophil peripheral blood granulocytes. Blood **81**, 2744-2749.

Metcalf D and Nicola NA (1983). Proliferative effects of purified granulocyte colony-stimulating factor (G-CSF) on normal mouse hemopoietic cells. J Cell Physiol **116**, 198-202.

Midorikawa Y, Yamashita T and Sendo F (1990). Modulation of the immune response to transplanted tumors in rats by selective depletion of PMN in vivo using a monoclonal antibody: Abrogation of specific transplantation resistance to chemical carcinogen-induced syngeneic tumors by selective depletion of PMN in vivo. Cancer Res **50**, 6243-6247.

Moore MAS, Welte K, Gabrilove J and Souza LM (1987). Biological activities of recombinant human granulocyte colony stimulating factor (rh-G-CSF) and tumor necrosis factor: In vivo and in vitro analysis. In "Hematology and Blood Transfusion. Modern Trends in Human

Leukemia VII" (eds R Neth, RC Gallo, MF Greaves and H Kabisch), pp 210-220. Springer-Verlag, Berlin.

Nicola NA, Metcalf D, Matsumoto M and Johnson GR (1983). Purification of a factor inducing differentiation in murine myelomonocytic leukemia cells: Identification as granulocyte colony-stimulating factor (G-CSF). J Biol Chem **258**, 9017-9021.

Nicola NA, Bergley CG and Metcalf D (1985). Identification of the human analogue of a regulator that induces differentiation in murine leukemia cells. Nature **314**, 625-628.

Okuda K, Tani K, Ishigatsobo Y, Yokota S and David CS (1980). Antigen pulsed neutrophils bearing I antigens can induce T-lymphocyte proliferative response to syngeneic or semi syngeneic antigen primed T-lymphocytes. Transplantation **30**, 368-372.

Perussia B, Kobayashi M, Rossi ME, Anegon I and Trinchieri G (1987). Immune interferon enhances functional properties of human granulocytes: Role of Fc receptors and effect of lymphotoxin, tumor necrosis factor, and granulocyte-macrophage colony-stimulating factor. J Immunol **138**, 765-774.

Stoppacciaro A, Melani C, Parenza M, Mastrocchio A, Bassi C, Baroni C, Parmiani G and Colombo MP (1993). Regression of an established tumor genetically modified to release granulocyte colony-stinulating factor requires granulocyte-T cell cooperation and T cell-produced interferon γ. J Exp Med, **178**, 151-161.

Tamura M, Hattori K, Nomura H, Oheda M, Kubota N, Imazeki I, Ono M, Ueyama Y, Nagata S, Shirafuji N and Asano S (1987). Induction of neutrophilic granulocytosis in mice by administration of purified human native granulocyte colony-stimulating factor (G-CSF). Biochem Biophys Res Commun **142**, 454-460.

Tepper RI, Pattengale PK and Leder P (1989). Murine interleukin-4 displays potent anti-tumor activity in vivo. Cell **57**, 503-512.

Tepper RI, Coffman RL and Leder P (1992). An eosinophil-dependent mechanism for the antitumor effect of interleukin-4. Science **257**, 548-551.

Tiku K, Tiku ML and Skosey J (1986). Interleukin 1 production by human polymorphonuclear neutrophils. J Immunol **136**, 3677-3685.

Tweardy DJ, Cannizzaro L, Palumbo A, Shane S, Hubner K, Vantuinen P, Ledbetter DH, Finan JB, Nowell PC and Rovera G (1987). Molecular cloning and characterization of a cDNA for human granulocyte colony-stimulating factor (G-CSF) from glioblastoma multiform cell line and localization of the G-CSF gene to chromosome band 17 q 21. Oncogene Res **1**, 209-220.

Wang JM, Chen ZG, Colella S, Bonilla MA, Welte K, Bordignon C and Mantovani A (1988). Chemotactic activity of recombinant human granulocyte colony-stimulating factor. Blood **72**, 1456-1460.

CLINICAL PERSPECTIVES

24

DIRECT GENE TRANSFER FOR IMMUNOTHERAPY IN CANCER

G. E. Plautz and G. J. Nabel
Department of Pediatrics and Communicable Diseases; Departments of Internal Medicine and Biological Chemistry, Howard Hughes Medical Institute, University of Michigan Medical Center, Ann Arbor, MI, USA

Introduction

Cancer arises from a series of genetic alterations in a clone of cells which allow uncontrolled growth and metastasis. Abnormal gene products generated during neoplastic transformation may be presented to the immune system as novel tumor antigens. Despite their potential immunogenicity, most tumors arise in immunocompetent hosts, and mechanisms to escape immune surveillance have been described. Appropriate stimulation of the host immune system, however, can often induce an antitumor response.

Gene transfer provides a powerful tool to introduce recombinant genes into cells to analyze their effects. Traditionally, genes have been transferred into target cells in vitro; recently, however, several approaches to direct gene transfer in vivo have been developed. A series of experiments has demonstrated that introduction of various recombinant genes into nonimmunogenic tumor cells can stimulate a strong immune response. Inoculation of these gene modified tumor cells into hosts can induce immune mediated destruction of tumor cells and can protect animals from rechallenge with nontransduced tumor cells. Because of their potent effects in stimulating an antitumor response, transfer of genes encoding cytokines and other immunostimulatory molecules offers promise as a cancer therapy.

To achieve a clinically applicable approach to gene therapy for malignancy, we have developed a method for direct gene transfer into tumors in vivo. Unlike modification of tumor cells in vitro and subsequent inoculation into a non tumor bearing host, this approach allows transduction of actively growing established tumors. Methods of augmenting the immune response to preexisting tumors can thus be evaluated, providing a more realistic model for clinical applications in human malignancies. In this chapter, we describe the introduction of an allogeneic class I MHC gene by direct gene transfer in vivo.

CYTOKINE-INDUCED TUMOR IMMUNOGENICITY
ISBN 0-12-261520-4

Immune recognition of malignant cells

Neoplastic cells are derived from normal precursors which undergo a series of genetic changes occurring in successive clones. The acquisition of a neoplastic phenotype involves abnormal gene expression. Exogenous viral genes, mutated versions of normal genes, and chimeric gene products which arise through chromosomal translocation can all potentially produce tumor specific, novel peptide fragments which can be presented as antigens to cellular components of the immune system. In addition, overexpression of normal genes or heterotopic expression of normally silent genes can alter the antigenic profile of tumor cells. Several mechanisms exist by which tumor cells can evade immune mediated destruction. Appropriate stimulation of the immune system, however, has provided an effective antitumor response in animal models and in some human tumors. Trials employing various combinations of systemic delivery of recombinant cytokine gene products, intratumoral delivery of immunogenic organisms, inoculation with tumor cell "vaccines", or administration of lymphocytes, activated ex vivo, have shown efficacy in animal models and in some human tumors.

In order to provide effective immunotherapy for malignant diseases, it is necessary to identify the relevant cellular effectors which mediate the immune response and signals which provide a sufficient stimulus for their activation. The immunodominant antigens on tumor cells must also be defined. It is necessary, therefore, to develop a reliable and clinically applicable system to deliver an adequate activation signal to the immune system without excessive toxicity.

The importance of the cellular immune system in the recognition and destruction of tumors has been demonstrated in animal tumor models. Lymphoid cells from tumor sensitized animals can specifically eliminate established tumors when adoptively transferred to secondary animals (Smith et al., 1977; Fernandez-Cruz et al., 1979; Cheever et al., 1980; Shu et al., 1985). Tumor draining lymph nodes contain tumor specific pre-effector lymphocytes. Stimulation and expansion of these cells with tumor and IL-2 (Shu et al., 1987), or with anti-CD3 plus IL-2 (Crossland et al., 1991; Yoshizawa et al., 1991) induces activity against established tumor upon adoptive transfer. Cellular effectors such as NK cells (Herberman, 1986) and LAK cells (Lotze et al., 1981; Grimm et al., 1982), which also participate in the antitumor immune response, have been defined. Such effector cells frequently induce severe toxicity, lack specificity for the tumor and fail to generate a memory response, limiting their efficacy. Although specific tumor antigens recognized by antibodies exist, and antibody mediated cytotoxicity with immunoconjugates have shown some promise, it has been difficult to demonstrate clinical responses in human tumors. Consequently, most strategies to augment the host immune response to tumors have concentrated on the identification and activation of T cells specific for tumor antigens.

In order to augment specific immune response against spontaneously arising human tumors the host must possess precursor T cells. CTL specific for autologous tumors have been isolated in tumor bearing humans from peripheral blood (DeVries and Spits, 1983; Mukherji and MacAlister, 1983; Hershey et al., 1986; Slovin et al., 1986; Anichini et al., 1989; Knuth et al., 1989), tumor draining lymph nodes (Cozzolino et al., 1987; Barnd et al., 1988; Mukherji et al., 1989), and within tumors (Whiteside et al., 1986; Muul et al., 1987; Anichini et al., 1989; Topalian et al., 1989). In some tumor types, particularly melanoma, effector cells are $CD8^+$ T cells and display class I MHC restricted lysis of autologous tumor. Presentation of tumor antigens through class II MHC by either tumor cells or APC such as macrophages (Fidler and Schroit, 1984; Brunda et al., 1991; Grabbe et al., 1991), is also very important. Melanomas frequently express class II MHC on their surface. $CD4^+$ T lymphocytes are present within tumors, and an increased CD4/CD8 ratio has been observed in regressing melanomas compared to nonregressing ones (Tefany et al., 1991). Tumor specific effector cells, either $CD4^+$ or $CD8^+$, can also release cytokines (Schwartzentruber et al., 1991; Hom et al., 1993a; Hom et al., 1993b) when specifically stimulated with autologous tumor antigens presented by class I or class II MHC, respectively. Through the specific antitumor response, the cellular immune system is capable of mediating direct cytolysis of tumor cells.

The nature of tumor antigens presented to cellular effectors from spontaneously arising human tumors is also under investigation. Animal models usually employ tumor cells derived from viral or chemically induced carcinogenesis, and it has been argued that most naturally arising tumors do not possess tumor antigens (Hewitt, 1982; Hewitt, 1983). Although it has been a very labor intensive process, tumor antigens recognized by CTL have now been cloned from murine tumors (Grabbe et al., 1991; Van den Eynde et al., 1991) and melanoma (Van der Bruggen et al., 1991). Several mutant or chimeric gene products have been tested for their ability to induce a $CD4^+$ T-cell response. Immunization of mice with synthetic peptides corresponding to a mutant ras protein (Jung and Schluesener, 1991), or the junction region of the translocation derived *bcr-abl* protein (Chen W et al., 1992), or the human melanoma associated antigen p97 (Kahn et al., 1991) generated $CD4^+$ specific T cells. Adoptive transfer of $CD4^+$ clones in mice bearing p97 transfected murine melanoma mediated tumor regression. There are several examples in which spontaneously arising poorly immunogenic murine tumors are capable of generating tumor specific immunity when admixed with the immunostimulant (*Corynebacterium parvum*) (Bursuker et al., 1991) or modified by gene transfer (Golumbek et al., 1991). These observations strengthen the concept that tumor antigens exist on spontaneous tumors although many human tumors have not yet been studied. Identification and characterization of tumor antigens presented by class I or class II MHC will improve our understanding of the immune response to malignant cells and may provide tools to more effectively stimulate a specific antitumor response.

One potential difficulty in extrapolating results derived from animal models to human neoplasia is that they usually involve inoculation of tumor cells or tumor fragments into otherwise healthy hosts. In contrast, spontaneously arising human tumors have co-evolved with the host immune system. There are several mechanisms by which lack of an effective immune response might occur. Advanced age is associated with a decreased ability to respond to antigens, and old mice fail to respond to immunotherapy which is protective against tumors in young mice (Dunn and North, 1991a; Dunn and North, 1991b). TGF-β1 has immunosuppressive properties and is secreted by several tumor types. Transfection of TGF-β1 into tumor cells can convert a highly immunogenic regressor tumor into a progressively growing tumor, and it can inhibit the tumor specific CTL response (Torre-Amione et al., 1990; Inge et al., 1992). Inadequate presentation of antigens by class I MHC on tumor cells is another mechanism by which they evade an effective immune response. Malignant cells are frequently observed to have low or absent expression of class I MHC molecules (Elliott et al., 1989). Often, there is a decrease in expression of class I MHC in metastatic foci compared to primary tumor lesions (Gopas et al., 1989; Cordon-Cardo et al., 1991; Pantel et al., 1991). Several mechanisms by which this could occur have been described, including altered binding of transcription factors to enhancer regions MHC genes (Blanchet et al., 1992), or abnormal expression of β2-microglobulin (Gattoni-Celli et al., 1992), or proteins involved in processing and transport of peptide fragments. In some tumor lines, tumorigenicity can be abrogated by transfection of transporter genes (Franksson et al., 1993) or by β2-microglobulin (Glas et al., 1992). Tumors which lack or have low level expression of class I MHC can acquire an immunogenic phenotype which induces rejection by transfection of the appropriate class I MHC gene (Hui et al., 1984; Tanaka et al., 1985; Wallich et al., 1985). In some cases, these immunogenic class I MHC expressing cells can protect against rechallenge with the MHC deficient parental tumor cells. These experiments, as well as others described below, suggest that despite poor immunogenicity or immunosuppressive properties of tumor cells, the host immune response to tumors can be augmented sufficiently to provide a therapeutic effect.

Gene transfer into tumor cells to provide immunotherapy

The presence of a cellular immune response against tumor cells provides the rationale for genetic manipulation of tumor cells to make them more immunogenic. Transfection of viral genes (Fearon et al., 1988) or viral infection of tumor cells enhances tumor cell immunogenicity and protects against rechallenge with parental nontransfected cells (Kobayashi et al., 1970; Kobayashi, 1979). In these experiments, presentation of a strong antigen by the tumor cells provides a stimulus sufficient to activate an immune response against associated weak tumor antigens. Localized delivery of cytokines mediated by gene modified tumor cells has also been used as a strategy for immunostimulation. Introduction of the IL-2 gene into several weakly immunogenic tumors using retroviral vectors (Fearon et al., 1990; Gansbacher et al., 1990b) or by transfection (Ley et al., 1991) abrogated their tumorigenicity. In these experiments the precursor frequency of specific CTL was increased and persisted at high levels, and CTL could provide protection in the absence of $CD4^+$ T cells but not in the absence of $CD8^+$ T cells. Interestingly, IL-2 secreting tumor cells provided protection against co-inoculated unmodified tumor cells (Gansbacher et al., 1990b) suggesting that the paracrine effects of local IL-2 secretion can augment a weak antitumor response. Tumor cells modified to secrete IL-4 are rejected with an infiltration of eosinophils and macrophages as a prominent component of the local response (Tepper et al., 1989; Li et al., 1990; Golumbek et al., 1991). These tumor cells grow in nude mice, demonstrating the requirement for T cells to mediate complete rejection. There was a significant delay in growth of these tumors in nude mice, however, suggesting that NK cells and macrophages provided partial protection in the rejection response induced by IL-4. In immunocompetent mice cured of IL-4 secreting tumors, $CD8^+$ T cells were necessary for rejection of unmodified parental cells upon rechallenge (Golumbek et al., 1991). This demonstrates that the relevant tumor antigens mediating a protective immune response are recognized by $CD8^+$ T cells.

Cytokines which alter the capacity of tumor cells to present antigens or deliver costimulatory signals to T cells have been introduced into tumor cells by gene transfer. IFN-γ induces the expression of class I and class II MHC molecules on target cells and activates macrophages and CTL. Introduction of the IFN-γ gene into a weakly immunogenic murine fibrosarcoma abrogated its tumorigenicity and generated a tumor specific immune response to nontransduced tumor cells protective against rechallenge (Gansbacher et al., 1990a). Introduction of IFN-γ into a nonimmunogenic tumor line induced TIL effective against the parental tumor cells upon adoptive transfer, a response dependent on $CD8^+$ T cells (Restifo et al., 1992). Activation of lymphocytes is mediated by interactions between adhesion molecules in addition to the specific binding of the TCR with the class I MHC/antigen complex. The B7 molecule, present on APC, interacts with the CD28 ligand on T cells to deliver an activation signal in the presence of TCR occupancy. Transfer of B7 into tumors caused their rejection mediated by $CD8^+$ lymphocytes and protection against rechallenge (Townsend and Allison, 1993) or established microscopic metastases (Chen L et al., 1992).

Cytokines have pleiotropic effects on hematopoietic and vascular cells; consequently, the mechanisms by which cytokine secreting tumor cells are rejected are not always identical. Several cytokines have antitumor effects which are dependent on macrophages and dendritic cells. Expression of TNF-α by tumor cells resistant to its direct effects on growth in vitro, nevertheless, suppressed tumor growth in animals (Asher et al., 1991; Blankenstein et al., 1991; Teng et al., 1991). This was associated with infiltration of tumors with macrophages (Blankenstein et al., 1991) and was independent of T cells, since it occurred in nude mice (Teng et al., 1991). Similarly, the rejection of IL-7 producing tumor cells was unaffected by depletion of $CD8^+$ T cells, but it required $CD4^+$ T cells and $CR3^+$ cells predominantly represented by macrophages (Hock et al., 1991).

Not all cytokines are equally effective in producing an antitumor response. For example, using an identical tumor line, adequate immunologic protection was conferred by TNF-α, but no response to IL-6 secretion was observed (Blankenstein et al., 1991). In another model using gene modified tumor cells, tumor inhibition was generated by IL-2 but not by TNF-α (Karp et al., 1993). The different histologic appearance within the tumor produced by various cytokines during the rejection response may be due to nonoverlapping effects of the particular cytokine expressed. Taken together, these experiments demonstrate the principle that modification of tumor cells by gene transfer can increase their immunogenicity and cause a tumor specific immune mediated rejection and protect against rechallenge. In some instances, the immune response elicited by gene modified tumor cells can eliminate previously established microscopic tumors (Golumbek et al., 1991; Chen L et al., 1992), providing the possibility that patients with preexisting tumors might be treated. The potential clinical applicability of cell mediated gene transfer approaches might, however, be limited by the time, cost and, in many cases, the inability to establish cell lines from human tumors. It may be possible to circumvent these difficulties for certain tumors if there are common tumor antigens which are presented by frequent HLA alleles. Such an approach for active immunotherapy, using IL-2 or IFN-γ expression by HLA-A1 or HLA-A2 melanoma cells for which HLA restricted common tumor antigens are known, is under development (Gansbacher et al., 1992). This strategy may not be widely applicable to many tumors for which the frequency of tumor antigen expression and distribution on normal tissues and HLA restriction are not characterized. To broaden the potential applicability of immunotherapy by recombinant gene expression in tumor cells, it would be advantageous to modify tumor cells directly in vivo.

Site specific gene transfer in vivo

Direct gene transfer provides a powerful tool to modify endogenous cells in order to examine the effects of recombinant gene expression in vivo. Several gene transfer techniques, including viral vectors, physical methods and chemical methods have been successfully utilized in vivo and may be applicable to tumor cell transduction. Viruses, modified to abrogate pathogenicity and to contain recombinant genes, have served as vectors for both ex vivo and in vivo gene transfer, including retroviruses (Price et al., 1987; Nabel et al., 1990; Culver et al., 1992), adenoviruses

(Rosenfeld et al., 1991; Jaffe et al., 1992; Rosenfeld et al., 1992), adeno-associated viruses (Samulski et al., 1991), and herpes viruses (Palella et al., 1988; Chiocca et al., 1990; Breakefield and DeLuca, 1991). Direct injection of DNA has allowed expression of recombinant genes in certain cell types, such as skeletal or cardiac myocytes and melanoma (Lin et al., 1990; Wolff et al., 1990; Acsadi et al., 1991; Vile and Hart, 1993). DNA has also been complexed with cationic liposomes to transfer recombinant genes directly into cells in vivo (Nabel et al., 1990; Plautz et al., 1993).

To introduce recombinant genes into specific target tissues we have delivered retroviral vectors and DNA-liposome complexes to anatomically localized sites. Transplantable murine tumors inoculated s.c. often exist as discrete masses with a pseudocapsule. Recombinant genes can thus be delivered to tumor cells in vivo, by direct intratumoral injection. This approach is an adaptation of previous work in which an intravascular catheter containing proximal and distal balloons was used to deliver recombinant genes to anatomically localized arterial segments. Recombinant genes were delivered into the protected space between the two balloons using either DNA-liposome complexes, or retroviral vectors (Nabel et al., 1990, Nabel et al., 1992c). Expression of the recombinant gene was limited to the site of introduction in both the intratumoral and intravascular gene transfer, with occasionally very low levels at other sites (Nabel et al., 1992c; Plautz et al., 1993).

Two liposome preparations have been used for site specific gene transfer. The cationic liposome, Lipofectin (Bethesda Research Laboratories, Gaithersburg, MD), a mixture of DOTMA (N-[2,3-dioleyloxy)propyl]-N,N,N-trimethylammonium chloride) and DOPE (dioleoyl phosphatidylethanolamine) (Felgner et al., 1987), has been used to deliver recombinant genes into vascular cells in peripheral (Nabel et al., 1990; Nabel et al., 1992c) and coronary (Lim et al., 1991) arteries in vivo. Another cationic liposome preparation, containing DC-cholesterol (3b[N-(N'N'-dimethylaminoethane)-carbamoyl] cholesterol) and DOPE (Gao and Huang, 1991), has also provided efficient gene transduction of vascular cells and malignant tumors in vivo. Cationic liposomes can be prepared in large batches by sonication to form particles of approximately 250 nm diameter, which retain activity in aqueous solution for several months. A complex is formed through an electrostatic interaction, highly dependent on the molar ratio between the DNA and liposomes (Stewart et al., 1992), which can traverse the plasma membrane of target cells. The DNA gains access to the nucleus to direct transcription through incompletely defined mechanisms (Felgner and Ringold, 1989). Cationic liposomes are significantly inactivated by serum (Felgner et al., 1989) and transduce cells most effectively at the site of delivery.

Several features of liposome mediated gene transfer enhance its application to direct gene transfer. In contrast to retroviral mediated gene transfer, replication of the recipient cell is not necessary; therefore, terminally differentiated somatic cells can serve as targets. Unlike viral vectors, there are few constraints on vector size or transcriptional signals, allowing the use of tissue specific enhancer/promoter elements. DNA-liposome complexes form rapidly in aqueous solution without extensive preparation, permitting rapid analysis of a series of vectors for activity in vitro or in vivo. The DC-cholesterol preparation is not toxic to cells over a wide range of concentrations in vitro, and it is metabolized in vivo (Gao and Huang, 1991). The DC-cholesterol liposome preparation has been demonstrated as safe in vivo after extensive toxicity testing (Stewart et al., 1992; Nabel et al., 1992b).

Direct gene transfer of allogeneic class I MHC genes

Genetic modification of tumor cells to increase their immunogenicity has shown promise in cell mediated gene transfer experiments. To determine whether a therapeutic immune response could be generated against preexisting macroscopic tumors, we introduced an allogeneic class I MHC gene by intratumoral injection of DNA-liposome complexes or retroviral vectors. Class I MHC proteins bind proteolytically derived peptide fragments from endogenously synthesized proteins (Jardetzky et al., 1991). The combination of the class I MHC molecule with its bound peptide fragment is displayed on the surface of nearly all somatic cells, in association with β2-microglobulin. This complex is recognized by T cells through steric interaction with unique TCR, and it is restricted to $CD8^+$ T cells by conserved interaction with class I MHC (Townsend and Bodmer, 1989). A combination of residues from the class I MHC in the peptide binding groove and the bound peptide fragment constitute the structure recognized by each individual TCR. MHC alleles are highly polymorphic, and most of the variation between alleles is concentrated in residues which comprise the peptide binding portion of the molecule (Parham et al., 1988). Consequently, transplanted tissue expressing allogeneic MHC molecules is highly immunogenic due to the altered shape of the MHC in the peptide binding region. A high precursor frequency of $CD8^+$ lymphocytes capable of reacting against allogenic cells exists (Skinner and Marbrook, 1976; Lindahl and Wilson, 1977). Therefore, introduction of an allogeneic class I MHC gene into tumors should stimulate a strong immune response. The murine class I MHC gene, H-$2K^s$, was introduced into tumors in vivo by direct intratumoral injection of a DNA-liposome complex using DC-cholesterol liposomes.

The weakly immunogenic murine tumor line CT26, a colon adenocarcinoma (H-$2K^d$), was inoculated s.c, and transduced with the H-$2K^s$ gene approximately 14 days later when tumors were 5-8 mm in diameter. Splenic lymphocytes obtained from these animals 14 days after gene transfer demonstrated lytic activity against CT26 targets expressing H-$2K^s$. More importantly, cytotoxic lymphocytes were also generated against the parental nontransduced CT26 cells.

Injection of a control gene, β-galactosidase, complexed with the liposomes, did not induce CTL. Cytolytic activity against both transduced and parental tumor cells could be blocked by antibodies against CD8+ but not CD4+ lymphocytes. This immune response was not protective, as the tumors were relatively large at the time of gene transfer and grew progressively during the generation of the antitumor response. Experiments using intratumoral transfer of the β-galactosidase gene suggest that the efficiency of gene transfer was approximately 1%.

Significant attenuation of tumor growth and, in some cases, complete regression of existing tumors was achieved following H-2K^s gene transfer when mice were presensitized to the H-2K^s gene. Mice were preimmunized against H-2K^s prior to tumor inoculation either by irradiated tumor cells expressing H-2K^s or by splenic lymphocytes from an H-2^s mouse strain. Two histologic types of tumors, the colon adenocarcinoma CT26 and the fibrosarcoma MCA 106 from two different mouse strains BALB/c and C57BL/6 respectively, responded to tumor transduction with the H-2K^s gene (Plautz et al., 1993). This demonstrated that the immunologic response to H-2K^s was not a peculiar strain specific response. The antitumor response was mediated by T lymphocytes since the growth of CT26 or MCA 106 tumors in nude mice treated with the H-2K^s gene was identical to β-galactosidase or untreated controls. Preimmunized mice which were depleted of CD8+ T cells prior to H-2K^s gene transfer did not respond to H-2K^s gene transfer, whereas depletion of CD4+ cells had no effect on the antitumor response (Plautz et al., 1993). Direct gene transfer of the H-2K^s gene into the poorly immunogenic murine melanoma B16BL/6 generated pre-effector cells in tumor draining lymph nodes in naive mice.

When stimulated ex vivo with the mAb anti-CD3 for 2 days, followed by culture in low dose IL-2 (10 U/ml) for 3 days, these effector cells had no direct cytolytic activity against B16BL6 targets. However, adoptive transfer of these cells to mice bearing 3-day-old lung metastases resulted in a marked decrease in lung metastases (Wahl et al., 1992). These studies demonstrate that actively growing macroscopic tumors can be transduced with recombinant genes in vivo by intratumoral gene transfer using DNA-liposome complexes. An appropriate immunologic response can be generated against an allogeneic MHC protein expressed within actively growing tumors, and this strong immune response can stimulate a host immune response to weak tumor antigens. Appropriate augmentation of this antitumor immune response can mediate regression of preexisting tumors.

Immunotherapy by direct gene transfer: preclinical studies

To determine whether recombinant genes can be delivered safely in vivo using DNA-liposome complexes, the acute and long term toxicity were analyzed extensively in animal models. DC-cholesterol liposomes complexed with DNA were injected into mice i.v. or intratumorally, and 2-3 weeks later the distribution of DNA was determined by PCR analysis of organs. The plasmid was detected in heart and lung tissue in most cases and at times in kidney when mice received i.v. DNA-liposome complexes. Occasionally, intratumoral injection resulted in isolation of plasmid in heart, lung and spleen. There were no histopathologic findings in these organs. Analysis of serum enzymes did not reveal evidence of liver, heart or kidney dysfunction. Because of the finding of injected DNA in heart tissue, electrocardiography was performed during liposome injection and at follow-up. These electrocardiograms were all normal (Stewart et al., 1992).

Expression of a foreign class I MHC gene on normal tissues could potentially induce autoimmunity. Mice which received the allogeneic MHC gene all generated a specific CTL response to the foreign gene product; however, there was no histopathologic or serologic evidence of organ toxicity (Nabel et al., 1992b). The potential for inadvertent introduction of recombinant genes into germline tissue is a concern for in vivo gene therapy. To address this question, 42 rabbits and pigs received i.v. or intraarterial injection of DNA-liposome complexes. Transfected DNA was never isolated from gonadal tissue using PCR, despite its presence in the transduced artery segments and occasionally in other tissues (Nabel et al., 1992b). These studies demonstrate that direct gene transfer and expression of recombinant genes, inclutsding a foreign MHC gene, do not induce generalized autoimmunity, toxicity, or transmission to germ cells. Direct gene transfer using DNA-liposome complexes, therefore, is suitable for human gene therapy.

Clinical trials of direct gene therapy in cancer

A phase I/II human gene therapy trial to provide immunotherapy for malignancy by direct transfer of an allogeneic MHC gene into tumors in vivo was approved by the Recombinant DNA Advisory Committee in January 1992 (Nabel et al., 1992a). Patients with metastatic melanoma and s.c. tumor nodules will be treated by intratumoral injection of a preparation of the human class I MHC gene, HLA-B7, complexed with DC-cholesterol liposomes. Four study groups will receive escalating doses of the DNA-liposome complex. The presence of transfected DNA, mRNA and surface expression of HLA-B7 protein will be analyzed in cells obtained from tumor biopsies two weeks after each injection. Peripheral blood lymphocytes and tumor infiltrating lymphocytes will be assayed by limiting dilution analysis for their reactivity to HLA-B7 and autologous tumor cells. The objectives of this ongoing clinical trial are to:

- Establish a safe and effective dose to introduce a recombinant gene, HLA-B7, into tumors by DNA-liposome transfection

- Confirm expression of this class I MHC gene by tumor cells following direct gene transfer in vivo
- Analyze the immune response against this antigen and other tumor antigens.

This protocol is the first clinical trial of direct gene transfer in vivo and represents the first step in the application of this promising technology. It is also the first utilization of a nonviral vector, DNA-liposome complexes. Previously, gene therapy efforts have relied on modification of cells ex vivo. This involves considerable effort, time and expense and is not applicable in situations in which cells fail to grow in culture. Direct gene transfer into patients obviates the cumbersome aspects of ex vivo gene therapy. Through anatomically localized direct gene transfer, using DNA-liposome complexes, specific tissues can transiently express recombinant genes. Because of the stability of the DNA and liposomes and the ease with which they can be reconstituted and administered to patients, such treatment could potentially be provided in a generalized outpatient setting.

References

Acsadi G, Jiao SS, Jani A, Duke D, Williams P, Chong W and Wolff JA (1991). Direct gene transfer and expression into rat heart in vivo. New Biol **3**, 71-81.

Anichini A, Mazzocchi A, Fossati G and Parmiani G (1989). Cytolytic T lymphocyte clones from peripheral blood and from tumor site detect intratumor heterogeneity of melanoma cells. J Immunol **142**, 3692-3701.

Asher AL, Mule JJ, Kasid A, Restifo NP, Salo JC, Reichert CM, Jaffe G, Fendly B, Kriegler M and Rosenberg SA (1991). Murine tumor cells transduced with the gene for tumor necrosis factor-α. J Immunol **146**, 3227-3234.

Barnd D, Kerr L, Metzgar R and Finn O (1988). Human tumor-specific T cell lines generated from invaded lymph nodes in patients with solid tumors. Transplant Proc **20**, 339-341.

Blanchet O, Bourge J-F, Zinszner H, Israel A, Kourilsky P, Dausset J, Degos L and Paul P (1992). Altered binding of regulatory factors to HLA class I enhancer sequence in human tumor cell lines lacking class I antigen expression. Proc Natl Acad Sci USA **89**, 3488-3492.

Blankenstein T, Qin Z, Uberla K, Muller W, Rosen H, Volk H D and Diamantstein T (1991). Tumor suppression after tumor cell-targeted tumor necrosis factor-α gene transfer. J Exp Med **173**, 1047-1052.

Breakefield XO and DeLuca NA (1991). Herpes simplex virus for gene delivery to neurons. New Biol **3**, 203-218.

Brunda MJ, Sulich V, Wright RB and Palleroni AV (1991). Tumoricidal activity and cytokine secretion by tumor-infiltrating macrophages. Int J Cancer **48**, 704-708.

Bursuker I, Petty BA, Neddermann KM and Keller LS (1991). Immunomodulation in an apparently non-immunogenic murine tumor. Int J Cancer **49**, 414-420.

Cheever MA, Greenberg PD and Fefer A (1980). Specificity of adoptive chemoimmunotherapy of established syngeneic tumors. J Immunol **125**, 711-714.

Chen L, Ashe S, Brady WA, Hellstrom I, Hellstrom KE, Ledbetter JA, McGowan P and Linsley PS (1992). Costimulation of antitumor immunity by the B7 counterreceptor for the T lymphocyte molecule CD28 and CTLA-4. Cell **71**, 1093-1102.

Chen W, Peace DJ, Rovira DK, You S-G and Cheever MA (1992). T-cell immunity to the joining region of p210BCR-ABL protein. Proc Natl Acad Sci USA **89**, 1468-1472.

Chiocca EA, Choi BB, Cai WZ, DeLuca NA, Schaffer PA, DeFiglia M, Breakefield XO and Martuza RL (1990). Transfer and expression of the lacZ gene in rat brain neurons mediated by herpes simplex virus mutants. New Biol **2**, 739-746.

Cordon-Cardo C, Fuks Z, Drobnjak M, Moreno C, Eisenbach L and Feldman M (1991). Expression of HLA-A,B,C antigens on primary and metastatic tumor cell populations of human carcinomas. Cancer Res **51**, 6372-6380.

Cozzolino F, Torcia M, Carossino A, Giordani R, Selli C, Talini G, Realli E, Novelli A, Pistoia V and Ferrarini M (1987). Characterization of cells from invaded lymph nodes in patients with solid tumors. J Exp Med **166**, 303-318.

Crossland KD, Lee VK, Chen W, Riddell SR, Greenberg PD and Cheever MA (1991). T cells from tumor-immune mice nonspecifically expanded in vitro with anti-CD3 plus IL-2 retain specific function in vitro and can eradicate disseminated leukemia in vitro. J Immunol **146**, 4414-4420.

Culver KW, Ram Z, Wallbridge S, Ishii H, Oldfield EH and Blaese RM (1992). In vivo gene transfer with retroviral vector-producer cells for treatment of experimental brain tumors. Science **256**, 1550-1552.

DeVries JE and Spits H (1983). Cloned human cytotoxic lymphocyte lines reactive with autologous melanoma cells. J Immunol **132**, 510-519.

Dunn PL and North RJ (1991a). Effect of advanced ageing on the ability of mice to cause tumor regression in response to immunotherapy. Immunology **74**, 355-359.

Dunn PL and North RJ (1991b). Effect of advanced aging on ability of mice to cause regression of an immunogenic lymphoma in response to immunotherapy based on depletion of suppressor T cells. Cancer Immunol Immunother **33**, 421-423.

Elliott BE, Carlow DA, Rodricks AM and Wade A (1989). Perspectives on the role of MHC antigens in normal and malignant cell development. Adv Cancer Res **53**, 181-245.

Fearon ER, Itaya T, Hunt B, Vogelstein B and Frost P (1988). Induction in a murine tumor of immunogenic tumor variants by transfection with a foreign gene. Cancer Res **48**, 2975-2980.

Fearon ER, Pardoll DM, Itaya T, Golumbek P, Levitsky HI, Simons JW, Karasuyama H,Vogelstein B and Frost P (1990). Interleukin-2 production by tumor cells bypasses T helper function in the generation of an antitumor response. Cell **60**, 397-403.

Felgner PL, Gadek TR, Holm M, Roman R, Chan HW, Wenz M, Northrop JP, Ringold GM and Danielsen M (1987). Lipofection: A highly efficient lipid-mediated DNA-transfection procedure. Proc Natl Acad Sci USA **84**, 7413-7417.

Felgner PL, Holm M and Chan H (1989). Cationic liposome mediated transfection. Proc Western Pharmacol Soc **32**, 115-121.

Felgner PL and Ringold GM (1989). Cationic liposome-mediated transfection. Nature **337**, 387-388.

Fernandez-Cruz E, Halliburton B and Feldman JD (1979). In vivo elimination by specific effectors of an established syngeneic rat Moloney virus-induced sarcoma. J Immunol **123**,1772-1777.

Fidler IJ and Schroit AJ (1984). Synergism between lymphokines and muramyl dipeptide encapsulated in liposomes: In situ activation of macrophages and therapy of spontaneous cancer metastases. J Immunol **133**, 515-518.

Franksson L, George E, Powis S, Butcher G, Howard J and Karre K (1993). Tumorigenicity conferred to lymphoma mutant by major histocompatibility complex-encoded transporter gene. J Exp Med **177**, 201-205.

Gansbacher B, Bannerji R, Daniels B, Zier K, Cronin K and Gilboa E (1990a). Retroviral vector-mediated interferon-γ gene transfer into tumor cells generates potent and long lasting antitumor immunity. Cancer Res **50**, 7820-7825.

Gansbacher B, Zier K, Daniels B, Cronin K, Bannerji R and Gilboa E (1990b). Interleukin 2 gene transfer into tumor cells abrogates tumorigenicity and induces protective immunity. J Exp Med **172**, 1217-1224.

Gansbacher B, Zier K, Cronin K, Hantzopoulos PA, Bouchard B, Houghton A, Gilboa E and Golde D (1992). Retroviral gene transfer induced constitutive expression of interleukin-2 or interferon-γ in irradiated human melanoma cells. Blood **80**, 2817-2825.

Gao X and Huang L (1991). A novel cationic liposome reagent for efficient transfection of mammalian cells. Biochem Biophys Res Commun **179**, 280-285.

Gattoni-Celli S, Kirsch K, Timpane R and Isselbacher KJ (1992). β2 microglobulin gene is mutated in a human colon cancer cell line (HCT) deficient in the expression of HLA class I antigens of the cell surface. Cancer Res **52**, 1201-1204.

Glas R, Sturmhofel K, Hammerling GJ, Karre K and Ljunggren HG (1992). Restoration of a tumorigenic phenotype by β2 microglobulin transfection to EL-4 mutant cells. J Exp Med **175**, 843-846.

Golumbek PT, Lazenby AJ, Levitsky HI, Jaffee LM, Karasuyama H, Baker M and Pardoll DM (1991). Treatment of established renal cancer by tumor cells engineered to secrete interleukin-4. Science **254**, 713-716.

Gopas J, Rager-Zisman B, Bar-Eli M, Hammerling G and Segal S (1989). The relationship between MHC antigen expression and metastasis. Adv Cancer Res **53**, 89-115.

Grabbe S, Bruvers S, Gallo RL, Knisely TL, Nazareno R and Granstein RD (1991). Tumor antigen presentation by murine epidermal cells. J Immunol **146**, 3656-3661.

Grimm EA, Mazumder, A, Zhang, HZ and Rosenberg, SA (1982) Lymphokine-activated killer cell phenomenon lysis of natural killer-resistant fresh solid tumor cells by interleukin-2-activated autologous human peripheral blood lymphocytes. J Exp Med **155**, 1823-1841.

Herberman RB (1986). Natural killer cells. Ann Rev Med **37**, 347-352.

Hershey P, MacDonald M, Schebeci S and Burns C (1986). Clonal analysis of cytotoxic T lymphocytes (CTL) against autologous melanoma. Cancer Immunol Immunother **22**, 15-23.

Hewitt HB (1982). Animal tumor models and their relevance to human tumor immunology. J Biol Resp Mod **1**, 107-119.

Hewitt HB (1983). Second point: Animal tumor models and their relevance to human tumor immunology. J Biol Resp Mod **2**, 210-216.

Hock H, Dorsch M, Diamantstein T and Blankenstein T (1991). Interleukin 7 induces $CD4^+$ T cell-dependent tumor rejection. J Exp Med **174**, 1291-1298.

Hom SS, Rosenberg SA and Topalian SL (1993a). Specific immune recognition of autologous tumor by lymphocytes infiltrating colon carcinomas: Analysis by cytokine secretion. Cancer Immunol Immunother **36**, 1-8.

Hom SS, Schwartzentruber DJ, Rosenberg SA and Topalian SL (1993b). Specific release of cytokines by lymphocytes infiltrating human melanomas in response to shared melanoma antigens. J Immunother **13**, 18-30.

Hui K, Grosveld F and Festenstein H (1984). Rejection of transplantable leukemia cells following MHC DNA-mediated cell transformation. Nature **311**, 750-752.

Inge TH, Hoover SK, Susskind BM, Barrett SK and Bear HD (1992). Inhibition of tumor-specific cytotoxic T-lymphocyte responses by transforming growth factor-β1. Cancer Res **52**, 1386-1392.

Jaffe HA, Danel C, Longenecker G, Metzger M, Setoguchi Y, Rosenfeld MA, Gant TW, Thorgeirsson SS, Stratford-Perricaudet LD, Perricaudet M, Pavirani A, Lecocq JP and Crystal RG (1992). Adenovirus-mediated in vivo gene transfer and expression in normal rat liver. Nature Genetics **1**, 372-378.

Jardetzky TS, Lane WS, Robinson RA, Madden DR and Wiley DC (1991). Identification of self peptides bound to purified HLA-B27. Nature **353**, 326-329.

Jung S and Schluesener J (1991). Human T lymphocytes recognize a peptide of single point-mutated, oncogenic ras proteins. J Exp Med **173**, 273-276.

Kahn M, Sugawara H, McGowan P, Okuno K, Nagoya S, Hellstrom KE, Hellstrom I and Greenberg P (1991). CD4+ T cell clones specific for the human p97 melanoma-associated antigen can eradicate pulmonary metastases from a murine tumor expressing the p97 antigen. J Immunol **146**, 3235-3241.

Karp SE, Farber A, Salo JC, Hwu P, Jaffe G, Asher AL, Shiloni E, Restifo NP, Mule JJ and Rosenberg SA (1993). Cytokine secretion by genetically modified nonimmunogenic murine fibrosarcoma. Tumor inhibition by IL-2 but not tumor necrosis factor. J Immunol **150**, 896-908.

Knuth A, Wolfel T, Klehmann E, Boon T and Meyer Zum Buschenfelde, KH (1989). Cytolytic T-cell clones against an autologous human melanoma: Specific study and definition of three antigens by immunoselection. Proc Natl Acad Sci USA **86**, 2804-2808.

Kobayashi H, Sendo F, Kaji H, Shirai T, Saito H, Takeichi N, Hosokawa M and Kodama T (1970). Inhibition of transplanted rat tumors by immunization with identical tumor cells infected with Friend virus. J Natl Cancer Inst **44**, 11-19.

Kobayashi H (1979). Viral xenogenization of intact tumor cells. Adv Cancer Res **30**, 279-297.

Ley V, Langlade-Demoyen P, Kourilsky P and Larsson-Sciard E (1991). Interleukin 2-dependent activation of tumor-specific cytotoxic T lymphocytes in vivo. Eur J Immunol **21**, 851-854.

Li W, Diamantstein T and Blankenstein T (1990). Lack of tumorigenicity of interleukin 4 autocrine growing cells seems related to the anti-tumor function of interleukin 4. Mol Immunol **27**, 1331-1337.

Lim CS, Chapman GD, Gammon RS, Muhlestein JB, Bauman RP, Stack RS and Swain JL (1991). Direct in vivo gene transfer into the coronary and peripheral vasculatures of the intact dog. Circulation **83**, 2007-2011.

Lin H, Parmacek MS, Morle G, Bolling S and Leiden JM (1990). Expression of recombinant genes in myocardium in vivo after direct injection of DNA. Circulation **82**, 2217-2221.

Lindahl KF and Wilson DB (1977). Histocompatibility antigen-activated cytotoxic T lymphocytes I. Estimates of the absolute frequency of killer cells generated in vitro. J Exp Med **145**, 500-522.

Lotze MT, Grimm EA, Mazumder A, Strausser JL and Rosenberg SA (1981). Lysis of fresh and cultured autologous tumor by human lymphocytes cultured in T-cell growth factor. Cancer Res **41**, 4420-4425.

Mukherji B and MacAlister TJ (1983). Clonal analysis of cytotoxic T cell response against human melanomas. J Exp Med **158**, 240-245.

Mukherji B, Guha A, Chakrabarty N, Sivandham M, Nafhed A, Sporn J and Ergin M (1989). Clonal analysis of cytotoxic and regulatory T cell responses against human melanoma. J Exp Med **169**, 1961-1967.

Muul LM, Spiess PJ, Director EP and Rosenberg SA (1987). Identification of specific cytolytic immune response against autologous tumor in humans bearing malignant melanoma. J Immunol **138**, 989-995.

Nabel EG, Plautz G and Nabel GJ (1990). Site-specific gene expression in vivo by direct gene transfer into the arterial wall. Science **249**, 1285-1288.

Nabel GJ, Chang A, Nabel EG, Plautz G, Fox BA, Huang L and Shu, S (1992a). Clinical protocols: Immunotherapy of malignancy by in vivo gene transfer into tumors. Hum Gene Ther **3**, 399-410.

Nabel EG, Gordon D, Xang Z-Y, Xu L, San H, Plautz GE, Gao X, Huang, L and Nabel GJ (1992b). Gene transfer in vivo with DNA-liposome complexes: Lack of autoimmunity and gonadal localization. Hum Gene Ther **3**, 649-656.

Nabel EG, Plautz G and Nabel GJ (1992c). Transduction of a foreign histocompatibility gene into the arterial wall induces vasculitis. Proc Natl Acad Sci USA **89**, 5157-5161.

Palella TD, Silverman LJ, Schroll CT, Homa FL, Levine M and Kelley WN (1988). Herpes simplex virus-mediated human hypoxanthine-guanine phosphoribosyltransferase gene transfer into neuronal cells. Mol Cell Biol **8**, 457-460.

Pantel K, Schlimok G, Kutter D, Schaller G, Genz T, Wiebecke B, Backmann R, Funke I and Riethmuller G (1991). Frequent down-regulation of major histocompatibility class I antigen expression of individual micrometastatic carcinoma cells. Cancer Res **51**, 4712-4715.

Parham P, Lomen CE, Lawlor DA, Ways JP, Holmes N, Coppin HL, Salter RD, Wan AM and Ennis PD (1988). Nature of polymorphism in HLA-A, -B and -C molecules. Proc Natl Acad Sci USA **85**, 4005-4009.

Plautz GE, Yang ZY, Wu B, Gao X, Huang L and Nabel GJ (1993). Immunotherapy of malignancy by in vivo gene transfer into tumors. Proc Natl Acad Sci USA **90**, 4645-4649.

Price J, Turner D and Cepko C (1987). Lineage analysis in the vertebrate nervous system by retrovirus-mediated gene transfer. Proc Natl Acad Sci USA **84**, 156-160.

Restifo NP, Spiess PJ, Karp SE, Mule JJ and Rosenberg SA (1992). A nonimmunogenic sarcoma transduced with the cDNA for interferon-γ elicits CD8+ T cells against the wild-type tumor: Correlation with antigen presentation capability. J Exp Med **175**, 1423-1431.

Rosenfeld MA, Siegfried W, Yoshimura K, Yoneyama K, Fukayama M, Stier LE, Paakko PK, Gilardi P, Stratford-Perricaudet LD and Perricaudet M (1991). Adenovirus-mediated transfer of a recombinant alpha 1-antitrypsin gene to the lung epithelium in vivo. Science **252**, 431-434.

Rosenfeld MA, Yoshimura K, Trapnell BC, Yoneyama K, Rosenthal ER, Dalemans W, Fukayama M, Bargon J, Stier LE and Stratford-Perricaudet L (1992). In vivo transfer of the human cystic fibrosis transmembrane conductance regulatory gene to the airway epithelium. Cell **68**, 143-155.

Samulski RJ, Zhu X, Xiao X, Brook JD, Housman DE, Epstein N and Hunter LA (1991). Targeted integration of adeno-associated virus (AAV) into human chromosome 19. EMBO J **10**, 3941-3950.

Schwartzentruber DJ, Topalian SL, Mancini M and Rosenberg SA (1991). Specific release of granulocyte-macrophage colony-stimulating factor, tumor necrosis factor-α and IFN-γ by human tumor-infiltrating lymphocytes after autologous tumor stimulation. J Immunol **146**, 3674-3681.

Shu S, Chou T and Rosenberg SA (1985). Adoptive immunotherapy of newly induced murine sarcomas. Cancer Res **45**, 1657-1662.

Shu S, Chou T and Rosenberg SA (1987). Generation from tumor-bearing mice of lymphocytes with in vivo therapeutic efficacy. J Immunol **139**, 295-304.

Skinner MA and Marbrook J (1976). An estimation of the frequency of precursor cells which generate cytotoxic lymphocytes. J Exp Med **143**, 1562-1567.

Slovin SF, Lackman RD, Ferrone S, Kiely PE and Mastrangelo M (1986). Cellular immune response to human sarcomas: Cytotoxic T cell clones reactive with autologous sarcomas. I. Development, phenotype and specificity. J Immunol **132**, 3042-3048.

Smith HG, Harmel RP, Hanna MG Jr, Zwelling BS, Zbar B and Rapp HJ (1977). Regression of established intradermal tumors and lymph node metastases in guinea pig after systemic transfer of immune lymphoid cells. J Natl Cancer Inst **58**, 1315-1321.

Stewart, MJ, Plautz GE, Del Buono L, Yang ZY, Xu L, Gao X, Huang L, Nabel EG and Nabel GJ (1992). Gene transfer in vivo with DNA-liposome complexes: Safety and acute toxicity in mice. Hum Gene Ther **3**, 267-275.

Tanaka K, Isselbacher KJ, Khoury G and Jay G (1985). Reversal of oncogenesis by the expression of H-2K antigen following H-2 gene transfection. Science **228**, 26-30.

Tefany FJ, Barnetson RS, Halliday GM, McCarthy SW and McCarthy WH (1991). Immunocytochemical analysis of the cellular infiltrate in primary regressing and non-regressing malignant melanoma. J Invest Derm **97**, 197-202.

Teng MN, Park BH, Koeppen HK, Tracey KJ, Fendly BM and Schreiber H (1991). Long-term inhibition of tumor growth by tumor necrosis factor in the absence of cachexia or T-cell immunity. Proc Natl Acad Sci USA **88**, 3535-3539.

Tepper RI, Pattengale PK and Leder P (1989). Murine interleukin-4 displays potent anti-tumor activity in vivo. Cell **57**, 503-512.

Topalian SL, Soloman D and Rosenberg SA (1989). Tumor-specific cytolysis by lymphocytes infiltrating human melanomas. J Immunol **142**, 3714-3725.

Torre-Amione G, Beauchamp RD, Koeppen H, Park BH, Schreiber H, Moses HL and Rowley DA (1990). A highly immunogenic tumor transfected with a murine transforming growth factor type b1 cDNA escapes immune surveillance. Proc Natl Acad USA **87**, 1486-1490.

Townsend A and Bodmer H (1989). Antigen recognition by class I-restricted T lymphocytes. Ann Rev Imm **7**, 601-624.

Townsend S and Allison JP (1993). Tumor rejection after direct costimulation of $CD8^+$ T cells by B7-transfected melanoma cells. Science **259**, 368-370.

Van den Eynde B, Lethe B, Van Pel A, De Plaen E and Boon T (1991). The gene coding for a major tumor rejection antigen of tumor P815 is identical to the normal gene of syngeneic DBA/2 mice. J Exp Med **173**, 1373-1384.

Van der Bruggen P, Traversari C, Choomez P, Lurquin C, De Plaen E, Van den Eynde B, Knuth A and Boon T (1991). A gene encoding an antigen recognized by cytolytic T lymphocytes on a human melanoma. Science **254**, 1643-1647.

Vile RG and Hart IR (1993). In vitro and in vivo targeting of gene expression to melanoma cells. Cancer Res **53**, 962-967.

Wahl WL, Plautz GE, Fox BA, Nabel GJ, Shu S and Chang AE (1992). Generation of therapeutic lymphocytes after in vivo transfection of tumor with a gene encoding allogeneic class I MHC antigen. Surgical Forum **63**, 476-478.

Wallich R, Bulbuc N, Hammerling G, Karzav S, Segal S and Feldman M (1985). Abrogation of metastatic properties of tumor cells by de novo expression of H-2K antigen following H-2 gene transfection. Nature **315**, 301-305.

Whiteside TL, Miescher S, Hulimann J, Moretta L and Von Hiedner U (1986). Clonal analysis and in situ characterization of lymphocytes infiltrating human breast carcinomas. Cancer Immunol Immunother **23**, 169-178.

Wolff JA, Malone RW, Williams P, Chong G, Acsadi G, Jani A and Felgner P (1990). Direct gene transfer into mouse muscle in vivo. Science **247**, 1465-1468.

Yoshizawa H, Chang AE and Shu S (1991). Specific adoptive immunotherapy mediated by tumor-draining lymph node cells sequentially activated with anti-CD3 and IL-2. J Immunol **147**, 729-737.

25

ADOPTIVE IMMUNOTHERAPY OF HUMAN MELANOMA WITH IL-2 AND LYMPHOCYTES: RESULTS AND PERSPECTIVES

L. Rivoltini, F. Arienti, F. Belli, N. Cascinelli and G. Parmiani
Divisions of Experimental Oncology D and Clinical Oncology B,
Istituto Nazionale Tumori, Milan, Italy

Introduction

A potential involvement of the immune system in the control of human melanoma has been suggested by a wealth of data (see Table 25.1). Spontaneous regression of primary lesions, which may occur in melanoma patients (Rumke, 1992), the presence of lymphocyte infiltrate in regressing lesions, and the specific reactivity of lymphocytes against autologous tumor cells (see Parmiani et al., 1990) have been considered as strong evidence for a relevant role of the immune system in this disease.

In recent years, the description of the biological activity and the availability of IL-2, which plays a major role in immunoregulation, allowed the expansion of lymphocytes both in vitro and in vivo, thus contributing to new studies of tumor immunology. A more recent landmark in the understanding of immune recognition of melanoma has been the identification of a melanoma specific antigen recognized by T cells (Van der Bruggen et al., 1991).

During the last decade many different strategies have been employed to define the role of biological therapy in the management of melanoma patients. One of these is represented by adoptive immunotherapy, which consists in the transfer into tumor bearing hosts of in vitro activated immune cells potentially capable of mediating antitumor effects. The major difficulty in the development of successful clinical protocols, however, has been the lack of identification of the subpopulation of immune cells responsible for antitumor activity in vivo. Leukocyte preparations derived from different immunological sites (peripheral blood, lymph nodes, tumor lesions) and consisting of different cell populations (MHC restricted T lymphocytes or LAK cell precursors) have been extensively investigated both in vitro and in vivo.

CYTOKINE-INDUCED TUMOR IMMUNOGENICITY
ISBN 0-12-261520-4

Table 25.1. *In vivo and in vitro evidence of the involvement of the immune system in melanoma*

In vivo	Occurrence of spontaneous regressions of primary tumors
	Detection of lymphocyte infiltrates in lesions regressed either spontaneously or after immunological treatments
	Expression of melanoma antigens by fresh tumor cells
	Clinical response after immunotherapy
In vitro	Detection of melanoma specific T lymphocytes (from peripheral blood, lymph nodes and TIL) after in vitro restimulation with autologous or HLA matched melanoma cells and low doses of IL-2
	Identification of T-cell recognized melanoma antigens (e.g. MAGE-1)

In spite of some encouraging results obtained using IL-2 and in vitro activated lymphocytes for adoptive immunotherapy, and the promising, but too early, approaches involving the use of gene modified T cells (Rosenberg, 1992), it is clear that significant improvements in the effectiveness of immunotherapy of human melanoma will come only when all the steps and factors involved in the immune recognition and rejection of tumor cells will have been delineated.

In this review we shall summarize the state of the art of adoptive immunotherapy and indicate which efforts are being made in our own and other laboratories to investigate the potential application of such an approach in melanoma patients.

In vitro evidence of the involvement of the immune system in melanoma patients

In animal models, immunogenic tumors can be successfully eradicated by active or adoptive immunotherapy with T cells (North et al., 1989). Specific recognition of target cells by T lymphocytes occurs through the contact between the TCR and a cell surface complex composed by an MHC molecule which presents in its groove an antigenic peptide. Thus, antigen recognition requires the presence of both functional MHC molecules and antigenic peptides; these are produced by the target cells through intracellular processing of protein antigens (Townsend and Bodmer, 1989).

The existence in melanoma patients of T lymphocytes able to recognize tumor antigens in vitro in an HLA restricted fashion has been clearly proven by several groups during the past ten years (reviewed in Parmiani et al., 1990; Platsoucas, 1991). In fact, lymphocytes that can elicit a cytotoxic activity against autologous melanoma cells (Anichini et al., 1987) or mediate cytokine release after stimulation with autologous tumor (Schwartzentruber et al., 1991) have been generated from different immunological districts, including PBL, lymph nodes and TIL (see Knuth et al., 1992). The antitumor activity of these cells is specific, since they do not recognize allogeneic HLA unmatched tumors or normal autologous cells (Parmiani et al., 1990). The in vivo presence of tumor specific effectors has been considered for several years as indirect evidence of the existence of melanoma antigens. A direct confirmation that such antigens do exist comes from the work of Thierry Boon and collaborators who succeeded in cloning and identifying T-cell recognized melanoma antigens (Van der Bruggen et al., 1991).

The analysis of melanoma specific T-cell reactions, at bulk or clonal level, has been performed mainly with PBL (Anichini et al., 1987) and, more recently, with TIL (Ito et al., 1988; Topalian et al., 1989). In these systems the presence of tumor specific T lymphocytes is assumed to indicate host response against tumor cells. MHC class I restricted $CD8^+$ CTL are considered to be the most important effector cells of tumor recognition and rejection (Platsoucas, 1991). However, for an optimal recruitment of CTL response, $CD4^+$ T helper cells are also needed. In some tumor systems, however, $CD4^+$ lymphocytes have been shown to mediate a direct tumor specific recognition (Forni et al., 1988), and $CD4^+$ cells recognizing human melanoma have been described (Fossati et al., 1988; Radrizzani et al., 1991).

Freshly isolated lymphocytes from cancer patients usually exert little or no cytotoxic activity against autologous tumor cells. Probably this is due to a low frequency of specific precursor PBL and TIL, to the in vivo presence of immunosuppressive factors (Knuth at al., 1992) or to an inherent anergy of T cells (Parmiani, 1993). One or all of these features may explain why autologous antitumor reactivity in metastatic melanoma patients is lower than in patients with primary lesions (Fossati et al., 1984). Thus, in vitro activation/restimulation of T lymphocytes is normally required to induce a significant antitumor activity. The use of irradiated autologous or HLA matched tumor cells (as antigen source) in conjunction with low doses of IL-2 has proven useful in initiating CTL cultures from PBL; this provides the right stimulus to elicit specific lymphocytes (Mukherji and MacAlister, 1983) and to selectively expand them. Since tumor specific CTL are believed to accumulate at the tumor site, the use of TIL has been recently preferred either for in vitro studies or for clinical trials (see below). Rosenberg's group reported that in approximately one third of melanomas, TIL cultures manifest in vitro cytolytic specificity for autologous tumor cells (Rosenberg, 1992). However, it is still not completely clear whether TIL are a population enriched for tumor specific CTL. Some limited comparative studies reported the obtainment of CTL specific for autologous tumor from TIL of draining lymph nodes and not from PBL of the same patient (Skornick et al., 1990). A recent study of our group, however, indicates a 200-fold higher accumulation of T cells, whose TCR recognize melanoma antigens, in TIL as compared with PBL of the same patient (Sensi et al., 1993).

The detection in vitro of tumor specific lymphocytes postulates the presence of tumor antigens on melanoma cells. Their identification, therefore, has become a key issue in this system and in the entire field of tumor immunology (see Boon in this book). Relevant progress has been made in recent years, with the cloning of the first T-cell recognized melanoma antigen by Boon and coworkers (Van der Bruggen et al., 1991). This is encoded by the MAGE-1 gene, whose product is recognized by melanoma specific T-cell clones in the context of HLA-A1, an HLA allele which shows a 20% frequency in the Caucasian population. MAGE-1 is expressed on about 40-50% of the melanoma lines tested, but it does not seem to be lineage restricted, since it can be detected also on some transformed B lymphocytes, breast carcinoma and other tumor cells (Van der Bruggen et al., 1991; Rimoldi et al., 1993; Zakut et al., 1993). Although MAGE-1 represents the first candidate for a tumor rejection antigen, its in vivo relevance has still to be proven. In fact, in a recent study by Rosenberg's group a panel of HLA-A1 restricted TIL cultures, some of which were effective in clinical trials, failed to recognize MAGE-1 products (Zakut et al., 1993).

An interesting issue is represented by the identification of melanoma antigens recognized by HLA-A2 restricted lymphocytes (Table 25.2.). The prevalence of the HLA-A2 haplotype is nearly 50% in melanoma patients (Marincola et al., 1992); furthermore, HLA-A2 restricted melanoma specific CTL can recognize allogeneic melanomas matched for HLA-A2 in a TCR, MHC restricted fashion, suggesting the existence of a common antigen (Darrow et al., 1989). Surprisingly, the same antigen appears to be detectable by HLA-A2 restricted lymphocytes on HLA-A2 constitutively negative melanoma cells made capable of expressing HLA-A2.1 molecules by gene transfer (Kawakami et al., 1992), suggesting that the lack of HLA class I expression, rather than that of tumor antigens, may be the limiting factor in the recognition of melanoma cells by T lymphocytes.

The postulated ubiquity of this antigen in melanoma makes this molecule an optimal candidate for vaccine therapy. Multiple antigens, however, could be expressed by the same lesion. In fact, using antigen loss variants of individual melanoma cell lines, the presence of at least three and six different epitopes for HLA-A2 and HLA-A1 restricted melanoma specific CTL, respectively, has been suggested (Van den Eynde et al., 1989; Wolfel et al., 1989). Recently, Engelhard and coworkers identified a minimum of six different peptides that can reconstitute melanoma specific epitopes for HLA-A2 restricted lymphocytes by analyzing the peptides bound to MHC molecules of HLA-A2$^+$ melanomas (Slingluff et al., 1993).

Table 25.2. *Data supporting the major role of HLA-A2 presented melanoma antigens as candidates for vaccine therapy*

HLA-A2 molecules are expressed on nearly 50% of melanoma patients
HLA-A2 restricted lymphocytes recognize the majority (67%) of HLA-A2+ melanoma lines (existence of a common antigen)
Naturally HLA-A2 negative melanoma lines can be specifically lysed by HLA-A2 restricted CTL after HLA-A2 gene transfection

One of the antigens expressed on HLA-A2 melanoma has been recently cloned by Boon's group and identified as tyrosinase (Brichard et al., 1993). Moreover, the finding that common antigens recognized by HLA-A2 specific lymphocytes are represented by tissue related protein could explain the lysis of HLA-A2+ melanocytes reported by Anichini et al. (1993a) using a panel of HLA-A2 restricted CTL clones from melanoma patients. However, the presence of predominant melanocyte lineage related antigens, if confirmed, may pose serious constraints in their use as immunogens for cancer therapy.

Since the recognition of antigenic peptides occurs via presentation by MHC molecules, a sufficient expression of these structures is the first condition required for efficient lymphocyte recruitment. Contrary to other human tumors, in which the defective expression of HLA molecules could explain the low number of T specific TIL and their inability to be expanded in vitro (Rivoltini et al., 1992), melanoma cells are believed to frequently express HLA class I molecules, at least when analyzed by mAb directed against a nonpolymorphic HLA-A,B,C determinant (Ruiter et al., 1991). More recent studies, however, using allele specific (e.g. HLA-A2) antibodies showed that an allelic loss in melanoma can occur spontaneously in vivo in 20-30% of cases (Natali et al., 1989) or after in vitro immunoselection (Knuth et al., 1989). Moreover, HLA-B and C loci, which can act as restriction elements, although with a lower frequency than the HLA-A locus (Hom et al., 1993), are not constitutively expressed by cultured melanoma cell lines but can be up-regulated in vitro by IFN-γ treatment (Marincola et al., 1993). Altogether these data underline the possibility that a down-regulation of MHC expression or an antigen processing defect could mediate an in vivo immunological escape, even in a potentially "immunogenic" tumor such as melanoma (Restifo, 1994).

To increase the complexity of the interactions between the immune system and melanoma cells is the requirement of accessory molecules, even for highly specific CTL. In fact, the affinity of the binding of the TCR to the MHC peptide complex can be modulated by the interaction with these molecules (or adhesion molecules), which play an additional role in triggering T-cell activation (Vanky et al., 1990). These molecules are also involved in the lytic mechanisms of other lymphocytes, such as CD3+ MHC unrestricted cells or CD3⁻ NK lymphocytes (Rivoltini et al., 1991). A clear correlation between tumor ICAM-1 expression and lysability by CTL has been reported at the clonal level by our group (Anichini et al., 1993b). In addition, the transfection of the costimulatory ligand B7 in murine melanoma cells was found to induce tumor rejection in vivo, which was mediated by CD8+ T cells, suggesting that B7 expression by tumor cells could improve the in vivo efficacy of immunological treatments (Townsend and Allison, 1993). Finally, another important issue in the characterization of melanoma T specific lymphocytes is the analysis of TCR usage. In fact, if a preferential V α and V β usage can be identified in specific CTL, the transfer of genes encoding TCR(s) responsible of tumor antigen recognition could be used to generate in vitro highly effective lymphocytes to be used in a clinical setting. Several groups have observed an oligoclonal pattern of V α and/or V β TCR in TIL derived T-cell lines (Platsoucas, 1991) or clones of melanoma patients (Sensi et al., 1992) exhibiting a specific antitumor cytotoxicity. Evidence for an accumulation of oligoclonal T cells at the tumor site has been recently provided in a melanoma regressing lesion (Ferradini et al., 1993) and, even more convincingly, T cells bearing a TCR recognizing specific melanoma antigens in an HLA-A2 restricted fashion have been found in TIL (Sensi et al., 1993). In another report, however, no clear evidence of a unique TCR rearrangement in TIL or clones recognizing the common antigen expressed on HLA-A2+ melanoma cells was found (Shilyansky et al., 1993).

Early studies of melanoma adoptive immunotherapy: the use of IL-2 and LAK cells

Because of the resistance of melanoma to chemo- and radiotherapy, and its potential susceptibility to immune recognition, different strategies aimed at potentiating the immune system have been attempted during the last several years. The first demonstration that a purely immunological approach could induce clinically documented regressions in a proportion of melanoma patients was obtained with the administration of high dose IL-2 with or without the adoptive transfer of ex vivo generated LAK cells (reviewed in Rosenberg, 1992). LAK cells are an heterogeneous population composed by both CD3+ and CD3- CD16+ NK lymphocytes that can be generated in vitro by short-term culture with high dose IL-2 and shown to mediate a MHC unrestricted cytotoxicity against a variety of fresh and cultured tumor cells. In patients with metastatic melanoma, the use of IL-2 as a single drug has shown some antitumor activity, but without CR. With the addition of LAK cells a response rate of about 20% has been reported by Rosenberg (1992), including CR, some of which lasted between 18 and 52 months. In a recent analysis of data from a prospective randomized trial comparing the administration of high dose IL-2 alone or in combination with LAK cells in 54 patients with advanced melanoma, the same NCI group from Bethesda observed a nearly 25% response rate both to IL-2 plus LAK cells and to IL-2 alone. Again, no CR was achieved with IL-2 alone, while half of the clinical regressions observed in IL-2 plus LAK cells treated patients were complete (Rosenberg et al., 1993). Moreover, a significant prolongation of survival was shown in the patients treated with the combination.

A variety of other groups were able to reproduce the data obtained by Rosenberg, though with a lower response rate (Dillman et al., 1993). The reasons for such variability may be found, among others, in the different frequency of immunogenic tumors in different series of patients (Parmiani, 1990).

The major limitation of this immunotherapeutic approach was represented by severe side effects associated with high dose IL-2 (Rosenberg, 1992). The toxicity appeared to be related to an increased membrane permeability leading to leakage of fluids into extravascular space. The relevant toxicity of IL-2 based therapies encouraged the modification in the schedules of IL-2 administration, adopting for example a 24 h constant infusion scheme (West et al., 1987).

After a phase I trial with IL-2 alone (Marolda et al., 1987), a pilot study including 15 stage IV melanoma patients treated with IL-2 and LAK cells was performed in our Institute (Cascinelli et al., 1989). Both studies proved ineffective, and only 2 minor and transient responses were noted in the IL-2 plus LAK cells treatment group of patients. The overall survival was not affected by either of the two treatment schedules, and most patients died during the first year after immunotherapy. Because of the severe toxicity experienced during IL-2 administration, the total amount of IL-2 delivered was significantly lower (3-6-fold) than the dosage reported in Rosenberg's studies (1992). This could partially explain the absence of therapeutic efficacy observed in our trial, even though no clear correlation exists between tumor regression and doses of IL-2 administered (Rosenberg et al., 1987).

Although LAK cells plus IL-2 have shown antitumor effects in both animal models and clinical trials, the mechanism of target recognition remains to be elucidated. Homing studies performed in melanoma patients by using ^{111}In labelled LAK cells did not show localization of these cells into tumor deposits (Fisher et al., 1989).

To overcome the limited tumor targeting ability of LAK cells in vivo and to reduce IL-2 toxicity of systemic administration, we have recently evaluated the treatment of melanoma metastases confined to an extremity with IL-2 plus LAK cells administered by isolated limb perfusion (Belli et al., 1992). In a pilot study including 6 patients treated with local IL-2 plus LAK cells, whose lesions relapsed after isolated perfusion with chemotherapy, 1 CR, 4 PR and 1 stable disease were obtained, suggesting that nonspecific immune cells can be effective if their concentration at the tumor site is optimized by locoregional perfusion. An update of this study, which now includes 10 patients, still shows a remarkable benefit compared to an historical control group (Table 25.3). This protocol, therefore, warrants further clinical investigation by a randomized clinical trial.

Therapeutic activity of TIL

In the attempt to overcome the limited clinical response of the IL-2/LAK cell therapy, in 1988 Rosenberg and coworkers carried out a clinical study using TIL in melanoma patients. In fact, TIL were shown to encompass T cells recognizing autologous melanoma in a MHC restricted fashion in at least 50% of cases (Topalian et al., 1989; Parmiani et al., 1990). In addition, homing studies showed that TIL, but not LAK cells, localize preferentially in metastatic lesions (Mukherji et al., 1988; Fisher et al., 1989). In vitro expanded TIL were administered with high doses of IL-2 to 20 patients with a remission rate of 55% (Rosenberg et al., 1988); such a result was, however, not confirmed by other smaller studies which reported remission rates ranging from 17% to 24% (Topalian et al., 1988; Kradin et al., 1989; Dillman et al., 1991). More recently, Rosenberg's group reported an update showing a remission rate of 40% in 55 evaluable patients (Aebersold et al., 1991). In both studies, TIL were expanded in the presence of large quantities of IL-2 (6000 IU/ml) with or without autologous irradiated melanoma cells. Data on the in vitro cytotoxicity of TIL against autologous melanoma cells were often unknown. This drawback is significant considering that the cytotoxic activity of TIL can play a central role in this type of therapeutic approach (Aebersold et al., 1991).

For this reason in our Institute we adopted a two phase schedule of TIL culture: initially with a low amount of IL-2 (120 IU/ml) in the presence of autologous melanoma cells for the first 2 weeks to select for CTL with high affinity IL-2 receptor (i.e. receptor induced by antigen stimulation) and to promote activation of tumor specific TIL, followed by 6000 IU/ml of IL-2 to expand them (Arienti et al., 1993). Fifty-three sequential melanoma patients were submitted to surgical resection of metastatic lesions and were evaluated for TIL growth. Of these patients, only 16 were treated by infusion of autologous TIL plus IL-2 according to the clinical protocol. In fact, between the initiation of TIL culture and the therapeutic delivery of TIL, a consistent number of patients developed brain metastases or died from progression of disease or were treated with different therapies in other hospitals. In addition, some skin lesions were completely necrotic or contaminated, and in 17 cases an insufficient TIL

Table 25.3. *Treatment, clinical results and follow-up of melanoma patients treated by locoregional perfusion with IL-2 and LAK*

Patient #	Total IL-2[a]	Total LAK[b]	Clinical response	Further treatment	Follow-up (months)[d]	Present status[e]
1	211	27	SD[c]	Chemotherapy	33	NED
2	136	26	CR	--	33	NED
3	108	16	PR	Excision	28	NED
4	126	16	PR	Excision	27	NED
5	138	17	PR	Local IFN	23	AWD
6	147	16	PR	Excision	19	AWD
7	133	16.3	PR	Excision	13	AWD
8	133	28	SD	Excision	13	NED
9	177	16	PR	Excision	12	NED
10	110	16	SD	Chemotherapy	10	AWD

Surgical removal of residual disease was considered according to the number and the site of lesions after a period of observation of no less than 2 months. Radical excision of nodules was possible in 6 cases in a period ranging from 2 to 3 months after isolated perfusion. New local metastases were documented in 4 cases. Two of them underwent a new isolated perfusion with TNF-α, IFN-γ and melphalan in hyperthermic condition. No clinical responses were observed. The other 2 patients were treated with surgical excision only. No correlation between clinical response and doses of IL-2 or LAK cell numbers was evident. The follow-up ranged from 10 to 33 months (mean 21.1)
[a]IU x 10^6
[b]x 10^9
[c]Stable disease
[d]At May 1993
[e]NED, no evidence of disease; AWD, alive with disease

growth was observed after the culture period with low doses of IL-2 (Arienti et al., 1993). These data show that the clinical utilization of autologous TIL in advanced cancer patients has severe limitations in terms of organization, selection of patients and cost.

In our study, only 12 of 16 patients treated with TIL and IL-2 were evaluable for clinical response; one CR and 3 PR were obtained, with a remission rate of 33%. Of the 4 responding patients, 3 received TIL showing an in vitro preferential lysis of autologous tumor cells (autologous melanoma cells of the fourth patient were not available).

Table 25.4. *Clinical response rates related to cytotoxicity of autologous melanoma cells by TIL*

Feature	Response	No response
Preferential lysis of autologous melanoma	3/3[a]	3/6[a]
Lysis of autologous melanoma (mean ± SD)[b]	2959±1892	155±109[c]

[a]No. of patients showing preferential lysis/no. of patients evaluated. Autologous melanoma cells of one responding and two nonresponding patients were not available
[b]$p < 0.003$, Student's unpaired t test
[c]LU/10^7

Table 25.5. *Clinical response rates related to different features in melanoma patients treated with TIL and IL-2*

Feature	Response	No response
Number of patients	4	8
Preferential lysis of autologous melanoma	3/3[a]	3/6[a]
Lysis of K562 cells (mean ± SD)[b]	316±205	830±1323
Lysis of Daudi cells (mean ± SD)[b]	146±179	65±107
Lysis of autologous cells (mean ± SD)[b]	2959±1892[a]	155±109[a,c]
Lysis of allogeneic cells (mean ± SD)[b]	141±166	99±124
Predominance of $CD8^+$ TIL	3/4	5/8
Mean of $CD8^+$ percentage	49±17	59±33
Mean of culture days	64±13	50±14
Number of TIL administered (x 10^9)	5.1±5.1	6.24±4.9
Total dose of IL-2 (IU x 10^6)	171±40	109±55

[a,b,c]As in Table 25.4

Nevertheless, injection of TIL with an in vitro preferential lysis of autologous melanoma cells did not appear to be sufficient to obtain a clinical response since no significant regression of tumor lesions was observed in 3 patients treated with this type of TIL. On the other hand, no clinical response was observed after infusion of TIL with an in vitro LAK-like activity.

A strong association between the level of lysis against autologous tumor cells by TIL at the end of the culture time and their therapeutic efficacy was observed. In fact, a statistically significant difference was found between the cytotoxicity against autologous melanoma cells exhibited by TIL infused in responding patients in comparison with the nonresponding ones (see Table 25.4). Other in vitro and in vivo parameters were analyzed in order to evaluate a possible association with the clinical outcome; neither the source of TIL (s.c. lesions or lymph nodal), the number of starting lymphocytes, the culture time, the phenotype and the number of TIL injected, nor the total dose of IL-2 infused appeared to be significantly associated with the clinical response (Table 25.5). Therefore, the concentration of tumor specific, MHC restricted T cells given to the patients is likely to be the major factor impacting on the clinical response in adoptive immunotherapy.

A further approach to the use of T cells for adoptive immunotherapy of melanoma is that proposed by Chang et al. (1993). These authors generated T cells obtained from the lymph nodes draining a site of active immunization of patients with irradiated autologous melanoma cells and BCG, and restimulated them in vitro with autologous tumor cells in the presence of IL-2. Such lymphocytes were then given to patients along with a concomitant administration of IL-2. Despite the induction of delayed-type hypersensitivity to autologous tumor cells in 7 out of 9 patients, the clinical response was marginal (1 PR), suggesting that induction of tumor immunity may not be associated with a clinical response.

Perspectives

We have illustrated how adoptive immunotherapy of melanoma has evolved in the recent years following the availability of recombinant IL-2 for clinical studies. Thanks to the exciting new developments in basic immunology and molecular biology, new approaches aimed at a genetic manipulation either of immune cells or of cancer cells are being attempted and have already led to the proposal of innovative clinical protocols.

New methods or strategies of selection and expansion of tumor specific T effectors could result in a better therapeutic response for melanoma patients infused with such cells. The use of allogeneic melanoma lines matched for MHC class I antigens and expressing common melanoma associated antigen(s) (Crowley et al., 1990), of gene modified tumor cells expressing the gene for cytokines (Gansbacher et al., 1992), and/or of well defined melanoma antigens like MAGE-1 could offer the possibility of obtaining more specific and numerous effector cells in vitro.

We are, therefore, witnessing a boom of proposals aimed at exploiting for therapeutic purposes the genetic modifications of tumor cells and/or host lymphocytes in order to modify the host-tumor relationship. Heeding the lesson of recent years, we should not be too eager to transfer into the clinic concepts which do not have a solid and convincing rationale at the preclinical level. The discovery of genes encoding human tumor antigens, although of exceptional importance, has not yet been followed by an adequate definition of the optimal conditions for immunizing tumor bearing patients with such antigens. Even in animal models such a task is far from being completed. In fact, it may turn out that tumor bearing patients, like those undergoing an active infectious disease, cannot be successfully treated with active immunization, given the subversion that the disease is causing in the body immune system.

Acknowledgements

This work has been supported by EuroCetus (Amsterdam) and by grants of the Italian Ministry of Health (Rome) and CNR (Rome). We thank Ms Grazia Barp for typing and editing.

References

Aebersold P, Hyatt C, Johnson S, Hines K, Korcak L, Sanders M, Lotze M, Topalian SL, Yang J and Rosenberg SA (1991). Lysis of autologous melanoma cells by tumor-infiltrating lymphocytes: Association with clinical response. J Natl Cancer Inst **83**, 932-937.

Anichini A, Fossati G and Parmiani G (1987). Clonal analysis of the cytolytic T cell response to human tumors. Immunol Today **8**, 385-389.

Anichini A, Macalli C, Mortarini R, Salvi S, Mazzocchi A, Squarcina P, Herlyn M and Parmiani G (1993a). Melanoma cells and normal melanocytes share antigens recognized by HLA-A2-restricted cytotoxic T cell clones from melanoma patients. J Exp Med **177**, 989-998.

Anichini A, Mortarini R, Alberti S, Mantovani A and Parmiani G (1993b). T cell receptor engagement and tumor ICAM1 upregulation are required to by-pass low susceptibility of melanoma cells to autologous CTL-mediated lysis. Int J Cancer **53**, 994-1001.

Arienti F, Belli F, Rivoltini L, Gambacorti-Passerini C, Furlan L, Mascheroni L, Prada A, Rizzi M, Marchesi E, Vaglini M, Parmiani G and Cascinelli N (1993). Adoptive immunotherapy of advanced melanoma patients with interleukin-2 (IL-2) and tumor-infiltrating lymphocytes selected in vitro with low doses of IL-2. Cancer Immunol Immunother **36**, 315-322.

Belli F, Arienti F, Rivoltini L, Santinami M, Mascheroni L, Prada A, Ammatuna M, Marchesi E, Parmiani G, Cascinelli N and Vaglini M (1992). Treatment of recurrent in transit metastases from cutaneous melanoma by isolation perfusion in extracorporeal circulation with IL-2 and lymphokine activated killer cells. A pilot study. Melanoma Res **2**, 263-271.

Brichard V, Van Pel A, Wölfel T, Wölfel C, De Plaen E, Lethe B, Coulie P and Boon T (1993). The tyrosinase gene codes for an antigen recognized by autologous cytolytic T lymphocytes on HLA-A2 melanomas. J Exp Med **178**, 489-495.

Cascinelli N, Belli F, Marchini S, Marolda R, Prada A, Sciorelli G, Villani F, Gambacorti-Passerini C, Galazka A and Parmiani G (1989). A phase II study of administration of recombinant interleukin-2 (IL-2) plus lymphokine activated killer (LAK) cells in stage IV melanoma patients. Tumori **75**, 233-244.

Chang AE, Yoshisawa H, Sakai K, Cambron MJ, Sondak VK and Shu S (1993). Clinical observation on adoptive immunotherapy with vaccine-primed T lymphocytes secondarily sensitized to tumor in vitro. Cancer Res **53**, 1043-1050.

Crowley NJ, Slingluff CL, Darrow TL and Seigler HF (1990). Generation of human autologous melanoma-specific cytotoxic T cells using HLA-A2-matched allogeneic melanomas. Cancer Res **50**, 492-498.

Darrow TL, Slingluff CL and Siegler HF (1989). The role of HLA class I antigens in recognition of melanoma cells by tumor-specific cytotoxic T lymphocytes: Evidence for shared tumor antigens. J Immunol **142**, 3329-3334.

Dillman RO, Oldham RK, Barth NM, Cohen RJ, Minor DR, Birch R, Yannelli JR, Maleckar JR, Sferruzza A, Arnold J and West WH (1991).

Continuous interleukin-2 and tumor-infiltrating lymphocytes as treatment of advanced melanoma. Cancer **68**, 1-8.

Dillman RO, Church C, Oldham RK, West WH, Schwartzberg L and Birch R (1993). Inpatient continuous infusion interleukin-2 in 788 patients with cancer. Cancer **71**, 2358-2370.

Ferradini L, Mackensen A, Genevee C, Bosq J, Duvillard P, Avril M-F and Hercend T (1993). Analysis of T cell receptor variability in tumor-infiltrating lymphocytes from a human regressive melanoma. J Clin Invest **91**, 1183-1190.

Fisher B, Packard BS and Read EJ (1989). Tumor localization of adoptively transferred indium-111 labeled tumor-infiltrating lymphocytes in patients with metastatic melanoma. J Clin Oncol **7**, 250-261.

Forni G, Fujiwara H, Martino F, Hamaoka T, Jemma C, Caretto P and Giovarelli M (1988). Helper strategy in tumor immunology: Expansion of helper lymphocytes and utilization of helper lymphokines for experimental and clinical immunotherapy. Cancer Met Rev **7**, 289-309.

Fossati G, Taramelli D, Balsari A, Bogdanovich G, Andreola S and Parmiani G (1984). Primary but not metastatic human melanomas expressing DR antigens stimulate autologous lymphocytes. Int J Cancer **33**, 591-597.

Fossati G, Anichini A, Squarcina P, Mazzocchi A and Parmiani G (1988). Proliferative and/or cytotoxic activity of lymphocyte clones to autologous human melanoma. Int J Cancer **42**, 239-245.

Gansbacher B, Zier K, Cronin K, Hantzopoulos PA, Bouchard B, Houghton A, Gilboa E and Golde E (1992). Retroviral gene transfer induced constitutive expression on interleukin-2 or interferon-γ in irradiated human melanoma cells. Blood **80**, 2817-2825.

Hom SS, Schwartzentruber DJ, Rosenberg SA and Topalian SL (1993). Specific release of cytokines by lymphocytes infiltrating human melanomas in response to shared melanoma antigens. J Immunother **13**, 18-30.

Ito K, Platsoucas CD and Balch CM (1988). Autologous tumor-specific cytotoxic T lymphocytes in the infiltrate of human metastatic melanoma. Activation by interleukin 2 and autologous tumor cells, and involvement of the T cell receptor. J Exp Med **168**, 1419-1442.

Kawakami Y, Zakut R, Topalian SL, Stotter H and Rosenberg SA (1992). Shared human melanoma antigens. I. Recognition by tumor-infiltrating lymphocytes in HLA-A2.1 transfected melanomas. J Immunol **148**, 638-643.

Knuth A, Wolfel T, Klehmann E, Boon T and Meyer zum Buschenfelde K-H (1989). Cytolytic T cell clones against an autologous human melanoma: Specificity study and definition of three antigens by immunoselection. Proc Natl Acad Sci USA **86**, 2804-2808.

Knuth A, Wolfel T and Meyer zum Buschenfelde K-H (1992). T cell responses to human tumors. Cancer Surv **13**, 39-52.

Kradin RL, Kurnick JT, Lazarus DS, Preffer FI, Dubinett SM, Pinto CE, Gifford J, Davidson E, Grove J, Callahan RJ and Strauss HW (1989). Tumour-infiltrating lymphocytes and interleukin 2 in treatment of advanced cancer. Lancet i, 577-580.

Marincola FM, Venzon D, White D, Rubin JT, Lotze MT, Simonis TB, Balkissoon J, Rosenberg SA and Parkinson DR (1992). HLA association with response and toxicity in melanoma patients treated with interleukin 2-based immunotherapy. Cancer Res **52**, 6561-6566.

Marincola FM, Shamamian P, Simonis TB, Abati A, Fetsch P, Restifo NP, Mule' JJ and Rosenberg SA (1993). Locus specific down-regulation of HLA class I in melanoma cell lines. Proc Am Ass Cancer Res, **453**.

Marolda R, Belli F, Prada A, Villani F, Gambacorti-Passerini C, Galazka A, Parmiani G and Cascinelli N (1987). A phase I study of recombinant interleukin-2 in melanoma patients. Toxicity and clinical effects. Tumori **73**, 575-584.

Mukherji B and MacAlister TJ (1983). Clonal analysis of cytotoxic T cell response against autologous melanoma. J Exp Med **169**, 1961-1976.

Mukherji B, Arnbjarnson O and Spitznagle LA (1988). Imaging pattern of previously in vitro sensitized and interleukin-2 expanded autologous lymphocytes in human cancer. Nucl Med Biol **15**, 419-423.

Natali PG, Nicotra MR, Bigotti A, Venturo I, Mercenaro L and Giacomini P (1989). Selective changes in expression of HLA class I polymorphic determinants in human solid tumors. Proc Natl Acad Sci USA **86**, 6719-6723.

North RJ, Awwad M and Dunn PL (1989). T cell mediated tumor regression in experimental system. Prog Immunol **7**, 1097-1103.

Parmiani G (1990). An explanation of the variable clinical response to interleukin 2 and LAK cells. Immunol Today **11**, 113-115.

Parmiani G, Anichini A and Fossati G (1990). Cellular immune response against autologous human malignant melanoma: Are in vitro studies providing a framework for a more effective immunotherapy? J Natl Cancer Inst **85**, 361-370.

Parmiani G (1993). Tumor immunity as autoimmunity: Tumor antigens include normal self proteins which stimulate anergic peripheral T cells. Immunol Today, **14**, 536-538.

Platsoucas CD (1991). Human autologous tumor-specific T cells in malignant melanoma. Cancer Met Rev **10**, 151-176.

Radrizzani M, Benedetti B, Castelli C, Longo A, Ferrara GB, Herlyn M, Parmiani G and Fossati G (1991). Human allogeneic melanoma-reactive T helper lymphocyte clones: Functional analysis of lymphocyte-melanoma interactions. Int J Cancer **49**, 823-830.

Restifo NP (1994). Antigen presentation by INF-γ releasing tumor cells. In "Cytokine Induced Tumor Immunogenicity" (eds G Forni, R Foa, A. Santoni and L Frati), pp 307-328. Academic Press, London.

Rimoldi D, Romero P and Carrel S (1993). The human antigen-encoding gene, MAGE-1, is expressed by other tumour cells of neuroectodermal origin such as glioblastomas and neuroblastomas. Int J Cancer **54**, 527-528.

Rivoltini L, Cattoretti G, Arienti F, Mastroianni A, Melani C, Colombo MP and Parmiani G (1991). The high lysability by LAK cells of colon-carcinoma cells resistant to doxorubicin is associated with high expression of ICAM1, LFA3 and NCA and a less differentiated phenotype. Int J Cancer **47**, 746-754.

Rivoltini L, Arienti F, Orazi A, Cefalo G, Gasparini M, Gambacorti-Passerini C, Fossati-Bellani F and Parmiani G (1992). Phenotypic and functional analysis of lymphocytes infiltrating paediatric tumors, with a characterization of tumor phenotype. Cancer Immunol Immunother **34**, 241-251.

Rosenberg SA, Lotze MT, Muul LM, Chang AE, Avis FP, Leitman S, Rarston Limenan W, Robertson C, Lee R, Rubin JT, Seipp CA, Simpson GC and White DF (1987). A progress report on the treatment of 157 patients with advanced cancer using lymphokine activated killer cells and interleukin 2 alone. N Engl J Med **316**, 889-897.

Rosenberg SA, Packard BS, Aebersold PM, Solomon D, Topalian SL, Toy ST, Simon P, Lotze MT, Yang JC, Seipp CA, Simpson C, Carter C, Bock S, Schwartzentruber D, Wei JP and White DE (1988). Use of tumor infiltrating lymphocytes and interleukin-2 in the immunotherapy of patients with metastatic melanoma. A preliminary report. N Engl J Med **319**, 1676-1680.

Rosenberg SA (1992). The immunotherapy and gene therapy of cancer. J Clin Oncol **10**, 180-199.

Rosenberg SA, Lotze MT, Yang JC, Topalian SL, Chang AE, Schwartzentruber DJ, Aebersold P, Leitman S, Linehan WM, Seipp CA, White DE and Steinberg SM (1993). Prospective randomized trial of high-dose interleukin-2 alone or in conjuction with lymphokine activated killer cells for treatment of patients with advanced cancer. J Natl Cancer Inst **85**, 622-632.

Ruiter DJ, Mattijssen V, Broecker EB and Ferrone S (1991). MHC antigens in human melanomas. Semin Cancer Biol **2**, 35-45.

Rumke PH (1992). Systemic therapy of disseminated melanoma. Sem Surg Oncol **8**, 392-399.

Schwartzentruber DJ, Topalian SL, Mancini M and Rosenberg SA (1991). Specific release of granulocyte-macrophage colony stimulating factor, tumor necrosis factor-α and IFN-γ by tumor-infiltrating lymphocytes after autologous tumor stimulation. J Immunol **146**, 3674-3681.

Sensi M, Castelli C, Salvi S, Mazzocchi A, Mortarini R, Nicolini G, Anichini A and Parmiani G (1992). Expansion of major histocompatibility complex-restricted antimelanoma cytotoxic T-lymphocyte clones with identical T-cell receptor from tumor-infiltrating lymphocytes. J Immunother **12**, 207-211.

Sensi M, Salvi S, Castelli C, Maccalli C, Mazzocchi A, Mortarini R, Nicolini G, Herlyn M, Parmiani G and Anichini A (1993). T cell receptor (TCR) structure of autologous melanoma-reactive cytotoxic T lymphocyte (CTL) clones: Tumor-infiltrating-lymphocytes overexpress in vivo the TCRβ chain sequence used by an HLA-A2-restricted and melanocyte-lineage-specific CTL clone. J Exp Med, **178**, 1231-1246.

Shilyansky J, Nishimura MI, Yannelli JR, Kawakami Y, O'Neil B, Mule JJ and Rosenberg SA (1993). T cell receptor variable gene usage in melanoma specific TIL clones and lines. Proc Keystone Symp "Cellular Immunity and Immunotherapy of Cancer", p 111.

Skornick Y, Topalian SL and Rosenberg SA (1990). Comparative studies of the long-term growth of lymphocytes from tumor infiltrates, tumor-draining lymph nodes and peripheral blood by repeated in vitro stimulation with autologous tumor. J Biol Resp Mod **9**, 431-438.

Slingluff CL, Cox AL, Henderson RA, Hunt DF and Elgelhard VH (1993). Recognition of human melanoma cells by HLA-A2 restricted cytotoxic T lymphocytes is mediated by at least six shared peptide epitopes. J Immunol **150**, 2955-2963.

Topalian SL, Solomon D, Avis FP, Chang AE, Freerksen DL, Linehan WM, Lotze MT, Robertson CN, Seipp CA, Simon P, Simpson CG and Rosenberg SA (1988). Immunotherapy of patients with advanced cancer using tumor-infiltrating lymphocytes and recombinant interleukin-2: A pilot study. J Clin Oncol **6**, 839-853.

Topalian SL, Solomon D and Rosenberg SA (1989). Tumor-specific cytolysis by lymphocytes infiltrating human melanomas. J Immunol **142**, 3714-3725.

Townsend A and Bodmer H (1989). Antigen recognition by class I restricted T lymphocytes. Ann Rev Immunol **7**, 601-624.

Townsend SE and Allison JP (1993). Tumor rejection after direct costimulation of $CD8^+$ T cells by B7-transfected melanoma cells. Science **259**, 368-370.

Van den Eynde B, Hainant P, Herin M, Knuth A, Lemoine C, Weynants P, Van der Bruggen P, Fauchet R and Boon T (1989). Presence on a human melanoma of multiple antigens recognized by autologous CTL. Int J Cancer **44**, 634-640.

Van der Bruggen P, Traversari C, Chomez P, Lurquin C, De Plaen E, Van den Eynde B, Knuth A and Boon T (1991). A gene encoding an antigen recognized by cytolytic T lymphocytes on a human melanoma. Science **254**, 1643-1647.

Vanky F, Wang P, Patarroyo M and Klein E (1990). Expression of the adhesion molecule ICAM-1 and major histocompatibility complex class I antigens on human tumor cells is required for their interaction with autologous lymphocytes in vitro. Cancer Immunol Immunother **31**, 19-27.

West WH, Tauer KW, Yannelli JR, Marshall GD, Orr DW, Thurman GB and Oldham RK (1987). Constant infusion recombinant interleukin-2 in adoptive immunotherapy of advanced cancer. N Engl J Med **316**, 898-905.

Wolfel TE, Klehmann C, Müller C, Schutt K-H, Meyer zum Buschenfelde K-H and Knuth A (1989). Lysis of human melanoma cells by autologous cytolytic T cell clones: Identification of human histocompatibility leukocyte antigen A2 as a restriction element for three different antigens. J Exp Med **170**, 797-802.

Zakut R, Topalian SL, Kawakami Y, Mancini M, Eliyahv S and Rosenberg SA (1993). Differential expression of MAGE-1-2 and -3 messanger RNA in transformed and normal human cell lines. Cancer Res **53**, 5-8.

26

IMMUNORECOGNITION OF HUMAN LEUKEMIC CELLS ELICITED BY IL-2

R. Foa, A. Guarini, A. Gillio Tos, L. Riera and A. Cignetti
Department of Biomedical Sciences and Human Oncology and CNR Center for Immunogenetics and Experimental Oncology, University of Turin, Turin, Italy

Introduction

Immunotherapy has been a long desired goal for the management of acute leukemia patients. Despite multiple approaches, largely based on vaccination protocols with irradiated leukemic cells and with *Corynebacterium parvum* or BCG vaccine, the overall results have been mostly disappointing (Foon et al, 1983; Stryckmans et al, 1983). Regardless of the negative clinical findings so far encountered, the rationale for attempting immunological maneuvers has nonetheless increased over these last few years in parallel with our better understanding of the mechanisms of immune recognition and surveillance. In turn, this has allowed the design and clinical implementation of new and less empirical therapeutic strategies for tumor patients.

In acute leukemia several clinical and biological considerations lead to the belief that if an optimal activation of the immune system of the host could be achieved, this might play a role in controlling the disease. On clinical grounds, it is well recognized that acute leukemia patients may relapse even after several years in apparent complete remission with a leukemic picture with morphological, cytochemical and immunological features identical to those observed at diagnosis. This strongly suggests that the neoplastic clone had not been eradicated, but rather that it had remained dormant during the entire remission period and undetectable according to conventional criteria. Should this be the case, it is conceivable that the few residual blasts may have remained stable under the control of the host immune system and that at some stage, for reasons at present uncertain, they may have escaped surveillance. Further clinical indications that the proliferative kinetics of leukemic cells may, in certain circumstances, be under an exogenous control is suggested by patients with myelodysplastic syndromes or with chronic myeloid leukemia in blast crisis who may show a proportion of neoplastic cells that can remain unchanged without treatment over a considerable period of time.

CYTOKINE-INDUCED TUMOR IMMUNOGENICITY
ISBN 0-12-261520-4

Biologically, numerous experimental evidences support a direct or indirect role of the immune system in the clinical course of hematologic malignancies. The marked immunodepression in patients with Hodgkin's disease at diagnosis has been known for many years (Levy and Kaplan, 1974), while in patients in long lasting remission the immune status may be restored (Lauria et al., 1986). In B-cell chronic lymphocytic leukemia the impairment of both the cellular and humoral immunity appears to correlate with the stage of the disease, with a partial sparing in early stage patients (for a review see Foa, 1992). T-cell depletion procedures in patients undergoing an allogeneic bone marrow transplant lead to an increased relapse rate (Champlin, 1990), as well as to a relatively high incidence of EBV associated malignant lymphomas occurring in the donor lymphocytes (Shapiro et al., 1988). The primary role played by the immune system is further strengthened by the evidence that these latter patients show a dramatic clinical response following infusion of donor lymphocytes, probably related to anti-EBV CTL (Papadopoulos et al., 1994). In acute leukemia, evidence has been accumulated of an impairment of the spontaneous cytotoxic activity (NK) in patients with active disease (Lotzova et al., 1985). Furthermore, it has been suggested that functional monitoring of the NK function in acute leukemia patients in remission may represent a relapse predictive variable (Pizzolo et al., 1988).

Taken together these data strongly suggest that, should a therapeutic strategy aimed at inducing a positive triggering of the host immune system be identified, this could conceptually lead to an amplified control over the neoplastic clone. Ideally, an optimal immunotherapeutic approach should aim at generating or amplifying CTL directed specifically against the autologous tumor. This requires that the cytotoxic compartment of the host can be adequately stimulated and that the tumor is immunogenic. While this goal is starting to be achieved in other tumors, particularly in melanoma, following the demonstration that CTL progenitors can be found within the circulating lymphocytes of the patients (Coulie et al., 1992) and the recognition of tumor associated antigens (Van der Bruggen et al., 1991; Brichard et al., 1993), as well as through novel technological strategies (see other chapters in the book), little information is available for acute leukemia.

Is there evidence for an immunogenic potential of human acute leukemia cells?

This still remains an uncertain issue. All antigens identified on human leukemic cells have proven to be also expressed by normal cells in some tissue area; thus, no leukemia associated or specific antigen has so far been identified. Recently, using irradiated leukemic cells as stimulators and cells from HLA genotypically identical sibling donors as responders, it has been suggested that specific alloreactive CTL clones may be generated in vitro (Faber et al., 1992). Furthermore, the possibility of triggering cytotoxic cells capable of lysing autologous tumor cells has been observed following coculture of the patients' lymphocytes with allogeneic AML blasts in the presence of IL-2 (personal data). This appears to occur less frequently in autologous coculture conditions, i.e. in the absence of a so far ill-defined exogenous (but leukemic) signal. The whole scenario of tumor immunogenicity, including leukemia, will have to be reassessed and possibly reevaluated on the basis of our advancements in knowledge and technology. This will include investigating the activation role of cytokines, evaluated both in vitro and in clinical practice, as well as gene transfer approaches aimed at introducing foreign material into given target cells. Through these technologies it will be possible to investigate whether the immunogenic potential of a tumor cell can be manipulated following transduction of different genes. This approach, which until recently would have appeared futuristic, is a rapidly evolving reality.

Is there a rationale for the use of immunotherapy in acute leukemia?

Over the last decade immunotherapy for cancer patients has been revitalized following the recognition that IL-2, among its many activities, is capable of generating a previously unrecognized cytotoxic function within normal lymphocytes (Grimm et al., 1982). These effector cells, named LAK, in a classic 4 h ^{51}Cr release assay can lyse the majority of tumor cells, while being largely ineffective against normal cells. LAK cells differ, therefore, from the classic effectors of natural cytotoxicity, NK cells, which by definition are lytic against few tumor cells, and the prototype of which is represented by the K562 cell line (Herberman, 1983). Studies in experimental models have convincingly documented the potential antitumor effectiveness of IL-2 and LAK cells (Mule et al., 1984; Lafraniere and Rosenberg, 1985). These findings have prompted researchers to conduct pilot studies in cancer patients with advanced stage disease no longer responsive to conventional chemo- or radiotherapy. These studies have shown that PR or, less frequently, CR may be recorded in a proportion of patients treated only with IL-2 based immunotherapy (Rosenberg et al., 1985; Rosenberg et al., 1987; West et al., 1987). Some of the remissions have been long lived. The overall results obtained over the last years indicate that one may expect to obtain an objective clinical response in 10-20% of patients with metastatic melanoma and renal cell carcinoma treated with IL-2 plus ex vivo generated LAK cells or with IL-2 alone (Rosenberg et al., 1989). In other tumors the clinical results have been less encouraging and less consistent.

The reasons for investigating the potential role of immunotherapy with IL-2/LAK cells in acute leukemia stem largely from some of the above considerations, and from the unfavorable natural course and overall poor prognosis of most adult acute leukemia patients. This would allow for the enrolment of patients with more limited disease and enable to assess the response to IL-2 over a relatively short period of time. The kinetics of solid tumor has made it impossible to evaluate the effect of IL-2, with or without LAK cells, as adjuvant therapy in patients with minimal disease.

The necessary preclinical studies carried out in vitro and in vivo in nude mice have documented that:

- murine leukemias may be eradicated by IL-2 plus/minus LAK cells
- human leukemic blasts may be lysed by normal LAK effectors
- acute leukemia patients in CR often show a good LAK activity against allogeneic and autologous targets
- IL-2 only rarely induces a proliferative signal on leukemic blasts
- in immunosuppressed nude mice normal LAK cells and IL-2 alone are capable of blocking the growth of human acute leukemia cells (Oshimi et al., 1986; Adler et al., 1988; Fierro et al., 1988; Lista et al., 1989; Foa et al., 1990a).

Clinical results with IL-2 based immunotherapy in acute leukemia

In 1986, as a follow-up to the above reported preclinical findings, in collaboration with the Chair of Hematology in Rome, we started to treat acute leukemia patients with IL-2 alone. The protocol employed is based on repeated 5 day cycles of IL-2 given i.v. by continuous infusion at daily escalating doses in order to reduce toxicity. Each cycle is followed by 2-3 days off therapy. The results obtained over the years have enabled us to draw several conclusions (Foa et al., 1990b; Foa et al., 1991c and reviewed by Foa, 1993). First of all, they allowed to show that high dose IL-2 can be administered to acute leukemia patients. Using a daily dose escalating protocol, toxicity is usually acceptable. If necessary, side effects can be further controlled following reduction or suspension of IL-2. On clinical grounds, patients with a large tumor mass show little response to IL-2 alone. On the basis of some clinical (Foa et al., 1991c) and in vitro observations (personal data), the combination of IL-2 and chemotherapy needs to be further investigated.

CR have, instead, been documented with IL-2 alone in acute myeloid leukemia (AML) patients with a limited proportion of residual bone marrow blasts (>5%, <30%) (Foa et al., 1990b; Foa et al., 1991c; Meloni et al., submitted). Some of these remissions have been long lived and have exceeded the longest remission in the natural history of the patient (Meloni et al, submitted).

Clinical responses have also been reported in other series of patients, particularly in AML with a small proportion of residual blasts (Maraninchi et al., 1991; Lim et al., 1992; Toze et al., 1993). The main results so far reported are summarized in Table 26.1.

The administration of IL-2 is coupled with marked hematological and clinical changes (Foa et al., 1991c). During the infusion of IL-2, patients disclose a profound neutrophilia associated with eosinophilia. Following suspension of IL-2, a more or less evident leukocytosis takes place, with a marked lymphocytosis and persistent eosinophilia. At this time, an increase in medium to large granular lymphocytes with prominent polar azurophilic granules is present. Repeated cycles of IL-2 induce in practically all patients a thrombocytopenia, which in some may require supportive measures. Marked thrombocytopenia has been recorded also in patients who started IL-2 with normal platelet values. Though thrombocytopenia may be contributed by peripheral sequestration due to organomegaly (Paciucci et al., 1990), we have shown that in acute leukemia patients a direct inhibitory effect of LAK effectors on the in vitro growth of autologous bone marrow megakaryocytic progenitors takes place, and that this is mediated by TNF-α (Guarini et al., 1991). As discussed below, IL-2 treated patients reveal an increase in endogenous LAK cells, as well as the in vivo release of TNF-α. A moderate degree of anemia, which normally does not require transfusion support, is often observed. Clinically, practically all patients exhibit a variable degree of hepato-splenomegaly (Foa et al., 1991c).

A clinical setting where immunotherapy is an obvious conceptual option is in the context of autologous bone marrow transplant in an attempt to enhance surveillance and potentially further decrease or eradicate the residual neoplastic clone. The likelihood that such an approach may prove beneficial derives, clinically, from the evidence that following an autograft patients should be at their lowest tumor burden and, biologically, from the demonstration that an enhanced cytotoxic progenitor cell compartment can be detected at this time (Higuchi et al., 1989; Reittie et al., 1989). The results so far reported indicate the feasibility of administering IL-2 after an autograft (Blaise et al., 1990; Higuchi et al., 1991; Meloni et al., 1992; Benyunes et al., 1993), though no response data are available at present. A major concern has been the potential thrombocytopenic effect of IL-2; this has prompted to starting IL-2 at a time when the platelet values were recovering and deemed safe, which meant initiating treatment at about 70-80 days after engrafting. In view of the kinetics of relapse of autografted acute leukemia patients, which often occurs early after transplant, and of the precocious presence of circulating cytotoxic progenitors, this delay could prove an important limiting factor. An alternative approach indicates the possibility of administering daily low doses of IL-2 by continuous i.v. infusion immediately after engrafting (Soiffer et al., 1992). This schedule appears to be feasible and

Table 26.1. *Clinical results in leukemic patients*

Author	Disease (No. of patients)	Clinical status	Response
Foa et al. (1991c)	AML (12)	Relapsed or resistant (8-90% BM blasts)	5 CR (only in patients with <20% BM blasts)
Maraninchi et al. (1991)	AML (10)	Relapsed or resistant (13-85% BM blasts)	2 CR 1 PR
Lim et al. (1992)	AML (13)	7 in CR 6 in relapse	4 CR 1 CR (patient with 10% BM blasts)
Toze et al. (1993)	RAEB[a] (1)	7% BM blasts	CR

[a]Refractory anemia with excess of blasts

capable of generating a marked enhancement of the host cytotoxic function.

Immunological changes in the IL-2 treated patients

This is clearly an important issue because it has made it possible to demonstrate that the known immunomodulating properties induced in vitro by IL-2 can be documented in vivo in the treated patients (Gottlieb et al., 1989; Foa et al., 1991b). In the IL-2 treated acute leukemia patients these changes were recorded after the high dose "induction' cycles 2-3 days after stopping IL-2, at a time when the greatest leukocytosis and lymphocytosis are observed. These changes are documented phenotypically by the expression of activation markers (HLA-DR and IL-2 receptor α chain) and by the amplification of a cytotoxic related phenotype (CD16 and CD56). Functionally, IL-2 induces an amplification of the NK and LAK functions; more relevantly, it is capable of activating endogenous LAK effectors. It is important to note that these phenotypic and functional modifications of the host immune system could be documented both on peripheral blood and on bone marrow lymphocytes

(Foa et al., 1991b). This is of relevance because it shows that the IL-2 administered by continuous infusion is capable of activating cells outside the blood stream. These data indicate that IL-2 generated cytotoxic effectors can either recirculate to reach other body sites or can be directly activated in different organs by IL-2. In the context of acute leukemia patients, the demonstration that bone marrow lymphocytes can be stimulated by IL-2 is relevant since the marrow is the primary site of residual disease and of relapse.

Activation of the host immune system occurs in all acute leukemia patients treated; similarly to what has been observed in solid tumors (Favrot et al., 1990), it does not appear to correlate with the clinical response to IL-2 (Foa et al., 1991b). Nonetheless, it should be noted that the amplification of the cytotoxic compartment is notably more evident, both in the blood and in the marrow, in acute leukemia patients with limited disease than in those with a large marrow blastosis (Foa et al., 1991c; Meloni et al., submitted). This indicates that patients with limited or minimal disease have a less compromised immune compartment which can be more effectively amplified following immunotherapeutic maneuvers.

In addition to these phenotypic and functional changes, the administration of IL-2 is associated with the in vivo release of different cytokines and growth factors (Heslop et al., 1989 and personal data). These include TNF-α, IFN-γ, IL-3, IL-5, GM-CSF. On the one hand, these may play a role in some of the side effects (TNF-α) and hematological changes recorded with IL-2 administration (IL-3 and GM-CSF on the leukocyte count, IL-5 on the eosinophilia), while on the other hand they may contribute to the clinical response (TNF-α, IFN-γ, IL-5) and, potentially, to a better tumor antigen presentation (IFN-γ, GM-CSF).

Patients who have received IL-2 after an autologous bone marrow transplant show an even greater activation of the cytotoxic compartment than do nongrafted patients (Meloni et al., 1992). These findings confirm that in autografted patients there is an increased proportion of circulating cytotoxic progenitors and demonstrate that this includes also potential effectors of LAK activity.

Despite the convincing evidence that the administration of IL-2 to acute leukemia patients is capable of inducing a marked modulation of the host immune system and, in detail, an overall amplification of the cytotoxic compartment, several issues closely connected with the clinical response of acute leukemia patients to IL-2 based immunotherapy still need to be clarified. More in detail, it is uncertain why patients with a larger tumor mass do not respond to IL-2 and why the clinical responses observed in patients with limited disease have been heterogeneous and, so far, unpredictable. Furthermore, the reasons for the lack of correlation between immune activation and clinical response to IL-2 are still nuclear. It is possible that some of these unresolved issues may be related to autologous recognition and specific killing capacities. In broader terms, what is the evidence that the cytotoxic activity boosted or generated by the in vivo infusion of IL-2 is directed against the autologous tumor cell population and not only against unrelated allogeneic targets? This is clearly an important question in order to define

- whether the clinical responses observed are at least partly mediated by LAK/CTL
- whether the heterogeneous responses may be due to a variable cytotoxic activation induced by IL-2
- whether a potential clinical response may be predicted in vitro.

Indirectly, this information could help to clarify the CTL precursor frequency in acute leukemia and, in turn, the immunogenicity of leukemic blasts. Subsequently, new technologies aimed at investigating the possibility of generating a specific recognition and thus at readdressing the issue of the immunogenicity of acute leukemia cells will need to be carefully approached.

The negative results observed in patients with a large neoplastic burden could be attributed simply to the insufficient numbers of cytotoxic progenitors. Though this may be a contributing factor, in vitro studies have shown that, in the great majority of acute leukemias at diagnosis and, to a further extent, at relapse, the patients' LAK cells are incapable of lysing autologous tumor cells (Foa et al., 1991a). This is unrelated to the fact that the blasts may be killed by allogeneic normal LAK effectors and that the patients' LAK cells may be active against classic NK resistant targets (HL-60, Daudi, etc) (Foa et al., 1991a). This clearly indicates that in patients with active disease there is often a defect in the autologous recognition machinery. On the other hand, patients in CR often reveal a restored autologous killing capacity (Foa et al., 1991a). These findings suggest that the lack of response in patients with a large leukemic disease may also be due to the inability of IL-2 to induce effectors capable of lysing the autologous tumor. Furthermore, it is conceivable that the heterogenous clinical responses observed in patients with apparently similar clinico-hematological pictures may be due to a different individual activation of LAK effectors directed against the autologous tumor. This possibility, that may also hold true for other tumors, has so far not been addressed. In our experience, however, a patient who achieved a IVth remission with IL-2 alone, and who persists in remission more than 5 years later, displayed a good LAK mediated recognition of his own leukemic blasts (Foa et al., 1990b), while this has not been the case for some nonresponding patients (personal data).

This entire issue of autologous recognition and killing is not readily approachable in solid tumors, since it would require the viable single cell freezing of the neoplastic cells at the time of diagnosis and ideally the enrolment of patients with limited or minimal disease in the immunotherapeutic protocol. However, this aspect should be better defined by the study currently under way in Italy on AML patients in IInd complete remission randomized to receive IL-2 or not after reinduction chemotherapy.

What new approaches can be employed to investigate the immunogenicity of human acute leukemia cells?

The possibility of transducing foreign material into the DNA of normal and neoplastic cells has opened new avenues in our understanding of different issues of basic science, as well as new therapeutic strategies. In cancer, great hopes rest on the feasibility of introducing cytokine and growth factor genes into tumor cells or into normal effector cells. These approaches have recently become a reality, largely through the use of retroviral vectors which allow a more efficient and stable DNA integration. While most of the information gathered over the last few years has been reviewed and discussed in several chapters of the present book, it is worth noting that through cytokine gene transfer into animal tumor cells some of the primary goals of immunotherapy have been fulfilled. Following retroviral vector mediated transduction of the IL-2 gene, it could be shown not only that the engineered tumor cells did not grow in mice but also that they could elicit a specific CTL response against the autologous tumor (Fearon et al., 1990; Gansbacher et al., 1990).

Furthermore, the IL-2 released constitutively by the transduced cells was capable of inducing an anti-tumor memory against subsequent challenges (Fearon et al., 1990; Gansbacher et al., 1990 and reviewed in Foa et al., 1992). These are in essence the findings which have prompted the activation of vaccination clinical trials for melanoma and renal cell carcinoma patients with advanced disease using IL-2 gene transduced allogeneic cell lines. In man, it has been reported that the IL-2 gene can be successfully transduced into melanoma and renal cell carcinoma cell lines, and that this leads to a decreased tumorigenicity in nude mice (Gansbacher et al., 1992; Gastl et al., 1992).

Taking advantage of autologous conditions, i.e. tumor cell lines from living patients, we succeeded in documenting that melanoma cells engineered with the IL-2 gene can induce the generation of cytotoxic effectors capable of specifically recognizing the autologous tumor cells (Guarini et al., 1994). Though theoretically conceivable in view of the knowledge that melanoma is an immunogenic tumor, that autologous CTL may be induced in vitro following incubation with low dose exogenous IL-2, and that melanoma tumor antigens have been recently identified (for a review see Boon, 1993), these findings are relevant to implementing gene therapy protocols in oncology since they demonstrate in a human setting the possibility of generating specific cytotoxic lysis by IL-2 gene transduced tumor cells.

The interest in exploring the possibility of inserting the IL-2 into human leukemia cells stems from different considerations. First of all, the encouraging clinical results described above were observed in AML patients with limited disease treated with high dose exogenous IL-2. As mentioned, however, the administration of IL-2 is associated with some limitations, which include the relevant toxicity, as well as the heterogeneous clinical response and lack of evidence that exogenous IL-2 can induce the generation of specific CTL. Furthermore, based on the results with engineered melanoma cell lines, gene transfer could offer a new and more refined approach to investigating the immunogenicity of human acute leukemia cells.

The IL-2 gene has been successfully transduced into a murine leukemic cell line (38C13). The engineered and IL-2 producing cells no longer grow in the syngeneic mice (Gansbacher et al., 1991). By using retroviral vectors we have recently succeeded in transducing the IL-2 gene into both myeloid and lymphoid human leukemia lines (Cignetti et al., 1994). The integration was shown by reverse transcriptase-PCR and the constitutive release using both a bioassay and an ELISA kit. Since, in general, the overall production of IL-2 appears lower compared to that observed with IL-2 gene transduced human melanoma and renal cell carcinoma lines (Foa et al., 1992; Gansbacher et al., 1992; Gastl et al., 1992), the engineered leukemic cells can be further subcloned in order to obtain clones which produce variable amounts of IL-2.

Taking advantage of this, we could also investigate whether the IL-2 released by the engineered cells could modify some of the features of the parental cells. This issue is relevant in view of the evidence that some leukemic cells may express the α and/or β chain of the IL-2 receptor (Armitage et al., 1986; Rosolen et al., 1989) and of the possibility that in some cases IL-2 could induce a proliferative signal on the neoplastic clone (Carron and Cawley, 1989). In our experience, the IL-2 gene transduced leukemic cells do not modify their morphologic, phenotypic and proliferative pattern, even in the high IL-2 producing subclones (Cignetti et al., 1994). Experiments carried out in immunosuppressed nude mice, a model that allows the growth of human leukemic cells (Caretto et al., 1989), demonstrated that IL-2 gene transduced acute leukemia cell lines show a reduced or abrogated tumorigenicity, which correlates with the quantities of IL-2 released by the engineered cells (Cignetti et al., 1994).

Taken together, the results of these studies show that using retroviral vectors human leukemic cell lines, of both myeloid and lymphoid origin, can be successfully transduced with the IL-2 gene; this leads to the stable and prolonged release of variable amounts of IL-2 which, in turn, can reduce the tumorigenic potential of the leukemic cells. A model is thus established through which the immunogenicity of leukemic cells may be readdressed in a more "physiologic" manner which mimics the scenario potentially occurring in vivo through the activation of vaccination protocols with IL-2 gene transduced leukemic cells.

Conclusions and perspectives

The clinical data so far accumulated using IL-2 as an immunotherapeutic agent in the management of acute leukemia have now extended beyond the phase I stage and have shown that in a proportion of

AML patients with advanced disease residual blasts may be eliminated by IL-2 alone. Since the amplification of the NK and LAK compartment, as well as in vivo release of TNF-α and IFN-γ, appears to occur in all treated patients, irrespective of the clinical response, it is likely that the heterogeneous behavior may be related to the immunogenicity of each individual case and thus to the possibility or otherwise of inducing CTL effective against the autologous tumor. The evidence that autologous lysis may occur in cases in remission and the favorable responses observed strongly suggest that the in vivo administration of IL-2 is capable in some AML patients of inducing an immune mediated process of leukemia blasts recognition and lysis. This implies that IL-2 can activate effective cytotoxic cells capable of lysing the autologous neoplastic cells. It is worth noting that while promising results have been recorded in AML, so far this has not been observed for ALL. Though fewer ALL patients have been treated with IL-2, it is nonetheless conceivable that in ALL the host immune system may be compromised by a neoplastic process that involves immunocompetent cells.

The availability of cytokines in recombinant form and the possibility of introducing cytokine genes into the genome of neoplastic cells has opened new possibilities to investigate the potential immunogenicity of neoplastic cells. Based on the above discussed data in experimental models, which have shown that IL-2 gene transduced tumor cells can elicit specific CTL, as well as an antitumor memory, and based on the fact that human melanoma cells engineered to secrete IL-2 can generate an in vitro specific recognition, the investigation of the immunogenicity of acute leukemia cells needs to be further addressed. In addition to IL-2, other cytokines and cytokine genes will need to be explored both alone and in combination. These include IFN-γ and GM-CSF, which are potentially capable of enhancing HLA class I expression and tumor antigen presentation. Particular attention should be devoted to GM-CSF, which could have a key role in amplifying putative tumor antigens. This has been convincingly documented with the malignant murine 38C13 B cells which express Ig variable regions which are tumor specific but very weakly immunogenic. By fusing the tumoral idiotype with GM-CSF and inoculating i.p. twice (with an interval of 2 weeks), specific anti-idiotypic antibodies could be generated 10 days after each inoculum. When 38C13 cells were inoculated 2 weeks after the second immunization, at day 120, 60% of mice were completely tumor free (Tao and Levy, 1993).

Through the combined analysis of different cytokine genes it may be possible to unravel differences in the process of recognition and lysis between AML and ALL and within different subgroups of cases or according to the timing of peripheral blood sample collection. In turn, these studies should allow a better indication of whether immunotherapeutic strategies, as well as gene therapy protocols, may play a role in the management of patients with acute leukemia.

Acknowledgments

This work was supported by grants from Associazione Italiana per la Ricerca sul Cancro (AIRC), Special Project on "Gene Therapy", Milan, Istituto Superiore di Sanita, Italy-USA Program on "Therapy of Tumors", and by Consiglio Nazionale delle Ricerche (CNR), Project on "Applicazioni Cliniche della Ricerca Oncologica", Rome, Italy. AGT is a PhD student in Oncology at the University of Turin. LR and AC are in receipt of fellowships from Fondazione "A. Bossolasco", Turin and AIRC, Milan, respectively.

References

Adler A, Chervenick PA, Whiteside TL, Lotzova E and Herberman RB (1988). Interleukin 2 induction of lymphokine-activated killer (LAK) activity in the peripheral blood and bone marrow of acute leukemia patients. I. Feasibility of LAK generation in adult patients with active disease and in remission. Blood **71**, 709-716.

Armitage RJ, Lai AP, Robert PJ and Cawley JC (1986). Certain myeloid cells possess receptors for interleukin-2. Br J Haematol **64**, 799-807.

Benyunes MC, Massumoto C, York A, Higuchi CM, Buckner CD, Thompson DA, Petersen FB and Fefer A (1993). Interleukin-2 ± lymphokine activated killer cells as consolidative immunotherapy after autologous bone marrow transplantation for acute myelogenous leukemia: A preliminary report. Bone Marrow Transplant **12**, 159-163.

Blaise D, Olive D, Stoppa AM, Viens P, Pourreau C, Lopez M, Attal M, Jasmin C, Monges G, Mawas C, Mannoni P, Palmer P, Franks C, Philip T and Maraninchi D (1990). Hematologic and immunologic effects of the systemic administration of recombinant interleukin-2 after autologous bone marrow transplantation. Blood **76**, 1092-1097.

Boon T (1993). Tumor antigens recognized by cytolytic T lymphocytes: Present perspectives for specific immunotherapy. Int J Cancer **54**, 177-180.

Brichard V, Van Pel A, Wölfel T, Wölfel C, De Plaen E, Lethe B, Coulie P and Boon T (1993). The tyrosinase gene codes for an antigen recognized by autologous cytolytic T lymphocytes on HLA-A2 melanomas. J Exp Med **178**, 489-495.

Caretto P, Forni M, D'Orazi G, Scarpa S, Feraiorni P, Jemma C, Modesti A, Ferrarini M, Roncella S, Foa R and Forni G (1989). Xenotransplantation in immunosuppressed nude mice of human solid tumors and acute leukemias directly from patients or in vitro cell lines. Res Clin Lab **19**, 231-243.

Carron JA and Cawley JC (1989). IL-2 and myelopoiesis: IL-2 induces blast cell proliferation in some cases of acute myeloid leukaemia. Br J Haematol **73**, 168-172.

Champlin RE (1990). T-cell depletion for bone marrow transplantation: Effects on graft rejection, graft-versus-host disease, graft-versus-leukemia, and survival. Cancer Treat Rep **50**, 99-111.

Cignetti A, Guarini A, Carbone A, Forni M, Cronin K, Forni G, Gansbacher B and Foa R (1994). Transduction of the IL2 gene into human acute leukemia cells induces tumor rejection without modifying cell proliferation and IL2 receptor expression. J Natl Cancer Inst, in press.

Coulie PG, Somville M, Lehmann F, Hainaut P, Brasseur F, Devos R and Boon T (1992). Precursor frequency analysis of human cytolytic T lymphocytes directed against autologous melanoma cells. Int J Cancer **50**, 289-297.

Faber LM, Van Luxemburg-Heijs SAP, Willemze R and Falkenburg JHF (1992). Generation of leukemia-reactive cytotoxic T lymphocyte clones from HLA-identical bone marrow donor of a patient with leukemia. J Exp Med **176**, 1283-1289.

Favrot MC, Combaret V, Negrier S, Philip I, Thiesse P, Freydel C, Bijmann JT, Franks CR, Mercatello A and Philip T (1990). Functional and immunophenotypic modifications induced by interleukin-2 did not predict response to therapy in patients with renal cell carcinoma. J Biol Resp Mod **9**, 167-177.

Fearon ER, Pardoll DM, Itaya T, Golumbek P, Levitsky HI, Simons JW, Karasuyama H, Vogelstein B and Frost P (1990). Interleukin-2 production by tumor cells bypasses T helper function in the generation of an antitumor response. Cell **60**, 397-403.

Fierro MT, Xin-Sheng L, Lusso P, Bonferroni M, Matera L, Cesano A, Lista P, Arione R, Forni G and Foa R (1988). In vitro and in vivo susceptibility of human leukemic cells to lymphokine activated killer activity. Leukemia **2**, 50-54.

Foa R, Caretto P, Fierro MT, Bonferroni M, Cardona S, Guarini A, Lista P, Pegoraro L, Mandelli F, Forni G and Gavosto F (1990a). Interleukin 2 does not promote the in vitro and in vivo proliferation and growth of human acute leukaemia cells of myeloid and lymphoid origin. Br J Haematol **75**, 34-40.

Foa R, Fierro MT, Tosti S, Meloni G, Gavosto F and Mandelli F (1990b). Induction and persistence of complete remission in a resistant acute myeloid leukemia patient following treatment with recombinant interleukin 2. Leuk Lymph **1**, 113-117.

Foa R, Fierro MT, Cesano A, Guarini A, Bonferroni M, Raspadori D, Miniero R, Lauria F and Gavosto F (1991a). Defective lymphokine-activated killer cell generation and activity in acute leukemia patients with active disease. Blood **78**, 1041-1046.

Foa R, Guarini A, Gillio Tos A, Cardona S, Fierro MT, Meloni G, Tosti S, Mandelli F and Gavosto F (1991b). Peripheral blood and bone marrow immunophenotypic and functional modifications induced in acute leukemia patients treated with interleukin 2: Evidence of in vivo lymphokine activated killer cell generation. Cancer Res **51**, 964-968.

Foa R, Meloni G, Tosti S, Novarino A, Fenu S, Gavosto F and Mandelli F (1991c). Treatment of acute myeloid leukaemia patients with recombinant interleukin 2: A pilot study. Br J Haematol **77**, 491-496.

Foa R (1992). Pathogenesis of the immunodeficiency in B-cell chronic lymphocytic leukemia. In "Chronic Lymphocytic Leukemia" (ed BD Cheson), pp 147-166. Marcel Dekker, New York.

Foa R, Guarini A and Gansbacher B (1992). IL2 treatment for cancer: From biology to gene therapy. Br J Cancer **66**, 992-998.

Foa R (1993). Does interleukin-2 have a role in the management of acute leukemia? J Clin Oncol **11**, 1817-1825.

Foon KA, Smalley RV, Riggs CW and Gale PG (1983). The role of immunotherapy in acute myelogenous leukemia. Arch Intern Med **143**, 1726-1731.

Gansbacher B, Zier K, Daniels B, Cronin K, Bannerji R and Gilboa E (1990). Interleukin 2 gene transfer into tumor cells abrogates tumorigenicity and induces protective immunity. J Exp Med **172**, 1217-1224.

Gansbacher B, Guarini A, Gilboa E and Golde D (1991). Lymphoma regression induced by gene transfer mediated localized IL2 secretion. Blood **78**, 384a.

Gansbacher B, Zier K, Cronin K, Hantzopoulos PA, Bouchard B, Houghton A, Gilboa E and Golde D (1992). Retroviral gene transfer induced constitutive expression of interleukin-2 or interferon-γ in irradiated human melanoma cells. Blood **80**, 2817-2825.

Gastl G, Finstad CL, Guarini A, Bosl G, Gilboa E, Bander NH and Gansbacher B (1992). Retroviral vector-mediated lymphokine gene transfer into human renal cancer cells. Cancer Res **52**, 6229-6236.

Gottlieb DJ, Prentice HG, Heslop HE, Bello-Fernandez C, Bianchi AC, Galazka AR and Brenner MK (1989). Effects of recombinant interleukin-2 administration on cytotoxic function following high-dose chemo-radiotherapy for hematological malignancy. Blood **74**, 2335-2342.

Grimm EA, Mazumder A, Zhang HZ and Rosenberg SA (1982). Lymphokine-activated killer cell phenomenon. Lysis of natural killer-resistant fresh solid tumor cells by interleukin 2-activated autologous human peripheral blood lymphocytes. J Exp Med **155**, 1823-1841.

Guarini A, Sanavio F, Novarino A, Gillio Tos A, Aglietta M and Foa R (1991). Thrombocytopenia in acute leukaemia patients treated with IL2: Cytolytic effect of LAK cells on megakaryocytic progenitors. Br J Haematol **79**, 451-456.

Guarini A, Gansbacher B, Cronin K and Foa R (1994). Specific T lymphocyte reactivity to IL-2 gene transduced human melanoma cells. In "Cytokine Induced Tumor Immunogenicity" (eds G Forni, R Foa, A Santoni and L Frati), pp 403-413. Academic Press, London.

Herberman RB (1983). Possible role of natural killer cells in host resistence against tumours and other diseases. Clin Immunol **3**, 479-483.

Heslop HE, Gottlieb DJ, Bianchi ACM, Meager A, Prentice HG, Mehta AB, Hoffbrand AV and Brenner MK (1989). In vivo induction of gamma interferon and tumor necrosis factor by interleukin-2 infusion following intensive chemotherapy or autologous marrow transplantation. Blood **74**, 1374-1380.

Higuchi CM, Thompson JA, Cox T, Lindgren CG, Buckner DC and Fefer A (1989). Lymphokine-activated killer function following autologous bone marrow transplantation for refractory hematological malignancies. Cancer Res **49**, 5509-5513.

Higuchi CM, Thompson JA, Petersen FB, Buckner CD and Fefer A (1991). Toxicity and immunomodulatory effects of interleukin-2 after autologous bone marrow transplantation for hematologic malignancies. Blood **77**, 2561-2568.

Lafraniere R and Rosenberg SA (1985). Successful immunotherapy of murine experimental hepatic metastases with lymphokine-activated killer cells and recombinant interleukin 2. Cancer Res **45**, 3735-3741.

Lauria F, Raspadori D, Foa R, Tazzari PL, Lusso P, Fierro MT, Matera L, Baccarani M and Tura S (1986). Normal T-lymphocyte function in patients with Hodgkin's disease in long-lasting remission. Tumori **72**, 75-80.

Levy R and Kaplan KS (1974). Impaired lymphocyte function in untreated Hodgkin's disease. N Engl J Med **24**, 181-186.

Lim SH, Newland AC, Kelsey S, Bell A, Offerman E, Rist C, Gozzard D, Bareford D, Smith MP and Goldstone AH (1992). Continuous intravenous infusion of high-dose recombinant interleukin-2 for acute myeloid leukaemia - A phase II study. Cancer Immunol Immunother **34**, 337-342.

Lista P, Fierro MT, Xin-Sheng L, Bonferroni M, Brizzi MF, Porcu PL, Pegoraro L and Foa R (1989). Lymphokine-activated killer (LAK) cells inhibit the clonogenic growth of human leukaemic stem cells. Eur J Haematol **42**, 425-430.

Lotzova E, Savary CA, Keating MJ and Hester JP (1985). Defective NK cell mechanism in patients with leukemia. In "Mechanisms for Cytotoxicity by NK cells" (eds RB Herberman and D Callewaert), pp 507-513, Academic Press, San Diego.

Maraninchi D, Blaise D, Viens P, Brandely M, Olive D, Lopez M, Sainty D, Marit G, Stoppa AM, Reiffers J, Gratecos N, Bertau-Perez P, Mannoni P, Mawas C, Hercend T, Sebahoun G and Carcassonne Y (1991). High-dose recombinant interleukin-2 and acute myeloid leukemias in relapse. Blood **78**, 2182-2187.

Meloni G, Foa R, Tosti S, Vignetti M, Mancini F, Guarini A, Marchis D, Gavosto F and Mandelli F (1992). Autologous bone marrow transplantation followed by interleukin-2 in children with advanced leukemia: A pilot study. Leukemia **6**, 780-785.

Mule JJ, Shu S, Schwarz SL and Rosenberg SA (1984). Successful adoptive immunotherapy of established pulmonary metastases with LAK cells and recombinant IL-2. Science **255**, 1487-1489.

Oshimi K, Oshimi Y, Akutsu M, Takei Y, Saito H, Okada M and Mizoguchi H (1986). Cytotoxicity of interleukin 2 activated lymphocytes for leukemia and lymphoma cells. Blood **68**, 938-948.

Paciucci PA, Mandeli J, Oleksowiz L, Ameglio F and Holland JF (1990). Thrombocytopenia during immunotherapy with interleukin-2 by constant infusion. Am J Med **89**, 308-312.

Papadopulos EB, Ladanyl M, Emanuel D, Mackinnon S, Boulad F, Carabasi MH, Castro-Malaspina H, Childs BH, Gillio AP, Small TN, Young JW, Kernan NA amd O'Reilly RJ (1994). Infusions of donor leukocytes as treatment of Eptstein-Barr virus associated lymphoproliferative disorders complicationg allogenenic marrow transplantation. N Engl J Med (in press).

Pizzolo G, Trentin L, Vinante F, Agostini C, Zambello R, Masciarelli M, Feruglio C, Dazzi F, Todeschini G, Chilosi M, Veneri D, Zanotti R, Benedetti F, Perona G and Semenzato G (1988). Natural killer cell function and lymphoid subpopulations in acute non-lymphoblastic leukaemia in complete remission. Br J Cancer **58**, 368-372.

Reittie J, Gotlieb D, Heslop H, Leger O, Hazlerhurst G, Drexler R, Hoffbrand A, Prentice H and Brenner M (1989). Endogenously generated activated killer cells circulate after autologous and allogeneic marrow transplantation. Blood **73**, 1351-1358.

Rosenberg SA, Lotze MT, Muul LM, Leitman S, Chang AE, Ettinghausen SE, Matory YL, Skibber JM, Shiloni E, Vetto JT, Seipp CA, Simpson C and Reichert CM (1985). Observations on the systemic administration of autologous lymphokine-activated killer cells and recombinant interleukin-2 to patients with metastatic cancer. N Engl J Med **313**, 1485-1492.

Rosenberg SA, Lotze MT, Muul LM, Chang AE, Avis FP, Leitman S, Linehan WM, Robertson CN, Lee RE, Rubin JT, Seipp CA, Simpson CG and White DE (1987). A progress report on the treatment of 157 patients with advanced cancer using lymphokine-activated killer cells and interleukin-2 or high-dose interleukin-2 alone. N Engl J Med **316**, 889-897.

Rosenberg SA, Lotze MT, Yang JC, Aebersold PM, Linehan WM, Seipp CA and White DE (1989). Experience with the use of high-dose interleukin-2 in the treatment of 652 cancer patients. Ann Surg **210**, 474-484.

Rosolen A, Nakanishi M, Poplack DG, Cole D, Quinones R, Reaman G, Cotelingman JD, Trepel JB, Sausville E, Marti GE, Jaffe E, Neckers LM and Colamonici OR (1989). Expression of interleukin-2 receptor β subunit in hematopoietic malignancies. Blood **76**, 1968-1972.

Shapiro RS, McClain K, Frizzera G, Gajl-Peczalska KJ, Kersey JH, Blazar BR, Arthur DC, Patton DF, Greenberg JS, Burke B, Ramsay NKC, McGlave P and Filipovich AH (1988). Epstein-Barr virus associated B cell lymphoproliferative disorders following bone marrow transplantation. Blood **71**, 1234-1243.

Soiffer RJ, Murray C, Cochran K, Cameron C, Wang E, Schow PW, Daley JF and Ritz J (1992). Clinical and immunologic effects of prolonged infusion of low-dose recombinant interleukin-2 after autologous and T-cell-depleted allogeneic bone marrow transplantation. Blood **79**, 517-526.
Stryckmans PS, Otten J, Delbeke J, Sucin S, Fiere D, Bury J, Solbu G and Benoit M for the EORTC Hemopathies Working Party (1983). Comparison

of chemotherapy with immunotherapy for maintenance of acute lymphoblastic leukemia in children and adults. Blood **62**, 606-615.

Tao MH and Levy R (1993). Idiotype/granulocyte macrophage colony-stimulating factor fusion protein as a vaccine for B-cell lymphoma. Nature **362**, 755-758.

Toze CL, Barnett MJ and Klingemann H-G (1993). Response of therapy-related myelodysplasia to low-dose interleukin 2. Leukemia **7**, 463-465.

Van der Bruggen P, Traversari C, Chomez P, Lurquin C, De Plaen E, Van Den Eynde B, Knuth A and Boon T (1991). A gene encoding an antigen recognized by cytolytic T lymphocytes on a human melanoma. Science **254**, 1643-1647.

West WH, Tauer KW, Yannelli JR, Marshall GD, Orr DW, Thurman GB and Oldham RK (1987). Constant-infusion recombinant interleukin-2 in adoptive immunotherapy of advanced cancer. N Engl J Med **316**, 898-905.

27

SPECIFIC T LYMPHOCYTE REACTIVITY TO IL-2 GENE TRANSDUCED HUMAN MELANOMA CELLS

A. Guarini, B. Gansbacher, K. Cronin and R. Foa
Department of Biomedical Sciences and Human Oncology and CNR Center for Immunogenetics and Experimental Oncology, University of Turin, Italy; Division of Hematologic Oncology, Memorial Sloan-Kettering Cancer Center, New York, NY, USA

Background

It is well established that human melanoma is an immunogenic tumor capable of inducing both in vitro and in vivo the generation of CTL reactive against autologous melanoma cells (De Vries and Spits, 1984; Anichini et al., 1985; Coulie et al., 1992). In fact, specific CTL have been found in the peripheral blood (Roberts et al., 1986) and in lymph nodes of patients with melanoma (Cozzolino et al., 1987), as well as in the tumoral masses (Topalian et al., 1989). Different authors have shown that specific CTL may be obtained in vitro, in both allogeneic and autologous conditions, by cocultivating PBL with irradiated melanoma cells and low doses of IL-2 (Herin et al., 1987; Crowley et al., 1990; Coulie et al., 1992). These CTL have been found to express a $CD3^+$, $CD8^+$ phenotype and to be HLA class I restricted. HLA class I seems to play an important role in the induction of a specific response by melanoma cells (Darrow et al., 1989). In particular, the A2 allele appears to be involved, in conjunction with the A1 allele, in the process of antigen presentation (Crowley et al., 1990; Crowley et al., 1991). In patients with HLA-A2 melanoma, CTL capable of lysing tumor cells derived from other patients which expressed HLA-A2 have been isolated (Darrow et al., 1989; Anichini et al., 1993).

Analysis of the induction of specific CTL enabled T. Boon's group to identify in 1991 the first human tumor antigen encoded for by the so-called MAGE-1 gene (Van der Bruggen et al., 1991), which appears to be expressed by the neoplastic melanoma cells in the context of the A1 allele of the HLA system (Traversari et al., 1992), as well as by thyroid, breast and prostate cancer and by the normal testis. More recently, the same group has identified in two melanoma patients specific CTL which

CYTOKINE-INDUCED TUMOR IMMUNOGENICITY
ISBN 0-12-261520-4

respond to the antigen coded by the tyrosinase gene presented by the HLA-A2 allele (Brichard et al., 1993).

Based on these considerations, the obvious question which arises is the following: if human melanoma cells express specific antigens capable of activating specific antitumor lymphocytes, why is the tumor not recognized as nonself by the immune system of the host and rejected? The answer that we propose is that the number of inducible CTL is numerically insufficient to counteract the disease and/or that an inadequate interconnection may prevent a satisfactory recognition of the tumor cells by the lymphocytes. Methodological strategies aimed at increasing the number of specific CTL in vivo clearly represent an approach which carries important therapeutic implications. In theory, cytokine gene therapy may be a potential solution to this problem.

Goals of therapy utilizing tumor cells transduced with the IL-2 gene

These approaches are aimed at the design and clinical activation of vaccination protocols utilizing allogeneic melanoma cells genetically transduced to produce constitutively IL-2 (for a review see Foa et al., 1992). In the management of melanoma patients, vaccination protocols employing irradiated melanoma cells have been extensively attempted with clinically contradictory results (Livingston et al., 1985; Livingston, 1991). More encouraging results have instead been obtained in melanoma patients following the i.v. administration of high doses of exogenous IL-2, with or without the addition of ex vivo activated LAK effectors (Rosenberg et al., 1988). Responses on the order of 15-20% of patients can be expected for melanomas with advanced disease. The systemic administration of IL-2, which has revealed a certain degree of anti-neoplastic activity in other tumors such as renal cell carcinoma (Rosenberg et al., 1985) and acute myeloid leukemia (Foa et al., 1991b), is capable of activating in vivo the NK and LAK compartment of the host (Sondel et al., 1988; Foa et al., 1991a). Both the clinical responses and the host immune activation have been observed when IL-2 is administered i.v. at doses which are always associated with relevant toxicity. However, the activation of the cytotoxic compartment and of other immunological parameters occurs in all treated patients and appears to be unrelated to the clinical response to IL-2 (Favrot et al., 1990; Foa et al., 1991a). Furthermore, there is no demonstration that the infusion of exogenous IL-2 can activate CTL directed specifically against the neoplastic cells of the tumor bearing patient.

These data have led to the design of a vaccination protocol based on the utilization of neoplastic cells which have been genetically modified ex vivo to produce constitutively IL-2 (Gansbacher, 1992). This innovative therapeutic approach is based on a number of different conceptual considerations:

- Irradiated melanoma cells present tumor associated antigens; when inoculated in the proximity of a lymph node they are exposed to the lymphocytes locally recruited
- The IL-2 released constantly by the engineered and irradiated neoplastic cells for a time period which ranges between 1 and 4 weeks, according to the tumor cell line studied (see below), may induce an activation or reactivation signal on the surrounding lymphocytes. This may overcome the inhibitory effects promoted by "paralyzing" factors released by the tumor cells themselves, as has been clearly documented for TGF-β (Inge et al., 1992).
- If genetically modified neoplastic cells can induce the generation or expansion of specific antitumor CTL, the latter can then reproduce themselves through the stimulation signal induced by the IL-2 released by the engineered tumor cells over a prolonged time period. In turn, this should allow the generation of a larger compartment of circulating cells capable of recognizing and killing neoplastic cells in different sanctuaries of residual disease.

Transduction of melanoma cells with the IL-2 gene

Transduction of the IL-2 gene has been successfully accomplished in different experimental models (Fearon et al., 1990; Gansbacher et al., 1990) (see also different chapters in this book). The neoplastic cells engineered to release IL-2 are capable of inducing a specific immunologic response and loose their capacity to form a tumor (Fearon et al., 1990; Gansbacher et al., 1990). Furthermore, they activate cytolytic $CD8^+$ T lymphocytes which block the growth of unmodified parental tumor cells at subsequent challenges, thus documenting that IL-2 gene transduced neoplastic cells can induce the generation of a compartment of memory cells capable of rejecting the tumor. This has been demonstrated also in the murine melanoma B16 (Fearon et al., 1990), a very aggressive and poorly immunogenic tumor which originates in C57BL/6 mice. Thus, two of the primary expectations of an optimal immunotherapeutic protocol, i.e. antitumor specificity and memory, have been fulfilled in different experimental models.

Utilizing retroviral vectors, several human melanoma cell lines have been transduced with the IL-2 gene (Gansbacher et al., 1992). The integration of the proviral DNA into the DNA of the engineered cells has been documented by Southern blot analysis, while the specific translation of the mRNA of the gene inserted can be demonstrated by Northern blot analysis or by reverse transcriptase-PCR. The constitutive release of IL-2 by the transduced cells is measured using a biological assay (Zier, 1982) to assess the real activity of the protein produced and confirmed with a quantitative evaluation using an enzyme linked immunosorbent assay.

Following in vitro cloning by limiting dilution, it is possible to obtain clones from individual tumor cell lines that produce highly heterogenous quantities of IL-2 (in our experience, between 0 and >600 U/ml per 10^6 cells). It could also be shown that the transduction of the gene and subsequent IL-2 production induced no notable effects on the morphology, pigmentation or proliferation of the cells compared to the parental unmodified cells, or to cells transduced with an irrelevant gene. Secretion of IL-2 was shown to remain stable over a time period of at least 6 months. Since vaccination programs which contemplate the in vivo administration of cytokine gene transduced tumor cells will, in our opinion, require that the cells are first irradiated (a maneuver which may also enhance the antigenic expression of the tumor cells), it becomes essential to document that gamma radiation does not block the production of IL-2. Following irradiation with 5000-10000 rad at a dose/rate of 100 rad per minute, several human melanoma cell lines transduced with the IL-2 gene maintain for several weeks their capacity to release variable amounts of IL-2 without, however, showing evidence of cell proliferation (Gansbacher et al., 1992). The quantity of IL-2 released is a function of cell viability.

It is very important to identify and standardize the number of rad necessary for each cell line in order to:

- obtain a block in cell proliferation
- document a persistent release of IL-2
- define the time course of IL-2 release (i.e. how much and for how long)
- allow cell viability to persist for more than 7-10 days.

These different features are of relevance for vaccination protocols which require that the transduced and inoculated cells do not proliferate, but must survive for a sufficient time to enable the cytokine released to activate the immune system of the host.

Human neoplastic cells transduced with cytokine genes have limited possibilities of being further studied preclinically in vivo, since the only possibility is represented by nude or scid mouse which can allow the growth of human tumor cell lines. Nude mice are, however, a limited model since immunologically they lack an efficient T-cell compartment. The growth in nude mice of IL-2 gene transduced human melanoma cell lines is abrogated, indicating that the IL-2 released by the engineered cells is biologically active and capable of inducing an activation signal on the NK, monocyte/macrophage and neutrophil compartments which, in turn, block tumor cell growth. The nude mouse model, however, is incapable of showing whether antitumor CTL can be generated and whether a specific antitumor memory can be obtained. The possibility of inducing both specificity and memory, which, as mentioned earlier, has been convincingly demonstrated in experimental tumor models, represents a primary goal for any vaccination protocol. This is a key issue towards transforming vaccination trials with tumor engineered cells into a truly new antitumor strategy.

Induction of specific recognition induced by tumor cells transduced with the IL-2 gene

In an attempt to obtain more preclinical information, we have utilized an in vitro culture model with effector cells and tumor cells which release IL-2. These experimental conditions are based on data accumulated earlier by other authors who have shown that it is possible in melanoma to generate a specific cytolytic response, HLA class I restricted and mediated by $CD3^+$ $CD8^+$ lymphocytes, both with autologous and with allogeneic patients' lymphocytes cultured for more than 20 days with irradiated tumor cells and low doses of IL-2 (5-250 U/ml) (Herin et al., 1987; Crowley et al., 1990; Coulie et al., 1992). In the establishment of this specific response, an important role appears to be played by the A2 allele of the HLA class I system present on the tumor cell; this allele is known to favor the process of antigen presentation (Crowley et al., 1990; Kawakami et al., 1992).

In two melanoma cell lines, one HLA-$A2^+$ and one HLA-$A2^-$, derived from patients from whom viable lymphocytes were available, the IL-2 gene has been transduced with a retroviral vector (Gansbacher et al., 1992). The engineered lines have not been further subcloned since, in agreement with other authors (Pardoll, 1993), we believe that therapeutically the use of bulk cell lines may be preferable in view of the fact that they are more likely to possess a wider antigenic profile which is more representative of the primary tumor. Following retroviral vector mediated gene transfer, the two melanoma cell lines produced 27 and 120 U of IL-2/ml per 10^6 cells in 48 h, respectively. After irradiation, they survived and continued to release IL-2 for a minimum of 10 days and a maximum of 21 days. Thereafter, the patients' lymphocytes were cultured, at a 10:1 ratio, with the autologous tumor cells transduced and irradiated for 28 days; tumor cells were renewed every 7 days. After 7 days of coculture, the lymphocytes developed a good NK function, as well as an "unspecific" antitumor activity (documented in a classic ^{51}Cr release assay). After 28 days of coculture, the lymphocytes incubated with the HLA-$A2^+$ melanoma generated a "specific" cytotoxic activity, i.e. they were capable of lysing the autologous tumor (the unmodified parental cell line), but they no longer lysed allogenic melanoma cells and the K562 cell line (target of NK cell lysis). The cytotoxic activity could be inhibited by antibodies against HLA class I and anti-CD8, thus proving that the lytic effectors were class I and CD8 restricted. On the other hand, lymphocytes incubated with the IL-2 producing HLA-$A2^-$ melanoma continued to reveal only an unspecific cytotoxic activity.

Taken together, these data indicate that human melanoma cells engineered to release IL-2 may, at least in vitro, induce the generation of specific cytotoxic lymphocytes. The evidence that this occurred when the tumor cells were HLA-A2+ further suggests that the A2 allele may play a relevant role in the process of tumor antigen presentation and, thus, in tumor recognition by the immune system of the host. Since our system is autologous, the activation signals must be ascribed to the tumor cells and to IL-2 only, and not to other stimuli induced by the histocompatibility system.

The same experiment has been carried out utilizing the melanoma cell line SK-MEL-29 (HLA-A2+) used in the gene therapy protocol recently activated for advanced melanoma patients at Memorial Sloan-Kettering Cancer Center, New York. We have incubated PBL obtained from two melanoma patients which were HLA-A2+ with the transduced and irradiated melanoma cell line that has been selected to be used as vaccine. Prior to incubation, the patients' lymphocytes showed, in a chromium release assay, a normal NK activity against K562 cells, but there was no lysis against the melanoma cells. After 23 days of incubation with the line transduced with the IL-2 gene, the patients' lymphocytes lost the capacity to lyse K562, while they acquired the capacity to kill the melanoma cells utilized to vaccinate (Table 27.1). These findings demonstrate, at least in vitro, that a human melanoma cell line engineered to release IL-2 can induce the generation of cytolytic effectors directed against the tumor cells.

Clinical trials with IL-2 gene transduced tumor cells

The design and activation of vaccination protocols utilizing IL-2 gene transduced melanoma cells seem a logic follow-up of the above described conceptual considerations and the experimental data accumulated. In the US, both the Food and Drug Administration and the Recombinant DNA Advisory Committee have approved phase I trials aimed at immunizing melanoma patients with advanced disease. The first patients were enrolled in 1993 and no study has so far been concluded. At Memorial Sloan-Kettering Cancer Center, New York the protocol is designed for HLA-A2+ stage IV patients to receive cells from a matched allogeneic melanoma cell line (SK-MEL-29) engineered to release IL-2 following transduction with the DC/AD/R/IL-2 retroviral vector (Gansbacher, 1992). After gamma radiation, tumor cells are inoculated s.c. three times, at intervals of 2 weeks, with a booster 2 months later. In this initial study the choice of limiting the entry to HLA-A2 patients is related both to the high frequency of this allele in the Caucasian population and to the evidence that from HLA-A2+ patients CTL directed against both the autologous tumor and against other HLA-A2+ allogeneic melanoma cells can be obtained (Darrow et al., 1989; Anichini et al., 1993).

Table 27.1. *Cytotoxicity of PBL from melanoma patients*

		% of lysis against	
Case	Day	SK-MEL-29	K562
# 1	0	1.6[a]	45.0
	23[b]	41.0	7.7
# 2	0	3.5	22.0
	23	16.3	2.1

[a]Effector:target cell ratio = 50:1
[b]PBL recovered after 23 days of coculture with irradiated SK-MEL-29 cells transduced with the IL-2 gene

Since this first activated protocol represents a feasibility and toxicity study, two levels of IL-2 gene transduced cells have been planned: the first one with only 10 x 10^6 cells and the second with 50 x 10^6 engineered cells. The results so far encountered demonstrate that the s.c. administration of the IL-2 gene transduced melanoma cells is devoid of systemic toxicity with both levels of cells inoculated. Most patients experience a local reactivity at the site of the vaccination which increases over a period of approximately 1 week from the injection. Several of the enroled patients have been unable to complete the planned scheme of four vaccinations owing to intervening disease progression.

Future prospects

The described suggestion that IL-2 gene transduced human melanoma cells may induce the generation of CTL directed against the autologous tumor further strengthens the rationale of utilizing vaccination protocols with genetically modified tumor cells in the management of patients with melanoma. In addition to clarifying the feasibility of such an approach and the number of engineered cells that can be inoculated, it will be important to define if and to what extent the overall small quantities of IL-2 released in vivo by the transduced cells are capable of activating the immune system of the host. It will be of even greater relevance to assess whether CTL directed against the patients' tumor cells can be amplified in vivo following inoculation with HLA matched engineered allogeneic cells. For a better design of the future trials, it will be important to correlate the eventual generation of specific CTL with the number of engineered cells inoculated. The planned activation of protocols based on the administration of gene modified autologous cell lines will permit the definition of possible differences in CTL generation and the assessment of whether much more cumbersome, but potentially more specific, approaches aimed at establishing and gene transducing autologous cell lines should be widely attempted.

These biological analyses appear to be crucial in an effort to expand cytokine gene therapy strategies, since it is most unlikely that clinical responses will be observed in the phase I protocols currently activated in view of the fact that only patients with advanced disease are recruited. Should the generation of specific antitumor CTL be demonstrated with given doses of IL-2 released by the engineered cells, the extension of this therapeutic strategy to patients with less advanced disease, in whom there is a greater likelihood of a clinical and immunological response, could appropriately be attempted.

Future prospects will be aimed at evaluating whether this same approach can be extended to HLA-A2$^-$ patients and whether strategies aimed at obtaining a greater immune response can be devised. These will be particularly addressed at immunizing directly with melanoma associated antigenic sequences (Boon, 1993), both alone and possibly in combination with the IL-2 gene, in an attempt to combine a specific activation signal induced by the antigen and an immunological modulation induced by IL-2.

Acknowledgments

Work supported by Associazione Italiana per la Ricerca sul Cancro (AIRC), Milan and Consiglio Nazionale delle Ricerche (CNR), Project on "Applicazioni Cliniche della Ricerca Oncologica", Rome, Italy.

References

Anichini A, Fossati G and Parmiani G (1985). Clonal analysis of cytotoxic T-lymphocyte response to autologous human metastatic melanoma. Int J Cancer **35**, 683-689.

Anichini A, Maccalli C, Mortarini R, Salvi S, Mazzocchi A, Squarcina P, Herlyn M and Parmiani G (1993). Melanoma cells and normal melanocytes share antigens recognized by HLA-A2-restricted cytotoxic T cell clones from melanoma patients. J Exp Med **177**, 989-998.

Boon T (1993). Tumor antigens recognized by cytolytic T lymphocytes: Present perspectives for specific immunotherapy. Int J Cancer **54**, 177-180.

Brichard V, Van Pel A, Wölfel T, Wölfel C, De Plaen E, Lethe B, Coulie P and Boon T (1993). The tyrosinase gene codes for an antigen recognized by autologous cytolytic T lymphocytes on HLA-A2 melanomas. J Exp Med **178**, 489-495.

Coulie PG, Somville M, Lehmann F, Hainaut P, Brasseur F, Devos R and Boon T (1992). Precursor frequency analysis of human cytolytic T lymphocytes directed against autologous melanoma cells. Int J Cancer **50**, 289-297.

Cozzolino F, Torcia M, Carossino AM, Giordani R, Selli C, Talini G, Reali E, Novelli A, Pistoia V and Ferrarini M (1987). Characterization of cells from invaded lymph nodes in patients with solid tumors. Lymphokine requirement for tumor-specific lymphoproliferative response. J Exp Med **166**, 303-318.

Crowley NJ, Slingluff CL Jr, Darrow TL and Seigler HF (1990). Generation of human autologous melanoma specific cytotoxic T-cells using HLA-A2-matched allogeneic melanomas. Cancer Res **50**, 492-498.

Crowley NJ, Darrow TL, Quinn-Allen MA and Seigler HF (1991). MHC-restricted recognition of autologous melanoma by tumor-specific cytotoxic T cells. J Immunol **146**, 1692-1699.

Darrow TL, Slingluff CL and Seigler HF (1989). The role of HLA class I antigens in recognition of melanoma cells by tumor-specific cytotoxic T lymphocytes. Evidence for shared tumor antigens. J Immunol **142**, 3329-3335.

De Vries JE and Spits H (1984). Cloned human cytotoxic T lymphocyte (CTL) lines reactive with autologous melanoma cells. J Immunol **132**, 510-519.

Favrot MC, Combaret V, Negrier S, Philip I, Thiesse P, Freydel C, Bijmann JT, Franks CR, Mercatello A and Philip T (1990). Functional and immunophenotypic modifications induced by interleukin-2 did not predict response to therapy in patients with renal cell carcinoma. J Biol Resp Mod **9**, 167-177.

Fearon ER, Pardoll DM, Itaya T, Golumbek P, Levitsky HI, Simons JW, Karasuyama H, Vogelstein B and Frost P (1990). Interleukin-2 production by tumor cells bypasses T helper function in the generation of an antitumor response. Cell **60**, 397-403.

Foa R, Guarini A, Gillio Tos A, Cardona S, Fierro MT, Meloni G, Tosti S, Mandelli M and Gavosto F (1991a). Peripheral blood and bone marrow immunophenotypic and functional modifications induced in acute leukemia patients treated with interleukin 2: Evidence of in vivo lymphokine activated killer cell generation. Cancer Res **51**, 964-968.

Foa R, Meloni G, Tosti S, Novarino A, Fenu S, Gavosto F and Mandelli F (1991b). Treatment of acute myeloid leukaemia patients with recombinant interleukin 2: A pilot study. Br J Haematol **77**, 491-496.

Foa R, Guarini A and Gansbacher B (1992). IL2 treatment for cancer: From biology to gene therapy. Br J Cancer **66**, 992-998.

Gansbacher B, Zier K, Daniels B, Cronin K, Bannerji R and Gilboa E (1990). Interleukin 2 gene transfer into tumor cells abrogates tumorigenicity and induces protective immunity. J Exp Med **172**, 1217-1224.

Gansbacher B (1992). A pilot study of immunization with HLA-A2 matched allogeneic melanoma cells that secrete interleukin-2 in patients with metastatic melanoma. Hum Gene Ther **3**, 677-690.

Gansbacher B, Zier K, Cronin K, Hantzopoulos A, Bouchard B, Houghton A, Gilboa E and Golde D (1992). Retroviral gene transfer induced constitutive expression of IL-2 and IFN-gamma in irradiated human melanoma cells. Blood **80**, 2817-2825.

Herin M, Lemoine C, Weynants P, Vessiere F, Van Pel A, Knuth A, Devos R and Boon T (1987). Production of stable cytolytic T-cell clones directed against autologous human melanoma. Int J Cancer **39**, 390-396.

Inge TH, Hoover SK, Susskind BM, Barrett SK and Bear HD (1992). Inhibition of tumor-specific cytotoxic T-lymphocyte responses by transforming growth factor β1. Cancer Res **52**, 1386-1392.

Kawakami Y, Zakut R, Topalian SL, Stotter H and Rosenberg SA (1992). Shared human melanoma antigens. Recognition by tumor-infiltrating lymphocytes in HLA-A2.1-transfected melanomas. J Immunol **148**, 638-643.

Livingston PO, Albino A and Chung TSC (1985). Serologic response of patients with stage II melanoma to allogenic melanoma cell vaccines. Cancer **56**, 2194-2200.

Livingston PO (1991). Active specific immunotherapy in treatment of patients with cancer. Immunol Allergy Clin Nort Am **11**, 3329-3333.

Pardoll DM (1993). Cancer vaccines. Immunol Today **14**, 310-316.

Roberts TE, Shipton U and Moore M (1986). Proliferative and cytotoxic responses of human peripheral blood lymphocytes to autologous malignant effusions. An analysis at the clonal level. Cancer Immunol Immunother **22**, 107-113.

Rosenberg SA, Lotze MT, Muul LM, Leitman S, Chang AE, Ettinghausen SE, Matory YL, Skibber JM, Shiloni E, Vetto JT, Seipp CA, Simpson C and Reichert CM (1985). Observations on the systemic administration of autologous lymphokine-activated killer cells and recombinant interleukin-2 to patients with metastatic cancer. N Engl J Med **313**, 1485-1492.

Rosenberg SA, Packard BS, Aebersold PM, Solomon D, Topalian SL, Toy ST, Simon P, Lotze MT, Yang JC, Seipp CA, Simpson C, Carter C, Bock S, Schwartzentruber D, Wei JP and White DE (1988). Immunotherapy of patients with metastatic melanoma using tumor infiltrating lymphocytes and interleukin-2: Preliminary report. N Engl J Med **319**, 1676-1680.

Sondel PM, Kohler PC, Hank JA, Moore KH, Rosenthal NS, Sosman JA, Bechhofer R and Storer B (1988). Clinical and immunological effect of recombinant interleukin 2 given by repetitive cycles to patients with cancer. Cancer Res **48**, 2561-2567.

Topalian SL, Solomon D and Rosenberg SA (1989). Tumor-specific cytolysis by lymphocytes infiltrating human melanomas. J Immunol **146**, 3714-3725.

Traversari C, Van der Bruggen P, Luescher IF, Lurquin C, Chomez P, Van Pel A, De Plaen E, Amar-Costesec A and Boon T (1992). A nonapeptide encoded by human gene MAGE-1 is recognized on HLA-A1 by CTL directed against tumor antigen MZ2-E. J Exp Med **176**, 1453-1457.

Van der Bruggen P, Traversari C, Chomez P, Lurquin C, De Plaen E, Van Den Eynde B, Knuth A and Boon T (1991). A gene encoding an antigen recognized by cytolytic T lymphocytes on a human melanoma. Science **254**, 1643-1647.

Zier K (1982). Functional and antigenic properties of cultured T cells in the cell mediated lympholysis (CML) assay. Human Immunol **4**, 147-154.

28

IMMUNE RECOGNITION AND REJECTION OF HEAD AND NECK SQUAMOUS CELL CARCINOMA ELICITED BY IL-2 INJECTION

G. Cortesina, A. De Stefani, E. Galeazzi, M. Sacchi and S. Vai
Otorhinolaryngology Clinic, University of Turin and CNR Center for Immunogenetics and Experimental Oncology, University of Turin, Italy

Introduction

The complex relationships established between the immune system and a cancer during the course of its pathogenesis and treatment are not yet understood in their entirety. One fundamental tenet of cancer immunology, however, is that the antigens expressed by tumor cells differ both qualitatively and quantitatively from those on normal cells. A fuller understanding of these relationships, therefore, could open the way to their modification in the hope of suppressing or defeating tumors.

In the case of squamous cell carcinoma of the head and neck (HNSCC), indeed, evidence of the part played by the immune system in its pathogenesis and the host response has been building up over the last ten years, and has provided the rationale for the use of BRM in its treatment.

This form of cancer offers a particularly advantageous model for the investigation of tumor immunology, since it is usually locoregional and responsible for a degree of immune impairment often related to its stage. Attention can thus be concentrated on the tumor area when performing both biological and therapeutic studies. In recent years, too, molecular biology has joined classic cellular immunology, and it has already been the source of stimulating insights into the subject of cancer as a whole.

Biological studies

The immune system's response to a tumor is usually heterogeneous. Several mechanisms are involved, such as regulatory functions, effector cells and target structures. Generally speaking, however, two responses

CYTOKINE-INDUCED TUMOR IMMUNOGENICITY
ISBN 0-12-261520-4

are evoked to control tumor growth: a nonspecific, NK response and a specific response.

The NK activity is defective in head and neck cancer (HNCa), both locally and peripherally (Cortesina et al., 1984a). This deficiency has been related to tumor stage in studies of peripheral blood lymphocytes (Cortesina et al., 1984b; Mickel et al., 1988), whereas in the regional lymph nodes it is the outcome of metastasis. It is also determined by the distance of a node from the tumor site (Cortesina et al., 1990; Wang et al., 1991), resulting in what Wang et al. (1991) call "hierarchical immunosuppression". Basal NK activity in HNCa is low in both involved and uninvolved lymph nodes (Letessier et al., 1990) and their hierarchical suppression can be demonstrated by lymphocyte activation with IL-2 in vitro: generation of LAK cells by lymphocytes from metastatic lymph nodes is less efficient compared with those from uninvolved normal lymph nodes or lymph nodes far from the tumor (Cortesina et al., 1982; Cortesina et al., 1990; Letessier et al., 1990). NK deficiency has also been related to the risk of regional lymph node and distant recurrence in HNCa after definitive therapy, though its predictability would seem to be confined to moderately or poorly differentiated forms as a possible consequence of their low expression of class I MHC antigen (Racz et al., 1989). Suppression of the production of lymphokines with chemotactic activity has also been described (Wolf et al., 1984). This is due to the presence of adherent cells (Cortesina et al., 1987). Defective secretion of cytokines able to activate an immune response, namely IL-1β, TNF-α and IFN-γ, observed in the absence of demonstrable suppressor cells, has been related to the possible presence of tumor derived immunosuppressive factors such as prostaglandins, TGF-β, and p15E-related peptides (Berlinger, 1984; Letessier et al., 1990). There is thus plenty of evidence of local impairment of the immune response in patients with HNCa. Reinstatement of both the nonspecific NK and the specific antitumor T-cell response has been observed after local administration of IL-2 (Rivoltini et al., 1990).

T-cell activation following preoperative IL-2 therapy has been demonstrated by in situ histochemistry (Valente et al., 1990). These studies revealed $CD8^+$ cells close to tumor cells and in a more aggressive position than in control tumors together with enhanced tumor infiltration by eosinophils. High grade tumor associated tissue eosinophilia is regarded as a favorable prognostic indicator in HNCa and is associated with the absence of distant metastatis (Goldsmith et al., 1992). A degree of local immune activation can be achieved with IL-2 (Rivoltini et al., 1990; Valente et al., 1990) or IFN-α (Vlock et al., 1991), as in other tumors (Rosenberg et al., 1989). The possibility that other factors may be involved in tumor response to cytokines, however, must not be overlooked. IFN-α, IFN-γ and TNF-α have all been shown to possess direct tumor cytotoxicity (Sacchi et al., 1991a), and the constitutive expression of their receptors on HNCa cells in vitro and in situ suggest that the same may be true of IL-2, though factors released in situ by activated cells, e.g. LAK or adherent LAK (A-LAK) cells, could also be responsible (Sacchi et al., 1991a;

Cortesina et al., 1992). Selection of cells for specific activation in HNCa is a current topic of interest since techniques are now available for generation of subpopulations of both NK cells (Sacchi et al., 1991b) and T cells (Letessier et al., 1991), and enrichment in HNCa-restricted effector cells has recently been achieved by positive selection on antibody coated flasks of $CD8^+$ cells from the uninvolved nodes of HNCa patients (Letessier et al., 1991). Employment of subpopulations of this kind offers a possible strategy for adoptive immunotherapy of these tumors.

Clinical experience

Poor results were obtained with nonspecific immunostimulators (BCG, *Corynebacterium parvum*, OK432, thymic hormones, levamisole) administered peritumorally or systemically in the management of HNCa (Olivari et al., 1979; Bier et al., 1981).

Interest in immunotherapy has now been revived by development of the DNA recombination technology and a new, expanding family of cytokines. The two classes of BRM are listed in Table 28.1.

Those that have been used clinically in the treatment of HNCa have produced variable and often poor results by contrast with the encouraging experimental findings (Forni et al., 1985; Forni and Giovarelli, 1987). Two main reasons for this failure can be adduced. In the first place, trials are often empirical with respect to their methods, choice of doses, administration route and type of tumor treated. Secondly, the tumors themselves are usually advanced and no longer open to conventional therapy. They thus provide a poor model for investigation of the effects of immunotherapy. In addition, the complexity of the immune system itself is a bar to any ready a priori assessment of how it will respond to a new modulator or stimulator.

Cytokines, on the other hand, rest on a different rationale and hold out the promise of better results. This is especially true for IL-2 and the IFN. Early work by Rosenberg et al. (1985 & 1987) showed for the first time that substantial regression of advanced, metastatic tumors could be obtained with high doses of IL-2 administered systemically with or without LAK cells. Their findings were confirmed in subsequent trials (Rosenberg et al., 1987; West et al., 1987; Rosenberg et al., 1989), but the severe associated toxicity of this treatment modality suggested that a local approach would be more appropriate, particularly in light of the experimental evidence (Forni et al., 1985; Forni and Giovarelli, 1987).

HNCa was chosen as a suitable target for the reasons already given. Several trials have been run over more than a decade (Table 28.2), though their clinical impact is still not well defined.

Table 28.1. *BRM used in cancer immunotherapy*

Immunomodulators	Immunostimulators
Thymic hormones	Bacterial products
Levamisole	BCG
Transfer factor	*Corynebacterium parvum*
Prostaglandin inhibitors	*Nocardia rubra*
Antitumor antibodies	OK 432
	Tumor antigen extracts
	Tumor vaccine
	Pyran copolymers (MVE-2)
	Interferons
	Interleukins
	Other cytokines
	Muramyldipeptide (MDP)
	Bestatine
	Azimexon

Clinical trials with IL-2

Many patients with locally advanced HNCa have a poor prognosis as a result of its early lymphatic spread. Immunodepression, too, is related to the severity of the prognosis. Tumors of this kind are suitable for local management as they can easily be reached by direct or perilymphatic injections. In our first trial, 20 patients with recurrent HNSCC received 10 daily injections of 200 U of natural IL-2 (nIL-2) in human serum albumin 1.5 cm below the tip of the mastoid, which is a "strategic" point for the cervical lymphatic circulation (Traissac, 1968). Three CR, 3 PR and 7 MR were obtained. These, however, were always followed by recurrences, and a second course was ineffective (Cortesina et al., 1988; Cortesina et al., 1991).

The second (phase II) trial was conducted with recombinant IL-2 on 31 patients with advanced rHNSCC (unpublished data). Four patients treated prior to surgery displayed a substantial inflammatory response within the tumor (Musiani et al., 1989). A subsequent histological study of 22 patients with oropharynx and oral cavity carcinoma treated with IL-2 preoperatively (Valente et al., 1990) revealed areas of necrosis in the tumor, enhanced osinophilic infiltration, and aggressive location of $CD4^+$ and $CD8^+$ cells close to nests of tumor cells. In addition the percentage of $CD25^+$ and $LAK1^+$ cells in the TIL was higher in treated as opposed to

Table 28.2. *Trials of HNSCC immunotherapy*

Tumor type	Treatment	No. of patients	Schedule	Results	Author
Recurrent SCCHN	nIL-2	20	200 U/day neck nodes	3 CR 3 PR 7 MR	Cortesina et al., 1988
	rIL-2 LAK	14	2400-1.8 x 10^6 U/day neck and tumor	3 PR 3 MR	Squadrelli-Saraceno et al., 1990
	rIL-2	15	6000 U/day neck nodes	15 NR	Mattijssen et al., 1991
Advanced SCCHN	rIL-2	12	200-20000 U/day neck and tumor	1 PR	Selvaggi et al., 1990
	rIL-2	20	8 x 10^5 U/day tumor	2 CR 1 PR 1 MR	Saito et al., 1991
	rIL-2	46	intra and peritumorally	2 CR 4 PR	Sawaki, 1990
Maxillary sinus SCC	rIL-2 LAK	4	intraarterial preoperative	histologic response	Ikawa et al., 1989
Recurrent HNSCC	rIL-2 CDDP+5FU	12	12 x 10^6 U/day i.v.	2 CR 4 PR	Valone et al.,1991
	rIL-2 CDDP+5FU	25	0.5-6 x 10^6 U/day i.v.	1 CR 6 PR	Dimery et al., 1989
	rIL-2 IFN-α	11	3 x 10^6 U/day i.v. 5 x 10^6 U/day s.c.	2 PR	Urba et al., 1991
HNSCC	IFN-α	30	3 x 10^5 U/day local	histologic response	Padovan et al., 1981
	IFN-α	14	10 x 10^2 U/day i.m.	3 CR 4 PR	Medenica and Slack, 1985
NP[a] SCC	IFN-α	13	10 x 10^6 U/day i.m.	2 PR	Connors et al., 1985

(continued)

Table 28.2. *(continued)*

Tumor type	Treatment	No. of patients	Schedule	Results	Author
Advanced NP[a] SCC	IFN-γ	14	0.25-0.5 μg/m²/day day i.m.	5 MR 2 SD[b]	Dimery et al., 1987
Advanced HNSCC	IFN-α	13	10×10^6 U/ day i.m.	1 PR 2 SD	Fierro et al., 1988
	IFN-γ	8	50 μg/m²	1 CR 3 MR	Richtsmeier et al., 1990
	IFN-α	28	6×10^6 U/day	1 CR 1 SD	Agarwala et al., 1991
	IFN-α	31	12×10^6 U/day i.m.	1 CR 9 SD	
	IFN-α	14	10×10^6	1 CR 1 SD 1 mixed response	Vlock et al., 1991
	IFN-α rIL-2	14	5×10^6 U/i.m. 3×10^6 U/i.v.	1 CR 2 CR	Schantz et al., 1992

[a]NP nasopharynx
[b]SD stable disease

untreated patients, while a parallel study of node lymphocytes (Cortesina et al., 1990) showed that spontaneous LAK activity was greater in nodes ipsilateral to the tumor and closest to the injection area. NK activity, on the other hand, was depressed. In two patients treated preoperatively, high levels of mRNA for the IL-2 receptor were observed. It was absent in untreated patients.

Bilateral induction of cytotoxicity against autologous tumor cells and the K562 and Daudi lines was observed by Rivoltini et al. (1990) in the cervical nodes of 12 recurrent HNSCC patients following the local administration of 1×10^5 U rIL-2/day for 20 days. Evidence of induction of an immune response was noted in the form of apparently specific autologous cytotoxicity on the part of $CD25^+$ cells, and LAK activity attributed to $CD16^+$ cells.

Squadrelli-Saraceno et al. (1990) obtained 3 PR and 3 MR (all in tumors smaller than 20 cm²) with 10 daily local injections of increasing doses (2500-1.8×10^6 U/day) of IL-2 plus one injection of autologous LAK cells on day 1, in 14 patients with advanced or recurrent HNSCC. No response was observed in cervical metastatic nodes. No clinical or immunopathological effects were observed, however, when Mattijssen et al. (1991) repeated this study in 15 patients.

Saito et al. (1991) treated 22 HNSCC patients for 4 weeks with 8 x 10^6 U/day of IL-2 injected directly into the tumor and obtained 2 CR, 1 PR and 1 MR. An immunopathological study showed that T-cell infiltration was related to the clinical response, and there was an increase in $CD25^+$ cells. Similar results have also been obtained in a multicentre phase II trial involving 46 patients treated systemically and locally (Sawaki, 1990): 2 CR and 4 PR (all in locally treated subjects).

Selvaggi et al. (1990) have reported 1 PR in 12 patients treated with peritumoral and local IL-2 injections. In an ongoing trial by the Eastern Cooperative Group, increasing doses of IL-2 are being injected peritumorally and in the cervical nodes of patients with advanced HNCa. A finding common to all these studies is the absence of substantial toxicity.

Systemic administration of IL-2 (alone or with LAK cells) has often been employed in the management of a variety of malignancies. This approach, however, has rarely been adopted for HNCa. Ikawa et al. (1989) administered autologous LAK cells via the maxillary artery to 4 patients with carcinoma of the maxillary sinus. A decrease in tumor mass was obtained, together with histological evidence of necrosis and infiltration by macrophages.

Schantz et al. (1992) have evaluated the efficacy of systemic IL-2 plus IFN-α in 14 patients with advanced HNSCC. Five responses were obtained: 2 PR and 3 transient (1 CR and 2 PR lasting less than 4 weeks). Examination of serum samples from these patients indicated that lymphocytes and multiple serum factors during this combined treatment regulated the induction of LAK activity.

We are currently running a pilot study on the effect of topical low-dose IL-2 and IFN-α (3 x 10^6 U on days 1, 3, 5 of each week for 2 weeks, s.c.) in the treatment of recurrent HNSCC.

Not many experiments have been conducted with other cytokines in HNCa. Padovan et al. (1981) obtained interesting pathological data with IFN-α administered intratumorally or topically prior to surgery. Systemic management with IFN-α was followed by 3 CR and 4 PR in 14 HNSCC patients treated by Medenica and Slack (1985), and 2 PR in 13 patients with nasopharyngeal cancer treated by Connors et al. (1985). Fierro et al. (1988) reported 1 PR and 2 SD (stable disease) in 13 HNSCC patients treated with IFN-α.

Mention may also be made of two recent IFN-α trials. Agarwala et al. (1991) obtained 1 CR and 1 SD in 28 patients treated with low dose IFN-α (6 x 10^6 U/day for 3 days a month), and 1 CR and 9 SD in 31 patients treated with high dose IFN-α (12 x 10^6 U/day for 3 days a week). Survival was the same in both groups. Systemic administration of 10 x 10^6 U/day was followed by 1 CR and 1 SD in 14 patients treated by Vlock et al. (1991).

IFN-γ has also been employed. Dimery et al. (1987) obtained 5 MR and 2 SD following its systemic administration in 14 patients with nasopharyngeal cancer, while Richtsmeier et al. (1990) have reported 1 CR and 3 MR in 8 patients with IFN-γ after systemic administration of 50 $\mu g/m^2$ in 24 h. TNF, on the other hand, has been followed by poor responses and high toxicity when used in the treatment of a variety of tumors (Chapman et al., 1986; Blick et al., 1987).

Conclusion

Immunotherapy is the fourth treatment modality for some tumors. The use of BRM in the management of HNCa, on the other hand, is quite recent. Their effectiveness can be judged from the responses achieved in patients with advanced cancers no longer open to conventional treatment. Local management may be more effective here than in other forms of tumor, since the natural history of HNSCC is marked by local recurrences. Systemic activation of the immune system following topical treatment has been demonstrated in many studies. We are now conducting a cooperative and randomized phase III trial of adjuvant IL-2 in oral cavity and oropharynx cancers following the observation of good responses from small tumors in these sites.

Control of minimum residual disease should optimize the use of BRM. Topical administration of IL-2 and IFN has never been followed by the severe toxicity associated with systemic treatments. Like that of chemotherapy, the toxicity of systemic immunotherapy is dose dependent, but it is also self limiting. Immunotherapy, initially nonspecific, is moving towards more specific goals. A future possibility, now in the course of investigation (Cavallo et al., 1992), is the establishment of tumor immunization through the use of cytokine gene transfected cells.

References

Agarwala S, Vlock D, Johnson J, Lynch E, Kirkwood J and Haselow R (1991). Phase II of interferon alpha (IFN) in locally recurrent or metastatic head and neck cancer (SCCHN). Results of ECOG trial P-Z386. Proc ASCO **10**, 205.

Berlinger NT (1984). Deficient immunity in head and neck cancer due to excessive monocyte production of prostaglandins. Laryngoscope **94**, 1407-1413.

Bier J, Rapp HJ, Borsos T, Zbar B and Linschuster ST (1981). Randomized study on intratumoral BCG-cell wall preparation (CWP) therapy in patients with squamous cell carcinoma in the head and neck region. Cancer Immunol Immunother **12**, 71-75.

Blick M, Sherwin SA, Rosenbloom M and Gutterman J (1987). Phase I study of recombinant tumor necrosis factor in cancer patients. Cancer Res **47**, 2986-2989.

Cavallo F, Giovarelli M, Gulino A, Vacca A, Stoppacciaro A, Modesti A and Forni G (1992). Role of neutrophils and $CD4^+$ T lymphocytes in the primary and memory response to nonimmunogenic murine mammary adenocarcinoma made immunogenic by IL-2 gene. J Immunol **149**, 3627-3635.

Chapman P, Lester E and Casper J (1986). Phase I study of recombinant tumor necrosis factor. Proc ASCO **5**, 905.

Clayman GL, Young G, Taylor DL, Savage ME, Lavedan P and Scantz SP (1992). Detection of regulatory factors of lymphokine activated killer cells activity in head and neck cancer patients treated with interleukin-2 and interferon alpha. Ann Otol Rhinol Laryngol **101**, 909-915.

Connors JM, Andiman WA, Howarth CB, Lin E, Merigan TC, Savage ME and Jacobs C (1985). Treatment of nasopharyngeal carcinoma with human leukocyte interferon. J Clin Oncol **6**, 813-817.

Cortesina G, Cavallo GP, Beatrice F, Sartoris A, Bussi M, Morra B, Di Fortunato V, Poggio E and Rendine S (1982). Production of leukocyte migration factor by lymphocytes of larynx cancer patients stimulated by laryngeal carcinoma solubilized membrane antigens. Tumori **68**, 39-43.

Cortesina G, Morra B, Beatrice F, Cavallo G, Bussi M, Di FortunatoV, Poggio E, Vercellino M and Sartoris A (1984a). Evaluation of blocking mechanisms against immunological response in patients with laryngeal carcinoma. Laryngoscope **94**, 6-11.

Cortesina G, Sartoris A, Di Fortunato V, Cavallo GP, Morra B, Bussi M, Beatrice F, Poggio E, Marcato P and Rendine S (1984b). Natural killer-mediated cytotoxicity in patients with laryngeal carcinoma. Ann Otol Rhinol Laryngol **93**, 189-191.

Cortesina G, Beatrice F, Morra B, Barioglio MG, Poggio E, Bussi M, Di Fortunato V, Sacchi M, Rendine S and Sartoris A (1987). Specific lymphokine responses in cervical nodes of patients with laryngeal cancer. Laryngoscope **97**, 843-847.

Cortesina G, De Stefani A, Giovarelli M, Barioglio MG, Cavallo GP, Jemma C and Forni G (1988). Treatment of recurrent squamous cell carcinoma of the head and neck with low doses of IL-2 injected perilymphatically. Cancer **62**, 2482-2485.

Cortesina G, De Stefani A, Galeazzi E, Bussi M, Giordano C, Cavallo GP, Jemma C, Vai S, Forni G and Valente G (1990). The effect of preoperative local interleukin-2 (IL-2) injections in patients with head and neck squamous cell carcinoma. An immunological study. Acta Otolaryngol **111**, 428-433.

Cortesina G, De Stefani A, Galeazzi E, Cavallo GP, Jemma C, Giovarelli M, Vai S and Forni G (1991). Interleukin-2 injected around tumor draining lymph nodes in head and neck cancer. Head and Neck **13**, 125-131.

Cortesina G, Sacchi M, Galeazzi E, Johnson JT and Whiteside TL (1992). Interleukin-2 receptors on squamous cell carcinomas of the head and neck. Characterization and functional role. Acta Otolaryngol **112**, 370-375.

Dimery I, Jacobs C, Tseng A, Saks S, Hong WK and Utterman JU (1987). Recombinant interferon gamma in the treatment of recurrent nasopharyngeal carcinoma. Proc ASCO **6**, 129.

Dimery I, Martin, Bradley E, Kramer A and Hong W (1989). Phase I trial of interleukin 2 plus cisplatinum and 5 fluorouracil in recurrent or advanced squamous cell carcinoma of the head and neck. Proc ASCO **8**, 170.

Fierro R, Johnson TJ, Myers E, Colao D, Pelch K, Rust D, Wagner R, Whiteside T and Vlock D (1988). Phase II trial of recombinant interferon alpha (IFN) in recurrent squamous cell carcinoma of the head and neck (SCCHN). Proc ASCO **7**, 605.

Forni G, Santoni A and Giovarelli M (1985). Lymphokine activated tumor inhibition in vivo. The local administration of interleukin-2 triggers non reactive lymphocytes from tumor bearing mice to inhibit tumor growth. J Immunol **134**, 1305-1311.

Forni G and Giovarelli M (1987). Tumor immunotherapy with interleukin 2 and leukocytes. In: "Tumor Immunology: Mechanisms, Diagnosis, Therapy" (eds W den Otter and EJ Ruitenberg), pp 266-282. Elsevier Science Publisher, Amsterdam.

Goldsmith MM, Belchis DA, Cresson DH, Merritt WD and Askin FB (1992). The importance of the eosinophil in head and neck cancer. Otolaryngol Head Neck Surg **106**, 27-33.

Ikawa T, Eura M, Fukiage T, Murakami H, Yamaraki S, Fukuda M, Arishima S, Machara T and Ishikawa T (1989). Adoptive immunotherapy by intraarterial infusion of ATLAK or Allo-TLAK cells in patients with head and neck cancer. Gan To Kagaku Ryoho **16**, 1438-1447.

Letessier EM, Sacchi M, Johnson JT, Herberman RB and Whiteside TL (1990). The absence of lymphoid suppressor cells in tumor-involved lymph nodes of patients with head and neck cancer. Cell Immunol **139**, 446-458.

Letessier EM, Heo DS, Okarma T, Johnson JT, Herberman RB and Whiteside TL (1991). Enrichment in tumor-reactive CD8+ T-lymphocytes by positive selection from the blood and lymph nodes of patients with head and neck cancer. Cancer Res **51**, 3891-3899.

Matijssen V, De Mulder PH, Shornagel HJ, Verwei J, Van den Bock P , Galazka A, Roy S and Ruiter DJ (1991). Clinical and immunopathological results of a phase II study of perilymphatically injected recombinant IL-2 in locally advanced, non pretreated head and neck squamous cell carcinomas. J Immunother **10**, 63-68.

Medenica R and Slack N (1985). Clinical results of leukocyte interferon-induced tumor regression in resistant human metastatic cancer resistant to chemotherapy and/or radiotherapy-pulse therapy schedule. Cancer Drug Deliv **2**, 53-76.

Mickel RA, Kessler DJ, Taylor JMG and Lichtenstein A (1988). Natural killer cell cytotoxicity in the peripheral blood, cervical lymph nodes tumor of head and neck cancer patients. Cancer Res **48**, 5017-5022.

Musiani P, De Campora E, Valitutti S, Castellino F, Calearo C, Cortesina G and Giovarelli M (1989). Effects of low doses of interleukin 2 injected perilymphatically and peritumorally in patients with advanced primary head and neck squamous cell carcinoma. J Biol Resp Modif **8**, 571-578.

Olivari AJ, Glait HM and Guardo A (1979). Levamisole in squamous cell carcinoma of head and neck. Cancer **63**, 983-990.

Padovan I, Brodarec I, Ikic D, Knezevic M and Soos E (1981). Effects of interferon in therapy of skin and head and neck tumors J Cancer Res Clin Oncol **100**, 295-310.

Racz T, Sacks PG, Taylor DL and Schantz SP (1989). Natural killer cell lysis of head and neck cancer. Arch Otolaryngol Head Neck Surg **115**, 1328-1331.

Richtsmeier WJ, Koch WM, Mc Guire WP, Poole ME and Chang EH (1990). Phase I-II study of advanced head and neck squamous cell carcinoma patients treated with recombinant human interferon gamma. Arch Otolaryngol Head Neck Surg **116**, 1271-1277.

Rivoltini L, Gambacorti-Passerini C, Squadrelli-Saraceno M, Grosso MI, Cantu G, Molinari R, Orazi A and Parmiani G (1990). In vivo interleukin 2-induced activation of lymphokine-activated killer cells and tumor cytotoxic T-cells in cervical lymph nodes of patients with head and neck tumors. Cancer Res **50**, 5551-5557.

Rosenberg SA, Lotze MT, Muul LM, Leitman S, Chang AE, Ettinghausen SE, Matory YL, Skibber JM, Shiloni E, Vetto JT, Seipp CA, Simpson C and Reichert CM (1985). Observation on the systemic administration of autologous lymphokine activated killer cells and recombinant interleukin 2 to patients with metastatic cancer. N Engl J Med **313**, 1485-1492.

Rosenberg SA, Lotze MT, Muul LM, Chang AE, Avis FP, Leitman S, Linehan WM, Robertson CN, Lee RE, Rubin JT, Seipp CA, Simpson CG and White DE (1987). A progress report on the treatment of 157 patients with advanced cancer using lymphokine-activated killer cells and interleukin-2 or high-dose interleukin-2 alone. N Engl J Med **316**, 889-897.

Rosenberg SA, Lotze MT, Yang JC, Aebersold PM, Linehan WM, Seipp CA and White DE (1989) Experience with the use of high-dose interleukin-2 in the treatment of 652 cancer patients. Ann Surg **210**, 474-484.

Sacchi M, Klapan I, Johnson JT and Whiteside TL (1991a). Antiproliferative effects of cytokines on squamous cell carcinoma. Arch Otolaryngol Head Neck Surg **117**, 321-326.

Sacchi M, Vitolo D, Sedlmayr P, Rabinowich H, Johnson JT, Herberman RB and Whiteside TL (1991b). Induction of tumor regression in experimental model of human head and neck cancer by human a-LAK cells and IL-2. Int J Cancer **47**, 784-791.

Saito T, Yoda J and Tabata T (1991). Local administration trials of interleukin 2 for head and neck cancer. Auris Nasus Larynx **18**, 169-178.

Sawaki S (1990). A phase II study of recombinant interleukin 2 (S-6820) for head and neck cancer. Gan No Rinsho **36**, 111-120.

Schantz SP, Dimery I, Lippman SM, Clayman GL, Pellegrino C and Morice R (1992). A phase II study of interleukin 2 and interferon alpha in head and neck cancer. Invest New Drug **10**, 217-223.

Selvaggi KJ, Vlock DR, Johnson JT, Snyderman CH, Rubin J, Kirkwood J, Haselow R, Letessier E, Whiteside T and Prescot K (1990). Phase Ib trial of peritumoral and intranodal injections of IL-2 in patients with advanced squamous cell carcinoma of the head and neck (SCCHN). Preliminary results. Proc ASCO **9**, 674.

Squadrelli-Saraceno M, Rivoltini L, Cantu G, Rovagnani F, Parmiani G and Molinari R (1990). Local adoptive immunotherapy of advanced head and neck tumors with LAK cells and interleukin 2. Tumori **76**, 566-571.

Traissac L (1968). Cervical lymphography: Difficulties and advantages. Rev Laryngol **898**, 81-90.

Urba S, Forastiere AA, Amrein PC, Woolf GT and Zeffren J (1991). Phase II pilot study of concomitantly administered recombinant human Interleukin 2 and Roferon A in patients with locally recurrent or metastatic head and neck cancer. Proc ASCO **10**, 695.

Valente G, De Stefani A, Jemma C, Giovarelli M, Geuna M, Cortesina G, Forni G and Palestro G (1990). Infiltrating leukocyte populations and T lymphocyte subsets in head and neck squamous cell carcinomas from patients receiving perilymphatic injections of recombinant interleukin 2. Modern Pathol **6**, 702-708.

Valone FH, Gandara DR, Deisseroth AB, Perez EA, Rayner A, Aronson FR, Luce J and Paradise L (1991). Interleukin 2, cisplatin and 5 Fluorouracil for patients with non-small cell lung and head and neck carcinomas. J Immunother **10**, 207-213.

Vlock DR, Johnson J, Myers E, Day R, Gooding WE, Whiteside TL, Pelch K, Sigler B, Wagner R, Colao D and Rust D (1991). Preliminary trial of non recombinant interferon alpha in recurrent squamous cell carcinoma of the head and neck. Head and Neck **13**, 15-21.

Wang MB, Lichtenstein A and Mickel RA (1991). Hierarchical immunosuppression of regional lymph nodes in patients with head and neck squamous cell carcinoma. Otolaryngol Head Neck Surg **105**, 517-527.

West WH, Tauer KW, Yannelly JR, Marshall GD, Orr DW, Thurmann GB and Oldham RK (1987). Constant-infusion recombinant IL-2 in adoptive immunotherapy of advanced cancer. N Engl J Med **316**, 898-905.

Wolf GT, Lovett EJ and Peterson KA (1984). Lymphokine production and lymphocyte subpopulations in patients with head and neck squamous carcinoma. Arch Otolaryngol **110**, 731-735.

29

LOCAL IMMUNOTHERAPY OF HUMAN BLADDER CANCER WITH RECOMBINANT IL-2: ACTIVATION OF IMMUNE RESPONSE AND POSSIBLE DIRECT ACTION OF IL-2 ON TUMOR CELLS

F. Velotti, M. Cippitelli, A. M. Giuffrida, A. Punturieri, F. Moretti, A. Stoppacciaro, A. Tubaro, R. De Maria and A. Santoni
Laboratory of Molecular Pathology, University of Tuscia, Viterbo; Department of Experimental Medicine, University La Sapienza, Rome; Institute of Human Pathology and History of Medicine, University of Chieti, Chieti; Urology Clinic, University of L'Aquila, L'Aquila and Laboratory of Pathophysiology, Regina Elena Cancer Institute, Rome, Italy

Introduction

Urinary bladder cancer is a frequent tumor, and approximately 90% of patients with this cancer have transitional cell carcinoma (TCC). It is usually of papillary type and can develop either as a superficial (confined to mucosa and lamina propria) or an invasive (spread through muscle tissues) bladder cancer. Although both superficial and locally invasive bladder cancer derive from transitional cell epithelium of the bladder, they are actually considered as two different diseases in terms of natural history, tendency to metastasize and generation of local inflammatory response (Stoppacciaro et al., manuscript in preparation).

Transurethral resection of bladder tumor (TURBT) is currently considered as the "curative" treatment for superficial bladder cancer. However, the tumor often recurs locally after endoscopic surgery. Neoadjuvant or adjuvant therapies with antiproliferative drugs or biological response modifiers are currently being used, either to prevent local relapses after TURBT or to induce regression of incompletely resected bladder neoplasms (Lumm and Torti, 1991).

Since urinary bladder is one of the anatomic regions amenable to local therapy, local chemotherapy and immunotherapy have been used extensively during the last decade to treat patients with bladder cancer (Kurth et al., 1984; Lamm et al., 1991). For local treatment of the bladder,

CYTOKINE-INDUCED TUMOR IMMUNOGENICITY
ISBN 0-12-261520-4

the ways of delivering drugs can be intralesional, intravesical or intraarterial.

Standard protocols are available for intravesical administration of chemotherapeutic drugs such as methotrexate, vinblastin, adriamycin and cisplatin (Raghavan et al., 1990). However, the efficacy of chemotherapy in preventing tumor relapses is still debated. A recent review by Lamm (1992) indicated that none of the currently used chemotherapeutic regimens resulted in a modification of the local relapse rate in long-term follow-up.

For the past 30 years, immunotherapy using microbial products has also been investigated for intravesical treatment with the aim of potentiating the host immune response against the tumor. Particular interest has been given to BCG. Since the first report by Morales et al. (1976) who showed the clinical efficacy of BCG immunotherapy in the prevention of bladder tumor recurrences, several investigations have reinforced the view that BCG is a therapeutic and prophylactic agent against recurrent tumors (Lamm, 1992). Currently, in many countries BCG is considered the treatment of choice against superficial bladder cancer in terms of reducing the local recurrence rate after TURBT and the regression of unresectable superficial neoplasms (Lamm et al., 1991; Lamm, 1992). However, local and systemic side effects such as cystitis, fever and tuberculosis are common (Lamm et al., 1986). The use of BCG is therefore restricted to patients with a high risk of recurrence and progression.

The mechanism underlying the antitumor activity mediated by BCG is not completely understood. Animal studies have shown that an intact T-cell system is required for the antitumor efficacy of BCG (Ratliff et al., 1987). Furthermore, lymphokines, including IL-2, have been found in the urine of patients treated with BCG (Haaf et al., 1986; Bohle et al., 1990).

The possible role of local release of IL-2 in BCG induced tumor regression and the recent assessment of immunotherapeutic activity of pure cytokines in tumors prompted us and other investigators to explore the possible therapeutic use of natural or recombinant IL-2 (nIL-2 or rIL-2) in the local treatment of bladder cancer.

Here we report our experience in local immunotherapy of superficial bladder cancer with rIL-2.

Local immunotherapy with rIL-2 in human superficial bladder cancer

Preliminary results suggested that perilesional injections of IL-2 containing supernatants were effective in patients with invasive bladder neoplasms (Pizza et al., 1984). More recently, it has been shown that intravesical continuous infusion of nIL-2 (Huland and Huland, 1989) could induce regression of locally advanced bladder tumors.

We designed two new neoadjuvant immunotherapeutic regimens with rIL-2 for local treatment of superficial (Ta-T1, N0, M0, stage G1 to G2) TCC of the bladder. Two phase I-II clinical trials were performed based on two different ways of delivering IL-2 to the bladder:

- intraarterial (trial 1) (through the hypogastric/internal iliac artery, which supplies blood to the bladder)
- intravesical (trial 2).

In trial 1, 12 patients (median age: 62 years, range 37-69) were treated with two 5 day cycles of continuous intraarterial infusion of rIL-2, separated by a 48 h interval. TURBT was performed 4 days after the end of the second IL-2 cycle. IL-2 dosages were escalated from 18 x 10^3 to 18 x 10^6 IU/m^2 daily in four groups of 3 patients (Table 29.1).

In trial 2, 25 patients (median age: 63 years, range 37-75) were treated with 10 daily bolus intravesical instillations of rIL-2 (referred to as induction course), using an interpatient dose escalation scheme from 3 to 18 x 10^6 IU/day (Table 29.2). TURBT was performed 10 days following the end of the induction course. Thereafter, four maintenance courses of 10 daily intravesical instillations of 6 x 10^6 IU/day of rIL-2 were started 1 month after surgery and carried out every 4 months.

In both trials, the administration of rIL-2 before tumor resection was performed with the aim of inducing an immune response against the tumor and of eliciting, possibly through the presence of a tumor antigen, a specific antitumor immune response and a long-term immunity capable of preventing tumor relapses. Under defined experimental conditions, induction of an early nonspecific immune reaction (mostly constituted by macrophages, NK cells and PMN) against the tumor was observed after administration of cytokines at the tumor site. Under these conditions, the level of tumor immunogenicity increased giving rise to a specific T-cell mediated immune response against the tumor, as well as against subsequent challenges with the parental tumor (Forni et al., 1987; Colombo et al., 1992).

The two main objectives of these clinical trials were:

- to define safety and toxicity of both local treatment approaches
- to evaluate their possible therapeutic effect, both on the tumor and on preventing local recurrences.

During trial 1, which is based on intraarterial IL-2 treatment, toxic effects consisting of grade II (WHO classification) hypotension and fever developed in 2 of the 12 patients treated with the two highest doses of rIL-2. However, these toxic effects rapidly disappeared after the infusion was stopped. In contrast, trial 2, based on intravesical injection of rIL-2, did not cause any local, systemic or laboratory toxicity within the dose range tested.

Objective clinical responses to IL-2 treatments were assessed before TURBT by bladder ultrasound sonography, computed tomography scan and cystoscopy. CR was defined as the disappearance of the tumor, clinically and histologically, in biopsies obtained from the previous tumor area. PR was defined as a decrease of tumor cross sectional area (product

Table 29.1. *Intraarterial administration of rIL-2*

Patient	rIL-2 dose (IU/m^2 daily)	Objective response	Tumor infiltration[a]		Local relapse (months)
			Before	After[b]	
1	18 x 10^3	NC[c]	ND[d]	ND	3
2	18 x 10^3	NC	ND	ND	-
3	18 x 10^3	NC	+	++	-
4	18 x 10^4	CR	+	++	-
5	18 x 10^4	CR	+	+++	-
6	18 x 10^4	NC	ND	ND	3
7	18 x 10^5	NC	+	++	-
8	18 x 10^5	PR	+	++	-
9	18 x 10^5	PR	+	+++	-
10	18 x 10^6	NC	++	++	-
11	18 x 10^6	PR	+	+	-
12	18 x 10^6	NC	+	+++	-

[a]+: 50-200 cells/mm^2; ++ : 200-500 cells/mm^2; +++ : >500 cells/mm^2
[b]4 days after the end of the second cycle of rIL-2
[c]NC: no change
[d]ND: not done

of the two major diameters of the tumor) superior or equal to 50% of the initial size.

Two CR and 3 PR occurred in 5 of 12 patients (overall response rate of 42%) treated with intraarterial administration of rIL-2 (Tubaro et al., 1991) (Table 29.1). A PR was observed in 5 out of 25 patients (overall response rate of 20%) treated with daily bolus intravesical instillations of rIL-2 (Tubaro et al., submitted) (Table 29.2). None of the patients experienced a progression of the initial tumor.

With respect to the recurrence rate, 2 out of 12 patients (17%) treated with the intraarterial treatment had a local recurrence. No tumor relapse was evident after a median follow-up of 45 months (mean 46.7, range 37-57) in the remaining patients (Table 29.1). Local tumor recurrences were found in 10 out of 25 patients (40%) treated with the intravesical treatment, after a median follow-up of 33 months (mean 34.5, range 24-45) (Table 29.2).

These findings demonstrated that both trials were safe and effective in terms of tumor regression. Our results with rIL-2 are therefore consistent with previous data obtained with nIL-2 preparations in locally advanced bladder cancer (Huland and Huland, 1989) and show that rIL-2 can induce the regression of Ta-T1 TCC of the bladder.

Table 29.2. *Intravesical administration of rIL-2*

Patient	IL-2 dose (1 x 10^6 IU/ day)	Objective response	Tumor inflammatory infiltration[a] Before	1 day after[b]	10 day after[c]	Local relapse (months)
1	3	NC[d]	+	+	+	-
2	3	NC	+	+	+	-
3	3	NC	++	++	++	18
4	3	NC	+	+	+	-
5	6	PR	++	ND[e]	+++	-
6	6	NC	++	ND	+	-
7	6	NC	+	ND	+	34
8	6	NC	+	ND	++	-
9	6	NC	+	ND	+	-
10	6	NC	++	ND	+++	4
11	9	PR	+	+++	++	18
12	9	NC	+	++	++	3
13	9	NC	++	+++	+++	-
14	9	NC	++	ND	+++	-
15	9	NC	++	ND	+++	-
16	9	PR	++	ND	+++	-
17	9	PR	+	+++	++	4
18	9	PR	++	ND	++	-
19	9	NC	+	ND	+	19
20	18	NC	+	++	+	18
21	18	NC	++	+++	++	3
22	18	NC	++	+++	+++	-
23	18	NC	++	ND	+++	-
24	18	NC	+	ND	+	-
25	18	NC	+	ND	+	5

[a]+: 50-200 cells/mm^2; ++ : 200-500 cells/mm^2; +++ : >500 cells/mm^2
[b]1 day after the end of the induction course of rIL-2
[c]10 days after the end of the induction course of rIL-2
[d]NC: no change
[e]ND: not done

Although in these trials the incidence of local recurrences appears to be lower than anticipated, the exact incidence has so far not been assessed because a 5 year follow-up is required (Lamm at al., 1991).

From the overall evaluation of the clinical results, no definitive conclusion can be drawn on which of the two trials is more effective. Contrary to the results of the intraarterial trial, no CR was obtained with the intravesical instillations of rIL-2. Whether or not a CR can be clinically relevant in this setting is, however, unknown. Nonetheless, there are some points to be considered. The intraarterial treatment can hardly be proposed for routine clinical use: bed rest is required because of the intraarterial catheter invasiveness. In contrast, daily intravesical instillation is a minimally invasive and non toxic approach; therefore, it is possible to give maintenance courses of rIL-2. This treatment could in the future be performed on an outpatient basis.

Activation of host immune response after local immunotherapy with rIL-2 in human bladder cancer

To investigate whether the local administration of rIL-2 could affect a local and/or systemic immune response, tumor and mucosa specimens, urine and peripheral blood samples were collected and analyzed for defined immunological parameters before and after treatment (Velotti et al., 1991a; Velotti et al., 1991b; Velotti et al., manuscript in preparation).

Immunohistochemical staining of biopsy specimens was performed on acetone fixed cryostatic sections by the peroxidase-anti-peroxidase method. Before treatment, a scanty or moderate inflammatory leukocyte infiltrate ($CD45^+$ cells) was present in the tumor connective stroma and in the submucosa of all patients. Treatment with rIL-2 was followed by a 2- to 20-fold increase of this inflammatory infiltrate in 8 out of 9 (88%) patients treated with intraarterial rIL-2 and in 13 out of 17 (70%) patients treated with intravesical rIL-2 (Tables 29.1 and 29.2). Immunohistochemical characterization of the inflammatory cells revealed that they were primarily composed of $CD3^+$ α/β T lymphocytes and of $CD68^+$ macrophages.

In both trials the analysis of lymphoid activation markers showed a dose dependent increase in the number of HLA-DR^+ and/or $CD25^+$ (p55 IL-2 receptor) lymphoid cells. The appearance of the p75 IL-2 receptor mRNA was also observed by in situ hybridization after IL-2 treatment.

The systemic administration of high doses of rIL-2 to cancer patients has led to the in vivo induction of cytokine production by circulating peripheral blood mononuclear cells (Kasid et al., 1989). In vitro studies also described cytokine production (e.g. TNF-α, IFN-γ) by TIL stimulated by rIL-2 (Belldegrun et al., 1989). We therefore analyzed whether the local administration of rIL-2 was capable of inducing cytokine production by TIL at the tumor site. The presence of cytokines in vivo was investigated by in situ hybridization, by immunocytochemistry on biopsy samples and by enzyme linked immunosorbent assays in the urine. Before treatment, few or no cells were positive for the expression of cytokine mRNAs. In most of the patients treated, IL-1α, IL-1β and TNF-α mRNA were induced or increased in macrophages, while IFN-γ and IL-2 mRNA appeared in lymphoid cells. The percentage of TIL positive for the expression of cytokine mRNA appeared to be higher following the intravesical treatment, though no statistical values can be proposed because of the small number of cases analyzed. The corresponding cytokines were also present, as shown by immunoreaction on biopsy sections. Furthermore, increased cytokine (IL-1β, TNF-α and IFN-γ) concentrations over baseline levels were detected in the urine of a number of patients.

These results show that the local administration of rIL-2 can trigger a cytokine cascade at the tumor site. It is highly probable that these cytokines may in turn have pleiotropic immunomodulatory effects and be cytotoxic/cytostatic for tumor cells. The specificity of T-cell lymphokine release or of a possible T-cell mediated cytotoxic activity could not be directly assessed because of the paucity of effector and target cells obtained from the biopsy specimens. We decided against using T-cell cultures, since these may select and amplify T-cell subpopulations with a minor representation in vivo. Therefore, we are currently analyzing the TCR repertoire of TIL by PCR on biopsy specimens. This approach is supported by recent studies in autoimmunity (Davies et al., 1991), allograft rejection (Miceli and Finn, 1989), as well as in malignant tumors (Nitta et al., 1990; Ferradini et al., 1993), indicating that effector cells may utilize a very restricted set of TCR V gene segments which might depend on specific tumor antigens being recognized. Analysis of the immune response at the systemic level has shown that the intraarterial administration of rIL-2 resulted in an increased percentage of $CD3^+/CD16^+$ lymphocytes and $CD14^+$ monocytes in the peripheral blood. The study of lymphocyte activation markers revealed an enhancement of $HLA\text{-}DR^+$ and $CD71^+$ cells at all doses employed, while $CD25^+$ cells increased only in patients treated with the highest doses of rIL-2. In contrast, intravesical instillations of rIL-2 affected neither the distribution of peripheral blood cell populations nor the percentage of lymphoid activation markers. Following both treatments, no increase over baseline levels of cytokines such as IL-1β, TNF-α and IFN-γ was found in the plasma. Taken together, these results indicate that both trials based on the local administration of rIL-2 were capable of increasing an inflammatory reaction and of activating an immune response at the tumor site. These results demonstrate that although these trials induced a similar immunomodulation action at the tumor site, they seem to differ in their capacity to affect a systemic immune response. Although the local administration of rIL-2 affected both the tumor mass and the immune response at the tumor site, no correlation was found between the level and quality of the local inflammatory reaction and the objective clinical responses or local relapses.

Does IL-2 interact with human TCC of the bladder?

Because the therapeutic efficacy of rIL-2 administered locally to patients with bladder cancer did not correlate with the activation of the immune response, we investigated whether rIL-2 could have a direct effect on the tumor.

Recent evidence supports the possibility that nonlymphoid cells including oligodendroglial and oligodendroglioma cells (Benveniste and Merrill, 1986; Okamoto et al., 1990), endothelial cells (Frazier-Scott et al., 1988), embryonic fibroblasts and fibrosarcomas (Carloni et al., 1989; Plaisance et al., 1992), squamous cell carcinoma (Weidmann et al., 1992) and intestinal epithelial cells (Ciacci et al., 1993) may respond to IL-2.

It is now well established that the IL-2 receptor (IL-2R) expressed on lymphoid cells is composed of at least three distinct polypeptides: the 55 kDa α chain (IL-2Rα; Tac, p55, CD25), the 70-75 kDa β chain (IL-2Rβ; p70/p75) and the 64 kDa γ chain (IL-2Rγ; p64); these last two chains combined correspond to the functional entity (Taniguchi and Minami, 1993).

In order to investigate whether IL-2 could directly act on bladder cancer cells, we analyzed the expression of the IL-2R chains in three human TCC cell lines: RT4 (G1 histological grade of the original tumor) (Rigby and Franks, 1970), RT112 (G2) (Marshall et al., 1977) and MGH-U1 or EJ (G3) (Evans et al., 1977).

The expression of the mRNA encoding for the IL-2R α, β and γ chains was analyzed in the three cell lines by RT-PCR (amplification of cDNA synthesized by reverse transcription of total cellular RNA). The band corresponding to the expected specific sequences for the IL-2R α chain was observed in RT4, RT112 and MGH-U1 cell lines. Positivity for the IL-2R γ chain was clearly evident for RT4 and RT112, while detectable only after a longer exposure in MGH-U1 (Fig. 29.1). Under our experimental conditions, specific sequences for the IL-2R β chain were not detected in any of the three cell lines.

These results demonstrate that the IL-2R α and γ chains are indeed transcribed in bladder cancer cells, thus providing the first evidence for IL-2R γ chain mRNA expression in nonlymphoid cells.

The expression of the IL-2R chains was also investigated at the protein level, using anti-CD25 (anti-p55) and anti-p70/75 (TU27 and TU11) (Takeshita et al., 1989; Takeshita et al., 1992) mAb. As shown by immunofluorescence and flow cytometric analysis, all mAb positively stained the three bladder cancer cell lines, although with a different degree of reactivity. Similar results were obtained in primary tumor cell lines derived from two patients with TCC of the bladder (T3 G3 and T2 G3, respectively). Our results suggest that bladder cancer cells constitutively express the IL-2R α, β and γ chains. Experiments are in progress to clarify the apparent discrepancy between expression of the IL-2R β chain at the mRNA and protein level.

To investigate whether bladder cancer cells could respond to IL-2,

Fig. 29.1. *Expression of the IL-2R γ chain mRNA by human TCC bladder cell lines.* Total RNA was isolated by single step guanidine thiocyanate extraction procedure (Chomczynsky and Sacchi, 1987). Using a standard method, 10 μg of total RNA from the Jurkat T-cell line and 20 μg from the RT4, RT112 and MGH-U1 (EJ) TCC cell lines were reverse transcribed in a total volume of 20 μl. To set up a 50 μl PCR reaction, 2 μl of Jurkat cDNA and 10 μl of the other cell lines cDNA were used. The IL-2R γ chain mRNA was amplified using the following primers: (sense) 5' CCATGTTGAAGCCATCATTAC and (antisense) 5' ACACTAGGCAAGGAGAACTTA. The expected product (579 bp length) was identified by Southern blot technique after transferring on nitrocellulose a 2% agarose gel loaded with different amounts of the PCR reactions. We used as a probe an identical PCR product obtained from the Molt-4 T-cell line, verified by sequencing identical to the described γ chain mRNA.
Lane 1: 5 μl of the Jurkat PCR reaction; lane 2: 50 μl of the RT112 PCR reaction; lane 3: 50 μl of the RT4 PCR reaction; lane 4: 50 μl of the MGH-U1 PCR reaction.

we analyzed the ability of rIL-2 to affect tumor growth. Under the experimental conditions used, no major changes in cell proliferation were found, as evaluated by ^{3}H-thymidine incorporation and colorimetric MTT based assays. This would indicate that IL-2 is not a growth promoting factor for bladder cancer cells.

At present, the biochemical pathways involved in the IL-2R signalling events remain largely undefined in lymphoid cells. IL-2 induced activation of NF-κB expression in the nucleus of lymphoid cells has been reported by Arima et al. (1992). Preliminary results obtained by Western blot analysis showed that, in contrast to lymphoid cells, IL-2 treatment of RT112 bladder tumor cells induced a transient decrease in the expression of the p50- and p65-NF-κB transcription factors in the nucleus, while no change was observed in the cytoplasm. A transient decrease in the translocation of the transcription factors from the cytoplasm to the nucleus could account for the IL-2 induced inhibition of NF-κB factor expression.

The biological significance of this observation is unknown; these results, however, suggest that the IL-2R expressed by tumor bladder cells may be functional. The fact that IL-2 provokes a negative regulation of NF-κB is consistent with recent observations showing IL-2 induced negative signals in lymphoid (Tsudo et al., 1989) and nonlymphoid cells (Weidmann et al., 1992).

Conclusions

Our studies have demonstrated that the two trials based on intraarterial and intravesical administration of rIL-2 are safe within the dose range used. Moreover, we have shown that both treatments can induce a total or partial regression of the tumor mass, though phase I-II studies have not documented a possible benefit to the patients. Although our results have provided an insight into two new therapeutic approaches, the use of rIL-2 in the treatment of superficial bladder cancer should be considered experimental until randomized trials can show a therapeutic advantage over the currently used drug regimens.

The mechanisms underlying the antitumor effect of rIL-2 are not yet completely understood. Our results have shown that both treatments are effective in inducing a local host immune activation and in triggering a cytokine cascade at the tumor site. Moreover, we have found the expression of the IL-2R on tumor bladder cancer cell lines and on some primary cultures of tumor bladder cancer cells.

Though many questions remain to be answered, our results raise the possibility that the antitumor effect mediated by IL-2 on bladder tumors may be the result of two different mechanisms: activation of an immune response, and a direct interaction with the tumor cell.

Acknowledgments

We thank Dr A.M. Jackson and S. Prescott (Department of Surgery/Urology, University of Edinburgh, UK) for the kind gift of long-term human bladder cancer cell lines, Dr G. Cusumano (Division of Urology, Regina Elena Cancer Institute, Rome) for providing human bladder cancer surgery specimens, Dr A. Nista (Laboratory of Pathophysiology, Regina Elena Cancer Institute, Rome) for flow cytometric analysis on bladder cancer cell lines and Prof K. Sugamura (Department of Microbiology, Tohoku University, Japan) for providing TU27 and TU11 mAb. We are grateful to P. Birarelli and D. Milana (Department of Experimental Medicine, University La Sapienza, Rome) for technical assistance. This work was supported by a grant from the Consiglio Nazionale Ricerche (CNR), progetto finalizzato "Applicazioni Cliniche della Ricerca Oncologica" (ACRO).

References

Arima N, Kuziel WA, Gradina TA and Greene WC (1992). IL-2 induced signal transduction involves the activation of nuclear NF-κB expression. J Immunol **149**, 83-91.

Belldegrun A, Kasid A, Uppenkamp M, Topalian SL and Rosenberg SA (1989). Human tumor-infiltrating lymphocytes. Analysis of lymphokine mRNA expression and relevance to cancer immunotherapy. J Immunol **142**, 4520-4526.

Benveniste EN and Merrill JE (1986). Stimulation of oligodendroglial proliferation and maturation by interleukin-2. Nature **321**, 610-613.

Bohle A, Nowc C, Ulmer A, Musehold J, Gerdes J, Hofstetter AG and Flad HD (1990). Elevations of cytokines interleukin-1, interleukin-2 and tumor necrosis factor in the urine of patients after intravesical Bacillus Calmette-Guerin immunotherapy. J Urol **144**, 59-64.

Carloni G, Paterson H, Mareel M., Augery-Bourget Y, Sahraoui J, Rubinstein E, Suarez H and Azzarone B (1989). N-ras dependent revertant phenotype in human HT 1080 fibrosarcoma cells is associated with loss of proliferation within normal tissues and expression of an adult membrane antigenic phenotype. Oncogene **4**, 873-880.

Chomczynski P and Sacchi N (1987). Single-step method of RNA isolation by acid guanydium thiocyanate-phenol-chloroform extraction. Ann Biochem **162**, 156-159.

Ciacci C, Mahida YR, Dignass A, Koizumi M and Podolsky DK (1993). Functional interleukin-2 receptors on intestinal epithelial cells. J Clin Invest **92**, 527-532.

Colombo M P, Modesti A, Parmiani G and Forni G (1992). Local cytokine availability elicits tumor rejection and systemic immunity through granulocyte-T-lymphocite cross-talk. Cancer Res **52**, 4853-4857.

Davies TF, Martin A, Conception ES, Graves P, Cohen L and Ben-Nun A (1991). Evidence of limited variability of antigen receptors on intrathyroidal T cells in autoimmune thyroid disease. N Engl J Med **325**, 238-244.

Evans DR, Irwin RJ, Havre PA, Bouchard JG, Kato T and Prout GR (1977). The activity of pyrimidine biosynthetic pathway in MGH-U1 transitional carcinoma cell growth in tissue culture. J Urol **117**, 712-719.

Ferradini L, Mackensen A, Genevée C, Bosq J, Duvillard P, Avril M-F and Hercend T (1993). Analysis of T cell receptor variability in tumor-infiltrating lymphocytes from human regressive melanoma. J Clin Invest **91**, 1183-1190.

Forni G, Giovarelli M, Santoni A, Modesti A and Forni M (1987). Interleukin-2 activated tumor inhibition in vivo depends on the systemic involvement of host immunoreactivity. J Immunol **138**, 4033-4041.

Frazier-Scott K, Hatzakis H, Seong D, Jones CM and Wu KK (1988). Influence of natural and recombinant interleukin-2 on endothelial cell arachidonate metabolism. J Clin Invest **82**, 1877-1883.

Haaf EO, Catalona WJ and Ratliff TL (1986). Detection of IL-2 in the urine of patients with superficial bladder tumors after treatment with intravesical BCG. J Urol **136**, 970-974.

Huland E and Huland H (1989). Local continuous high dose interleukin-2: A new therapeutic model for the treatment of advanced bladder carcinoma. Cancer Res **49**, 5469-5474.

Kasid A, Director EP and Rosenberg SA (1989). Induction of endogenous cytokine-mRNA in circulating peripheral blood mononuclear cells by IL-2 administration to cancer patients. J Immunol **143**, 736-739.

Kurth KH, Schroder FH, Tunn U, Reginald AY, Pavone Macaluso M, Debruyne F, De Paw M, Dalesio O and Tenkate F (1984). Adjuvant chemotherapy of superficial transitional cell bladder carcinoma: Preliminary results of a European organization for research on treatment of cancer. J Urol **132**, 258-262.

Lamm DL, Stogdill VD, Stogdill BJ and Crispen RG (1986). Complications of Bacillus Calmette Guerin immunotherapy in 1,278 patients with bladder cancer. J Urol **135**, 272-274.

Lamm DL, Blumenstein BA, Crawford ED, Montie JE, Scardino P, Grossman HB, Stanisic TH, Smith JA Jr, Sullivan J, Sarosdy MF, Crissman JD and Coltman CA (1991). A randomized trial of intravesical doxorubicin and immunotherapy with Bacillus Calmette-Guerin for transitional-cell carcinoma of the bladder. N Engl J Med **325**, 1205-1209.

Lamm DL (1992). Long-term results of intravesical therapy for superficial bladder cancer. Urol Clin North Am **3**, 573-580.

Lumm BL and Torti FM (1991). Adjuvant intravesicular pharmacotherapy for superficial bladder cancer. J Natl Cancer Inst **83**, 682-694.

Marshall CJ, Franks LM and Carbonell AW (1977). Markers of neoplastic transformation in epithelial cell lines derived from human carcinomas. J Natl Cancer Inst **58**, 1743-1748.

Miceli MC and Finn OJ (1989). T cell receptor β-chain selection in human allograft rejection. J Immunol **142**, 81-86.

Morales A, Eidinger D and Bruce AW (1976). Intracavitary bacillus Calmette-Guerin in the treatment of superficial bladder tumors. J Urol **116**, 180-183.

Nitta T, Oksenberg JR, Rao NA and Steinman L (1990). Predominant expression of T cell receptor Va7 in tumor-infiltrating lymphocytes in uveal melanoma. Science **249**, 672-674.

Okamoto Y, Minamoto S, Shimizu K, Mogami H and Taniguchi T (1990). Interleukin-2 receptor β-chain expressed in an oligodendroglioma line binds interleukin-2 and delivers growth signal. Proc Natl Acad Sci USA **87**, 6584-6588.

Pizza G, Severini G, Menniti D, De Vinci C and Corrado F (1984). Tumor regression after intralesional injection of interleukin-2 (IL-2) in bladder cancer. Preliminary report. Int J Cancer **34**, 359-367.

Plaisance S, Rubinstein E, Alileche A, Sahraoui Y, Krief P, Augery-Bourget Y, Jasmin C, Suarez H and Azzarone B (1992). Expression of the interleukin-2 receptor on human fibroblasts and its biological significance. Int Immunol **4**, 739-746.

Raghavan D, Shipley WU, Garnick MB, Russell PJ and Ritchie JP (1990). Biology and management of bladder cancer. N Engl J Med **322**, 1129-1139.

Ratliff TL, Gillen D and Catalona WJ (1987). Requirement of a thymus dependent immune response for BCG-mediated antitumor activity. J Urol **137**, 155-158.

Rigby CC and Franks LM (1970). A human tissue culture cell line from a transitional cell carcinoma of the urinary bladder: Growth, chromosome pattern and ultrastructure. Br J Cancer **24**, 746-754.

Takeshita T, Goto Y, Tada K, Nagata K, Asao H and Sugamura K (1989). Monoclonal antibody defining a molecule possibly identical to the p75 subunit of the interleukin-2 receptor. J Exp Med **169**, 1323-1332.

Takeshita T, Ohtani K, Asao H, Kumaki S, Nakamura M and Sugamura K (1992). An associated molecule, p64, with IL-2 receptor β-chain. J Immunol **148**, 2154-2158.

Taniguchi T and Minami Y (1993). The IL-2/IL-2 receptor system: A current overview. Cell **73**, 5-8.

Tsudo M, Karasuyama H, Kitamura F, Nagasaka Y, Tanaka T and Miyasaka M (1989). Reconstitution of a functional IL-2-receptor by the β-chain cDNA. A newly acquired receptor transduces negative signal. J Immunol **143**, 4039-4043.

Tubaro A, Velotti F, Stoppacciaro A, Santoni A, Vicentini C, Bossola PC, Galassi P, Pettinato A, Morrone S, Napolitano T, Frati L, Ruco L, Franks RF, Palmer PA, Pourreau CN and Miano L (1991). Continuous intra-arterial administration of recombinant interleukin-2 in low-stage bladder cancer. Cancer **68**, 56-61.

Velotti F, Stoppacciaro A, Ruco L, Tubaro A, Pettinato A, Morrone S, Napolitano T, Bossola P, Franks CR, Palmer P, Pourreau CN, Piccoli M, Baroni CD, Frati L, Miano L and Santoni A (1991a). Local activation of immune response in bladder cancer patients treated with intra-arterial infusion of recombinant interleukin-2. Cancer Res **51**, 2456-2462.

Velotti F, Stoppacciaro A, Tubaro A, Giuffrida A, Morrone S, Ruco L, Vicentini C, Tonelli VH, Miano L and Santoni A (1991b). Immune response in bladder cancer patients (pts) treated with two different recombinant interleukin-2 (rIL-2) local/regional regimens. Proc AACR, 246.

Weidmann E, Sacchi M, Plaisance S, Heo DS, Yasumura S, Lin W-C, Johnson JT, Herberman RB, Azzarone B and Whiteside TL (1992). Receptors for interleukin-2 on human squamous cell carcinoma cell lines and tumor in situ. Cancer Res **52**, 5963-5970.

30

ADOPTIVE THERAPY OF HUMAN VIRAL DISEASES WITH VIRUS SPECIFIC CD8+ CYTOLYTIC T-CELL CLONES

P. D. Greenberg, M. Gilbert, B. Nelson, S. Nagoya, C. Yee, L. Finch, B. Walter and S. R. Riddell
Departments of Medicine/Oncology and Immunology, University of Washington, Seattle, WA, USA

Introduction

Numerous studies in murine models have demonstrated the potential utility of adoptive transfer of antigen specific T cells as a means of treating viral and malignant diseases (Byrne and Oldstone, 1984; Lukacher, et al., 1984; Reddehase, et al., 1985; Greenberg, 1991). These studies have identified critical issues that must be resolved for each human disease to most effectively apply the principles derived from murine adoptive therapy models, including:

- the requisite effector cell: which subpopulation(s) of T cells mediate(s) the desired effect, and which effector functions participate
- the target antigen: what is the nature, immunogenicity and stability of the protein eliciting the response, and need it be expressed in every target cell
- principles of cell transfer: what are the therapeutic limitations of adoptive therapy with regard to stage and localization of disease, and what potential toxicities do the transferred T cells have to normal host tissues.

Several laboratories have already pursued the use of adoptive T-cell therapy for the treatment of human malignant disease. In general, these studies have isolated T cells from the lymphoid infiltrate in a tumor mass or in a tumor draining lymph node, and have expanded the cells by continuous exposure to pharmacologic doses of IL-2 as a T-cell growth factor (Rosenberg et al., 1988). The results have affirmed the hypothesis that effector cells are present in the expanded population which have the capacity to mediate antitumor responses in vivo, but significant clinical responses have been observed in only a minority of the treated patients, and these responses have tended to be of short duration. There are many

CYTOKINE-INDUCED TUMOR IMMUNOGENICITY
ISBN 0-12-261520-4

factors that could contribute to the limited efficacy observed, but the precise problems are difficult to characterize in this complex therapeutic setting. The obstacles to elucidating these factors and resolving the problems limiting therapeutic efficacy include difficulties identifying:

- the precise effector T cell present in these mixed cell populations that is responsible for the antitumor activity
- the actual number of such effector cells being infused in an individual patient
- the target antigen being recognized on the suseptible tunor cells and the antigenic status of nonsusceptible tumors (i.e. is the protein not expressed or mutated?)
- the requirements for effective cell transfer, particularly since the use of high dose IL-2 as growth signal during in vitro effector cell expression generally necessitates that exogenous IL-2 is administered to patients in order for the transferred T cells to persist and function in vivo.

In light of these obstacles, our laboratory elected to evaluate initially the application of principles of adoptive T-cell transfer previously elucidated in murine models to the treatment of human viral infection in immunocompromised hosts, rather than to the therapy of malignant disease. This provides a setting in which it should be possible to identify the nature of the effector T cell required to eliminate the pathogen, the viral antigen that is being expressed in target cells and recognized by the effector population, and the principles of cell transfer including the persistence, function and toxicity of specific effector cells. It is hoped that the results of such studies can form the conceptual framework for the development of analogous approaches to the treatment of human malignancies.

CMV infection in allogeneic bone marow transplant recipients

Patients undergoing allogeneic bone marrow transplantation (BMT) are transiently rendered immunoincompetent as a consequence of the ablative high dose chemoradiotherapy that is administered to treat their underlying malignancy. The period of immunoincompetence persists until lymphoid progenitors from the donor bone marrow mature in the host and reconstitute the host immune system. During this time patients are at high risk for potentially serious infections with organisms such as cytomegalovirus (CMV) that are minimally pathogenic in normal hosts. CMV has historically represented the most common cause of infectious mortality in patients undergoing allogeneic BMT. The majority of BMT patients have latent CMV infection as a consequence of previous exposure to CMV, and reactivation of this virus occurs post-BMT in 70-80% of cases. Prior to the development of anti-viral therapy, up to 50% of patients excreting virus developed CMV pneumonia or enteritis (Meyers et al., 1986). Combined treatment with ganciclovir and Ig has improved survival for these patients but, even with combined therapy, 50% of patients developing CMV pneumonia still die (Emanuel et al., 1988; Reed et al., 1988). Thus, additional therapeutic modalities are necessary for treatment of this opportunistic infection in immunocompromised patients. Studies in murine models employing a related virus, mCMV, have demonstrated that reconstitution of T-cell immunity by adoptive transfer of CMV related T cells can protect immunosuppressed mice from developing pneumonia (Reddehase et al., 1985; Reddehase et al., 1987). Therefore, we have evaluated if adoptive therapy by transfer of T cells specific for CMV would be of therapeutic benefit for BMT patients at high risk of developing CMV disease.

Based on the need to satisfy the principles required for efficacy elucidated in murine models, we first attempted to identify the nature of the effector T cell responsible for resistance to CMV disease. Most BMT patients at our center are administered a cellular bone marrow aspirate that has not been selectively depleted of lymphocytes, and thus receive as contaminants of the bone marrow inoculum a small number of donor B lymphocytes, as well as T cells of both CD4 and CD8 phenotype. However, the total number of such mature "contaminating" cells that are specific for an individual pathogen such as CMV is variable from patient to patient, and it is generally too low to promptly reconstitute immunity. Thus, significant differences are observed between patients in the tempo of immunologic recovery post-BMT, with the variation dependent on: the number of mature naive and memory lymphocytes infused, the post-transplant immunosuppressive regimen employed, the development of graft-versus-host disease complicating normal immunologic reconstitution, and the development of other potentially immunosuppressive illnesses, such as renal failure.

Previous studies have demonstrated that the presence and titer of CMV specific antibodies does not correlate with protection. Similarly, analysis of PBL of BMT recipients for lymphoproliferative $CD4^+$ Th responses has demonstrated that the presence of CMV specific $CD4^+$ T-cell responses does not correlate strongly with the development of CMV infection (Meyers et al., 1980; Quinnan et al., 1982; Meyers et al., 1987).

These results, as well as studies in a murine model, suggested that the $CD8^+$ response to the virus might be necessary and sufficient for protection (Reddehase et al., 1987). To evaluate the temporal recovery after BMT of $CD8^+$ CMV specific CTL responses, an in vitro stimulation system was developed that efficiently detected the presence of CMV reactive $CD8^+$ precursor T cells (Riddell et al., 1991). PBL were obtained at 30, 60 and 90 days post-transplant, and the presence of cytolytic responses to CMV correlated with the occurrence of CMV disease. These time points were chosen since the peak incidence of CMV disease post-BMT occurs during this time, with a median at day 50-60 post-BMT. Our initial evaluation of 15 patients strongly suggested that the $CD8^+$ response to this virus has a protective role (Reusser et al., 1991), and these findings have now been affirmed in a larger analysis of 56 patients. Twenty-five of 56 patients demonstrated recovery during the first 100 days post-BMT of specific CD8+ reactivity to CMV, and none of these patients developed CMV pneumonia. By contrast, 31/56 patients failed to demonstrate recovery of CMV specific CD8+ reactivity during the first 100 days post-transplant, and 15 of these patients developed evidence of CMV pneumonia. Thus, the presence of a specific cytolytic T-cell response to CMV correlated strongly with resistance to the development of CMV disease, and the absence of such a response placed patients at high risk for the development of CMV pneumonia.

Analysis of possible target antigens expressed in CMV infected cells

The above studies strongly suggested that CD8+ CTL were the essential effector cells responsible for protection from CMV infection. However, such studies did not clarify if CD8+ T cells specific for any CMV protein can mediate protection, or if the protective response reflects the reactivity of a subset of T cells specific for a particular CMV protein. CMV is the largest member of the herpes virus family and contains approximately 200 open reading frames that potentially encode viral proteins (Griffiths and Grundy, 1987). Expressing each of these genes individually in target cells to analyze the relative frequency of T cells among the virus reactive population that are specific for the protein would represent an onerous task, and we therefore focused our initial efforts on characterizing the responses to three classes of proteins that represented likely targets of a protective response. These are:

- immediate-early (IE) proteins, which are regulatory proteins required to transactivate the other CMV genes and are expressed in target cells within hours of infection
- envelope proteins, such as gB, which form the outer coat of the virus
- matrix proteins, which are present in abundance in viral particles and can be detected in the nucleus and cytoplasm of infected cells shortly following viral penetration and uncoating.

Prior studies evaluating murine CMV had demonstrated that the protective response in BALB/c mice was specific for the major IE protein, and CMV specific CTL reactive with IE had been reported in humans (Reddehase et al., 1987; Borysiewicz et al., 1988). Thus, IE was evaluated as a promising candidate for an immunodominant protective response. However, studies in our laboratory have demonstrated that, despite abundant expression of IE in CMV infected target cells, the frequency of IE specific CTL was very low relative to the frequency of other CTL reactive to CMV (Gilbert et al., 1993). Moreover, these IE specific CTL efficiently lysed target cells expressing only the CMV derived IE gene, but were very inefficient at lysing CMV infected target cells expressing equivalent or greater levels of IE.

The basis for this discrepancy in lytic efficiency has only recently become clarified, with studies from our laboratory demonstrating that another protein in the CMV virion selectively interferes with access of IE proteins to the antigen processing machinery. From the perspective of the virus, this novel mechanism for evading immune recognition probably evolved as a means to assure that initiation of transcription of viral proteins can occur in infected cells but, from the standpoint of T-cell therapy, these results suggest that IE specific CTL clones would likely be ineffective at recognizing virally infected targets.

A second class of proteins evaluated has been the viral envelope proteins, as represented by gB. As observed with the response to IE, CTL reactive to gB represent a small percentage of the total CTL response to CMV (Gilbert et al., 1993). However, gB specific CTL clones, unlike the IE specific clones described above, do lyse CMV infected target cells. The envelope genes such as gB belong to the class of CMV late genes, and as a result gB is not synthesized until late time points following viral infection. Analysis of the temporal susceptibility of CMV infected targets to lysis by gB specific CTL is consistent with the data on protein synthesis from this late gene, since these CTL show no activity at early time points post infection, and lysis is not efficient until 48-72 h following infection. Consequently, the lysis of virally infected cells by gB specific clones would not be expected to occur until after viral replication has occurred in the target cell. Thus, this effector response would not efficiently prevent the release of infectious virions, and would not likely be optimal as a protective response.

The third class of proteins we have evaluated are matrix proteins, as represented by pp150 and pp65. Viral particles contain large amounts of these proteins, suggesting a potentially important role for them in the early stages of viral infection. Analysis of the specificity of CMV reactive $CD8^+$ CTL in healthy seropositive people has suggested that responses to matrix proteins reflect the immunodominant response (Riddell et al., 1991; McLaughlin-Taylor et al., in press). The processing of these proteins is quite unique, in that the matrix proteins introduced into the cell by the infectious virus are rapidly processed and presented. Thus the proteins are presented even in the absence of viral gene expression, and CTL specific for matrix proteins lyse virally infected targets efficiently within the first 4 following infection (Gilbert et al., 1993; McLaughlin-Taylor et al., 1994).

Moreover, CTL specific for these proteins were capable of lysing clinical isolates that were divergent by Southern blot analysis, suggesting that the epitopes are relatively conserved among distinct strains. Thus, CTL specific for these matrix proteins meet most of the criteria for a protective CTL response to the virus in that the cells are present in high frequency in individuals who have protective immunity; CTL specific for these proteins lyse virally infected cells very shortly following the initiation of viral infection; the target proteins appear relatively stable among viral strains. Consequently, we have evaluated whether CTL specific for these proteins can be effective in adoptive T-cell transfer.

Evaluation of T-cell transfer

The restoration of viral immunity by the adoptive transfer of T-cell clones has required the development of techniques that permit the isolation and rapid expansion of virus specific T-cell clones under conditions that permit the retention of function and specificity. Methods were developed in our laboratory in which CMV specific $CD8^+$ cell clones could be isolated by limiting dilution following stimulation with autologous CMV infected fibroblasts, and the reactive clones then expanded by weekly cycles of stimulation with fresh autologous infected stimulator cells and the addition of IL-2 (Riddell and Greenberg, 1990; Riddell et al., 1992b). Such repetitive cycles in which the TCR of the clone was first ligated, and the cells subsequently exposed to IL-2 as a proliferative signal, were found to be obligate for retention of normal phenotype and function. $CD8^+$ T-cell clones grown in this fashion can be rested in vitro with no antigen or IL-2, and will morphologically revert to small resting lymphocytes. Moreover, using these in vitro stimulation conditions, it has been possible to expand T-cell clones to greater than 10^9 cells in approximately 2-3 months. Therefore, employing this culture technology, a study was designed to evaluate the transfer of T-cell clones in CMV seropositive patients with hematologic malignancies who were undergoing BMT from an HLA matched sibling CMV seropositive donor. CMV specific $CD8^+$ CTL clones were isolated from the HLA matched donor, beginning 2-3 weeks prior to BMT; the T-cell clones, were expanded in vitro and T-cell infusions initiated approximately 35 days post-BMT. Each patient received four doses of cells at weekly intervals, beginning at an initial cell dose of 3.3×10^7 T cells/m^2, and increasing in half log increments so that the fourth dose was 10^9 T cells/m^2. This time period was chosen for the four infusions since it represents the period of maximal risk for CMV infection (Meyers et al., 1987).

Seven CMV seropositive patients have completed their cell infusions. Following administration of even the lowest dose of T cells, $3.3 \times 10^7/m^2$, transferred $CD8^+$ responses to CMV were recoverable in the peripheral blood. With subsequent infusions of the higher cell doses, the magnitude of $CD8^+$ responses in the recipients increased, and it was possible to achieve responses equivalent to or greater than those detected in healthy seropositive individuals who are resistant to CMV disease (Riddell et al., 1992b).

The persistence of transferred immunity has been evaluated by two approaches. By functional analysis, donor derived reactivity to CMV was readily detectable at 6 weeks post-transfer. After this time point responses were still evident but, since many patients begin recovering endogenous responses to CMV, it is not possible to readily distinguish by functional analysis the portion of the response that is of donor origin. Therefore, we have performed a molecular analysis using the TCR gene expressed in the transferred clones as a unique marker. CMV reactive $CD8^+$ T cells were isolated from treated patients, and cloned in vitro. The mRNA of the patient derived clones was then reverse transcribed, and the TCR gene was amplified by PCR with appropriate primers. The gene was then sequenced for comparison with the TCR sequence utilized in the clones that were administered.

With this latter molecular approach, it has been possible to demonstrate that the transferred clones persist and retain function for at least 12 weeks following T-cell administration.

Of the seven patients who have completed treatment so far, no patients have developed evidence of CMV viremia or pneumonia. One patient who was excreting CMV from his throat prior to T-cell therapy became culture negative following T-cell administration, and one patient who had CMV enteritis prior to the T-cell infusions exhibited resolution following T-cell therapy. T-cell administration was not associated with any toxicity in these patients. In particular, patients did not experience episodes of fever, changes in blood pressure or heart rate, or falls in oxygen saturation. Indeed, as the study progressed, most T-cell infusions were administered over a 30 minute period in the outpatient clinic.

In summary, the results of this study have demonstrated that

- $CD8^+$ CMV specific CTL clones can be isolated from virtually all healthy CMV seropositive donors
- infusion of greater than 2.2×10^9 $CD8^+$ CMV specific T-cell clones is safe, nontoxic and can be administered in an outpatient setting
- transfer of $CD8^+$ CMV specific T-cell immunity is detectable following infusion of cell doses as low as 3.3×10^7 T cells/m^2
- following adoptive therapy, patients can acquire $CD8^+$ CTL responses equivalent to those of healthy donors
- transferred $CD8^+$ T-cell immunity exhibits a gradual decline following T-cell administration but persists for at least 12 weeks.

These results provide a rational basis for expanding this approach to the treatment of other viral and malignant diseases.

Genetic modification of T-cell clones

The CMV adoptive therapy study has demonstrated that it is feasible to transfer antigen specific T cells in humans, and that these T cells can retain function and provide a biologic effect in the recipient. However, the requirements for protection from CMV infection in the BMT setting represents a unique situation in which the period of host immunoincompetence is well defined and relatively short lived. The broad application of this approach to the treatment of other diseases would be greatly facilitated if repetitive cell infusions were not necessary to maintain the response at therapeutic levels.

One method by which this might be activated is to introduce genes into the CD8+ population prior to cell transfer that render the cells capable of proliferating independent of CD4+ T-cell help or exogenous IL-2. Therefore, using retroviral mediated gene transfer, we are exploring three strategies to genetically modify CD8+ T cells to produce a regulated autocrine growth circuit.

In all three strategies, the growth signal is tightly linked to TCR activation so that proliferation is not autonomous but only specifically follows recognition of an appropriate target cell.

The dependency of CD8+ T cells on exogenous IL-2 reflects in part a failure, under normal conditions, of CD8+ T cells to express the endogenous IL-2 gene. Therefore, our first approach has been to introduce into CD8+ T cells a signalling receptor that might induce expression of the endogenous IL-2 gene following specific activation. Previous studies from our laboratory have demonstrated that bifunctional CD8+ T-cell clones, that are both cytolytic and produce IL-2, can be isolated from mice, and that such CD8+ T cells differ from helper dependent CD8+ T cells in that they express an IL-1 receptor (Klarnet et al., 1989). Moreover, IL-2 production and proliferation of these bifunctional CD8+ cells requires coligation of the TCR and the IL-1 receptor. Therefore, in collaboration with Bob Overell (Targeted Genetics Corporation, Seattle, WA), the gene encoding the IL-1 receptor has been inserted into a retroviral vector, and this vector has been used to transduce CD8+ T-cell clones.

Preliminary studies have demonstrated that CD8+ T-cell clones expressing this transduced receptor will proliferate if both the TCR and IL-1 receptor are triggered, but they will not proliferate in response to either signal alone. The precise intracellular events responsible for this effect are currently being evaluated.

A second approach being examined is the introduction of an IL-2 gene under the control of a regulated promoter that is activated following ligation of the TCR. In collaboration with Steve Lupton at Targeted Genetics Corporation, a variety of enhancer/promoter regions is being evaluated. For example, CD8+ T cells produce a panel of cytokines following target recognition, such as IFN-γ and GM-CSF. Thus, hybrid genes containing the IFN-γ or GM-CSF promoter fused to the IL-2 coding region have been constructed and inserted into retroviral vectors. T cells are currently being transduced with these vectors to determine if adequate amounts of IL-2 are produced to support autocrine growth.

The third approach being examined is the construction and introduction of a chimeric cytokine receptor into CD8+ T cells that have the potential to deliver an autochthonous growth signal following specific activation of CTL. This approach takes advantage of cytokines which CD8+ CTL do produce following target recognition, but modifies the receptors expressed on the cell surface so that these cytokines deliver a growth signal. For example, chimeric cytokine receptors are being made that contain the extracellular binding regions of the α and β chains of the GM-CSF receptor fused to the β and γ chains of the IL-2 receptor. Preliminary studies from our laboratory have already demonstrated that similar chimeric IL-2 receptors can deliver an authentic IL-2 growth signal to T cells following binding of the appropriate cytokine to the extracellular binding region.

This strategy has several possible advantages over the previously discussed alternatives. In particular, the chimeric receptor can be introduced under control of a constitutive promoter, since the regulated response is provided by the cytokine produced by the $CD8^+$ T cell. Such constitutive expression is generally easier to achieve than introduction of a regulated gene, as with the hybrid IL-2 genes. Secondly, unlike the conditions created by the inserted IL-1 receptor, signalling through the chimeric receptor only requires target recognition by the TCR and does not require delivery of a signal to the T cell from a second cell.

Future prospects

The studies reported here have outlined some of our accomplishments in developing adoptive T-cell therapy for the treatment of CMV infection. The use of gene transfer technologies to introduce genes into T-cell clones promises to improve the safety and efficacy of adoptive T-cell therapy and to decrease the technical efforts required to generate and maintain an adequate response in the host. Indeed, we have already initiated a trial for the treatment of HIV infection with gene modified $CD8^+$ T-cell clones (Riddell et al., 1992a). As discussed in other chapters in this book, novel methods of generating T-cell responses to human tumors and to characterize human tumor antigens are being increasingly explored. The preliminary successes with many of these approaches suggest that it should in the near future be possible to begin evaluating therapy of human malignant disease with tumor reactive T-cell clones. The principles defined in the therapy of viral infection should facilitate success in this endeavor.

References

Borysiewicz LK, Hickling JK, Graham S, Sinclair J, Cranage MP, Smith GL and Sissons JG (1988). Human cytomegalovirus-specific cytotoxic T cells. Relative frequency of stage-specific CTL recognizing the 72-kD immediate early protein and glycoprotein B expressed by recombinant vaccinia viruses. J Exp Med **168**, 919-931.

Byrne JA and Oldstone MB (1984). Biology of cloned cytotoxic T lymphocytes specific for lymphocytic choriomeningitis virus: Clearance of virus in vivo. J Virol **51**, 682-686.

Emanuel D, Cunningham I, Jules EK, Brochstein JA, Kernan NA, Laver J, Stover D, White DA, Fels A, Polsky B, Castro-Malaspina H, Peppard JR, Bartus P, Hammerling U and O'Reilly RJ (1988). Cytomegalovirus pneumonia after bone marrow transplantation successfully treated with the combination of ganciclovir and high-dose intravenous immune globulin. Ann Intern Med **109**, 777-782.

Gilbert MJ, Riddell SR, Li CR and Greenberg PD (1993). Selective interference with class I major histocompatibility complex presentation of the major immediate-early protein following infection with human cytomegalovirus. J Virol **67**, 3461-3469.

Greenberg PD (1991). Adoptive T cell therapy of tumors: Mechanisms operative in the recognition and elimination of tumor cells. Adv Immunol **49**, 281-355.

Griffiths PD and Grundy JE (1987). Molecular biology and immunology of cytomegalovirus. Biochem J **241**, 313-324.

Klarnet JP, Kern DE, Dower SK, Matis LA, Cheever MA and Greenberg PD (1989). Helper-independent $CD8^+$ cytotoxic T lymphocytes express IL-1 receptors and require IL-1 for secretion of IL-2. J Immunol **142**, 2187-2191.

Lukacher AE, Braciale VL and Braciale TJ (1984). In vivo effector function of influenza virus-specific cytotoxic T lymphocyte clones is highly specific. J Exp Med **160**, 814-826.

McLaughlin-Taylor E, Tanamache B, Pande H, Li C-R, Zaia J, Forman S, Greenberg P and Riddell S (1994). Identification of the major late human cytomegalovirus matrix protein pp65 as a target antigen for $CD8^+$ virus-specific CTL. J Med Virol (in press).

Meyers JD, Flournoy N and Thomas ED (1980). Cytomegalovirus infection and specific cell-mediated immunity after marrow transplant. J Infect Dis **142**, 816-824.

Meyers JD, Flournoy N and Thomas ED (1986). Risk factors for cytomegalovirus infection after human marrow transplantation. J Infect Dis **153**, 478-488.

Meyers JD, Bowden RA and Counts GW (1987). Infections after bone marrow transplantation. In "Infections in Transplant Patients" (ed H Lode,

D Huhn and M Melzahn), pp 17-32. Georg Thieme Verlag, Stuttgart and New York.

Quinnan GVJ, Kirmani N, Rook AH, Manischewitz JF, Jackson L, Moreschi G, Santos GW, Saral R and Burns WH (1982). Cytotoxic T cells in cytomegalovirus infection: HLA-restricted T-lymphocyte and non-T-lymphocyte cytotoxic responses correlate with recovery from cytomegalovirus infection in bone-marrow-transplant recipients. N Engl J Med **307**, 7-13.

Reddehase MJ, Weiland F, Munch K, Jonjic S, Luske A and Koszinowski UH (1985). Interstitial murine cytomegalovirus pneumonia after irradiation: Characterization of cells that limit viral replication during established infection of the lungs. J Virol **55**, 264-273.

Reddehase MJ, Mutter W, Munch K, Buhring HJ and Koszinowski UH (1987). CD8-positive T lymphocytes specific for murine cytomegalovirus immediate-early antigens mediate protective immunity. J Virol **61**, 3102-3108.

Reed EC, Bowden RA, Dandliker PS, Lilleby KE and Meyers JD (1988). Treatment of cytomegalovirus pneumonia with ganciclovir and intravenous cytomegalovirus immunoglobulin in patients with bone marrow transplants. Ann Intern Med **109**, 783-788.

Reusser P, Riddell SR, Meyers JD and Greenberg PD (1991). Cytotoxic T-lymphocyte response to cytomegalovirus after human allogeneic bone marrow transplantation: Pattern of recovery and correlation with cytomegalovirus infection and disease. Blood **78**, 1373-1380.

Riddell SR and Greenberg PD (1990). The use of anti-CD3 and anti-CD28 monoclonal antibodies to clone and expand human antigen-specific T cells. J Immunol Methods **128**, 189-201.

Riddell SR, Rabin M, Geballe AP, Britt WJ and Greenberg PD (1991). Class I MHC-restricted cytotoxic T lymphocyte recognition of cells infected with human cytomegalovirus does not require endogenous viral gene expression. J Immunol **146**, 2795-2804.

Riddell SR, Greenberg PD, Overell RW, Loughran TP, Gilbert MJ, Lupton SD, Agosti J, Scheeler S, Coombs RW and Corey L (1992a). Phase I study of cellular adoptive immunotherapy using genetically modified CD8+ HIV-specific T cells for HIV seropositive patients undergoing allogeneic bone marrow transplant. Hum Gene Ther **3**, 319-338.

Riddell SR, Watanabe KS, Goodrich JM, Li CR, Agha ME and Greenberg PD (1992b). Restoration of viral immunity in immunodeficient humans by the adoptive transfer of T cell clones. Science **257**, 238-241.

Rosenberg SA, Packard BS, Aebersold PM, Solomon D, Topalian SL, Toy ST, Simon P, Lotze MT, Yang JC, Seipp CA, Simpson C, Carter C, Bock S, Schwartzentruber D, Wei JP and White DE (1988). Use of tumor-infiltrating lymphocytes and interleukin-2 in the immunotherapy of patients with metastatic melanoma. A preliminary report. N Engl J Med **319**, 1676-1680.

31

HERPES THYMIDINE KINASE SUICIDE GENE TRANSFER: A REVIEW OF PRECLINICAL MODELS AND PROPOSALS TO TREAT BRAIN TUMORS

C. A. Mullen
Metabolism Branch, Division of Cancer Biology, Diagnosis and Centers, National Cancer Institute, National Institutes of Health, Bethesda, MD, USA

Introduction

Oncologists seek agents which will selectively kill tumor cells without fatally damaging normal tissues. A variety of drugs have been developed for cancer treatment which have preferential toxicity for tumors but, in general, most produce severe dose limiting hematopoietic and mucosal toxicities. The therapeutic index of most anticancer drugs is small and, because of drug transport problems, even smaller for brain tumors. As a result, chemotherapy for brain tumors is usually unsuccessful.

An alternate approach for drug development is rendering the tumor sensitive to agents that can enter the central nervous system. One theoretical approach to this is the transfer of suicide genes to brain tumors. Suicide genes encode proteins which cause a cell to produce lethal intracellular toxins. This conditional lethality is accomplished by using nonmammalian metabolic enzymes and prodrugs which can be activated by only such enzymes.

Suicide genes

The basic principle underlying metabolic suicide gene systems is intracellular conversion of a relatively non toxic prodrug to a toxic drug by an enzyme that is not normally present in the cell. Viruses, bacteria and fungi often utilize unique metabolic pathways not used by mammalian cells, and they contain genes for enzymes that perform

CYTOKINE-INDUCED TUMOR IMMUNOGENICITY
ISBN 0-12-261520-4

metabolic conversions that mammalian cells do not perform. Such distinctive enzymes have often been the target of drugs developed for the treatment of infections. Such agents are lethal or inhibitory for the infecting microbe but do not harm the host cell because it lacks the enzyme system needed to activate the drug. Transfer of genes for such enzymes to mammalian cells can confer upon these altered cells the same distinctive metabolic sensitivity. If such genetically altered cells are produced in or placed in a host, treatment of the host with the prodrug will produce toxicity confined to the altered cell and/or its microenvironment without generation of significant systemic toxicity. This is so for two reasons. First, normal cells will not express the necessary activating suicide gene. Second, the active toxins generated by the suicide gene enzyme interfere with nucleic acid metabolism, and thus only cycling cells will be susceptible to damage. Several enzyme/prodrug systems have been developed and have been successfully used in vivo. Herpes simplex type 1 thymidine kinase has been most extensively tested and used, although other metabolic suicide genes are being developed. These include *Escherichia coli* cytosine deaminase (Mullen et al., 1992) and varicella-zoster thymidine kinase (Huber et al., 1991)

Herpes thymidine kinase

Herpes simplex virus type 1 (HSV-1) is a common human pathogen. It produces a thymidine kinase (HSV-1-TK) whose substrate specificity is distinct from normal cellular thymidine kinase. This enzyme has been the target of drugs like the guanosine analogues acyclovir and ganciclovir (Hirsch and Kaplan, 1990; Elion, 1982). Cells expressing the HSV-1-TK enzyme phosphorylate these drugs producing acyclovir-monophosphate organciclovir-monophosphate, while normal cells do not. Normal cellular guanylate kinase will then further metabolize the monophosphate to the diphosphate form (Miller and Miller, 1980). Activation is completed by other kinases (phosphoglycerate kinase, pyruvate kinase and phosphoenolpyruvate kinase), and ultimately the highly toxic triphosphate forms of the prodrugs are generated. Production of the monophosphate is the rate limiting step in the generation of the active form of the drugs. The triphosphate forms cause some inhibition of cellular α-DNA polymerase, and they produce additional toxic effects through incorporation into DNA and chain termination (Mar et al., 1985). Ganciclovir is more potent than acyclovir in this regard; it is a better substrate for HSV-1-TK (Biron et al., 1985).

Cloning and transfer of the HSV-1-TK gene to normal cells confers the same selective sensitivity to these nucleic acid analogues. In the absence of the prodrug, constitutive expression of the HSV-1-TK gene is not harmful. Altered cells grow normally in vitro and in vivo. In fact, transgenic mice expressing the HSV-1-TK gene have been produced

(Borrelli et al., 1988). In vitro cells expressing the HSV-1-TK gene are inhibited by ganciclovir in the 1-50 μM range, levels which can be achieved in patients treated with the drug.

Preclinical models of HSV-1-TK mediated tumor cell killing

The observation of selective killing of suicide gene modified cells by the appropriate prodrugs suggests that if tumors can be manipulated to express these suicide genes, it is theoretically possible that systemic treatment of the host with the prodrug may have a significant impact on the tumor.

This possibility has been tested by a number of investigators working with the HSV-1-TK gene in a variety of tumor models. In such studies tumor cells have been manipulated in vitro to express the HSV-1-TK gene and then reinjected into syngeneic hosts. Animals have then received parenteral treatment with acyclovir or ganciclovir and tumor growth was monitored. HSV-1-TK positive clones of a murine sarcoma line transfected by calcium phosphate precipitation were shown to be sensitive in vitro to 0.1-1 μM ganciclovir.

When these tumor lines were injected into animals receiving a 5 day course of ganciclovir (twice daily at a dose of 150 mg/kg) begun about 1 week after tumor injection, treated mice consistently failed to grow tumors while untreated mice exhibited progressive tumor growth. HSV-1-TK negative tumors were unaffected by prodrug treatment (Moolten, 1986). Similar results were seen in cloned sarcoma lines transduced in vitro with retroviral vectors, although about one-third of mice exhibited tumor recurrences once drug therapy was discontinued (Moolten and Wells, 1990).

The majority of recurrent tumors no longer expressed the HSV-1-TK gene. Murine adenocarcinoma clones transduced with an HSV-1-TK retroviral vector regressed in animals receiving oral ganciclovir 75-100 mg/kg daily for 14 days. Even animals whose drug treatment began more than 1 month after tumor inoculation and whose tumors were greater than 4 cm in diameter regressed (Plautz et al., 1991). A variety of other solid tumor types have been shown to be susceptible to HSV-1-TK/ganciclovir mediated killing (Golumbek et al., 1992; Panis et al., 1992; Hasegawa et al., 1993).

Glioma cells have also been shown to be susceptible to HSV-1-TK/ganciclovir mediated killing (Ezzeddine et al., 1991). Cloned rat C6 glioma cells expressing HSV-1-TK were sensitive to ganciclovir 2-5 μM in vitro. In nude mice these xenogeneic gliomas were suppressed by parenteral treatment with ganciclovir 50 mg/kg daily for 28 days by continuous infusion or by 25 mg/kg twice daily for 12 days.

In vivo transfer of the HSV-1-TK gene

The demonstration that tumors transduced with a HSV-1-TK gene ex vivo and reinjected into a host can be eliminated by the ganciclovir is encouraging, but it does not accurately model clinical cancer. The important therapeutic question is whether tumor in a host can be rendered sensitive in situ to the prodrug. Thus, gene delivery in vivo becomes a very important issue in metabolic suicide gene schemes.

Methods effective in vitro for transferring genes such as calcium phosphate precipitation, electroporation and exposure to viral vectors are not efficient in vivo. Direct injection of DNA, DNA-liposome complexes, DNA-metal complexes or retroviral vectors have all been shown to transfer DNA into cells, but at efficiencies of less than 1% of the target cells. To increase exposure of cells to the vector alternate methods have tried to deposit within tumors cells which can continuously manufacture and secrete retroviral vectors.

"Producer lines" are cell lines (usually fibroblast derived) which have been genetically engineered to produce replication defective retroviral vectors (Miller, 1990). The vectors themselves are defective viruses which lack genes (i.e, the retroviral *gag, pol* and *env* genes) needed to manufacture viral structural and regulatory proteins (Miller and Rosman, 1989). The absence of these genes in the vector itself prevents it from replicating in its target cells and propagating the infection. These important viral genes have been replaced by the sequences one wishes to transfer to target cells. Unlike the target cells, the vector producer line makes the needed viral proteins, complementing the genetic defect in the vector. Thus only in these cells can vectors reproduce, while in other target cells the vectors cannot reproduce.

One approach to in vivo tumor transduction is the injection of retroviral vector producer cells into a tumor mass. The hope is that the producer cells within the tumor mass would continuously secrete the vector for at least several days, increasing the exposure of the tumor cells to the vector. This approach was validated by experiments in which a producer line making a β-galactosidase vector was inoculated into gliomas growing in rat brain (Short et al., 1990). Histochemical evidence of β-galactosidase gene transfer was seen in 1-10% of the glioma cells, while normal brain was not transduced. The tumor specificity of the gene transduction is attributable to the fact that retrovirus vectors can only integrate into replicating cells while noncycling cells are resistant. Direct injection of vector preparations without the producer lines produced very little tumor transduction. The HSV-1-TK gene has been transferred with a similar technique, and efficiencies of 10-70% have been claimed (Culver et al., 1992; Ram et al., 1993). Small intracerebral gliomas in rats were stereotaxically injected with HSV-1-TK vector producer cells. It is important to note that 10-100-fold more producer cells were injected compared to the number of glioma cells inoculated 7 days earlier. Five-seven days later the rats received a course of ganciclovir. Under these conditions about 75% of tumors regressed as assessed by autopsy roughly 1 month after tumor inoculation. Long term survival data were not reported, and thus tumor recurrence rates after such treatment cannot be estimated.

Another approach to transduction of tumor cells in vivo involves introduction of live, wild type, replication competent retrovirus along with the HSV-1-TK vector (Takamiya et al., 1992; Takamiya et al., 1993). The presence of wild type "helper virus" has the effect of propagating infection in the tumor itself. This is because cells infected with the wild type virus express the needed *gag, pol* and *env* genes. Tumor cells infected with the wild type virus and HSV-1-TK vector can produce more wild type virus and also more HSV-1-TK vector, while HSV-1-TK vector targeted cells not infected by helper virus cannot produce more vector. In nude mice injected with a mixture of both tumor cells lacking the HSV-1-TK gene and tumor cells infected with both the wild type virus and HSV-1-TK vector, regression of the mixed tumor was seen after treatment with ganciclovir. This observation would be consistent with spread of the HSV-1-TK vector throughout the tumor. However, the contribution of metabolic "bystander killing" cannot be excluded.

"Bystander killing" of wild type tumor cells

A population of HSV-1-TK expressing cells will die in the presence of the ganciclovir, while a population of unmodified wild type cells will continue to grow normally. In a mixed population only the genetically modified cells will activate the prodrug, but it is possible that wild type cells in the immediate vicinity of the modified cell may be affected by the toxic metabolite generated by the suicide gene harboring cell. Such "bystander killing" of unmodified cells has been observed in the HSV-1-TK system. The results of the studies discussed above in which gene transfer in vivo was attempted point to the phenomenon. Gene transfer efficiency was reported as 10-70%, yet in many animals complete tumor regression was observed. Presumably 30-90% of the cells in these regressing tumors did not harbor the suicide gene and yet were inhibited when their HSV-1-TK positive neighbors were treated with ganciclovir. Inhibition of unmodified tumor cells by HSV-1-TK positive tumor cells has also been observed in mixed tumors in which no vector producing cells have been inoculated (Culver et al., 1992). In a fibrosarcoma tumor model in which only 50% of the cells were HSV-1-TK positive, regression was as complete as in 100% HSV-1-TK positive tumors. Even in mixed tumors in which only 10% of the cells were HSV-1-TK positive, over one-half of tumors regressed with ganciclovir treatment. This apparent "bystander killing" of unmodified tumor cells was tentatively explored in vitro in the earliest reports of the HSV-1-TK system (Moolten, 1986). There it was observed that when mixtures of HSV-1-TK positive and negative cells were plated at very low density in vitro, ganciclovir treatment eliminated only the HSV-1-TK positive cells. When the same cell mixture was plated at high density, nearly all cells, both HSV-1-TK positive and negative, were killed. It was speculated that some type of metabolic cooperation between adjacent cells allowed transport of activated drug. This phenomenon is under investigation in a number of laboratories. Preliminary evidence has suggested several possibilities ranging from phagocytosis of toxin-laden apoptotic vesicles produced by suicide gene expressing dying cells to intracellular transport of phosphorylated drug through intercellular junctions.

While the mechanism of "bystander killing" in neighboring tumor cells has not been fully elucidated, it is clear that normal tissues in the host are not damaged by the suicide gene system. While toxin concentrations may be high in the microenvironment, they seem to be diluted effectively such that systemic toxic concentrations are not reached. The many studies of animals bearing HSV-1-TK positive tumors agree that mice do not suffer systemic toxicity when treated with ganciclovir.

While systemic toxicity has not been observed, one can still ask if normal tissues in the microenvironment of the HSV-1-TK positive tumor would be damaged. This has not been observed to date, either with inoculation of HSV-1-TK positive cells alone or with addition of vector producing cells. Nonreplicating normal tissues are resistant to both retroviral transduction and the antimetabolites generated by the suicide gene.

Theoretical barriers to practical human application

While these observations in animal models are encouraging, enthusiasm for possible use of suicide genes in human therapies should be tempered by consideration of their limitations. First, there remain formidable technical obstacles to efficient delivery of genes to tissues in vivo. Even under nearly optimal circumstances, in preclinical models the majority of target cells are not successfully transduced. Second, delivery of a vector to a cell still does not guarantee efficient expression of its genes. Populations of cells transduced with retroviral vectors will include some cells which exhibit high levels of gene expression and many others which contain little gene product. The potency of the suicide gene systems is generally directly related to the amount of enzyme activity produced. Third, clinical tumors generally have a lower growth fraction or mitotic index than experimental murine tumors. Tumor cells not actively cycling at the time of gene transfer or metabolic activation may be resistant to both genetic transduction and the toxic effects of the activated drugs. Fourth, tumor cells are inherently genetically unstable. It is likely that within a large population of suicide gene transduced tumor cells (even assuming initially efficient gene transfer), variants will be generated which cease to express the suicide gene adequately. Such cells could easily escape and produce recurrent tumor.

Clinical trials

Despite the aforementioned limitations, several schemes for human gene therapy employing the HSV-1-TK suicide vector have been proposed. One group of clinical trials attempts to use HSV-1-TK to treat intracerebral tumors, focusing on in vivo gene transduction and relying heavily upon "bystander killing" of nontransduced tumor cells (Oldfield et al., 1993). In this scheme HSV-1-TK retroviral producer lines will be injected into intracerebral tumors (either primary brain tumors or cerebral metastases from other tumors), and 7 days later patients will receive a 14 day course of ganciclovir (5 mg/kg twice daily). Before and during the period of treatment the patients will be treated with high dose corticosteroids, for both standard treatment of cerebral edema associated with brain tumors and for control of local inflammation induced by the surgical manipulation of the brain and injection of xenogeneic cells.

The injection of vector producer lines into human brain has provoked some concerns about safety. The producer lines are derived from murine fibroblasts. It is expected that humans will mount a brisk immune response against these tissues which could in principle have two major problems. First, there is the risk of an intense inflammatory response to the xenogeneic antigens which could worsen cerebral edema. In preclinical safety studies in rats and primates there was no serious neurological toxicity in animals treated with dexamethasone during this period. However, it remains possible that if humans receive a second treatment with these xenogeneic cells a more powerful anamnestic immune response would be elicited. A second potential problem with the immune response to xenogeneic cells is elimination of the cells before effective vector secretion and gene transfer can take place. Preclinical studies in rats and primates indicate that the producer cells are eliminated within several days to a week of injection, but despite this gene transfer to tumor cells can occur.

A second major concern is the production of replication competent retrovirus and the development of secondary malignancy. In preclinical safety studies (Donahue et al., 1992), primates developed lymphomas after undergoing bone marrow transplantation with cells transduced with retroviral vector preparations contaminated with helper virus. The replication competent virus produced a productive infection in the animals and the persistent viremia resulted in multiple insertional events or "genetic hits" in lymphoid cells, ultimately producing a malignant clone. The vector producer lines to be used in the brain tumor trials will be extensively screened for the presence of helper virus, and thus it is unlikely that retrovirally mediated insertional mutagenesis and malignancy is a serious risk. However, if therapies based on live, replication competent virus gene transfer are developed (as in one of the preclinical models of in vivo gene transfer discussed above), the concern about retrovirus induced malignancies will be greater.

Other toxicity concerns include effects on hematopoietic and lymphoid cells. Ganciclovir is to be used for a relatively brief period and thus is unlikely to produce bone marrow suppression. Fatal infections were reported in 2 of 5 monkeys used in preclinical safety testing. Similarly, a number of rats died of infection after treatment. The deaths were likely due to immunosuppression induced by the corticosteroids used to minimize inflammation. It is likely that the supportive care available to patients will minimize serious infections.

It is hoped that these human brain tumors will regress like the rat glioma tumors used in the preclinical models. However, significant differences between the experimental model and these may limit the success of these initial human experiments. The human tumors will be larger, their cells will significantly outnumber the injected producer cells (the converse of the rat model), and they will have a lower growth fraction. Initial reports from one of these trials indicate that some tumors demonstrate some response, but that local recurrences regularly occur.

Conclusion

The poor prognosis of patients with brain tumors has been used to justify human trials of HSV-1-TK gene transfer based on a relatively modest amount of preclinical data. Numerous basic questions remain including the mechanism of HSV-1-TK mediated cell killing, the efficiency of gene transfer into established tumors and the probability of local relapse after initial tumor regression. Critics of these trials contend that more of these issues should be worked out experimentally before patients are treated. Proponents of early clinical trials argue that therapies should not wait for the resolution of all preclinical questions since the clinical urgency of the problem is so great. This latter argument seems to have won the day. One hopes that this human experimentation will be conducted not only to find out if tumors can shrink if treated with gene transfer, but also to contribute to the fund of basic knowledge which ultimately is the source of all rational treatment.

References

Biron KK, Stanat SC, Sorrell JB, Fyfe JA, Keller PM, Lambe CU and Nelson DJ (1985). Metabolic activation of the nucleoside analog 9-{[2-hydroxy-1-

(hydroxymethyl)ethoxy]methyl} guanine in human diploid fibroblasts infected with human cytomegalovirus. Proc Natl Acad Sci USA **82**, 2473-2477.

Borrelli E, Heyman R, Hsi M and Evans RM (1988). Targeting of an inducible toxic phenotype in animal cells. Proc Natl Acad Sci USA **85**, 7572-7576.

Culver KW, Ram Z, Wallbridge S, Ishii H, Oldfield EH and Blaese RM (1992). In vivo gene transfer with retroviral vector-producer cells for treatment of experimental brain tumors. Science **256**, 1550-1552.

Donahue RE, Kessler SW, Bodine D, McDonagh K, Dunbar C, Goodman S, Agricola B, Byrne E, Raffeld M, Moen R, Bacher J, Zsebo KM and Nienhuis AW (1992). Helper virus induced T cell lymphoma in nonhuman primates after retroviral mediated gene transfer. J Exp Med **176**, 1125-1135.

Elion GB (1982). Mechanism of action and selectivity of acyclovir. Am J Med **73**, 7-13.

Ezzeddine ZD, Martuza RL, Platika D, Short MP, Malick A, Choi B and Breakefield XO (1991). Selective killing of glioma cells in culture and in vivo by retrovirus transfer of the herpes simplex virus thymidine kinase gene. New Biol **3**, 608-614.

Golumbek PT, Hamzeh FM, Jaffee EM, Levitsky H, Lietman PS and Pardoll DM (1992). Herpes simplex-1 virus thymidine kinase gene is unable to completely eliminate live, nonimmunogenic tumor cell vaccines. J Immunother **12**, 224-230.

Hasegawa Y, Emi N, Shimokata K, Abe A, Kawabe T, Hasegawa T, Kirioka T and Saito H (1993). Gene transfer of herpes simplex virus type I thymidine kinase gene as a drug sensitivity gene into human lung cancer cell lines using retroviral vectors. Am J Respir Cell Mol Biol **8**, 655-661.

Hirsch MS and Kaplan JC (1990). Antiviral agents. In "Virology, Second Ed." (eds BN Fields and DM Knipe), pp 441-468. Raven Press Ltd, New York.

Huber BE, Richards CA and Krenitsky TA (1991). Retroviral-mediated gene therapy for the treatment of hepatocellular carcinoma: An innovative approach for cancer therapy. Proc Natl Acad Sci USA **88**, 8039-8043.

Mar EC, Chiou JF, Cheng YC and Huang ES (1985). Inhibition of cellular DNA polymerase alpha and human cytomegalovirus-induced DNA polymerase by the triphosphates of 9-(2-hydroxyethoxymethyl) guanine and 9-(1,3-dihydroxy-2-propoxymethyl) guanine. J Virol **53**, 776-780.

Miller AD (1990). Retrovirus packaging cells. Hum Gene Ther **1**, 5-14.

Miller AD and Rosman GJ (1989). Improved retroviral vectors for gene transfer and expression. Biotechniques **7**, 980-990.

Miller WH and Miller RL (1980). Phosphorylation of acyclovir (acycloguanosine) monophosphate by GMP kinase. J Biol Chem **255**, 7204-7207.

Moolten FL (1986). Tumor chemosensitivity conferred by inserted herpes thymidine kinase genes: Paradigm for a prospective cancer control strategy. Cancer Res **46**, 5276-5281.

Moolten FL and Wells JM (1990). Curability of tumors bearing herpes thymidine kinase genes transferred by retroviral vectors. J Natl Cancer Inst **82**, 297-300.

Mullen CA, Kilstrup M and Blaese RM (1992). Transfer of the bacterial gene for cytosine deaminase to mammalian cells confers lethal sensitivity to 5-fluorocytosine: A negative selection system. Proc Natl Acad Sci USA **89**, 33-37.

Oldfield EH, Ram Z, Culver KW, Blaese RM, De Vroom HL and Anderson WF (1993). Gene therapy for the treatment of brain tumors using intra-tumoral transduction with the thymidine kinase gene and intravenous ganciclovir. Hum Gene Ther **4**, 39-69.

Panis Y, Caruso M, Houssin D, Andreoletti M, Khayat D, Salzmann JL and Klatzmann D (1992). Treatment of experimental liver tumors by in vivo suicide gene transfer in rats. C R Acad Sci III **315**, 541-544.

Plautz G, Nabel EG and Nabel GJ (1991). Selective elimination of recombinant genes in vivo with a suicide retroviral vector. New Biol **3**, 709-715.

Ram Z, Culver KW, Walbridge S, Blaese RM and Oldfield EH (1993). In situ retroviral-mediated gene transfer for the treatment of brain tumors in rats. Cancer Res **53**, 83-88.

Short MP, Choi BC, Lee JK, Malick A, Breakefield XO and Martuza RL (1990). Gene delivery to glioma cells in rat brain by grafting of a retrovirus packaging cell line. J Neurosci Res **27**, 427-439.

Takamiya Y, Short MP, Ezzeddine ZD, Moolten FL, Breakefield XO and Martuza RL (1992). Gene therapy of malignant brain tumors: A rat glioma line bearing the herpes simplex virus type 1 thymidine kinase gene and wild type retrovirus kills other cells. J Neurosci Res **33**, 493-503.

Takamiya Y, Short MP, Moolten FL, Fleet C, Mineta T, Breakefield XO and Martuza RL (1993). An experimental model of retrovirus gene therapy for malignant brain tumors. J Neurosurg **79**, 104-110.

CONCLUSIONS

32

NEW CLINICAL PROSPECTS

D. L. Longo
Biological Response Modifiers Program, NCI-Frederick Cancer Research and Development Center, Frederick, MD, USA

The rational mind tells us that the relief of human suffering from disease is attainable through a series of logical steps. First science detects the cause of disease or its pathophysiology and then interventions to prevent or treat the disease follow logically from this knowledge. The basic premise is that disease processes should yield to logical thought processes. Indeed, from this premise arises the current trend of funding agencies to target particular diseases for attack.

The success of this premise in the physical and engineering sciences has made us slow to question its validity in biological sciences. Indeed, the National Aeronautics and Space Administration's (NASA's) $14 billion a year budget gives testimony to the persuasive power of setting and then reaching scientific goals. NASA is remarkable in its chutzpah; it manages to appear successful even when it fails. I would love to see the NIH director make a billion dollar mistake such as NASA's launching of a poorly operating Hubble telescope and then congratulate himself and the agency when he spends another billion dollars to correct the initial error. Yet that is precisely what NASA has just done.

Setting scientific goals in the biological sciences is very dangerous. Freeman Dyson, the great Princenton University physicist, has written that there is no illusion more dangerous than the notion that scientific progress is predictable. Yet goal setting has been used conspicuously by some leaders in biomedical research as a short term strategy to earn Congressional support and funding. This has led to some strange results. For example, the National Heart, Lung and Blood Institute (NHLBI) has enjoyed reasonable financial support in recent years and the rate of deaths from heart disease seems to be declining dramatically. However, there is no evidence that the application to the population of the many insights derived from research supported by the NHLBI is responsible for decline in heart disease mortality. Heart disease just seems to be declining. Similarly, Francis Collins is an eloquent spokesman for the value of the human genome project, citing the many human disease genes that will be identified in the course of sequencing the human genome. However, there is absolutely nothing about knowing the genetic cause of disease that leads logically to a successful treatment or cure. We have known the

CYTOKINE-INDUCED TUMOR IMMUNOGENICITY
ISBN 0–12–261520–4

precise molecular cause of sickle cell anemia since 1949 and still cannot treat it.

What the genetic information does is create enormous practical problems for patients and for medicine. There are now women in breast cancer prone families who are opting for mutilating surgery to prevent this dreaded disease, a situation that reverses the old adage about an ounce of prevention since indeed the vast majority of patients who actually develop breast cancer do not receive surgical treatment as radical as bilateral mastectomy as the preferred treatment. Similarly, the prevention of certain familial forms of colon cancer is total colectomy. The main alternative to such radical preventions is frequent surveillance. In the case of colon cancer, one wonders whether annual colonoscopy is sufficient and if so, who will pay for many colonoscopies to detect a few early cases. No health care plan is prepared for widespread screening and prevention programs that are more costly than treatments. We do not yet have medicines that have been proven capable of preventing these cancers (though some interesting ideas are being tested such as tamoxifen for breast cancer and nonsteroidal anti-inflammatory agents for colon cancer).

It is clear that our basic science is progressing much faster than our clinical science. And if steps are not taken to address the disparity (e.g. more training of clinical scientists and more funding for clinical science and providing clinical researchers with a more respected position in the research hierarchy), the medicine of the 21st century will be as frustrating as the medicine of the 19th century must have been. We shall know the molecular cause precisely, but shall be able to do little to intervene medically and alter the natural course of things. If we detect an organ at risk, we will have to cut it out or watch helplessly as the inevitability of a genetic risk takes its psychologic toll on the still healthy victim.

Those of us engaged in clinical research in cancer treatment are a particularly hardy (some might say foolhardy) lot. For there is no other specialty in medicine in which rational thought has contributed less to the existing armamentarium. The vast majority of agents and regimens in use today were not the product of rational drug design; they were discovered empirically or even accidentally. Yet we persist in trying to think rationally about the disease and about the agents we have at our disposal and to come up with clever hypotheses about how to use them better. The science of pharmacology (i.e. molecular pharmacology, pharmacokinetics, pharmacodynamics, pharmacogenetics) has contributed only modestly to the way we manage patients. Most progress has come by trial and error, and if there are underlying scientific principles in the development of successful treatment programs, they have not yet been clearly discerned and articulated. That makes predicting success little more than crystal ball gazing.

Indeed the first ten years of biological therapeutics (I am arbitrarily dating the beginning of this era to 1980 when Lee Nadler and his colleagues reported the first in vivo administration of a murine mAb in man; the advent of mAb technology and gene cloning made it possible to work with biological agents of pharmacologic purity; little of what came before using poorly characterized and poorly standardized preparations has yielded useful results, in my view) has failed to produce a single curative regimen in a previously incurable disease. There are at least two components contributing to this failure. One is that the expectations for success are so high both among physicians and among patients. Chemotherapy and radiation therapy have a terrible reputation with patients because of the anticipated side effects and when they learn that they may be eligible for biological treatment, they jump at the opportunity because they anticipate that biological approaches must be less toxic and more rational. Clearly this favorable a priori view of biological therapy is unwarranted.

The second major factor responsible for the failure of biological therapy is that we are developing agents according to the wrong paradigm, namely that used for chemotherapeutic agents. Chemotherapeutic drug development assumes that all drugs are toxic above a certain dose and phase I testing is aimed at identifying the dose that is the maximum tolerated dose. Phase II testing is aimed at detecting whether there is a therapeutic index for the drug. Is it active in any disease when given at its maximum tolerated dose? Are the toxic effects tolerable in the context of the observed therapeutic effect? Phase III testing compares the new agent to standard therapy in a randomized trial. Phase IV is the incorporation of an active agent into combinations with other agents and modalities to maximize the therapeutic effect. Applying this strategy to biological drug development is analogous to the drunk who searches for his lost wallet under the lamppost because that is the only place he can see. The only alteration in this scheme that has been made for biological agents is a phase Ib study in which one attempts to ascertain the dose of the agent at which it exerts its maximum biological effects. Unfortunately, with rare exceptions (e.g. the colony stimulating factors), we do not know which of the many effects of a biologic agent are critical to maximize and we do not know how to measure it in the tissues. Instead, we settle for measuring a variety of epiphenomena detectable in the peripheral blood, which is certainly not an immunologic end organ.

Treating cytokines and antibodies as though they are drugs has not been very fruitful. I.v. injection of more IL-2 than the patient has made in his entire life in the course of his many physiologic immune responses to environmental agents is several logs short of genius. Nevertheless, though it is easy to criticize this paradigm, we lack the tools and knowledge to replace it with a better one. If Star Trek's Bones could send back in the time some of his noninvasive diagnostic equipment, it might be possible to monitor a patient's immune reactivity in particular tissues as a function of a therapeutic intervention. However, lacking the sophistication of the 25th century, we are forced to make do with what we have. And what we have has not yet led us to a successful biological therapy.

One of the most exciting features of the work presented in this book is that it largely ignores the paradigm of cytokine as drug. The ideas presented here place the cytokine in a more physiologic context, produced locally in the tissues to initiate and amplify a local immune response acutely the hope that systemic immunity will be the ultimate outcome.

Chapter 2 through 6 explore the immunologic concepts involved in eliciting immunity to tumors. Professor Mitchinson reviewed in detail the data that would argue against an important role for the immune system in detecting and fighting cancer. There is a legitimate concern that the processes that maintain self tolerance may not be sufficiently different (in an antigenic sense) to arouse the immune response. In other words, are there tumor specific antigens in human cancer?

Boon and his colleagues report data from their studies of human melanomas that make us think there are tumor restricted if not tumor specific antigens and that different intracellular targets may be processed and presented on the cell surface depending upon the MHC type of the person. HLA-A1 can present a peptide of the MAGE-1 protein and HLA-A2 can present a peptide of tyrosinase to cytotoxic melanoma specific T cells. One of these two HLA types is expressed in about 75% of persons in North America and thus, it is conceivable that an appropriate vaccine based upon just these two antigens could be effective components of treatment for the majority of patients. Of course, it is not yet clear that there are similar antigenic peptides associated with other histologic types of cancer. However, there is now evidence coming from Boon's group and from Robert Fenton at the NCI's Biological Response Modifiers Program that there is a family of MAGE genes and their expression in cancers is more widespread than just melanoma; MAGE genes have been found to be expressed in breast and lung cancers, among others. Furthermore, active research is under way to identify other tumor antigens by obtaining human tumor samples, eluting from the surface HLA molecules the peptides they are presenting, characterizing the peptides by high pressure liquid chromatography and identifying the proteins from which the peptides were derived. Even without knowing the identity of the proteins from which the peptides are derived, it might be possible to devise a vaccine on the basis of the predominant peptide detected in association with a particular type of MHC molecule. In light of the physical constraints for peptide presentation on any individual MHC molecule, it seems quite likely that there will be immune response gene defects associated with certain haplotypes caused by the inability of a particular peptide to be presented by a particular MHC allele. It is hoped that there are enough antigenic peptides that some tumor related peptide will be presented by one of the various (at least six: two alleles each of HLA-A, B and C) types of class I molecule expressed on a tumor in a particular patient. However, there are no data on this point at the present time.

Thus, melanoma and, it is hoped, other types of cancer can express antigenic peptides that can be recognized by autologous CTL. Interestingly, both MAGE-1 and tyrosinase are expressed in normal human tissues, MAGE-1 in testis and tyrosinase in normal melanocytes. The detection of T cells capable of recognizing these antigens could be thought of as evidence that self tolerance has been broken, perhaps related to the method of expansion of melanoma reactive clones in vitro (i.e.

chronic antigen exposure plus IL-2). Certainly, without stimulation and amplification, the existence of these T cells is not sufficient to induce tumor regression, as noted by Parmiani and his colleagues.

One of the crucial recent insights about the nature of tumor antigens is that they need not be natively expressed on the tumor cell surface. Although, as reviewed by Houghton and Lewis, there are a number of cell surface antigens expressed on tumor cells that are under investigation for their ability to elicit tumor immunity, including serological defined carbohydrate blood group antigens (Lewis T, Tn, sialosyl-Tn) and gangliosides (GM2 and GD3), T cells recognize intracellular protein targets that become cell surface antigens when they are processed intracellularly into peptides and displayed in the groove of the HLA class I protein expressed on the cell surface. This critical observation greatly expands the potential menu of targets for immune attack.

Since mutated oncogenes are rather distinctively expressed in neoplastic cells, they may be excellent targets for an immune response. Indeed, Bob Fenton and his colleagues have already been able to obtain in vivo rejection of *ras* transformed tumors in mice immunized against the appropriate mutated *ras* protein and human T cells with specificity for mutated *ras* proteins have been isolated. There is no reason to think that other mutated oncogenes (p53, *myc*, *bcr-abl*) cannot also be used to elicit immunity. Clinical trials with these antigens are just beginning, but hold major promise for success, particularly if strategies can be developed to use the vaccine in a minimal residual disease setting. Clear difficulties exist in using the immune response to control macroscopic tumor and other problems have been associated with trying to elicit an immune response in patients receiving cytotoxic therapy. Thus, there are a number of clinical problems to solve before vaccination against protein antigens becomes a standard approach to cancer treatment.

There are several levels at which tumors may avoid immune detection. First, tumors may be heterogeneous in the expression of a particular antigen. Second, tumors may down-regulate their MHC molecules and thus fail to present appropriate targets for T-cell recognition. Third, tumors may not express important costimulatory molecules required for inducing T-cell response (e.g. B7 or cell adhesion molecules). Fourth, tumors may produce immunosuppressive factors such as TGF-β, prostaglandin E2, or down-regulatory cytokines such as IL-4 or IL-10 that lead to both generalized and specific inhibition of immune responses. Finally, tumors may be effective at inducing tolerance in the cells capable of recognizing them. There may also be other evasive actions that tumors take that have not yet been fully characterized.

In addition to these features of tumors that are important, it must also be recognized that there are attributes of the immune system that may favor rather than inhibit tumor growth and spread. TNF has been shown to act as a tumor promoter in experimental models and IFN-γ can be associated with angiogenesis and can make tumor cells resistant to cytotoxic T-cell or NK-cell lysis. There are numerous other cytokines and factors made by cells of the inflammatory/immune system that may counteract the hopes for antitumor effect.

Thus, the relationship between the tumor and its host is exceedingly complicated. Undoubtedly the relationship is different for different tumors in different sites in different hosts, and there are undoubtedly changes in the host-tumor relationship over time in an individual host. The process of neoplasia has the capacity (i.e. genetic instability) to adapt to changes in its environment; however, unlike an organism undergoing evolutionary change, the short sighted tumor makes adaptations that permit it to survive and kill its host. Since this invariably also leads to the death of the tumor, it cannot be said that tumors gain any ultimate advantage. However, in our role as physicians, we seek to understand all of the ways that tumors and hosts interact and to intervene in specific ways to shift the balance of power in favor of the host.

This book is filled with strategies for overcoming the many problems that exist in the host tumor relationship. Parmiani's group discusses overcoming the problem of lack of MHC expression and poor T-cell triggering through the use of bispecific antibodies with one Fab directed against the tumor and the other against CD3. They have shown such antibodies to result in tumor cell killing in vitro. My colleagues, Stephen Creekmore and John Ortaldo, have devised an alternative approach to the same problem using murine IgG3 antibodies that have the property of self aggregating via their Fc regions. Thus, antitumor IgG3 antibodies can be bridged to triggering anti-CD3 IgG3 antibodies via Fc-Fc linkage, and tumor cell killing is promoted. We have observed the efficacy of this approach both in vitro and in vivo in man using the R24 antibody directed against the GD3 ganglioside, a structure shared by melanomas and by a subset of T cells. Further studies are underway to evaluate the clinical efficacy of combinations of IgG3 antibodies, one directed against the tumor and one aimed at triggering T cells.

However, the major emphasis of the work described in this book is the use of cytokines to boost the immune response to a tumor. The initial rationale for the use of cytokines was the stimulation of T-cell responses. IL-1 and IL-2 have been administered systemically with substantial toxicity and very little success at inducing tumor regression. About 15% of patients with melanoma and renal cell cancer respond to systemic IL-2 administration with 1-2% of patients achieving a CR. Data with IL-1 are considerably less mature, but preliminary results suggest that it may be comparable to IL-2 in its antitumor efficacy. Response rates are higher in patients receiving IL-2 regionally. I.p. IL-2 and IFN have been used in patients with cancer restricted to the peritoneal cavity with response rates 2-3-fold higher than systemic administration of those cytokines. Furthermore, Forni and his colleagues have administered IL-2 regionally into s.c. tissue in the lymphatic drainage of primary head and neck cancers. Again, response rates were higher and IL-2 doses were lower than when IL-2 was administered systemically; however, the response durations were short.

There have also been some attempts to inject cytokines directly into tumor cells. In general, these efforts have been successful at inducing a local immune response or tumor regression. My colleague Brendan Curti has documented why this is so. Tumor nodules have elevated interstitial pressure, sometimes as high as 100 mm Hg. This enormous pressure gradient acts as a barrier to the penetration of cytokines or antibodies or even chemotherapeutic agents into the middle of tumors when injected systemically or even peritumorally. Furthermore, when proteins are injected directly into such tumor nodules, they are actively pumped out of the tumor and into surrounding tissue in a matter of minutes. Given the very short half life of the cytokine in the tissue site where we want it to work, there is little wonder that intralesional injection of cytokines has been ineffective.

An alternative approach to delivering cytokines to tumors is to directly inject gene transfected tumor cells (either autologous or allogeneic), gene transfected normal fibroblasts (again either autologous or allogenic), or liposome encapsulated cytokines into tumors. The major portion of this book includes descriptions of the outcome of experiments along these lines. The rationale for such experiments is grounded in both immunologic and pharmacologic principles. If the host immune response is blind to the tumor, let's make the tumor visible. If the tumor fails to present antigens to the host, let's supply the missing help.

And so a large number of experiments have been performed using a very large number of very different genes and cytokines including IL-1, IL-2, IL-3, IL-4, IL-5, IL-6, IL-7, IL-8, IL-9, IL-10, IL-11, IL-12, IFN-α, IFN-γ, TNF, GM-CSF, G-CSF, allogenic MHC molecules, T-cell costimulatory molecules such as B7, and perhaps by now, others. And *wowwee*! They all work, a fact that I find both encouraging and discouraging. The fact that something works in animal models encourages us to hope that such an approach might actually work in people. However, the fact everything seems to work undermines the credibility of the entire field. The only gene transfected tumor cells that do not elicit a florid host response in animals are those carrying the immunosuppressive TGF-β.

In general, the interested scientists have chosen to test the efficacy of a particular transfected gene under unrealistic and nonphysiologic conditions. Usually a transfected tumor is injected into an animal and its natural history is compared to another animal injected with parental tumor. A few investigations have mixed transfected and nontransfected tumor cells for injection. The usual result of such experiments is that transfected cells need to outnumber nontransfected cells 5-20:1 to maintain a detectable therapeutic effect. There have been no successful experiments in which an animal bearing an established macroscopic tumor has had that tumor successfully eradicated by any injection of a gene transfected tumor.

Of course, it is possible that it is too much to ask for a treatment to produce tumor regression in a murine model. The natural history of murine tumors is usually about 1 or, at most, 2 months. Even the most aggressive human cancers generally have at least 6 months between detection and death. More rapidly growing animal tumors with growth fractions of 50-100% may be a poor model for human tumors with growth fractions of 1-3%.

On the other hand, the successful treatment that we seek to develop really should work to produce significant cytoreduction of an established tumor. Animals and people bearing tumors have been documented to have a variety of abnormalities in their immune responses. One of the challenges of a new immunologic based treatment approach must be that it works in the face of the immunologic compromise associated with the cancer bearing state. I personally do not feel that it is a sufficiently ambitious goal for the field of immunotherapy to develop treatment approaches that are only active in the adjuvant setting.

Despite these considerations, some of the more careful investigators have learned a great deal from their animal models. From their work, it is quite clear that the in vivo effects of cytokines are often more diverse than researchers might imagine based upon in vitro effects on purified cell populations. The cytokines that are most effective at eliciting a host antitumor response may vary from tumor to tumor, but the ones that appear to be most effective on the basis of head-to-head comparison within a particular model system are GM-CSF and IL-2. Both of these cytokines have the attributes of eliciting a two-phase attack: the first is acute inflammation and the second is chronic inflammation. The first is dominated by granulocytes and macrophages; the second by T cells and APC. The first phase is antigen nonspecific; the second is antigen specific. It appears important that the first phase should not be too effective. If the acute inflammation completely eradicates the local tumor, the second phase of the response does not develop and memory is never established. On the other hand, if the acute inflammation is not adequate, the animal dies before the T cells can come in and help. The first phase is crucially dependent on granulocytes, the second on T cells and APC. In general, both $CD4^+$ and $CD8^+$ T cells participate in the process.

It is also clear that the cell carrying the cytokine gene can be just about anything and perhaps need not even be a cell. Pardoll and his colleagues have put GM-CSF in large, cell sized liposome bags and have achieved success in a murine tumor model. This is important for several practical reasons. First, it may mean that one does not need to create a unique therapeutic reagent for each patient based upon his or her own tumor. It may be possible to create a bank of transfected fibroblasts that can be injected into a patient's tumor. That the fibroblasts are not completely HLA matched to the recipient may actually be of benefit in eliciting the response. Thus, the individual patient's tumor may not be an essential ex vivo component of the vaccine. Second, the ability to use a standard cell line or cytokine preparation may eliminate the difficult technical problems associated with retroviral transfection and gene expression in tumor cells. It is notoriously difficult to get tumor explants to proliferate in vitro, yet retroviral mediated gene transfer is completely dependent upon target cell proliferation to be successful.

These generalizations gloss over many potentially important details. For example, Lollini and colleagues have documented dose response effects when the same tumor expresses different amounts of a particular cytokine. They found that animals given cells expressing large amounts of IFN-γ did not develop pulmonary metastases, but those whose tumors expressed small amounts developed significantly more pulmonary metastases than animals bearing a nontransfected tumor. Thus, at some concentrations, IFN-γ enhanced the metastatic spread of the tumor. Furthermore, my colleagues Robert Fenton and Robert Wiltrout have documented that some transfected tumors undergo epigenetic changes, such as methylation of the transfected gene, that can result in variable or no expression of the transfected cytokine.

Another important consideration is whether the tumor cell is capable of eliciting a memory T-cell response. There are several levels at which the tumor might fail to do so. First, it may contain an antigen that is not properly recognized as an antigen on the basis of its inability to be presented in a recognizable fashion by the MHC alleles of the patient. For example, there may be no T cells in an HLA-A1 bearing patient that are capable of recognizing the valine-12 *ras* mutation. This would be analogous to the so-called immune response gene defects of mice. Second, the tumor may fail to express MHC molecules, either through a down-regulation of class I α chain expression or an abnormality in β2-microglobulin expression. Third, the tumor may have a defect in the intracellular machinery involved in antigen processing. Several such instances have been outlined in the chapter by Restifo, including loss of chaperonins or proteases. Fourth, the tumor may fail to express costimulatory molecules needed to activate rather than anergize T cells. Each of these levels of defect may be correctable. However, it would seem impractical to develop a set of molecular diagnostic tests on every patient's tumor to assess its ability to present antigen and discern the level at which the tumor is defective in antigen presentation.

Of course, it is possible that defective antigen presentation by a tumor cell will not be a critical deficiency in the process of vaccine development. The requirements for eliciting a primary T-cell response are more stringent than the requirements for eliciting a memory T-cell response. Low levels of class I MHC expression or the expression of minor histocompatibility antigens may be sufficient for a cell to act as a target for a secondary T-cell response. In this scenario, all that would be required for the tumor to generate a primary response would be for some tumor cells to die and have their antigens picked up and processed by dendritic cells or some other professional APC that would prime antigen specific T cells. It appears that GM-CSF and IL-2 are both capable of eliciting a cellular response that includes professional APC.

It is fascinating that acute inflammation in general and granulocytes in particular seem to be crucial to the observed antitumor effects of locally produced cytokines. This observation raises important questions about how immune responses are regulated by granulocytes. Although much attention has been given to the nature of the enzymes in granulocyte granules, one would guess from these results that there must also be a number of chemotactic and activating factors produced by granulocytes that propagate the response and recruit lymphocytes to the site. Undoubtedly, more research in this area will permit a finer dissection of the physiologic process.

It is a bit surprising that there have not yet been reports of combinations of cytokines inserted into tumor cells or fibroblasts. Since it appears that Th1 based responses are more effective than Th2 based responses, it would seem logical to use a combination of IL-2 and IFN-γ, or IL-2 and IL-12, or GM-CSF and IL-2. It may also be reasonable to include an activating cytokine plus a chemokine to boost the directional migration of particular subsets of T cells and/or macrophages. Such experiments are undoubtedly under way and will certainly have some impact on the clinical development of the treatment approach.

However, it is now time to work out most of the remaining questions in human subjects. It is always difficult to know when an idea is ready for clinical testing; there are examples in biological therapy of making the transition from mouse to man too early. But the development of an effective cancer vaccine depends on understanding the regulation of acute and chronic inflammation in vivo in normal and tumor bearing man. That information cannot be fully revealed by more animal experiments. Certainly there is more to learn from continued animal based study; but such work should be conducted in parallel with, rather than in lieu of, human clinical trials.

We are embarking on a clinical trial in patients with melanoma using a melanoma cell line that expresses both HLA-A1 and HLA-A2 and expresses both MAGE-1 and tyrosinase, the melanoma specific antigens associated with these class I molecules. About 75% of patients with melanoma will express either HLA-A1, HLA-A2, or both. We have transfected the cell line with the B7 gene that encodes for the costimulatory molecule that triggers T cells through CD28. We have also transfected the cell line with GM-CSF. We will need to determine the appropriate number of cells to inject, the number of injections and their timing, the sites of injection, the requirement for adjuvants, the effect of tumor burden on efficacy, and dozens of other facts that will be essential to develop an effective treatment approach.

Such clinical trials will require a different kind of clinical monitoring of the immune system than is seen in the current state of the art. It is rare for current clinical trials to assess T-cell function in a sophisticated way. However, assays for antigen specific T-cell function, identical or analogous to those outlined by Professor Kourilsky and colleagues, will be required. We are interested in what is happening in the tissues, not in the peripheral blood. Peripheral blood LAK activity and NK cell numbers, though relatively simple to measure, have no meaning in assessing the efficacy of vaccine approaches to cancer treatment. More work will be required in this arena. But we must now get on with it.

Another interesting strategy for eliciting an immune response against a tumor is to do something to the tumor to make it appear more foreign. The term for this process is xenogenization and it includes the use of chemical (e.g. mutagen), physical (e.g. UV light or X-irradiation), and biological (e.g. IFN, viral infection) agents to alter the tumor in a way that would make it more readily visible to the host. An interesting form of xenogenization that is being developed by Nabel and his colleagues is the direct introduction into the tumor in vivo of DNA encoding a foreign class I MHC molecule. Tumor cells that successfully take up and express this foreign HLA-B7 molecule should elicit an immune response similar to the rejection of a histoincompatible organ graft. Key unknowns in this process are the extent of HLA-B7 expression that one will get, i.e. the fraction of tumor cells that will take up and express the injected DNA and the number of HLA-B7 molecules that will be expressed on each cell. The hope is that some T cells specific for a tumor related peptide plus self will be nonspecifically recruited by the allogeneic response and once such cells see the tumor, a tumor specific (rather than an allospecific) memory response will be elicited. In that way, tumor cells not expressing HLA-B7 will also be recognized and destroyed at distant sites.

Once it is clear what variables lead to optimal stimulation of T-cell immunity in vivo in man, it may be possible to harvest draining lymph nodes and specifically propagate and expand the antigen specific T-cell population. Adoptive cellular therapy may have improved therapeutic effects if one can transfer tumor specific cells, provided that the cells do not lose their specificity or homing capacity during in vitro culture and expansion.

Another interesting application of gene therapy techniques has been explored by Mullen and Blaese and their colleagues at the National Cancer Institute. They reasoned that if they could get the herpes simplex thymidine kinase gene expressed in tumor cells, such cells might become susceptible to killing by ganciclovir. Herpes simplex thymidine kinase, but not mammalian thymidine kinase, converts ganciclovir to a monophosphate and other intracellular enzymes convert it to a triphosphate that can inhibit DNA polymerase. Without the herpes simplex enzyme, the first phosphate cannot be added and ganciclovir is relatively nontoxic. Surprisingly, in preclinical brain tumor models, in situ injection of the herpes simplex thymidine kinase gene caused not only injected cells but also noninjected cells to become susceptible to ganciclovir. The mechanism for the bystander effect is not yet clear. Clinical trials of this approach to treating brain tumors have been started.

The ability to use ganciclovir to selectively kill cells expressing herpes simplex thymidine kinase has led to another interesting idea. One of the most effective immunologic antitumor effects that we know about is the graft-vs-tumor effect that occurs in the setting of allogeneic bone marrow transplantation. There is no question that allogeneic immunocompetstent cells are capable of killing tumor in the adoptive host. The major difficulty with this reaction is that it is often accompanied by graft-vs-host disease where the donor T cells do damage to various host normal organs and tissues. We reasoned that it might be possible to elicit a beneficial antitumor effect over a few weeks and then selectively kill the allogeneic donor T cells with ganciclovir if they had been transfected with herpes simplex thymidine kinase. In preclinical experiments, we have shown that functional human T cells expressing herpes simplex thymidine kinase retain their function and are selectively killed by ganciclovir without a bystander effect on noninfected cells. Clearly the major unanswered question is whether the herpes simplex gene will be expressed in the progeny of the adoptively transferred alloreactive cells in the face of the in vivo expansion of that population that is likely to occur. The difficulty in transfecting murine T cells makes it nearly impossible to address these issues in a preclinical setting.

The take home message from this book is that we are finally making a departure from the pharmacology paradigm of biologic therapy development. Cytokines are not drugs. Their physiologic action, in general, is over a short distance in the tissue in which they were produced. Our general understanding of immunology has improved to the point where, despite concerns over the ability of the immune system to recognize spontaneously arising human cancers, reasonable strategies are now available that must be tested. Indeed, the vast majority of human cancers are either carcinogen induced (e.g. lung cancer) or express one or more altered protein products that may be suitable targets for immune attack even though they are normally expressed only intracellularly (e.g. mutant p53, mutant *ras*). Harnessing the power of the host defense system and marshalling it against cancer is not a pipe dream. The tools and the ideas to make it happen are at hand. We must realistically assume that there is much more to learn as we begin to apply what we have learned to human beings. Our sophisticated understanding of molecular biology has not been accompanied by comparable development of understanding of human immune physiology. Our current challenge is to develop a sophisticated understanding of how to manipulate human immune physiology.

Given the general tendency to randomness in the universe described quantitatively as the second law of thermodynamics, the ordered and organized molecules and functions that define life must be viewed as miracles. As Lewis Thomas wrote, "Statistically, the probability of any one of us being here is so small that you'd think the mere fact of existing would keep us all in contented dazzlement of surprise". Cancer, then, can be viewed as in keeping with the natural increase in entropy, an escape from order and a tendency to randomness. Our struggle to understand it is a battle of wits. Thomas H. Huxley described our efforts to comprehend nature as a chess game: "The chessboard is the world, the pieces are the phenomena of the universe, the rules of the game are what we call the laws of Nature. The player on the other side is hidden from us". The battle against cancer is but one little skirmish in the game. But our ignorance is frustrating and our helplessness is depressing. As we struggle, we discern the nature of cancer and its weaknesses as through a glass darkly. With sufficient time and insight, perhaps some day we shall see it face to face.

SUBJECT INDEX

Introductory note: location references given in *italics* refer to illustrations or tables.

CYTOKINE-INDUCED TUMOR IMMUNOGENICITY
ISBN 0–12–261520–4